High-Performance and Specialty Fibers

The Society of Fiber Science and Technology, Japan
Editor

High-Performance and Specialty Fibers

Concepts, Technology and Modern Applications of Man-Made Fibers for the Future

Springer

Editor
The Society of Fiber Science and Technology, Japan
Tokyo
Japan

ISBN 978-4-431-56645-8 ISBN 978-4-431-55203-1 (eBook)
DOI 10.1007/978-4-431-55203-1

Softcover reprint of the hardcover 2nd edition 2016

Printed on acid-free paper

This Springer imprint is published by Springer Nature
The registered company is Springer Japan KK

Preface

The Society of Fiber Science and Technology, Japan (SFSTJ) was established in 1943 by merging the Cellulose Association (founded in 1924) and the Society of Industrial Fiber Technology (founded in 1935). Since then, SFSTJ has been contributing not only to research and education on fiber science and technology but also to the development of the fiber and textile industries. In December 2013, SFSTJ celebrated its 70th anniversary and held a commemorative ceremony for the anniversary on September 29, 2014. The ceremony was followed by the International Symposium for Fiber Science and Technology (ISF2014), looking back at its previous activities and looking ahead to the fiber science and technology of the future. For this purpose, many prominent scientists and important leaders from government, academia, and industry around the world were invited. In addition, SFSTJ made plans to publish this book commemorating the 70th anniversary by compiling the whole of fiber science and technology that have been developed thus far. In particular, new technologies for manmade fibers that have recently been industrialized in Japan are highlighted to shed light on the future technologies that will be needed in the fiber and textile industry of this century.

Manmade fibers have a long history, and their current developments ought to have their own historical backgrounds. We therefore succeeded in reviewing the historical origins of the newly developed fibers in addition to the concepts and processes for their development, as well as their key technologies and excellent characteristics. Recently, the production center of many low-cost manmade fibers such as polyester has shifted from Japan to China after the past drift from the Western countries, including the United States, to Japan. Such transition has forced Japanese fiber companies to change their business strategy from mass production to small-size production of high-quality, high-performance fibers. Currently, many of the specialty fibers are industrialized in Japan. For example, Zylon and Dyneema from Toyobo are the strongest organic and super high-elastic-modulus fibers, respectively. Technora from Teijin and Vectran from Kuraray are high-strength fibers. Torayca is the most famous carbon fiber produced by Toray. These fibers having outstanding properties constitute the main subjects of this book.

We are proud to have been able to invite the inventors or specialists from both industry and academia as authors who write about the various fibers. They describe in detail the design concepts and key technologies used for the respective manmade fibers in the chapters for which they were responsible. These chapters have been effectively compiled into a book that can be used not only as a good textbook for university teachers and students to learn basic concepts of modern fibers, but also as a review or a reference book for researchers in academic organizations and companies to overview the recent statuses of manmade fibers, particularly specialty fibers that have been developed in Japan to date. This book is therefore presented as a record of the outcome of important research by Japanese scientists and engineers who have been involved in the development of manmade fibers in Japan. It also has the distinction of being the first English book authorized by The Society of Fiber Science and Technology, Japan.

In the initial publication plan for this book, Springer explained to us that they now publish not only printed books but also ebooks, which readers can access through the Internet. This on-line publishing system is convenient for readers in that the books they need are readily available whether in the office or at home. Readers who are interested in having a hard copy of a book can order a printed version which will be mailed to them. On-line publications in principle keep on sailing forever, and no book ever goes "out of print". In other words, if a book is continuously revised, it will not lose contemporary value as a reference book and as a data and information source. This feature is especially attractive for authors and editors although continued revision may not always be easy. Therefore, revisions must be made in an organized manner from generation to generation by leaving it to scientific societies or organizations where the most capable authors can be nominated from among many eminent members at the time of revision in order to update the descriptions and data. Based on this concept, SFSTJ decided to edit this book and to continue its revision to maintain the contemporary quality of the book in the future. The present volume is therefore to be considered only a first version leading to a comprehensive treatise on fiber science and technology that will be accomplished after several revision steps.

Here, we express our deep appreciation to the many authors who devoted themselves to writing their respective chapters, sacrificing their valuable time. We also acknowledge the section directors who arranged the important sections by inviting excellent authors and sometimes editing the manuscripts submitted.

Finally, Ms. Taeko Sato and Dr. Shin'ichi Koizumi at Springer Japan are gratefully acknowledged for their valuable support for the publication of this book. Without their initial proposal and their follow-up arrangements, this book would not have appeared.

On behalf of SFSTJ, the associate editors hope that all readers enjoy this book, which we believe is highly interesting and instructive.

Managers

Yoshiharu Kimura (Chief)
Emeritus Professor, Kyoto Institute of Technology

Akira Tsuchida (Sub)
Professor, Gifu University

Kaname Katsuraya (Sub)
Professor, Wayo Women's University

Editor The Society of Fiber Science and Technology, Japan

List of the Editorial Staff

Editor
The Society of Fiber Science and Technology, Japan

Managers
Yoshiharu Kimura (Chief), Akira Tsuchida (Sub), Kaname Katsuraya (Sub)

Section Directors
Part I: Takeshi Kikutani
Part II: Kazuyuki Yabuki (Chief), Kohji Tashiro (Sub), Hiroki Murase (Sub)
Part III: Togi Suzuki
Part IV: Akihiko Tanioka
Part V: Makoto Endo
Part VI: Masatsugu Mochizuki
Part VII: Kaname Katsuraya

Advisory Brains
Masatsugu Mochizuki, Hiroki Murase

Superviser
Tsuneo Okubo

Authors
Part I

Chapter 1: Toshiji Kanaya and Keisuke Kaji
Chapter 2: Kohji Tashiro
Chapter 3: Takeshi Kikutani
Chapter 4: Masatoshi Aoyama and Yoichiro Tanaka

Part II

Chapter 5: Hiroki Murase and Kazuyuki Yabuki
Chapter 6: Kohji Tashiro
Chapter 7: Yasunori Fukushima, Hiroki Murase, and Yasuo Ohta
Chapter 8: Takeshi Kikutani
Chapter 9: Shigeru Hayashida
Chapter 10: Hideki Hoshiro, Ryokei Endo, and Forrest E. Sloan
Chapter 11: Yoshihiko Teramoto and Fuyuhiko Kubota

Part III

Chapter 12: Togi Suzuki
Chapter 13: Hiroshi Takahashi
Chapter 14: Togi Suzuki and Sonoko Ishimaru
Chapter 15: Sonoko Ishimaru

Part IV

Chapter 16: Akihiko Tanioka and Mitsuhiro Takahash
Chapter 17: Yuhei Maeda and Masato Masuda
Chapter 18: Akira Isogai
Chapter 19: Hidetoshi Matsumoto

Part V

Chapter 20: Makoto Endo
Chapter 21: Yutaka Arai
Chapter 22: Tetsuyuki Kyono
Chapter 23: Yoshitaka Kageyama

Part VI

Chapter 24: Osamu Yaida
Chapter 25: Masatsugu Mochizuki and Nobuhiro Matsunaga
Chapter 26: Eiji Shiota
Chapter 27: Hiroyasu Shimizu

Part VII

Chapter 28: Akira Yonenaga

Contents

Part I
Advancement of Fiber Science and Technology

Chapter 1
History of Fiber Structure

Toshiji Kanaya and Keisuke Kaji

Abstract The early history of fiber structure in molecular level is closely related to the history of development of the so-called macromolecular theory or the concept of macromolecules. After a long debate, it ended in victory of the macromolecular theory in 1936, concluding that the substances consist of many monomers connected by covalent bonds. Through the discussions of the macromolecular theory, the study of fiber structure was greatly activated. In this article we review the history of fiber structure. In the first section we describe the research history of structure of natural fibers, especially natural cellulose fiber, starting from a micelle model to a fringed-micelle model. In the second section, we focus on the history of the structural study of synthetic fibers from fringed-micelle microfibril model to shish-kebab model.

Keywords History of fiber structure • Natural cellulose fiber • Synthetic fiber • Micelle model • Fringed-micelle model • Fringed-micelle microfibril model • Shish-kebab model

1.1 Introduction

The early history of fiber structure in molecular level, which began in the 1870s, is closely related to the history of development of the so-called macromolecular theory or the concept of macromolecules. In 1920 Hermann Staudinger proposed the concept of macromolecules on the basis of many experimental results on polymerization by his research group, reporting a paper entitled "Über Polymerisation" [1]. This paper triggered a big controversy between the macromolecular hypothesis and the low molecular hypothesis for substances such as rubber,

T. Kanaya (✉)
Institute of Material Structure Science J-PARC, Material and Life Science Division, High Energy Accelerator Research Organization (KEK), 203-1 Shirakata, Tokai-mura, Naka-gun, Ibaraki 319-1106, Japan
e-mail: tkanaya@post.kek.jp

K. Kaji
Institute for Chemical Reserach, Kyoto University, Uji, Kyoto-fu 611-0011, Japan

Society of F.S. Technology, Japan (ed.), *High-Performance and Specialty Fibers*,
DOI 10.1007/978-4-431-55203-1_1

cellulose, and resins [2–4]. After a long debate, it ended in victory of the macromolecular theory in 1936, concluding that the substances consist of many monomers connected by covalent bonds. Through the discussions of the macromolecular theory, the study of fiber structure was greatly activated.

In the Sect. 1.2, "From Micelle Model to Fringed-Micelle Model for Natural Fibers," we describe the research history of structure of natural fibers, especially natural cellulose fiber, starting from a micelle model to a fringed-micelle model. In the Sect. 1.3, "From Fringed-Micelle Microfibril Model to Shish-Kebab Model for Synthetic Fibers," we focus on the history of the structural study of synthetic fibers which began to appear in the middle of 1930s.

1.2 From Micelle Model to Fringed-Micelle Model for Natural Fibers

Researches on fiber structure were mainly performed on natural cellulose fibers in the early stage. The research period almost corresponds to the term of controversy between the low molecular theory and the macromolecular theory. Fiber structure models clarified in this period gave a great impact on the subsequent studies of polymer physical structure.

1.2.1 Micelle Model

In the latter half of the nineteenth century, Nägeli [5] (1817–1891), the Swiss botanist, introduced a concept of "micelle," the origin of which came from Latin mica-ella meaning small particle (or grain), to account for the structure of biological systems such as starch granules, cellulose membranes, and gelatin (see Fig. 1.1). From polarized optical microscopic observations, he found some structural units indicating birefringence, named "micelles," in these substances, concluding that they should be crystal particles. Thereafter, the micelle theory was supported by their detailed optical anisotropic investigations including form birefringence and form dichroism measurements [6–8].

In the 1920s Herzog of the Kaiser-Wilhelm-Institut für Faserstoffchemie performed X-ray diffraction study on natural cellulose fibers such as cotton and ramie in cooperation with Jancke [9, 10] and Polanyi [11]. They found that the cellulose fiber exhibited the fiber diagram of a crystalline form with a fiber identity period (FIP) of 10.3 Å [11]. Although this fact had already been discovered in 1913 by Nishikawa and Ono [12], it was not noticed at that time in Western countries. On the basis of their results, Herzog considered that their finding supported "the crystal micelle theory" of Nägeli and concluded that the molecular chain length of cellulose was "equal to the FIP" or smaller because it was then considered that crystals

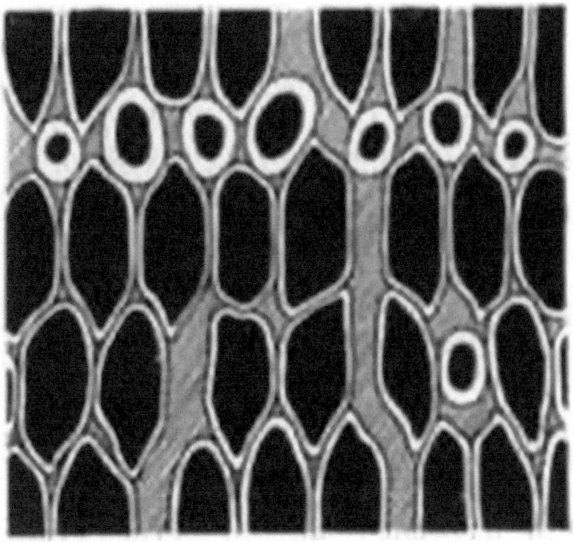

Fig. 1.1 Schematic diagram of "micelle," proposed for the structure of cellulose fibers by Nägeli. Extended black particles indicate the crystalline micelles (Nägeli [5])

should consist of small molecules, that is to say, molecules could not be larger than the crystallographic unit cell. Such an idea was then widespread over the crystallographers. When Staudinger gave a lecture on his macromolecular theory at a meeting of the Chemische Gesellschaft Zürich in 1925, the famous Swiss mineralogist Paul Niggli (1888–1953) stood up and shouted "Such a thing does not exist!"

Sponsler and Dore [13] and Meyer and Mark [14, 15] revealed that the unit cell of cellulose crystal contains four glucose units $C_6H_{10}O_5$. Furthermore, Hengstenberg and Mark [16] evaluated the crystallite size of cellulose along the fiber axis from the widths of the X-ray diffraction peaks using Scherrer's equation [17], resulting that the estimated sizes were at least 600 Å (60 nm). The lateral dimension perpendicular to the fiber axis was as well obtained to be 50–60 Å (5–6 nm) from the equatorial diffraction peaks. Hence the crystallites would be rodlike and the number of chain molecules in one crystallite would be 40–50 in the case of ramie. On the basis of these results, Meyer and Mark [18] proposed a "new micellar theory" or an "isolated micellar model" as shown in Fig. 1.2 by improving the Nägeli's micellar model. In the new model, a brick represented a crystal micelle where the molecular length was equal to the micellar length in the fiber-axial direction. In each micelle 40–60 cellulose molecules, each of which had about 60 glucose or 30 cellobiose units, were assembled in parallel. This structural model could explain the swelling behavior of the cellulose fibers: swelling agents probably entered the gaps among the bricks or micelles, resulting in increase of the fiber diameter. They considered that the forces acting between micelles were strong cohesive forces due to many OH groups, which was called "micellar forces." However, such intermolecular forces were supposed to as well operate in the crystalline micella, and so it was unclear what the difference was between them. Furthermore, the X-ray diffraction experiments revealed that the swelling agents

Fig. 1.2 "New micelle model" proposed by Mayer-Mark (Meyere and Mark [18])

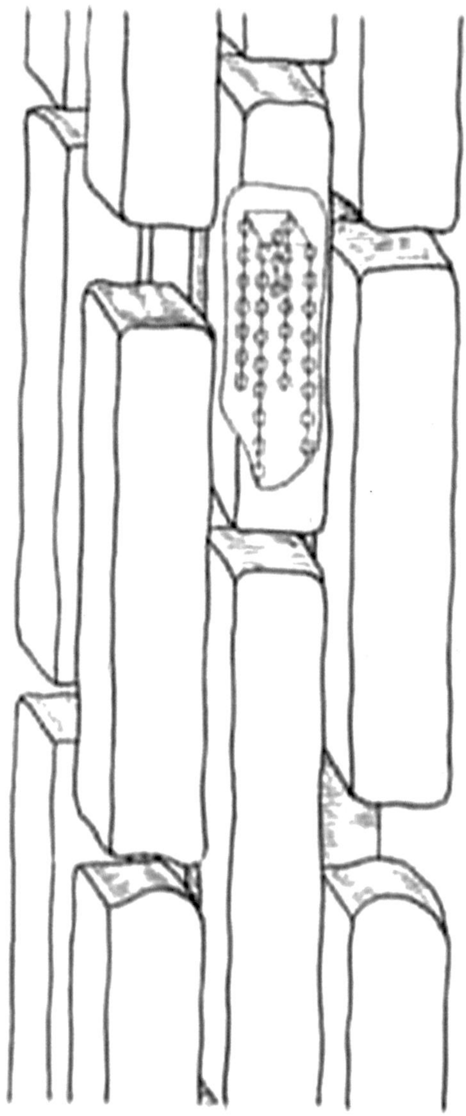

did not enter the crystalline regions or the micelles. If the swelling of fiber occurs only in the gap between the micelles, it would be difficult to understand why the micelle force due to OH groups is maintained. In order to resolve the question, some intermicellar materials capable of swelling, called "Kittstoff" (cement materials), was assumed as an interlinking substance between micelles, but this assumption was later denied since artificial fibers, made from highly refined cellulose, indicated a similar swelling behavior.

1.2.2 Fringed-Micelle Model for Macromolecules

In 1930 Gerngross and Herrmann [19, 20] proposed a fringed-micelle model for gelatin gel as shown in Fig. 1.3 where a network structure was formed by connecting micelles with molecules making the micelles themselves. On the other hand, the isolated micellar model or the low molecular micelle model in Fig. 1.2 could not explain the binding force between the micelles.

Frey-Wyssling [21, 22] studied the swelling behavior of cellulose fibers by introducing colloid particles of noble metal into the fibers and determining the size and distribution of the metal particles by small-angle X-ray scattering and optical measurements. The results showed that well-organized cavities exist in all native fibers. Based on the result, he proposed a cavity model for a micellar system as shown in Fig. 1.4. In this model, fibrils about 70 Å (7 nm) wide consisting of alternate amorphous and crystalline regions aligned along the fiber direction and the fibrils are connected by crosslinking molecules in several places, which Frey-Wyssling called "junction points," forming a network structure. In addition, long cavities about 10 Å (1 nm) wide are present between the fibrils. This model well explained the swelling behavior of cellulose fiber. He also visualized the molecular structure for this model as shown in Fig. 1.5, which was called a "continuous micelle model."

After the establishment of the macromolecular theory in 1935–1936, Kratky and Mark [23] submitted a fiber structure model as represented in Fig. 1.6a, referring to the micelle models in Figs. 1.3 and 1.5 proposed by Gerngross and Frey-Wyssling, respectively. In this model, long covalently bonded molecular chains or polymer

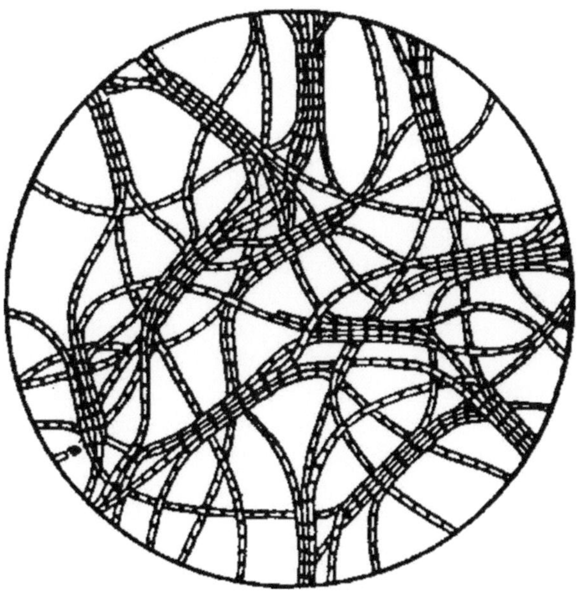

Fig. 1.3 Fringed-micelle model for gelatin gel by Gerngross et al. [19]

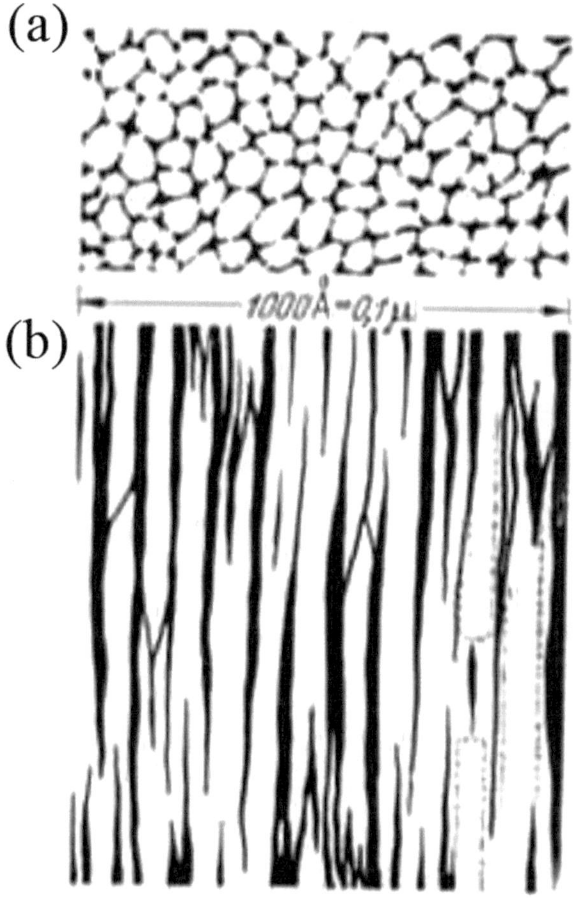

Fig. 1.4 Diagram of micellar texture of cellulose fibers by Frey-Wyssling. The diagram shows a longitudinal cross section (**a**) and a normal cross section (**b**) of the fibers. Black spaces represent the cavities (Frey-Wyssling [21, 22])

chains pass through several micelles to connect their micelles. This model is called "the fringed-micelle model" because the structural unit of this model consists of a micelle with "fringe-like" amorphous chains on both the end surfaces of the micelle (see Fig. 1.6b). This model as well as the Frey-Wyssling model in Figs. 1.4 and 1.5 solved the problem of intermicellar forces either; the interlinking substance "Kittstoff" is unnecessary. Furthermore, it explains the reasons why the fibers do not swell in the fiber-axial direction but only in thickness of the fiber and why the elastic modulus of natural cellulose fibers well agrees with the theoretical crystalline modulus calculated by Mayer and Lotmar [24].

The abovementioned "fringed-micelle model" for natural cellulose is supported also from the biosynthetic mechanism of its fiber [25]. Figure 1.7 shows how the cellulose microfibrils are formed; the microfibrils, each consisting of six molecules in this case, are continuously synthesized on the outer surface of the plasma membrane. The large cellulose synthase complexes at the growing ends are pushed ahead in the plane of the membrane; they run parallel to each other, but their

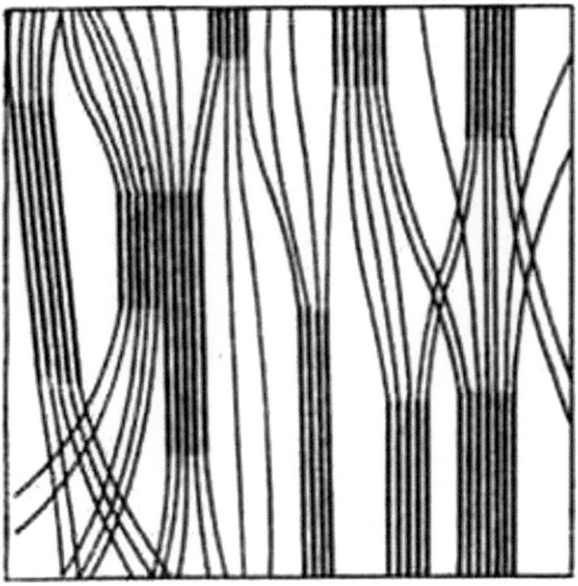

Fig. 1.5 Continuous micelle model proposed by Frey-Wyssling (Frey-Wyssling [21, 22])

directions are not precise though they are confined in some extent between the microtubules attached to the plasma membrane. Therefore, the fibers composed of these microfibrils may not be considered to crystallize completely.

1.2.3 Conclusion

In this part, we have mainly described the fiber structure of natural cellulose and concluded that the fiber structure of natural cellulose could be described in terms of the fringed-micelle structure model. Here, it should be emphasized that the cellulose chain molecules pass through both the crystalline and amorphous regions of which distributions or the correlation between the crystal regions is not regular. This idea was supported not only by small-angle X-ray scattering (SAXS) but also by small-angle neutron scattering (SANS) on natural cellulose fibers; the so-called long-period reflections being usually observed in synthetic fibers were not observed for natural cellulose fibers by SAXS and even for the sample deuterated only in the less ordered or amorphous region by SANS [26]. The latter technique utilizes the difference in coherent scattering lengths of H and D to raise the scattering contrast in SANS.

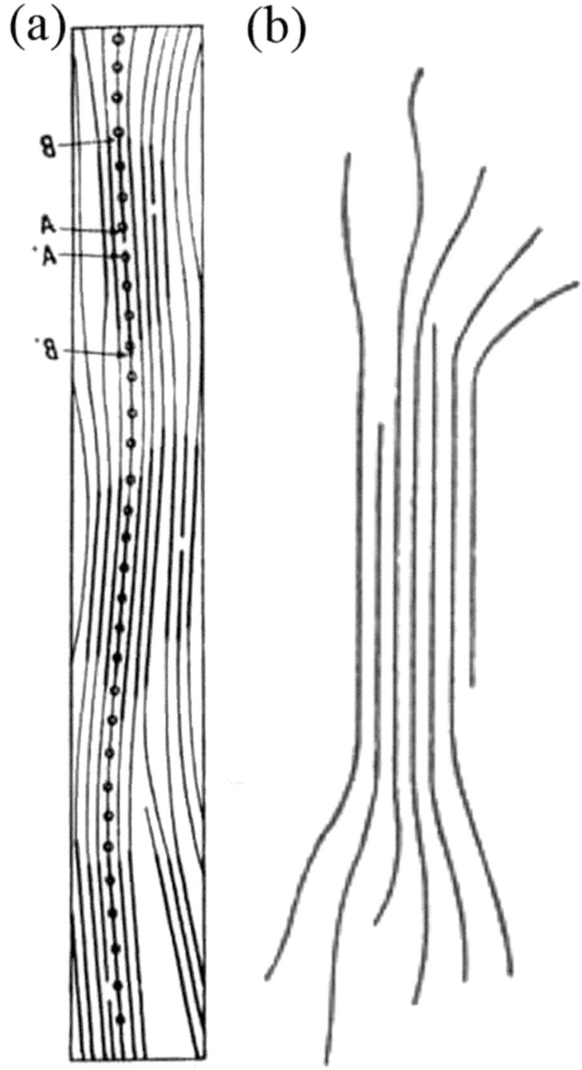

Fig. 1.6 Fringed-micelle model by Kratky and Mark (**a**) and its structure unit (**b**) (Kratky and Mark [23])

1.3 From Fringed-Micelle Microfibril Model to Shish-Kebab Model for Synthetic Fibers

As mentioned in the first section, the early studies on the fiber structure were carried out using mainly natural cellulose fibers, and the fringed-micelle structure model was proposed as the most reliable one. In order to explain the fiber strength, this model assumes that the molecular chains pass through both the crystalline and amorphous regions while the distribution of the two regions is irregular. Then, the trend of the structure studies moved to synthetic fibers such as polyamides and

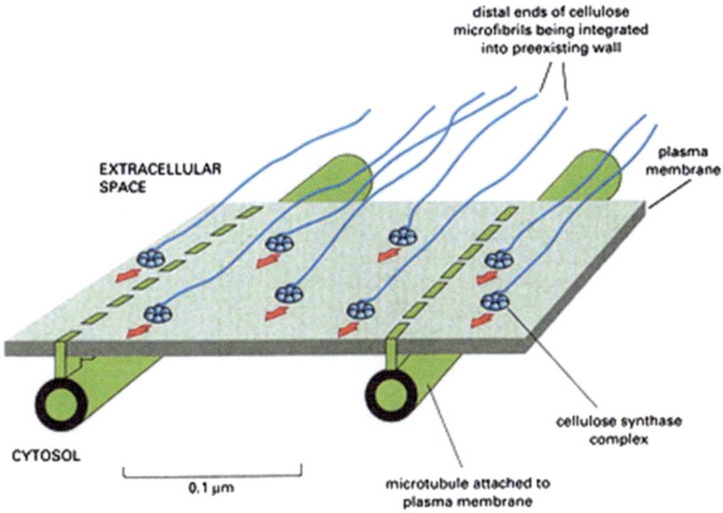

Fig. 1.7 Biosynthetic mechanism of cellulose microfibrils (Alberts et al. [23])

polyesters because manufactured synthetic fibers began to appear in the middle of 1930s.

Though the natural cellulose fiber does not show a long period as described in the first section, synthetic fibers as well as the regenerated cellulose fiber have a well-defined long-period structure in which the crystalline and amorphous regions are alternately arranged along the fiber axis. Therefore, we need to distinguish synthetic fibers from natural fibers. In this part, we describe the research history on the structure of synthetic fibers.

1.3.1 Fringed-Micelle Microfibril Model (Hess-Kiessig Model)

Structural studies of synthetic fibers began in earnest after the discovery of the long-period structure by small-angle X-ray scattering (SAXS). In 1942 Hess and Kiessig [27, 28] first observed the long-period diffraction peaks in the fiber-axial direction using a SAXS camera developed by Kiessig [29] and obtained the long repeating periods of 70–200 Å for polyamides (nylon) and polyesters. From these observations, they visualized a periodic structure in which the crystalline and amorphous regions were alternately arranged along the fiber axis. Furthermore, the large broadening of the long-period reflection on the layer line suggested that the width of the crystallite was not very large [30]. On the basis of these two observations, i.e., (i) the existence of the long period and (ii) the small width of the crystallite, they proposed a "fringed-micelle microfibril model" as shown in Fig. 1.8 by referring the fringed-micelle model for the natural cellulose fiber. This new model is different

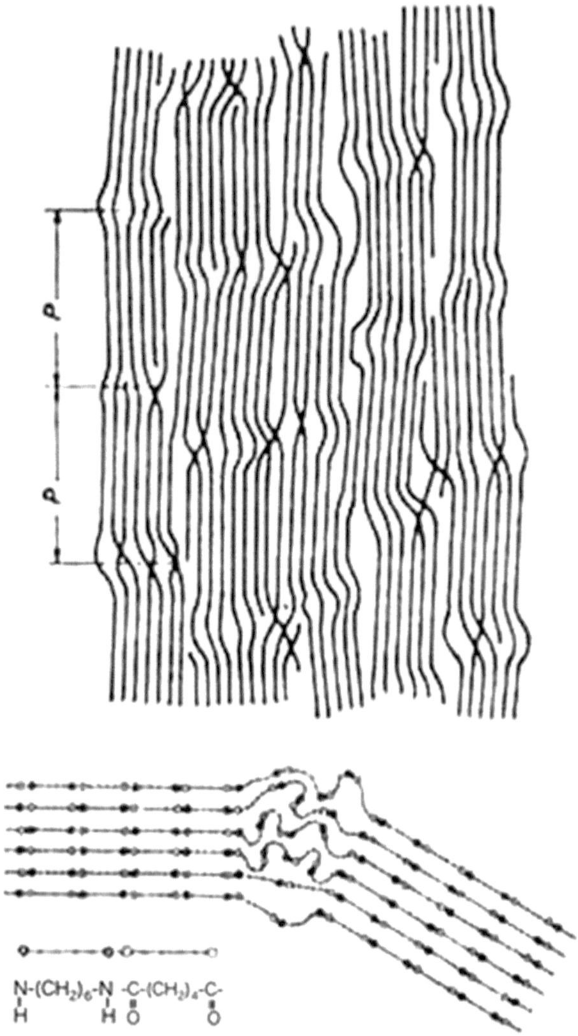

Fig. 1.8 Fringed-micelle microfibril model for synthetic fibers (Hess-Kiessig model); *P* denotes the long period. Lower part shows enlarged details of the model for an example of nylon-66 (Hess and Kiessig [28])

from the "fringed-micelle model" in two aspects; (i) the fibers are made of microfibrils, and (ii) the alternate periodic arrangement of the crystalline and amorphous regions in the microfibril gives the long period. Here the importance is that the microfibril should form a bulge in the amorphous or disordered region because of the lower density. If the fibril is very thick, bulges would become too big to be accommodated within the fiber. This leads to the concept of microfibrils; the small bulges of the amorphous part can be accommodated within the fiber by shifting the relative positions of the microfibrils in the fiber-axial direction. In order to validate this model, they performed electron microscopic observations on the synthetic fibers stained with iodide after cleaving the samples to succeed in observing microfibrils with diameter of about 100–300 Å as well as a long-period structure of the microfibril [30, 31].

1.3.2 Paracrystalline Layer Lattice-Microfibril Model (Hosemann-Bonart Model)

The fringed-micelle microfibril model mentioned above was a rather good model since it explained various features of synthetic fibers. Nevertheless, some problems were pointed out even on this model; it could neither explain completely the SAXS pattern nor the elastic modulus and strength of fibers. The first problem was raised by Bonart and Hosemann [32]. If the fringed-micelle microfibril model shown in Fig. 1.8 is right, some diffraction peaks would appear on the equator because it has the inter-microfibril periodicity in the direction normal to the fiber, which is due to the lower density of the bulges. However, no expected peaks were actually observed. In addition, if the broadening of the long-period reflection on the layer line is originated from the thickness of microfibril, the tails of scattering intensity at the center of diffraction pattern could be observed on the equator. However such tails were not observed experimentally.

Bonart and Hosemann [32] solved the first problem by introducing the so-called chain folding in the amorphous region. This idea was based on the discovery of the chain-folding structure in polymer single crystals by Keller [33, 34] in 1957 as well as in the crystalline lamella of bulk crystallized polymers by Fischer [35]. In other words, they succeeded in lowering the density of the amorphous regions without swelling. They also solved the latter problem by introducing the so-called paracrystalline layer structure model [36] as shown in Fig. 1.9, which was based on the concept of paracrystalline layer lattice presented by Hosemann [37, 38]. Here the distance in the fiber direction is a constant, but the lateral coordination statistic has a certain degree of freedom. These features are represented as wavy crystalline layers in Fig. 1.9. For this model the center reflection does not spread, but the broadening of high-order reflections on the individual layer lines increases with increasing the reflection order. Combining the features of the paracrystalline layer lattice model and the microfibril model, Bonart and Hosemann [36, 39] proposed a fiber structure model as shown in Fig. 1.10. This model should be referred to as a "paracrystalline layer lattice – folded chain microfibril model," but it is abbreviated as a "paracrystalline layer lattice microfibril model." This model corresponds to the one obtained by dividing the model of Fig. 1.9 in the fiber-axial direction into microfibrils. In other words, the microfibrils consist of the alternate arrangement of the folded chain crystalline region (black part) and the amorphous region including chain-folded parts and relaxed tie chains (white part). The characteristic feature of the model is that the crystalline regions in the neighboring microfibrils correlate with each other, forming a wavy layer lattice in the paracrystalline manner.

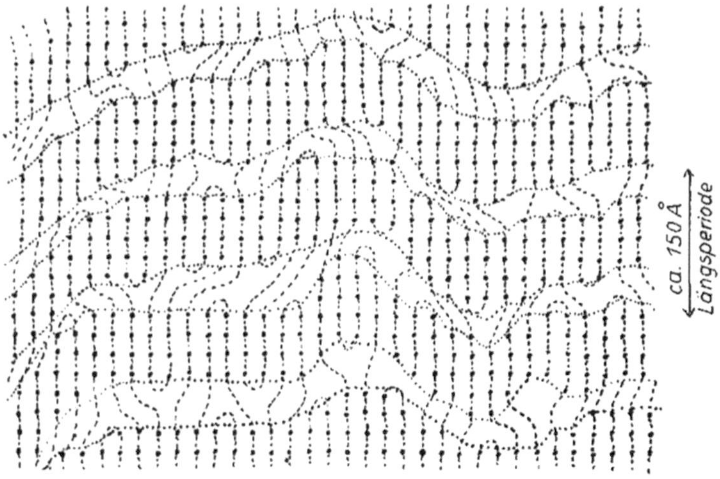

Fig. 1.9 Paracrystalline layer lattice model in which the crystalline lamellar surface is wavy. The density of noncrystalline regions is lowered by chain folds on the surfaces of crystals (Bonart and Hosemann [36])

Fig. 1.10 Paracrystalline layer lattice microfibril model for synthetic fibers (Hosemann-Bonart model) (Bonart and Hosemann [36])

1.3.3 Folded Chain Microfibril Model (Peterlin Model)

The problem about the physical properties of fibers mentioned before was discussed by Meinel and Peterlin [40, 41]. In order to explain the elastic modulus and strength of fiber, they suggested that some extended tie chains should be introduced in the amorphous region in addition to relaxed tie chains and chain folds, having estimated their fraction to be at most 30 % for highly stretched polyethylene fibers. Taking into account these considerations, Peterlin [42] proposed a fiber structure model as shown in Fig. 1.11, which contained the extended tie chains in the amorphous regions as well as all the features of the paracrystalline layer structural model by Bonart-Hosemann. Thus, some extended tie chains pass through the amorphous regions within each folded chain microfibril, and the crystallites in neighboring microfibrils make a wavy layer to take the paracrystalline layer structure. Such a model describes various properties of the synthetic fibers, and it has therefore been cited as a model of synthetic fibers until quite recently. However, a new discovery has shown that this model was essentially different from the real fiber structure.

1.3.4 Shish-Kebab Structure Model (Pennings-Keller Model)

The so-called shish-kebab structure was first found by Mitsuhashi in 1963 [43] and then by Pennings in 1965 [44]. This was obtained as fibrillar polymer crystals from

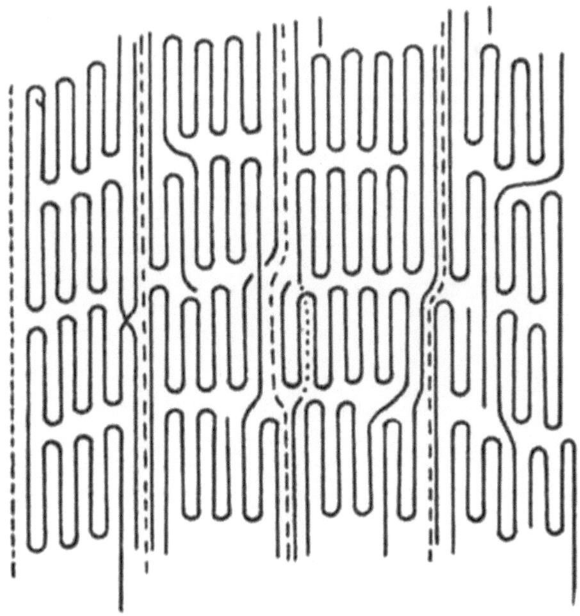

Fig. 1.11 Folded chain microfibril model (Peterlin model). In order to explain the fiber strength, some extended tie chains are introduced into the amorphous region of the Hosemann-Bonart model in Fig. 1.10 (Peterlin [42])

flowing solutions by vigorously stirring dilute solutions at high temperatures. Electron microscope observations on the fibrillar polymer crystals revealed the structures consisting of a central thread covered with folded chain platelets at regular intervals. These structures are called as shish kebabs after P. H. Lindenmeyer because they are similar to a Turkish dish "shish kebab" consisting of pieces of meat roasted on skewers. The thread and the platelets correspond to the shish and the kebabs, respectively. Subjected to ultrasonic cleaning, the kebabs were easily removed from the shish. This suggested that the kebabs would be crystallized epitaxially on the shish.

While the shish-kebab structure was discovered in the flowing solutions, similar shish-kebab structure was also found in the spinning and drawing process or the injection molding process of polymer melts by several researchers. For example, polarizing microscopic observations on an extruded nylon-66 filament revealed that the sample had the so-called row structure where many flow lines overgrown with many lamellae at nearly regular intervals developed along the extruding direction [45]. Furthermore, a clear shish-kebab structure was also observed in strain-induced crystals of rubber under the electron microscope by Andrews [46]. Keller et al. [47, 48] discussed the structure from the viewpoint of orientation-induced crystallization and referred to it as the "row structure" because the crystalline fibrils were aligned in the extrusion direction. However, they later renamed it "interlocking shish kebab" because the name "shish kebab" has been more popular in the scientific community. Hence, we also refer to as "shish kebab" in order to avoid misunderstanding.

Afterwards, Keller et al. [49–51] often found the shish-kebab structure in the research on the development of ultra-high modulus polyethylene fibers. Then, they also noticed that higher molecular weight components tended to become shish more easily, and hence the shish-kebab structure could be observed more clearly when a high molecular weight component was added. Figure 1.12 shows a transmission

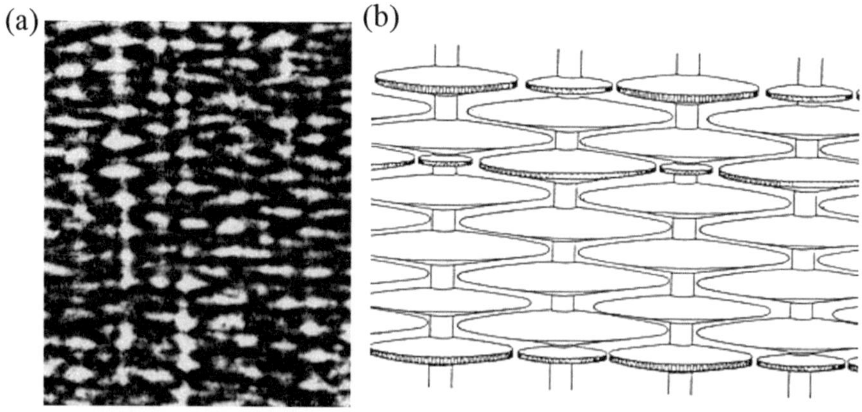

Fig. 1.12 Transmission electron micrograph of interlocking shish kebabs in melt-extruded polyethylene filaments (**a**) and a corresponding computer-generated perspective view (**b**) (Bashir et al. [51])

electron microscopic image of the thus-obtained shish-kebab structure (a) and its computer simulation model (b). They also found that similar shish kebab was obtained using polymers with a proper molecular weight distribution even if a high molecular weight component was not added [52]. Figure 1.12b shows how the lamellae of adjacent shish kebabs are interlocked. In this picture, the average thickness of the lamellae is 300 Å, the average diameter of the shish is about 250 Å, and the shish often reaches to 500 μm in length. It may be considered that the lamellae are folded chain crystals and the shish is an extended chain crystal. The number density of shish's per unit area increased with the draw ratio. Especially, drawn polyethylene fiber prepared from gel spinning [53] has an extremely high number density [54], so that the lamellar width is reduced significantly and the lamellae on the adjacent shish's are out of place, leading to fibrillation. For further reference, the evaluation methods of the volume fraction and the elastic modulus of shish were developed by Ruland et al. [55].

Comparing the folded chain microfibril model in Fig. 1.11 with the shish-kebab model in Fig. 1.12, it is obvious that the tie chains in the former model, which were introduced to explain the strength and elasticity of the fiber, correspond to the extended chain crystals or shish. It is also evident that the paracrystal layer structure introduced by Bonart-Hoseman is caused by the interlocked lamellae between the adjacent shish. In other words, the trajectory of lateral traversing along the crystalline lamellar surfaces fluctuates vertically and forms a wavy line. In addition, Ward et al. [56] found that when the fiber was highly stretched, the crystal size along the fiber axis reached more than twice of the long period though the long period hardly changed. This result also can be explained in terms of the shish-kebab model as follows. The long period corresponds to the spacing between the interlocked folded chain lamellae on the shish, which depends on the crystallization temperature but not on the draw ratio. On the other hand, the crystals in the shish must be stretched chain crystals, which are much larger than the folded chain lamellae. With increasing the draw ratio, the fraction of the shish increases, resulting in the contribution of the extended chain crystals in the shish which increases more largely. The long-period reflection intensity is reduced significantly when the fiber is highly drawn. This can be understood as a result of a reduction in the lamellar width and the lamellar fraction.

1.4 Conclusion

As described above, we can conclude that the shish-kebab model is the most reliable model for synthetic fibers. However, this model was developed based on the experimental results mainly on polyethylene and polyamides. It should be confirmed in future if the shish-kebab model is also valid for fibers of other polymers such as poly(vinyl alcohol) and polyester. There are still many unsolved problems in the molecular structure in the shish kebab. For example, one question is if the structure of shish would be similar to the fringed-micelle model by Hess-Kiessig. It is also a question if the kebabs are connected by tie chains and/or entangled chains. In any case, the structure of synthetic fibers differs significantly from the ideal structure in natural cellulose fibers. In order to develop a fiber having a super high performance such as natural fiber, it is necessary to clarify the formation mechanism of the shish-kebab structure. Such research has just started, and it is expected that the formation process is closely related with micro-phase separations during the incubation period before crystallization [57–63].

References

1. H. Staudinger, Berichte **53**, 1073 (1920)
2. H. Staudinger, *Arbeitserinnerungen* (Huthing Verlag, Heidelberg, 1961)
3. C. Priesner, *H. Staudinger, H. Mark und K. H. Meyer: Thesen zur Größe und Struktur der Makromoleküle* (Verlag Chemie, Weinheim, 1980)
4. Y. Furukawa, *Inventing Polymer Science: Staudinger, Carothers, and the Emergence of Macromolecular Science* (University of Pennsylvania Press, Philadelphia, 1998)
5. C. Nägeli, "Dei Micellartheorie", Ostwalds Klassiker Nr. 227, Akademie Verlag, Leipzig, 1928. Cf: His full name was Karl Wilhelm von Nägeli, but Karl was often spelled as Carl
6. H. Ambronn, Ber. Dtsch. Bot. Ges. **6**(85), 225 (1888)
7. H. Ambronn, Kolloid- Z. **18**(70), 273 (1916)
8. H. Ambronn, Kolloid- Z. **20**, 173 (1917)
9. R.O. Herzog, W. Jancke, Ber. Dtsch. Chem. Ges. **53**, 2162 (1920)
10. R.O. Herzog, Ber. Dtsch. Chem. Ges. **58**, 1254 (1925)
11. M. Polanyi, *Z. Physik*. **7**, 149 (1921); **9**, 123 (1922)
12. S. Nishikawa, S. Ono, Proc. Math. Phys. Tokyo **7**, 131 (1913)
13. O.L. Sponsler, W.H. Dore, Coll. Symp. Monog. **4**, 174 (1926)
14. H. Mark, Ber. Dtsch. Chem. Ges **59**, 2982 (1926)
15. H. Mark, K.H. Meyer, Z. Phys. Chem. **2B**, 115 (1929)
16. J. Hengsten, H. Mark, Z. Krist. **69**, 271 (1928)
17. P. Scherrer, Gottinger Nachr. **2**, 98 (1918)
18. K.H. Meyer, H. Mark, Ber. Dtsch. Chem. Ges., **61**, 593, 1932 (1928)
19. O. Gerngross, K. Herrmann, W. Abitz, Z. Phys. Chem. **10**, 371 (1930)
20. W. Abitz, O. Gerngross, K. Herrmann, Naturwissenchaftliche **18**, 754 (1930)
21. A. Frey-Wyssling, Protoplasma **25**, 261 (1935)
22. A. Frey-Wyssling, Protoplasma **26**, 45 (1936)
23. O. Kratky, H. Mark, Z. Phys. Chem. **B36**, 129 (1937)
24. K.H. Mayer, W. Lotmar, Helv. Chem. Acta. **19**, 68 (1935)
25. B. Alberts et al., Molecular Biology of the Cell. Garland Science, New York (2002)
26. E.W. Fischer, O. Herchenroder, R.S.J. Manley, M. Stamm, Macromolecules **11**, 213 (1978)
27. K. Hess, H. Kiessig, Naturwissenschaften **31**, 171 (1943)
28. K. Hess, H. Kiessig, Z. Phys. Chem. **A193**, 196 (1944)
29. H. Kiessig, Kolloid- Z. **98**, 213 (1942)
30. K. Hess, H. Mahl, Naturwissenschaften **41**, 86 (1954)
31. K. Hess, H. Mahl, E. Guetter, Kolloid- Z. **155**, 1 (1957)
32. R. Bonart, R. Hosemann, Makromol. Chem. **39**, 105 (1960)
33. A. Keller, Phil. Mag. **2**, 1171 (1957)
34. A. Keller, Nature (London) **180**, 1289 (1957)
35. E.W. Fischer, Z. Naturforsch. **12a**, 753 (1957)

36. R. Bonart, R. Hosemann, Kolloid- Z. Z. Polym. **186**, 16 (1962)
37. R. Hosemann, Acta. Cryst. (London) **4**, 520 (1951)
38. R. Hosemann, Kolloid- Z. **125**, 149 (1952)
39. R. Hosemann, Polymer **3**, 349 (1952)
40. G. Meinel, A. Peterlin, J. Polym. Sci. **B5**, 197 (1967)
41. G. Meinel, A. Peterlin, J. Polym. Sci. **A-2**(6), 587 (1968)
42. A. Peterlin, J. Polym. Sci. **A-2**(7), 1151 (1969)
43. S. Mitsuhashi, Bull. Text. Res. Inst. **66**, 1 (1963)
44. A. Pennings, A.M. Kiel, Kolloid- Z. Z. Polym. **205**, 160 (1965)
45. A. Keller, J. Polym. Sci. **21**, 363 (1956)
46. E.H. Andrews, Proc. R. Soc. (London) **A277**, 562 (1964)
47. A. Keller, M.J. Machin, J. Macromol. Sci. (Phys.) **B1**, 41 (1967)
48. M.J. Hill, A. Keller, J. Macromol. Sci. (Phys.) **B3**, 153 (1969)
49. J.A. Odell, D.T. Grubb, A. Keller, Polymer **19**, 617 (1978)
50. Z. Bashir, J.A. Odell, A. Keller, J. Mater. Sci. **19**, 3713 (1884)
51. Z. Bashir, M.J. Hill, A. Keller, J. Mater. Sci. Lett. **5**, 876 (1986)
52. Z. Bashir, J.A. Odell, A. Keller, J. Mater Sci. **21**, 3993 (1986)
53. P. Smith, P.J. Lemstra, J.P.L. Pijpers, A.M. Kiel, Colloid Polym. Sci. **259**, 1070 (1981)
54. J. Brady, E.L. Thomas, J. Polym. Sci. Phys. Educ. **26**, 2358 (1988)
55. L. Fischer, W. Ruland, Colloid Polym. Sci. **261**, 717 (1983)
56. A.G. Gibson, G.R. Davies, I.M. Ward, Polymer **19**, 683 (1978)
57. T. Hashimoto, H. Murase, Y. Ohta, Macromolecules **43**, 6542 (2010)
58. T. Hashimoto, H. Murase, Y. Ohta, Macromolecules **44**, 7335 (2011)
59. T. Kanaya, G. Matsuba et al., Macromolecules **40**, 3650 (2007)
60. T. Kanaya, G. Matsuba et al., Macromolecules **46**, 3031 (2013)
61. Y. Hayashi, G. Matsuba, Y. Zhao, K. Nishida, T. Kanaya, Polymer **50**, 2095 (2009)
62. G. Matsuba, C. Ito, Y. Zhao et al., Polym. J. **45**, 293 (2013)
63. S. Kimata et al., Science **316**, 1014 (2007)

Chapter 2
Progress in Structure Analysis Techniques of Fibers

Kohji Tashiro

Abstract The historical progress in structure analytical techniques developed for the fiber science and engineering has been reviewed. Wide-angle X-ray diffraction (WAXD) and small-angle X-ray scattering (SAXS) methods have contributed remarkably in the clarification of detailed information of molecular chain conformation, chain aggregation in the crystal lattice, and aggregation of crystalline lamellae in the fiber. The increasing role of the simultaneous measurement system of WAXD and SAXS is emphasized, which is helpful for the trace of the structural evolution process in the melt crystallization, melt spinning, etc. Neutron scattering method has played also an important role in the structural study of fiber. Several examples including polyethylene, polyoxymethylene, cellulose, and silk fiber have been reviewed for demonstrating the progress in structural analysis of fibers. Scanning electron microscopes (SEMs) and transmission electron microscopes (TEMs) are useful to give the direct images of fibers and the inner structures up to the level of the atomic array in the crystal lattice in some cases. The technical developments in FTIR and Raman vibrational spectroscopy, NMR spectroscopy, AFM and STM, thermal analyses, and computer simulations have been also reviewed.

Keywords Structure analysis • Fiber • Crystal structure • Chain conformation • Hierarchical structure

2.1 Introduction

Since the scientific establishment of polymer concept by H. Staudinger in 1920s, the structure analysis of polymers has been recognized to be the most important for the clarification of the images of polymer solids. The "structure" covers a wide hierarchical level, that is, the stereoregularity of polymer chains, the chain conformation, the aggregation state of these chains, the stacking of crystalline lamellae

K. Tashiro (✉)
Department of Future Industry-oriented Basic Science and Materials, Toyota Technological Institute, Tempaku, Nagoya 468-8511, Japan
e-mail: ktashiro@toyota-ti.ac.jp

Society of F.S. Technology, Japan (ed.), *High-Performance and Specialty Fibers*,
DOI 10.1007/978-4-431-55203-1_2

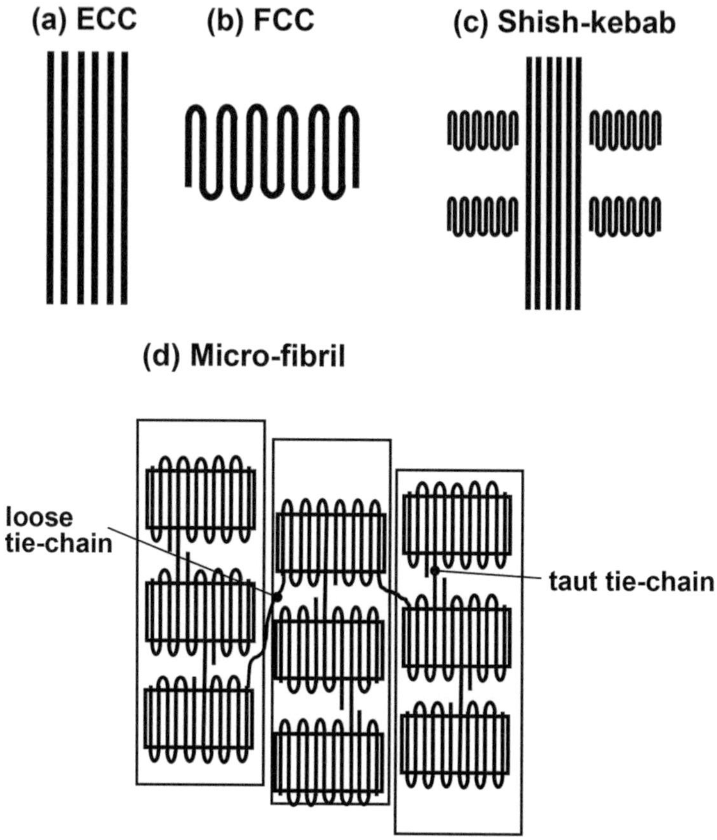

Fig. 2.1 Illustrative models of (**a**) extended chain crystal, (**b**) folded chain crystal, (**c**) shish-kebab and (**d**) micro fibril. The taut and loose tie chains are also indicated

including the amorphous regions, the outer and inner structure of spherulite, and so on. As for the fibers to be focused here, they show the various characteristic structures or morphologies, for which the various concepts were proposed including the microfibril structure, tie chains, fully extended chain crystal (ECC), shish-kebab consisting of ECC and FCC (folded chain crystal), etc. (Fig. 2.1). A monofilament or a single fiber shows the different structural feature between the skin and core parts. The characterization of inner structure of fibers is quite important for understanding the characteristic features of fibers from the macroscopic to microscopic points of view. The fiber science has been developed remarkably on the basis of these structural images necessary for the improvement of fiber property.

In this chapter, the development of analytical techniques to study the microscopically viewed structure of fibers will be described. That is to say, the development of the scattering methods includes X-ray, electron, neutron, and light scattering techniques to reveal the aggregation state of chains, the vibrational spectroscopic methods or the infrared and Raman spectroscopies to know the conformation and interatomic interactions of polymer chains, the NMR spectroscopy to reveal the molecular motion and stereoregularity of chains, and the transmission electron microscope (TEM), atomic force microscope (AFM), and scanning tunneling microscope (STM) to see the direct images of inner structure and/or surface structure of fibers. The visible and ultraviolet spectroscopy, extended X-ray absorption edge fine structure (EXAFS), and X-ray photon spectroscopy (XPS) might give us the details of information from the electron energy level, which will not be treated here.

2.2 Development of Structure Analysis of Fibers

2.2.1 Wide-Angle X-ray Scattering Technique for Crystal Structure Analysis

The X-ray scattering method played an important role in the establishment of polymer science in 1930s [1–3]. The fiber period along the chain axis was estimated for the various kinds of synthetic and natural fibers, which is only several to several tens angstrom. This order of the period was assumed at first to be impossible to include a giant polymer chain into a unit cell totally. But a concept of a repetition of monomeric units along the fiber axis allowed us to understand the X-ray scattering data quite reasonably [4]. That is to say, the long polymer chain is extended and can be stored into the crystal lattice. The array of monomeric units along the chain axis and the parallel arrangement of these chains create the crystal lattice, which gives the corresponding X-ray diffraction pattern. Since then the various trials of the crystal structure analysis were made for the various kinds of synthetic polymers including polyethylene, polyoxymethylene, poly(ethylene oxide), cellulose, and so on [5–7]. In 1953, the double-stranded helical structure model was proposed for DNA fiber, which was a great success in the crystal structure analysis of fiber substance and opened a molecular biology [8, 9]. The development of X-ray diffraction measurement technique has made it possible to collect more accurate and more number of the observed reflections, which lead to the structure analysis with higher reliability. The regular helical structure model of *isotactic* polypropylene established the concept of stereoregularity of synthetic polymers obtained from the stereospecific polymerization reaction by an organometallic catalyst in 1950s [10]. Not only these crystal structure analyses but also the quantitative analysis of individual reflection profile was found quite useful for the estimation of crystallinity and chain orientation of fibers. For example, the utilization of Scherer's equation (or the analysis of reflection width) made it possible to estimate the size of crystallites [6]. The data treatment of such a series of diffractions as 100, 200, 300, etc. allowed the analysis of the so-called second-kind disorder of polymer crystals, in which the unit cells are not packed regularly but the size fluctuates with a statistical correlation between the neighboring unit cells [11].

In addition to the wide-angle X-ray diffraction (WAXD), the small-angle X-ray scattering (SAXS) has given the details of higher-order structure of fibers. The so-called long period is an average of repeating period of the stacked lamellae [12]. The two-dimensional SAXS pattern gives the more detailed information about the spatial aggregation state of tilted lamellae in the fiber. The application of the theory of the second-kind disorder allowed us to know the more detailed spatial arrangement of lamellae. The development of SAXS theory revealed the size and shape of aggregated chains, the correlation of the neighboring lamellae, the roughness of lamellar surface, and the analysis of voids existing in the fiber [11, 12].

In this way, the development of the measurement technique and theory of X-ray scattering has led us to the deeper understanding of the fiber structure. The wide-angle X-ray scattering from the fiber substance, however, does not give very clear spots, different from the case of single crystals of the low-molecular-weight compounds. The total number of diffraction spots is quite small, only a few tens in number at most for many polymer substances, highly contrast to the several thousands of sharp spots in the low-molecular-weight compounds. For getting reliable structural information from such small number of broad reflections, the various techniques were developed in the X-ray diffraction measurements [13]. The two-dimensional X-ray diffraction diagram is indispensable for the estimation of the crystalline regularity as a whole, which can be measured using a flat camera and a cylindrical camera. The brilliance of X-ray beam should be as high as possible. Starting from the vacuum-tube type, the development of rotating anode combined with the focal mirror and the appearance of synchrotron X-ray source (1970s) give a highly brilliant diffraction pattern which can be useful for the quantitative structure analysis. At the same time, the detector became progressed remarkably: the Geiger–Müller tube, the scintillation counter, the 1D- and 2D-position-sensitive proportional counter (PSPC), the imaging plate, and the silicon pixel detector (Pilatus). The combination of highly brilliant X-ray beam with these highly sensitive detectors has made it possible to take the diffraction data only in seconds to minutes. As a result, the time-resolved measurement of the crystallization from the melt and the phase transition during the stretching of fiber, for example, can be made even in the laboratory [14].

The simultaneous measurement of WAXD and SAXS data is useful for tracing the structural change in the various phenomena. Some systems are as shown in Fig. 2.2 [13, 14]. The case (a) utilizes an imaging plate for the WAXD measurement, and the SAXS component can pass through the central hole of this first detector and reach the second detector positioned at a distance far from the sample. The case (b) utilizes only one detector onto which both of WAXD and SAXS signals can be recorded at the same time. As shown in a later section, the setting of a laser probe around the sample position makes it possible to measure the Raman scatterings simultaneously in addition to the WAXD and SAXS data. Another system is the simultaneous measurement of the transmission infrared spectra and the WAXD and SAXS data [15, 16]. In this way, the simultaneous measurement system is quite useful for the thin film and fiber materials subjected to the various external conditions.

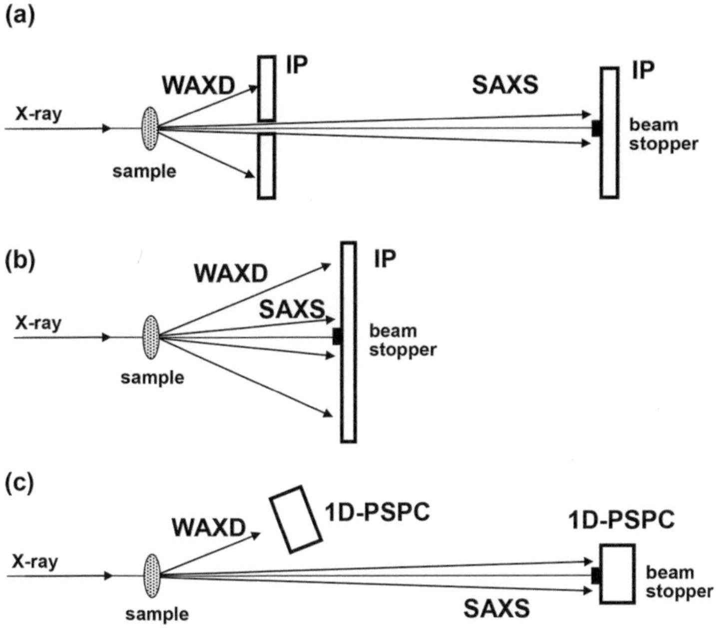

Fig. 2.2 The simultaneous measurement systems of wide-angle and small-angle X-ray scatterings [13]

2.2.2 *Neutron Scattering Method*

Neutron scatterings contain the various types [17, 18]. The neutron is scattered by the atomic nuclei and gives the different information from the X-ray scattering by electrons. Besides the scattering cross section or neutron scattering power of nucleus is not systematic but depends on the atomic species. For example, H atom has a negative cross section for coherent scattering, while D atom has a positive value and comparable to that of carbon atom. The coherent elastic neutron scattering allows us to analyze the crystal structure. The coherent inelastic neutron scattering is useful for the detection of vibrational frequency-phase angle relation of the solid materials. Quasi-elastic scattering reveals the diffusive motion of atoms in the solid. These various techniques are useful for the study of polymer materials. The small-angle elastic coherent neutron scattering (SANS) leads to the success of the experimental confirmation of Flory's theory that the amorphous chains behave like an ideal random coil with the relation of $R_g \propto M^{1/2}$ where R_g is a radius of gyration of a polymer chain and M is its molecular weight [19, 20]. In this experiment, an isolated hydrogenous polymer chain is surrounded by deuterated polymer species to increase the contrast of the single chain from the basic matrix. This type of polymer blends between the H and D species was used in many cases [19]. The SANS study about the shish-kebab structure of polyethylene, which is obtained in the crystallization from the melt under shear stress, was performed by using the blend samples consisting of high-molecular-weight D–PE chains and low-molecular-weight H–PE chains [21, 22].

The wide-angle neutron diffraction (WAND) method is useful for the crystal structure analysis of polymers. The physical property of polymer crystal may be assumed as that of ultimate state of bulk polymer substance. The physical property can be predicted theoretically if we can get the information of atomic coordinates in the crystal lattice [23]. However, the anisotropy in the mechanical property is governed mainly by the anisotropic interactions between the hydrogen atoms branching out of the skeletal chains. In other words, the accurate estimation of hydrogen atomic positions is indispensable for the precise prediction of anisotropic physical property of polymer crystal [24]. The X-ray scattering from the H atom with only one electron is too weak to extract the H position clearly by the X-ray crystal structure analysis. Rather, the utilization of neutron diffraction data measured for the deuterated sample may be more useful because of the almost comparable scattering power of deuterium to that of carbon atom. The measurement of neutron diffraction was previously difficult because of the too weak neutron source from the nuclear reactor. The introduction of imaging plate detector for the neutron experiment has made it possible to collect the diffraction data available enough for the quantitative analysis of polymer crystal including the determination of hydrogen (deuterium) positions. The crystal structure analysis by neutron diffraction technique was made by Tashiro et al. for the various kinds of polymer substances [24]. In more recent years, the stronger neutron pulses, which are generated from the nuclear fission of mercury atoms by bombardment of highly accelerated protons, can be used for getting the diffraction data of high signal-to-noise ratios (J–PARC) [25].

2.2.3 *Electron Diffraction Technique*

The electron diffraction (ED) method contributed greatly in the discovery of chain folding concept of polymer chain by the structure analysis of single crystal grown from the solution in 1950s [26–30]. The transmission electron microscope (TEM) showed the morphology of polymer single crystal. The corresponding ED data revealed that the chain axis is perpendicular to the surface of the single crystal, indicating clearly the short-chain stems stand up vertically along the normal to the single crystal plane. The TEM gave the various types of morphology of polymer crystals: the ECC, the shish-kebab structure, the aggregation of crystalline lamellae in the spherulites, and even the image of atomic arrays in the crystal lattice by an ultrahigh-resolution TEM technology [31–34]. The ED is useful also for the extraction of hydrogen atomic positions in the crystal lattice or, more exactly speaking, the spatial distribution of electron field potentials [35, 36]. The quantitative analysis using ED data gives sometimes a crystal structure with stereochemically curious geometry of molecular chains due to the modification of diffraction intensity by the so-called multiple-scattering effect in the single crystal

[35, 37]. The utilization of electron beam of higher power reduces such a dynamical effect, but the polymer sample is more easily damaged by electron beam [35].

2.3 Case Studies of Crystal Structural Analysis of Fibers

2.3.1 Synthetic Fibers

2.3.1.1 Polyethylene

Figure 2.3a–c shows the 2D WAXD diagrams measured for a uniaxially oriented polyethylene sample [33]. The evaluation of distance between the equatorial ($l = 0$) and the lth layer line gives the most basically useful information of fiber or the so-called fiber period or the repeating period along the chain axis [5–7]. It is 2.53 Å for polyethylene crystal, which is almost the same as the value expected for the all-transplanar-zigzag conformation, 2.54 Å. The analysis of the positions and intensities of all the observed diffraction spots allows us to determine the unit cell parameters and the space group symmetry of the crystal lattice. Polyethylene takes an orthorhombic unit cell of $a = 7.43$ Å, $b = 4.93$ Å, and c (fiber period) = 2.54 Å [38, 39]. The two chains are included in the unit cell as known from the density of the sample. The positions and orientation or the setting parameters of these zigzag chains were determined by a trial and error method so that the observed X-ray data can be reproduced as satisfactorily as possible. The utilization of X-ray beam of shorter wavelength has given us more number of observed diffraction spots as shown in Fig. 2.3b, from which the carbon atoms could be extracted using a so-called direct method. The Fourier synthesis $Fo - Fc$ has made it possible to extract the hydrogen atomic positions reasonably, where Fo and Fc are the observed structure factor and that calculated for the carbon atom, respectively. The extraction of H positions can be made more clearly by using an X-ray fiber pattern taken by a high-energy synchrotron X-ray beam, which gives several hundred of observed reflection spots as shown in Fig. 2.3a [40]. Figure 2.3d shows the thus-obtained crystal structure of the orthorhombic polyethylene.

As mentioned above, the X-ray scattering power of hydrogen atom is relatively weak compared with that of carbon atom. Figure 2.4a shows the fiber diffraction pattern of the oriented deuterated polyethylene sample taken by neutron experiment [37]. Figure 2.4b shows the case of hydrogenous sample. Figure 2.5a, b shows the hydrogen and carbon atomic positions derived from the quantitative analysis of WAND data of PE–h [-$(CH_2CH_2)_n$-] and PE–d [-$(CD_2CD_2)_n$-] samples, respectively. The H atomic positions were obtained as the peaks of negative density (Fig. 2.5b), while the D atomic positions were detected as the positive peaks because of the difference in negative and positive scattering cross sections of the H and D nuclei, respectively.

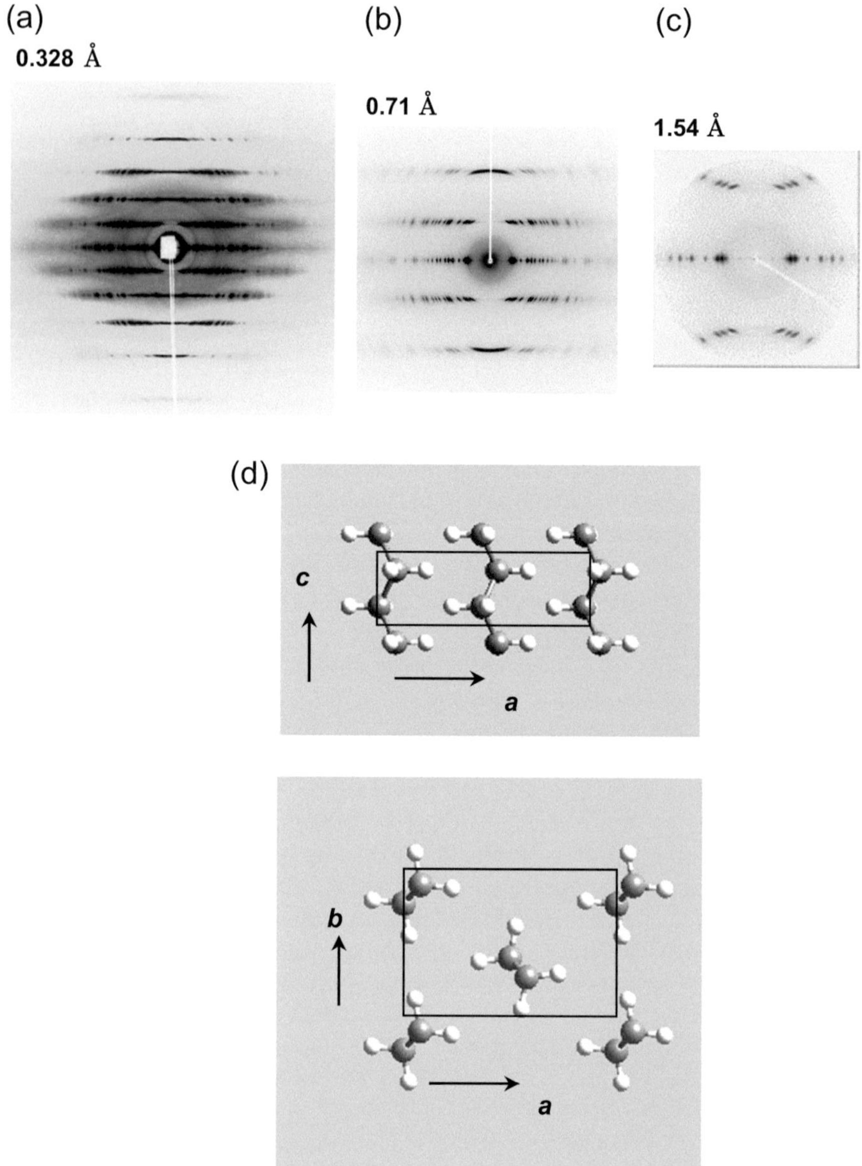

Fig. 2.3 Wide-angle X-ray diffraction diagrams of oriented polyethylene sample measured with an incident X-ray beam of wavelength (**a**) 0.32 Å, (**b**) 0.71 Å, and (**c**) 1.54 Å [40]. (**d**) The crystal structure of orthorhombic polyethylene [38, 39]

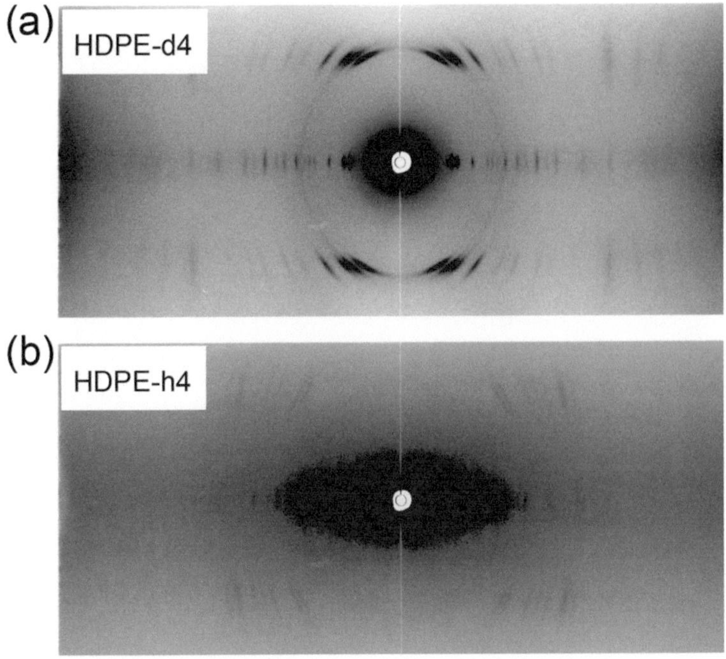

Fig. 2.4 Wide-angle neutron diffraction diagrams of uniaxially oriented (**a**) deuterated and (**b**) hydrogenous polyethylene samples [37]

2.3.1.2 Polyoxymethylene

Figure 2.6a–c shows the X-ray diffraction patterns of polyoxymethylene (POM), which was prepared by solid-state polymerization reaction of single crystals of tetraoxane monomers [39, 41–49]. The irradiation of high-energy synchrotron X-ray beam of 0.32 Å gave us more than 200 reflection spots, from which the C, O, and H atomic positions were extracted with high accuracy as shown in Fig. 2.6d [49]. The measurement of WAND patterns, as shown in Fig. 2.7a, b for hydrogenous and deuterated POM samples, respectively, gave also the positions of all these atoms accurately (Fig. 2.8). Such an accurate structural information could predict the anisotropic mechanical property of POM crystal with high reliability. Figure 2.9 shows the anisotropic curves of Young's modulus and linear compressibility of POM crystal in the plane perpendicular to the chain axis in comparison with those obtained for poly(L-lactic acid) α and δ forms. [50] The slim POM chains are packed compactly in the unit cell and give the relatively higher Young's modulus in all the directions compared with the poly(L-lactic acid) cases, making POM to be one of the most excellent engineering plastics.

The crystal structure analysis has been performed for many kinds of synthetic polymers, as described in Alexander and Tadokoro's textbooks [6, 7]. See also the references [24].

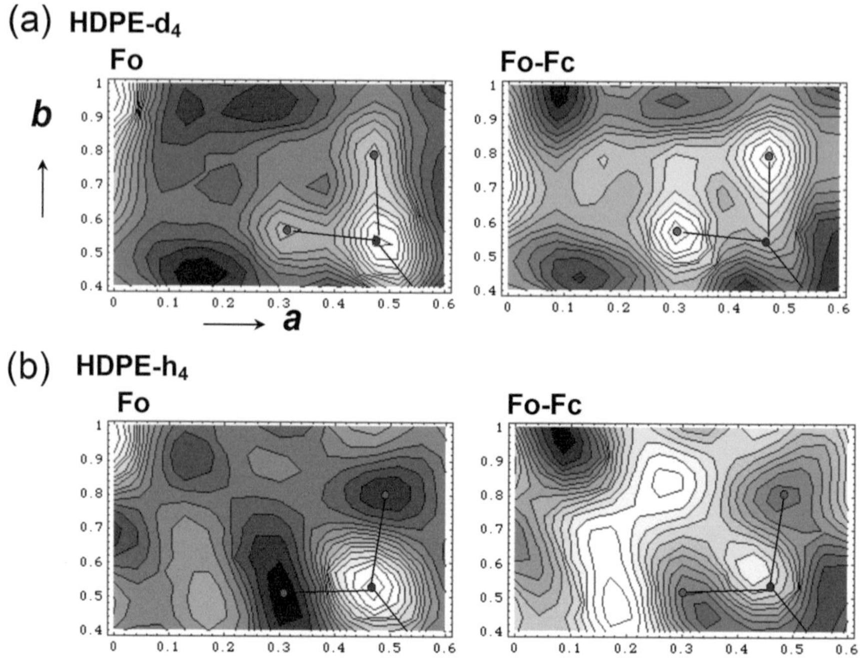

Fig. 2.5 Atomic density maps of orthorhombic polyethylene crystal [37]. (**a**) deuterated sample and (**b**) hydrogenous sample. In **a** the positive peaks were detected to show the positions of D atoms, and in **b** the negative peaks correspond to the H atoms

2.3.2 Natural Fibers

The crystal structures of natural fibers or cellulose, silk, etc. were analyzed by many researchers, which have not yet enough satisfactorily established. Some examples are reviewed here.

2.3.2.1 Cellulose

Cellulose exhibits the various crystal modifications depending on the preparation conditions. Among them the crystal forms I and II are important. Historically, the native cellulose fiber was analyzed by assuming as a pure crystal form I, but, more strictly speaking, it contains both of the different modifications, cellulose I_α and I_β [51]. The crystal structure of form I (I_β) consists of the parallel packing of cellulose chains along the fiber axis, which is a kind of extended chain crystal [52–54]. The chains are connected by the intermolecular hydrogen bonds (Fig. 2.10a). More important hydrogen bonds are created between the neighboring glucose rings along the chain axis. These intramolecular hydrogen bonds are the origin of the high Young's modulus of cellulose fiber [23, 55]. On the other hand, the crystal form II is

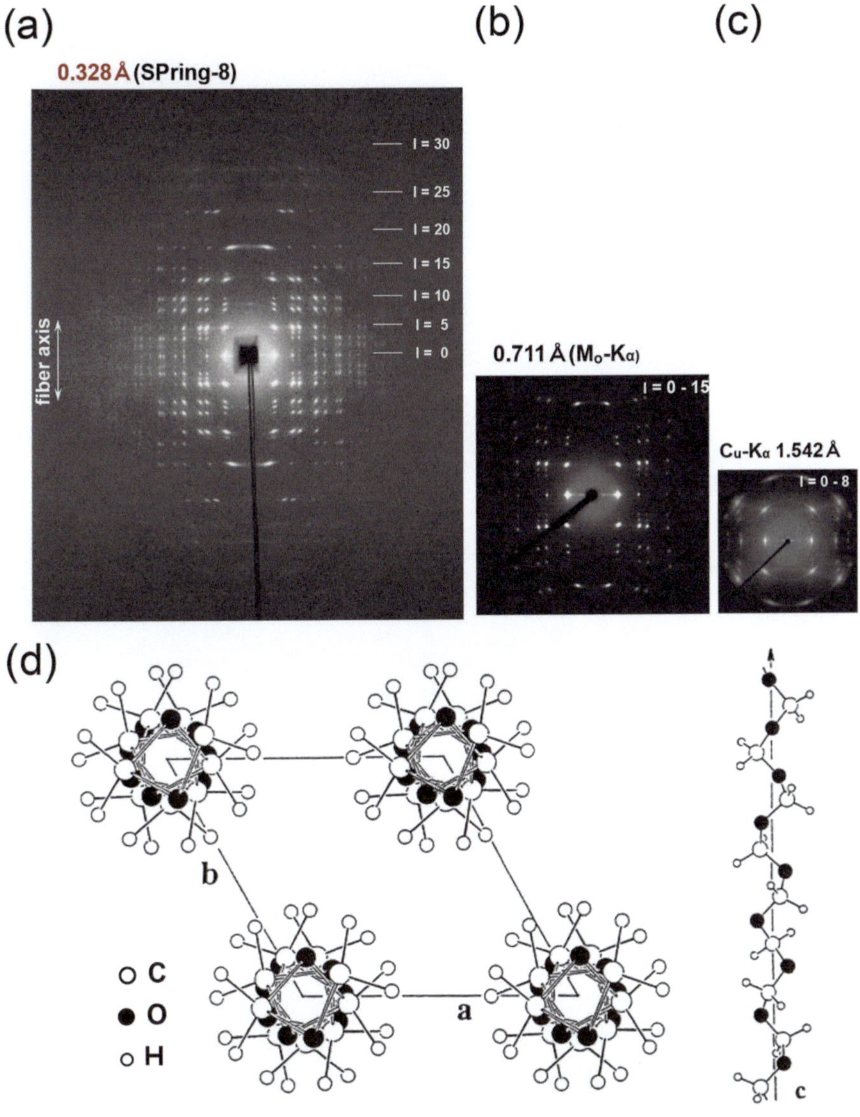

Fig. 2.6 Wide-angle X-ray diffraction diagrams of polyoxymethylene crystal obtained by photo-induced solid-state polymerization reaction, which were measured with an incident X-ray beam of wavelength (**a**) 0.32 Å, (**b**) 0.71 Å, and (**c**) 1.54 Å [49]. (**d**) The crystal structure of trigonal polyoxymethylene

a so-called regenerated cellulose and consists of antiparallel packing of fully extended molecular chains along the fiber axis as shown in Fig. 2.10b [53]. The crystal form II is prepared from the swollen cellulose sample containing the crystal form I. The transition mechanism from the parallel chain packing to the antiparallel packing was proposed, in which the crystallites are swollen and the molecular

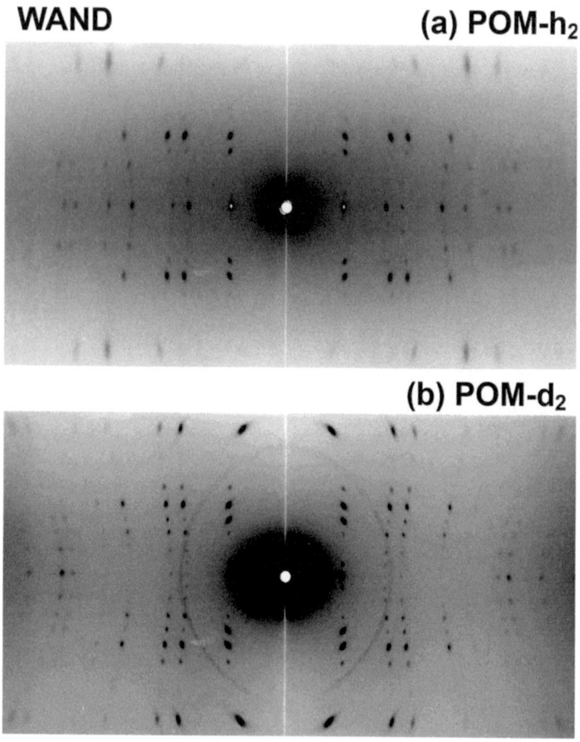

Fig. 2.7 Wide-angle neutron diffraction diagrams of **a** deuterated and **b** hydrogenous polyoxymethylene samples [49]

chains migrate from one crystallite to another and rebuild the new crystallites when the sample is dried up again [56].

2.3.2.2 Silk Fiber

In 1913, Japanese scientists Nishikawa and Ono measured the X-ray diffraction pattern of Bombyx mori silk fiber for the first time [57], which was almost the same timing as the starting of X-ray structure analysis [58–67]. Since then, the structural studies of silk have been performed by many researchers. Among them, in 1955, Marsh, Corey, and Pauling proposed the structure model shown in Fig. 2.11 [68]. The molecular chains were assumed to have a chemical formula of $-(G{-}A)_n-$ where G and A are glycine and alanine unit, respectively, although the actual silk fiber has more complicated chemical structure as shown below:

$$Y - G - A - G - A - (S - G - A - G - A - G)_8 - S - G - A - A - Y$$

where Y is tyrosine residue and S is serine unit. In such a sense, the chemical formula more suitable for the X-ray structure analysis might be $-(G{-}A{-}G{-}A{-}G{-}S)_n-$ [69]. The fiber period estimated from the X-ray fiber diagram is 6.98 Å.

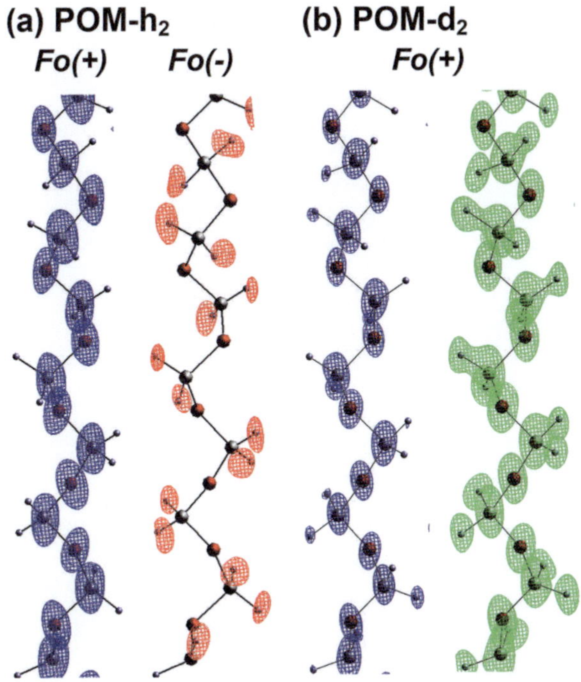

Fig. 2.8 Molecular conformation of polyoxymethylene chain with (**a**) H and (**b**) D atoms shown, which were derived from the quantitative analysis of wide-angle neutron diffraction data given in Fig. 2.7 [49]

According to the model proposed for –(G–A)$_n$– sequence by Marsh et al. [68], the molecular chain takes a zigzag shape, where the C–amide–C parts are almost straight but the chain is bent at the CH_2 and $CH(CH_3)$ positions because of their sp ^{3}-orbital characters. These chains are packed together to form a pleated sheet structure along the *a* axis. The neighboring molecular chains in a sheet are packed alternately upward and downward along the *c* axis or in an antiparallel mode, where the CO–NH vector direction is used for the indication of upward and downward chain orientation. This antiparallel packing mode is the most preferable for the formation of as many intermolecular hydrogen bonds as possible. These sheets are stacked together by the weaker van der Waals interactions along the *b* axis. The methyl groups of alanine residues orient to the normal to the sheet plane and in the same direction along the *b* axis. The orientation of the neighboring sheets or the orientation of methyl groups change alternately between the b+ and b− directions, which are symmetrically related by the 2_1 screw axis along the *c* axis. As a result, the steric repulsion between the methyl groups of the adjacent sheets can be avoided well. This mode regarding the CH_3 group orientation is called here the "polar" mode. In this way, the crystal structure model of silk proposed by Marsh et al. is expressed as a stacked sheet structure of polar and antiparallel mode.

In Fig. 2.11, it is seen that the unit cell contains the four chains. The two chains forming a sheet structure are symmetrically independent of each other as pointed out by Marsh et al. [68]. These two chains are aligned in an opposite direction along the *c* axis so that the intermolecular hydrogen bonds are created effectively.

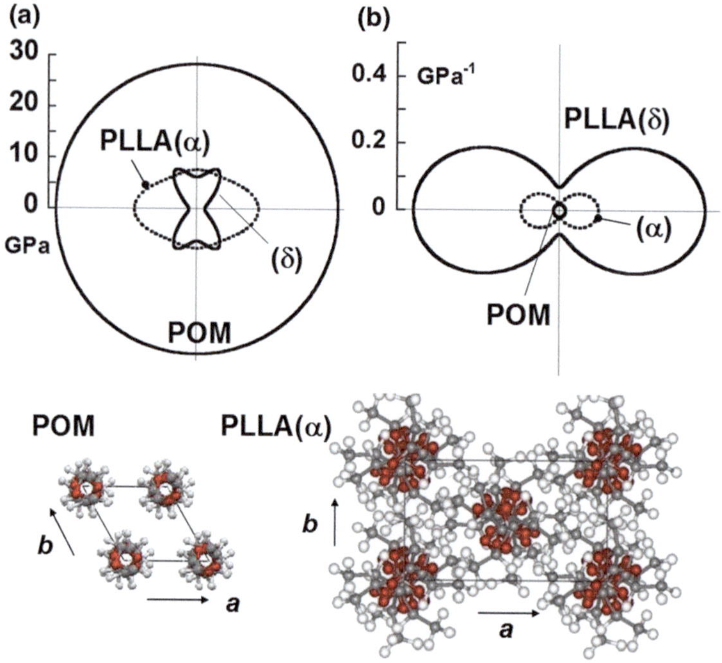

Fig. 2.9 Anisotropic curves of (**a**) Young's modulus and (**b**) linear compressibility in the plane perpendicular to the chain axis, which were calculated for poly(L-lactic acid) α and δ forms and polyoxymethylene crystals [50]

However, the orientation of methyl groups is not necessarily common. The methyl groups in the neighboring sheets may orient in the same direction (b+) or in the alternately opposite direction (b+ and b−). The latter is the model by Marsh et al. In the actual crystal lattice, the stacking of these sheets might occur more or less irregularly. For the description of structural disorder, it may be convenient to use the notation about the orientations of skeletal chain and methyl group as illustrated in Fig. 2.12. The methyl group orientation along the positive *b* direction is denoted as M^{b+}, and that along the negative *b* direction is M^{b-}. The molecular chain orientation along the positive *c* axis direction (or upward direction) is represented as C^{up}, and the opposite case of downward direction is C^{dw}. As a result, we have four possible types of orientation; $M^{b+}C^{up}$, $M^{b-}C^{up}$, $M^{b+}C^{dw}$, and $M^{b-}C^{dw}$ as shown in Fig. 2.12. The four chains included in the unit cell may take one of these four possible orientations. Of course, the basic sheet structure with strong hydrogen bonds must be created; in other words, the upward and downward chains must be arrayed alternately along the sheet plane. Figure 2.13 illustrates some possible structures of stacked sheets. For example, Fig. 2.13a is the model proposed by Marsh et al. (polar antiparallel mode). Figure 2.13b is the case of polar orientation of methyl groups in the antiparallel chain arrays, which may be energetically

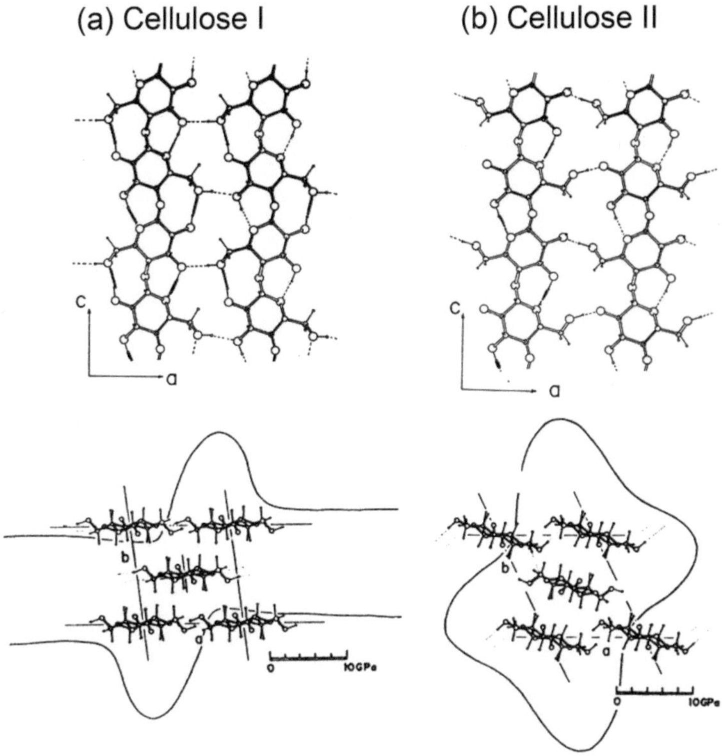

Fig. 2.10 Crystal structures of cellulose (**a**) I and (**b**) II. In the structure projected along the chain axis, the calculated anisotropic curve of Young's modulus in the plane perpendicular to the chain axis is shown for these two crystal modifications [55]

not very stable. Figure 2.13c is the model of anti-polar and antiparallel stacking of the sheets.

Takahashi et al. proposed the statistically disordered stacking model of these various regular models as illustrated in Fig. 2.14 [70]. For example, Fig. 2.14a is the disorder occurring in the regular model of Fig. 2.13c. The upward and downward chains in the sheet plane are inversed in some sheets. In the X-ray structure analysis, the statistically disordered structure may be extracted using an occupancy or probability of existence at a site, as shown in the right-side figure: the upward and downward chains with the same methyl group orientation exist at the probabilities of p and q. On the other hand, if the disorder occurs in Marsh model, Fig. 2.14b might be realized at the probabilities of p and q. In the X-ray structure analysis, these stacking disorder models are needed to be taken into consideration as discussed by Takahashi et al. [70]. However, even at present, the crystal structure of silk fiber is still unsolved well enough.

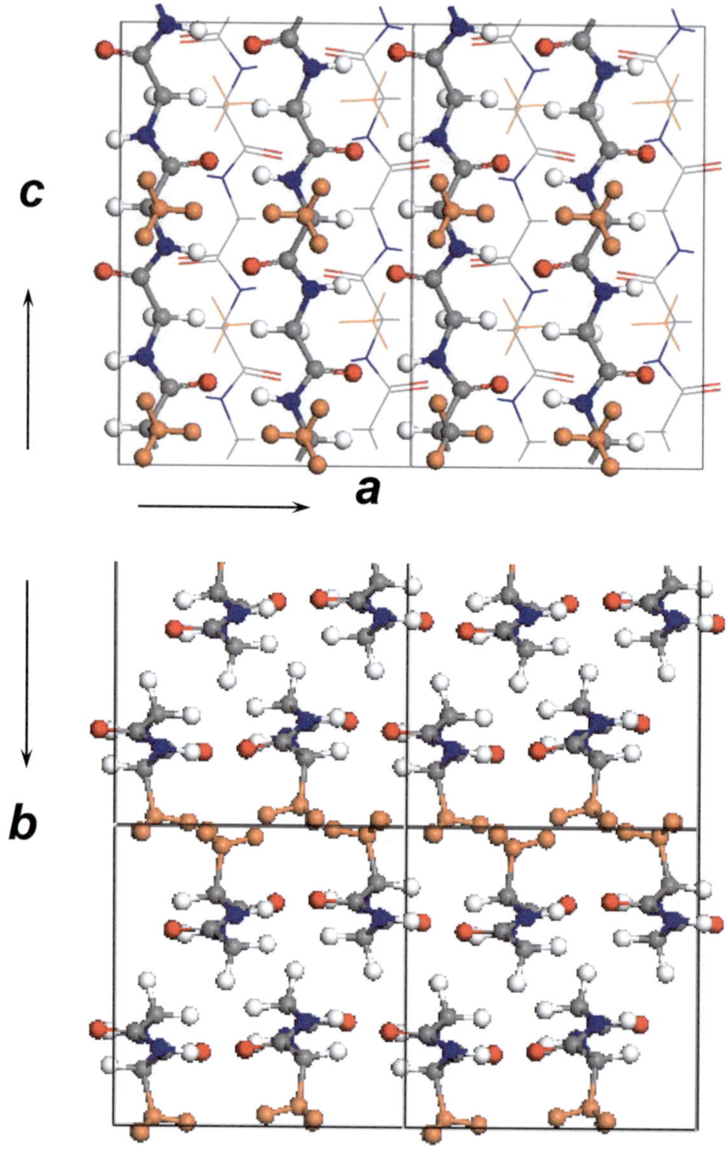

Fig. 2.11 Crystal structure model of Bombyx mori silk proposed by Marsh et al. [68]

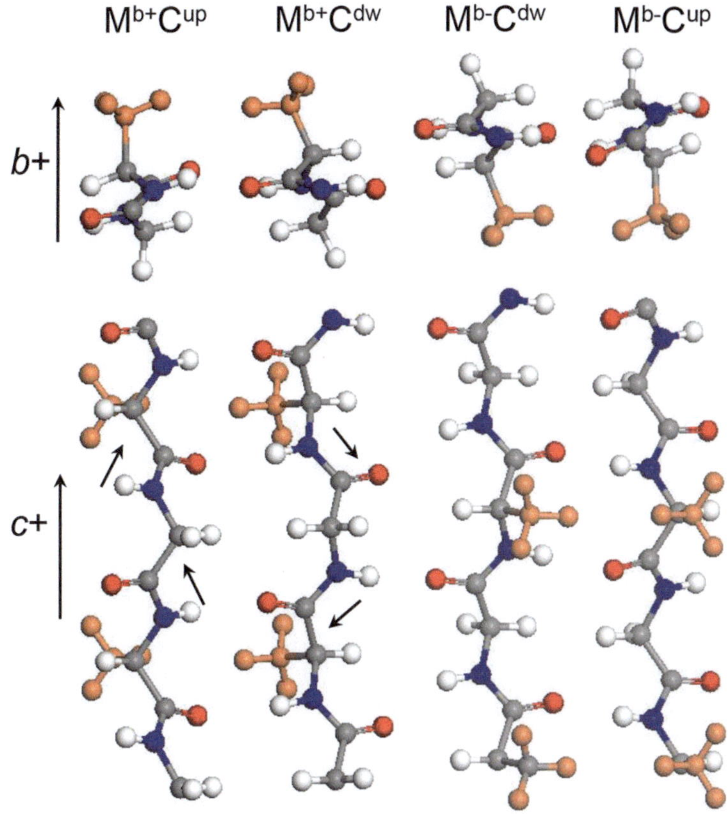

Fig. 2.12 Models to show the various possible orientations of silk chain along the *b* and *c* axes

2.4 Vibrational Spectroscopic Technique

2.4.1 Historical Development

The remarkable progress is seen in the development of vibrational spectroscopy for the study of polymer substance. In particular, the spectrometer has been made a great progress in this half century [71]. In the earlier stage of the development of infrared spectrometer, the dispersion-type monochromator of prism or grating was used for the separation of the signals of the different frequencies. The detector was a thermocouple. The total time for getting the whole spectral region of 400–4,000 cm^{-1} was about several tens of minutes. The transmission-type measurement is more quantitative to analyze the band profile, but the quite thin film is needed, about several to several tens micrometer. The reflection spectra, for example, the attenuated total reflection (ATR) spectra, are useful for the semiquantitative analysis of thicker film, but the band profile is sometimes deformed more or less due to the effect of abnormal dispersion of

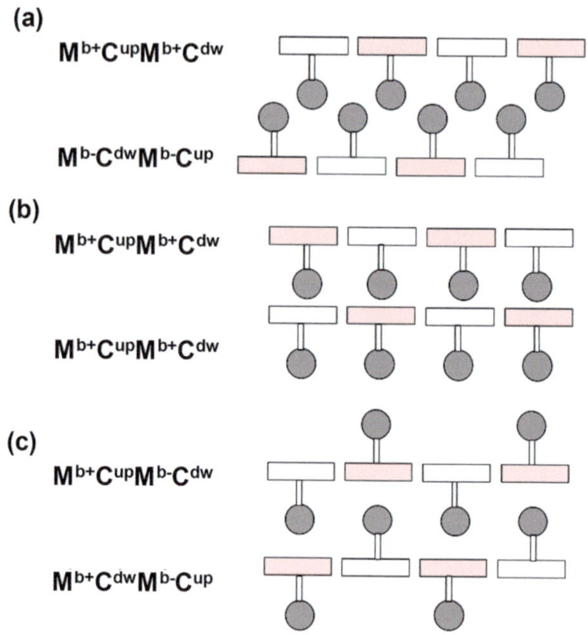

Fig. 2.13 Packing models of silk chains with the various orientations (Refer to Fig. 2.12)

refractive index. The development of Fourier transform infrared spectrometer (1969) allows us to perform more quantitatively accurate and highly time-resolved measurement of infrared spectra by using a brighter light source and more highly sensitive semiconductor detector. As a result, the more complicated experiments have become possible by using the various tools such as the heater, the stretcher, the high pressure anvil, etc. The higher brightness allows to measure the infrared spectra for a sample of 5–10 μm radius using an infrared microscope. The 2D imaging is also made relatively easy by scanning the sample stage or by using a multi-detector system. The fiber sample is measured also using the infrared microscope, but the round cross section makes the measurement difficult. The cross-sectional cutout of the single fiber in the lateral direction can be useful for the microscopic measurement. The infrared spectral measurement of the fiber can be made also using an ATR method. However, the contact between the fiber and highly refractive element crystal is not necessarily good. The hard element crystal must be pressed to the fiber strongly, causing the deformation of the round fiber.

For the purpose of structural study of fibers, the Raman spectroscopy might be more helpful. The Raman scattering phenomenon was discovered in 1928 [72], and the Raman spectrometer was commercialized in the 1940s. In 1966, the introduction of laser beam enhanced the Raman scattering intensity remarkably. This method was applied also to the polymer study. However, the strong fluorescence originated from the impurity contained in the polymer material makes it difficult to perform the normal Raman spectral measurement at all. The Fourier transform Raman spectrometer that was developed in 1986 [73], which uses a near-IR laser as

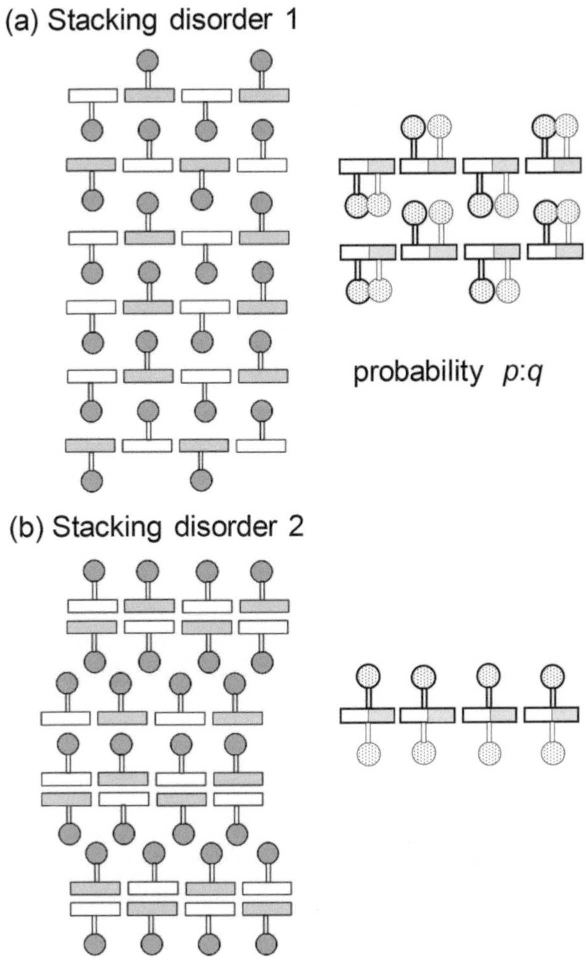

Fig. 2.14 Some illustrations of stacking disorders in the crystal structure of silk

an incident light, may give the normal Raman scattering without any generation of the fluorescence spectra. But the scattering power, which is proportional to the fourth power of laser frequency, is appreciably lower, making the measurement difficult more or less.

The bands in the low-frequency region of several tens cm^{-1} are generally difficult to measure even using a far-infrared spectrometer with a highly sensitive detector. The evacuation of the inner part of the spectrometer is needed for erasing many rotational bands of water vapors. The Raman spectrometer can measure the bands in such a low-frequency region relatively easily. The discovery of longitudinal acoustic mode (LAM or an accordion mode) may be one of the typical results in the polymer field.

2.4.2 Progression Bands

The infrared spectra of n-alkanes show many bands in the fingerprint region than our expectation from the simple chemical structure. They are called the progression bands and observed in 1,450–1,300 cm^{-1} (CH_2 bending-wagging mode), 1,300–1,100 cm^{-1} (CH_2 twisting-rocking mode), 1,200–1,000 cm^{-1} (skeletal CC stretching mode), and 1,000–700 cm^{-1} (CH_2 rocking-twisting mode) as shown in Fig. 2.15c for n-$C_{18}H_{28}$'s case [74, 75]. The origin of these progression bands comes from the vibrational frequency-phase angle curves of polyethylene (PE) chain with infinitely extended zigzag conformation (Fig. 2.15a) [76–81]. The infinitely long PE chain gives the observable bands at the phase angle δ of 0 and π for the vibrational modes of neighboring CH_2 units. Once the PE chain is cut shortly or for the n-alkane molecule with finite number of CH_2 units, the vibrational modes at the $\delta = k\pi/(n + 1)$ with $k = 1, 2, \ldots, n$ become optically active (of course, it is dependent on the odd and even numbers of n, the constraining condition at the end groups, and so on) [82]. As a result, many bands are observed in the spectra at the vibrational frequencies predicted for the phase angles on the dispersion curves as illustrated in Fig. 2.15b, c. In other words, the positions of the observed progression bands make it possible to estimate the effective trans-zigzag lengths of methylene segments included in a polymer chain.

Typical example is seen in the case of aliphatic nylon pq (-[NH$(CH_2)_p$-NHCO-$(CH_2)_{q\text{-}2}$-CO$]_x$-), which contains methylene segments of finite lengths p and q-2 and give the many progression bands in the infrared spectra [83–85]. The phase angle assignment of the observed bands was made reasonably as shown in Fig. 2.16, where the infrared spectra of a model compound of nylon 1010 [N1010CC, $CH_3(CH_2)_9NHCO(CH_2)_8CONH(CH_2)_9CH_3$] are shown. It was found that the n value in the equation $\delta = k\pi/(n + 1)$ must be p-2 and q-4 for NH$(CH_2)_p$-NH and CO-$(CH_2)_{q\text{-}2}$-CO sequences, respectively. In other words, the effective segments are shorter by two CH_2 units adjacent to the amide groups. This idea can be applied to the study of the conformational change in the phase transition of aliphatic nylon. Once the nylon is heated up to the so-called Brill transition temperature region, the progression bands observed in the lower-temperature region disappear, and the new but broad progression bands are detected at the different positions as shown in Fig. 2.16. These bands are assigned to the methylene segments of shorter trans-zigzag length, indicating that the methylene segments of long trans-zigzag form are disordered in the Brill transition region. The intermolecular hydrogen bonds are not totally broken as known from the analysis of the amide bands. In a higher temperature region near the melting point, nylon was found to show another phase transition, in which the methylene segments are totally conformationally irregular, and the hydrogen bonds are almost broken and so the molecular chains rotate violently around the chain axis [86]. In this way, the analysis of progression bands was quite useful in a study of phase transition of aliphatic nylons, in which the crystal lattice shows the order-to-disorder phase transition in the

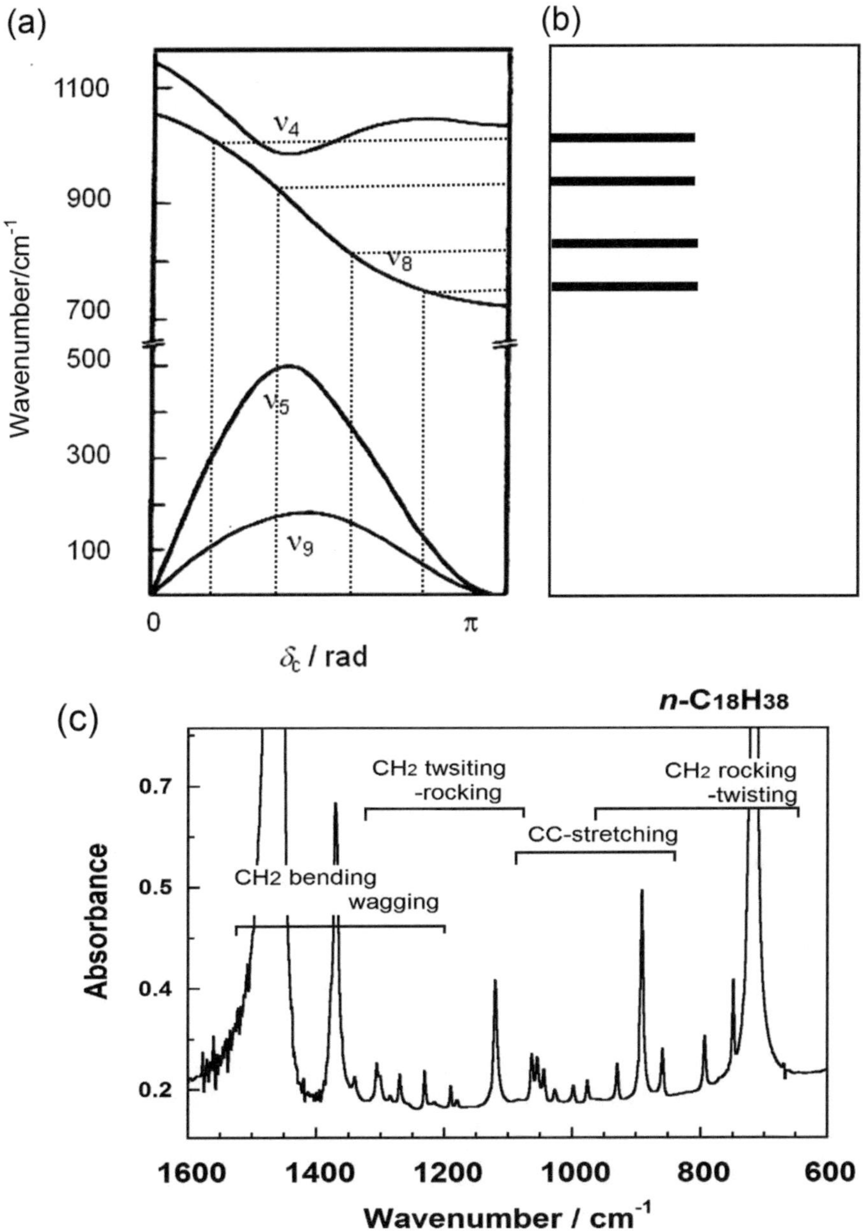

Fig. 2.15 (**a**) Vibrational frequency-phase angle dispersion curves calculated for an isolated fully extended polyethylene chain and the illustration of the progression bands observed at the several phase angle positions on the dispersion curve [81]. (**b**) The observed infrared spectrum of n-C_{18} compound to show the many progression bands corresponding to the various vibrational modes

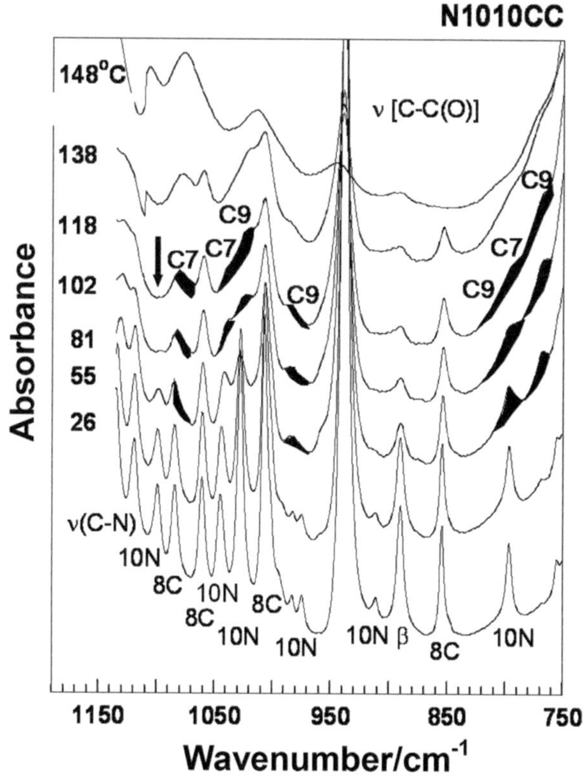

Fig. 2.16 Temperature dependence of infrared spectra of the model compound of nylon 1010 to show the drastic change in the progression bands. A symbol 8C or C7, for example, given for a band indicates that this band corresponds to that predicted for *n*-alkane molecule with C_n zigzag length (e.g., the band of 8C comes from the mode predicted for *n*-C_8H_{18} molecule). In other words, this band corresponds to the mode originated from the $(CH_2)_6$ trans-zigzag segment in the CO $(CH_2)_8$CO part of nylon 1010 chain. N and C indicate the methylene segments belonging to the N$(CH_2)_p$N and CO$(CH_2)_{q-2}$CO parts, respectively [83–85]

high-temperature region and the methylene bands are conformationally disordered by keeping the hydrogen bonds around the amide groups.

2.4.3 *Longitudinal Acoustic Mode Bands*

In addition to the progression bands, *n*-alkanes of finite chain length show the longitudinal acoustic mode (LAM) bands in the low-frequency region of Raman spectra. The band position is inversely proportional to the chain length *L*. The plot of the band position against the methylene chain length gives the dispersion curve corresponding to that predicted for the longitudinal acoustic mode of PE chain (see Fig. 2.15a, ν_5 and ν_9 branches). This discovery was made by Japanese researchers Mizushima and Shimanouchi in 1949 [87]. The acoustic mode frequency is approximately expressed by the equation $\nu_m = m(E/\rho)^{1/2}/(2Lc)$ where *E* is the Young's modulus of methylene chain, ρ is a density of a single chain, *L* is the chain length, *c* is a light velocity, and *m* is an integer. If *E* is known, then the effective chain length *L* can be estimated from the LAM band position [88, 89]. The application was made for the various kinds of PE annealed at the various temperatures. This was possible since the molecular chains are folded repeatedly at the top and bottom surfaces of the lamella, and the stem of effective trans-zigzag conformation has a finite length (Fig. 2.1). The *L* value estimated from the LAM band frequency was found to be in good correspondence to the lamellar thickness estimated from the SAXS long period [90, 91].

2.5 Solid-State NMR Spectroscopy

In addition to the molecular vibrations treated by the vibrational spectroscopy as mentioned above, the broad line NMR method is useful for the estimation of large-amplitude molecular motion in the solid state.

The NMR phenomenon was discovered in 1946 [92, 93]. Since then the high-resolution NMR spectrometer had been developed mainly, where the magnetic field strength was increased drastically to 60 MHz, 400 MHz, and 1 GHz now. The application of Fourier transform technique and the development of 2D NMR method (1975) made the analysis of complicated polymer chain configuration accurately [94]. The NMR study of polymer solids was limited to the estimation of relaxation time since the spectra are so broad. The high-resolution solid-state NMR method was developed in 1960–1970 through the introduction of magic angle rotational technique, fast Fourier transform method, etc. [95]. At present the solid-state NMR method is useful for getting the information of molecular shape and molecular aggregation in the crystalline and amorphous regions, chain folding structure on the lamellar surface, and so on. The structure information is obtained also for natural fibers including silk, cellulose, natural rubber, etc. [96–98] The simulation technique based on molecular orbital theory made it possible to predict the NMR spectra. In these years, the 3D imaging method was also developed as seen typically in the MRI images of the medical field.

2.6 AFM and STM

As already mentioned, the development of TEM and SEM made it possible to observe the direct image of crystalline and amorphous regions of polymer materials in nanoscale although the resolution was not very high [98]. The emission of characteristic X-rays from the electron-irradiated part tells us the distribution of the various elements in there. The 3D image can be also produced using a tomographic technique [99]. The STM (scanning tunneling microscope) technique developed in 1982, and the atomic force microscope (AFM) in 1986 made it possible to observe the arrangement of polymer chains on the surface, although the well-resolved image of atomic scale is still difficult [100, 101].

2.7 Thermal Analysis

The thermal analysis is useful also for the structural study of fibers. DSC was developed in 1962, which is quite useful for the estimation of equilibrium melting temperature of polymer crystals and the study of crystallization behavior from the melt [102]. The degree of crystallinity is easily evaluated by measuring the enthalpy change in the melting process. The constrained fiber was found to increase the melting point remarkably because of the smaller entropy change due to the mechanical constraint, suggesting an existence of tie chains, an important concept for the industrial consideration of mechanical property. The detailed analysis of thermal behavior of polymer has been made by measuring the heat capacity under a quasi-static condition, which can be obtained as the thermodynamic database. The thermomechanical analysis (TMA) is useful for the fiber study, which gives the thermal expansion and contraction of fibers under a constant load. Thermal gravimetric analysis (TGA) tells us the thermal degradation of fibers in the high temperatures. The recent thermometers can measure the DSC and TGA data simultaneously.

2.8 Computer Simulations

The stress distribution of a fiber is simulated by the finite element analysis (FEA) in the micrometer scale. In the experimental side, the X-ray diffraction measurement of the peak shifts at the various points of a single fiber using an X-ray micro-beam tells us the heterogeneous distribution of stress in the fiber. For example, the skin and core parts show the different deformation when the tensile force is applied to the fiber.

Computer simulation method is useful for the structural study of crystalline and amorphous regions of fibers. The molecular mechanics (MM) and the molecular dynamics (MD) allow us to predict the molecular shape and molecular aggregation state in these regions. In these calculations, the reliability of potential function parameters must be high. The molecular orbital theory had been said to be more accurate in the prediction of structure and properties, but it must be noted that the atomic orbitals used in the MO calculations are assumed as a combination of Gaussian functions, for example. The semiempirical and ab initio MO methods and the density functional theory are now being used for the study of interactions of polymer chains. But, the total number of atoms treatable in the calculation is still limited to the order of several hundreds. In other words, it is impossible now to predict the behavior of the bulk fiber sample from the atomic scale. Rather, the application of a continuous mean-field theory gives the prediction of micro-phase separations of polymer blends in nanoscale.

2.9 Conclusions

In this chapter, the development of various characterization techniques useful for the study of fiber materials has been reviewed briefly. About one century has passed already since the establishment of polymer science. The detailed analysis of inner structure of fibers has made it possible to control the processing from the microscopic point of view. However, it might be difficult to say clearly that we are now at the stage of satisfactory prediction of physical properties of fibers from the molecular level. The structure of fibers is still too complicated to do so. In 1930s, it was already pointed out that the purpose of characterization of polymer materials is to clarify the structure-property relationship from the microscopic point of view [4]. The development of the various techniques is now leading us to this ultimate purpose, but the speed is not very high, and the goal is still far beyond us.

References

1. K. Tashiro, Special issue. Rep. Progr. Polym. Phys. Jpn. **43**, 219 (2000)
2. H. Staudinger, *Der Aufbau der Hochmolecularen Organischen Verbindungen*. die Naturwissenschaften. (Springer, Berlin, 1932), pp. 84–89.
3. H. Staudinger, H. Johner, P. Signer, G. Mie, H. Hengstenberg, Z. Physik. Chem. **A126**, 425 (1927)
4. C.S. Fuller, Chem. Rev. **26**, 143 (1940)
5. B.K. Vainstein, *Diffraction of X-rays by Chain Molecules* (Elsevier, New York, 1966)
6. L.E. Alexander, *X-ray Diffraction Methods in Polymer Science* (Wiley-Interscience, New York, 1969)
7. H. Tadokoro, *Structure of Crystalline Polymers* (Wiley-Interscience, New York, 1979)
8. W. Cochran, F.H.C. Crick, V. Vand, Acta Crystallogr. **5**, 581 (1952)
9. J.D. Watson, F.H.C. Crick, Nature **171**, 737 (1953)
10. G. Natta, P. Corradini, Makromol. Chem. **16**, 77 (1955); **15**, 40 (1960)
11. R. Hosemann, S.N. Bagchi, *Direct Analysis of Diffraction by Matter* (North-Holland, Amsterdam, 1962)
12. L.A. Feigin, D.I. Svergun, *Structure Analysis by Small-Angle X-Ray and Neutron Scattering* (Plenum Press, New York, 1987)
13. K. Tashiro, S. Sasaki, Prog. Polym. Sci. **28**, 451 (2003)
14. R. Hirose, T. Yoshioka, H. Yamamoto, K.R. Reddy, D. Tahara, K. Hamada, K. Tashiro, J. Appl. Crystallogr. **47**, 922 (2014)
15. K. Tashiro, H. Yamamoto, T. Yoshioka, T. Hai Ninh, S. Shimada, T. Nakatani, H. Iwamoto, N. Ohta, H. Masunaga, Kobunshi Ronbunshu. Jpn. J. Polym. Sci. Tech. **69**, 213 (2012)
16. K. Tashiro, H. Yamamoto, T. Yoshioka, T.H. Ninh, M. Tasaki, S. Shimada, T. Nakatani, H. Iwamoto, N. Ohta, H. Masunaga, Macromolecules **47**, 2052–2061 (2014)
17. C.C. Wilson, *Single Crystal Neutron Diffraction from Molecular Materials*. Series on Neutron Techniques and Applications, vol. 2 (World Scientific Publishing Co, New York, 2000)
18. R.-J. Roe, *Methods of X-ray and Neutron Scattering in Polymer Science* (Oxford University Press, New York, 2000)
19. J. Higgins, H. Benoit, *Polymers and Neutron Scattering* (Clarendon, Oxford, 1994)
20. J. Schelten, C.D. Wignall, D.G. Ballard, W. Schmatz, Colloid Polym. Sci. **252**, 749 (1974)

21. S. Kimata, T. Sakurai, Y. Nozoe, T. Kasahara, N. Yamaguchi, T. Karino, M. Shibayama, T.A. Koenfield, Science **316**, 1014 (2007)
22. T. Kanaya, G. Matsuba, Y. Ogino, K. Nishida, H.M. Shimizu, T. Shinohara, T. Oku, J. Suzuki, T. Otomo, Macromolecules **40**, 3650 (2007)
23. K. Tashiro, Prog. Polym. Sci. **18**, 377 (1993)
24. K. Tashiro, M. Hanesaka, H. Yamamoto, K. Wasanasuk, P. Jayaratri, Y. Yoshizawa, I. Tanaka, N. Niimura, K. Kusaka, H. Takaaki, T. Ohhara, K. Kurihara, R. Kuroki, T. Tamada, S. Fujiwara, K. Katsube, K. Morikawa, Y. Komiya, T. Kitano, T. Nishu, T. Ozeki, Kobunshi Ronbunshu. Jpn. J. Polym. Sci. Tech. **71**, 508 (2014)
25. Homepage of J-PARC (http://j-parc.jp/)
26. A.P. Yundt, Tappi **34**, 89 (1951)
27. W. Schlesinger, H.M. Leeper, J. Polym. Sci. **11**, 203 (1953)
28. A. Keller, Philos. Mag. **2**, 1171 (1957)
29. E.W. Fischer, Z. Naturforsch. **12A**, 753 (1957)
30. K. Kobayashi, *4th Int. Cong. Electron Microscopy* (Berlin, 1958); Special issue of *Kagaku (Chem.)*, **8**, 203 (1962)
31. C.W. Bunn, A.J. Cobbold, R.P. Palmer, J. Polym. Sci. **28**, 365 (1958)
32. B. Wunderlich, T. Arakawa, J. Polym. Sci. A **2**, 3697 (1964)
33. A.J. Pennings et al., Colloid Polym. Sci. **253**, 452 (1975)
34. P.J. Lemstra et al., J. Mat. Sci. **15**, 505 (1980)
35. D.L. Dorset, *Structural Electron Crystallography* (Plenum Press, New York, 1995)
36. T. Ogawa, S. Moriguchi, S. Isoda, T. Kobayashi, Polymer **35**, 1132 (1994)
37. K. Tashiro, I. Tanaka, Y. Oohara, N. Niimura, S. Fujiwara, T. Kamae, Macromolecules **37**, 4109 (2004)
38. C.W. Bunn, Trans. Faraday Soc. **35**, 482 (1939)
39. K. Tashiro, K. Ishino, T. Ohta, Polymer **40**, 3469 (1999)
40. K. Tashiro, M. Hanesaka, T. Ozeki, Polym. Prepr. Jpn. **57**, 735 (2008)
41. S. Okamura, K. Hayashi, Y. Kitanishi, J. Polym. Sci. **58**, 925 (1962)
42. J. Hengstenberg, Z. Phys. Chem. **A126**, 435 (1927); Ann. Physik. **84**, 245 (1927)
43. E. Sauter, Z. Physik. Chem. **B18**, 417 (1932); **B21**, 186 (1933)
44. M.L. Huggins, J. Chem. Phys. **13**, 37 (1945)
45. H. Tadokoro, J. Polym. Sci. **44**, 266 (1960)
46. T. Uchida, H. Tadokoro, J. Polym. Sci. **A-2**(5), 63 (1967)
47. Y. Takahashi, H. Tadokoro, J. Polym. Sci. Polym. Phys. Ed. **16**, 1219 (1978)
48. G.A. Carazzolo, S. Leghissa, M. Mammi, Makromol. Chem. **60**, 171 (1963)
49. K. Tashiro, M. Hanesaka, T. Oohara, T. Ozeki, T. Kitano, T. Nishu, K. Kurihara, T. Tamada, R. Kuroki, S. Fujiwara, I. Tanaka, N. Niimura, Polym. J. **39**, 1253 (2007)
50. K. Wasanasuk, K. Tashiro, Macromolecules **45**, 7019 (2012)
51. J. Sugiyama et al., Macromolecules **24**, 4168 (1991)
52. K.H. Gardner, J. Blackwell, Biopolymers **13**, 1975 (1974)
53. A. Sarko, R. Muggli, Macromolecules **7**, 486 (1974)
54. C. Woodcock, A. Sarko, Macromolecules **13**, 1183 (1980)
55. K. Tashiro, M. Kobayashi, Polymer **32**, 1516 (1991)
56. T. Okano, A. Sarko, J. Appl. Polym. Sci. **30**, 325 (1985)
57. M. Nishikawa, S. Ono, Proc. Math. Phys. Soc. Tokyo **7**, 131 (1913)
58. W. C. Röntgen, Sitzgsber. Würzburg.Phys.-Med. Ges. **137** (1895)
59. W. Friedrich, P. Knipping, M.V. Laue, Münchner Ber. **303**, 363 (1912); Ann. Phys. **41**, 971 (1913)
60. P.P. Ewald, Phys. Z. **14**, 465 (1913); Ann. Phys., **54**, 519, 577 (1917)
61. W.L. Bragg, Proc. Camb. Phil. Soc. **17**, 43 (1913)
62. W.H. Bragg, W.L. Bragg, Proc. R. Soc. **A88**, 428 (1913)
63. T. Terada, Proc. Math. Phys. Soc. Tokyo **7**, 60 (1913)
64. P. Debye, Ann. Phys. **46**, 809 (1915)

65. P. Ehrenfest, Akad. Amst. Versl. **23**, 1132 (1915)
66. P. Debye, P. Scherrer, Nachr. Götting. Ges. **1u**, 16 (1916)
67. W.H. Bragg, Proc. Phys. Soc. **33**, 304 (1921)
68. R.E. Marsh, R.B. Corey, L. Pauling, Biochim. Biophys. Acta **16**, 1 (1955)
69. F. Lucas, J.T.B. Shaw, S.G. Smith, Biochem. J. **66**, 468 (1957)
70. Y. Takahashi, M. Gehoh, K. Yuzuriha, Int. J. Biol. Macromol. **24**, 127 (1999)
71. *Handbook of Vibrational Spectroscopy*, ed. by J.M. Chalmers, P.R. Griffiths (Wiley, 2002) **4**, 2437
72. C.V. Raman, Ind. J. Phys. **2**, 37 (1928)
73. D.B. Chase, J. Am. Chem. Soc. **108**, 7485 (1986)
74. R.G. Snyder, J.H. Schachtschneider, Spectrochim. Acta **19**, 85 (1963)
75. J.H. Schachtschneider, R.G. Snyder, Spectrochim. Acta **19**, 117 (1963)
76. M. Tasumi, T. Shimanouchi, T. Miyazawa, J. Mol. Spectrosc. **9**, 261 (1962); **11**, 422 (1963)
77. M. Tasumi, S. Krimm, J. Chem. Phys. **46**, 755 (1967)
78. M. Kobayashi, H. Tadokoro, J. Chem. Phys. **66**, 1258 (1977)
79. M. Kobayashi, J. Chem. Phys. **70**, 4797 (1979)
80. M. Kobayashi, H. Tadokoro, Macromolecules **8**, 897 (1975)
81. M. Kobayashi, J. Chem. Phys. **70**, 509 (1979)
82. H. Yamamoto, K. Tashiro, N. Nemoto, Y. Motoyama, Y. Takahashi, J. Phys. Chem. B **115**, 9537 (2011)
83. Y. Yoshioka, K. Tashiro, C. Ramesh, J. Polym. Sci. B: Polym. Phys. Ed. **41**, 1294 (2003)
84. Y. Yoshioka, K. Tashiro, Polymer **44**, 7007 (2003)
85. Y. Yoshioka, K. Tashiro, C. Ramesh, Polymer **44**, 6407 (2003)
86. K. Tashiro, Chinese J. Polym. Sci. **25**, 73 (2007)
87. S. Mizushima, T. Shimanouchi, J. Am. Chem. Soc. **71**, 1320 (1949)
88. R.F. Schaufele, T. Shimanouchi, J. Chem. Phys. **47**, 3605 (1967)
89. T. Shimanouchi, S. Mizushima, J. Chem. Phys. **23**, 707 (1955)
90. W.L. Petcolas, G.W. Hibler, J.L. Lippert, A. Peterlin, H.G. Olf, Appl. Phys. Lett. **18**, 87 (1971)
91. H.G. Olf, A. Peterlin, W.L. Peticolas, J. Polym. Sci. Polym. Phys. Ed. **12**, 359 (1974)
92. E.M. Purcell, H.C. Torrey, R.V. Pound, Phys. Rev. **69**, 37 (1946)
93. F. Bloch, W.W. Hansen, M. Packard, Phys. Rev. **69**, 127 (1946)
94. R.R. Ernst, G. Bodenhansen, A. Wokaun, *Principles of Nuclear magnetic Resonance in One and Two Dimensions* (Clarendon, Oxford, 1989)
95. F.A. Bovey, *High Resolution NMR of Macromolecules* (Academic, New York, 1972)
96. W.M. Pasika (ed.), *Carbon-13 NMR in Polymer Science, ACS103* (American Chemical Society, Washington, DC, 1979)
97. R.A. Komoroski (ed.), *High Resolution NMR of Synthetic Polymers in Bulk* (VCH Publications, Florida, 1986)
98. L. Sawyer, D. Grubb, G.F. Meyers, *Polymer Microscopy* (Springer, New York, 2008)
99. H. Jinnai, R.J. Spontak, Polymer **50**, 1067 (2009)
100. R. Wiesendanger, *Scanning Probe Microscopy and Spectroscopy* (Cambridge University Press, Cambridge, 1994)
101. P. Eaton, P. West, *Atomic Force Microscopy* (Oxford University Press, New York, 2010)
102. B. Wunderlich, *Thermal Analysis* (Academic, New York, 1990)

Chapter 3
Progress in Fiber Spinning Technology

Takeshi Kikutani

Abstract The aim for the progress of fiber spinning technology is to analyze the phenomena occurring in the industrial fiber-producing processes and seek the way to produce fibers which would fulfill the required properties. Fiber formation behavior needs to be investigated through two different approaches, i.e., theoretical analysis of the spin-line dynamics and in situ measurement of the spin-line. Regarding theoretical analysis, after the introduction of the fundamentals of spin-line dynamics, various approaches for analyzing the spin-line behavior including the melt spinning of noncircular cross-section fibers, non-steady-state spinning, higher-order structure development, and the control of the state of molecular entanglement are discussed. Regarding the in situ measurement, analyses on the structure development behavior in the spin-line through the measurements of molecular orientation and crystallization are emphasized. In addition, detailed necklike deformation behavior occurring in the high-speed melt spinning process is discussed. Lastly, recent development of various distinctive fiber spinning technologies such as the utilization of low-temperature extrusion, mimicking of the bio-spinning process applying microchip technology, fabrication of fibers with various cross-sectional configurations, and structure control through multicomponent melt spinning process are introduced.

Keywords Spin-line dynamics • In situ measurement • Structure development • Necklike deformation • Cross-sectional configuration

3.1 Introduction

In the Society of Fiber Science and Technology, Japan (*Sen'i Gakkaishi*), a research committee for studying fiber spinning technology was established in 1962 [1, 2]. The committee was one of a series of research committees established in

Note: Main part of this article is based on the review published in *Sen'i Gakkaishi*.[79]

T. Kikutani (✉)
Department of Organic and Polymeric Materials, Graduate School of Science and Engineering, Tokyo Institute of Technology, Tokyo, Japan
e-mail: kikutani.t.aa@m.titech.ac.jp

Society of F.S. Technology, Japan (ed.), *High-Performance and Specialty Fibers*,
DOI 10.1007/978-4-431-55203-1_3

the same year to promote the research activity in the society for specialized fields in fiber and textile engineering. In the opening lecture given by Prof. K. Kanamaru in the first meeting [3], he named Prof. A. Ziabicki [4] as one of the pioneering founders of this research field.

Since then, review articles on fiber spinning technology were published in the journal of the Society of Fiber Science and Technology, Japan (*Sen'i Gakkaishi*), every 5–10 years [5–9] with the last one by Matsui in 2004 [9]. On the other hand, the journal of the Japan Society of Polymer Processing (*Seikei-Kakou*) has been publishing special issues on research review every year. Review articles on fibers and film processing has been included in these special issues [10–19]. In addition, in the special issue commemorating the 20th anniversary of the establishment of the Japan Society of Polymer Processing, a review article on fibers and film processing was contributed by the author in 2008 [20]. In addition to the above, a review article on the detailed industrial technologies on polyester fibers by Fukuhara [21], some review articles on fiber structure development [22, 23], and books including some chapters in the field of fiber spinning technology [4, 24–28] were published.

Considering the above, research activities in the field of fiber spinning technology will be reviewed in this chapter with a rather long time span. Recent notable progress in the field also will be included.

3.2 Definition of Fiber Spinning Technology

The aim of researches on "fiber spinning technology" is to analyze the phenomena occurring in the industrial fiber-producing processes and seek the way to produce fibers which would fulfill the required properties.

As shown schematically in Fig. 3.1, fiber spinning process is consisting of the following processes: (1) giving fluidity to material through heating or with the aid of solvent, (2) shaping the material into thin and long thread, and (3) solidifying the material either by cooling or removal of solvent. In the production of polymeric synthetic fibers, characteristics of the fibers would be highly dependent on the higher-order structure developed during the fiber-forming process. Therefore, in the analysis of fiber spinning technology, combination of two totally different viewpoints, i.e., transfer phenomena and materials science, is necessary. The transfer phenomena includes the concepts of mass, energy, and momentum conservation occurring in the spinning process, whereas the materials science is necessary for understanding the mechanism of higher-order structure development. Combination of these two different disciplines is important for the complete understanding of the process. It is important to note that the macroscopic spin-line dynamics and microscopic higher-order structure formation interact with each other. For example, crystallization causes heat evolution and change in flowability of polymer melt. Crystallization rate is affected by the tensile stress applied to the spin-line, whereas the spin-line dynamics itself can be affected by crystallization. Concept of the macroscopic and microscopic analyses and their interactions in the fiber spinning process is summarized in Fig. 3.2.

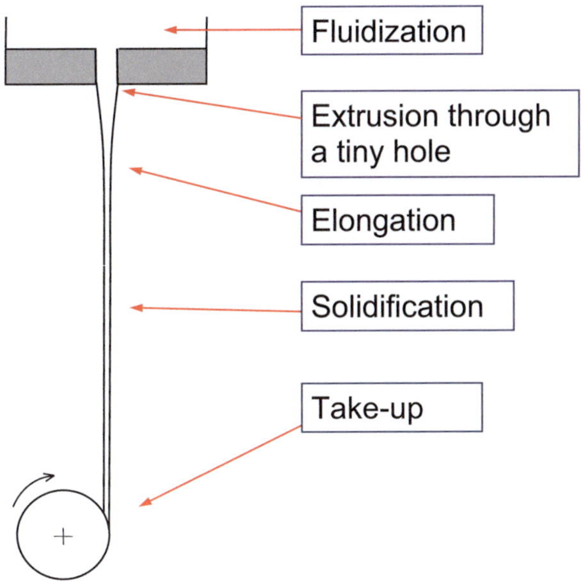

Fig. 3.1 Basic concept of fiber spinning process

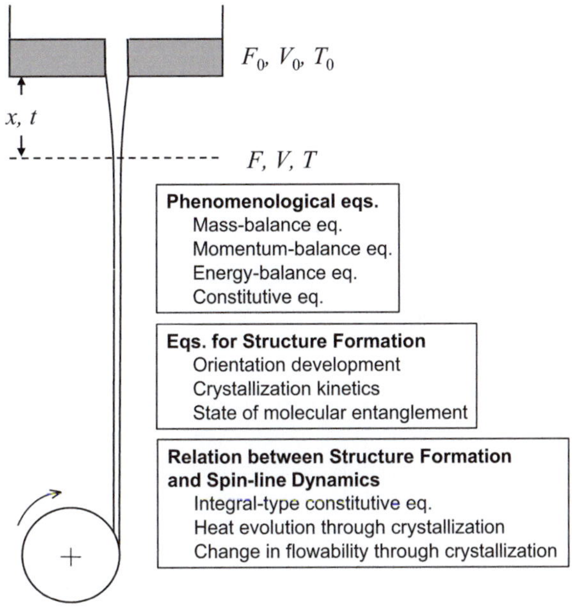

Fig. 3.2 Two important factors and their interactions in spin-line dynamics

A first approximation for the fiber spinning process is single-component fiber of circular cross section, thin thread with no variation of variables along radial direction, and steady-state process with constant thickness of resultant fibers. Nevertheless, fibers of various unique characteristics can be produced by deviating from such fundamental concepts. Those include engineered cross-sectional shape, multicomponent fibers, control of fiber characteristics along the fiber length, etc.

3.3 Theoretical Analysis of Fiber Spinning Dynamics

Fiber formation behavior needs to be investigated through two different approaches, i.e., theoretical analysis of the spin-line dynamics and in situ measurement of the spin-line. Systematic theoretical analysis for the fundamental aspect of fiber spinning dynamics was developed by Ziabicki [4]. Based on his theory, Kase and Matsuo conducted the numerical analysis of both steady-state and non-steady-state melt spinning processes in early 1960 [29, 30]. This achievement is known as the first successful trial for the numerical analysis of the polymer processing. For the melt spinning of circular cross-section fibers, one-dimensional analysis was applied to obtain the variations of velocity, diameter, temperature, and tension of the spin-line as a function of time and distance from the spinneret. For the steady-state condition, velocity can be converted to diameter based on the mass conservation principle. In this kind of analysis, constitutive equation for the flowability of polymer melt and various empirical equations describing the cooling and air-friction behaviors of the spin-line play dominant roles.

3.3.1 Melt Spinning of Noncircular Cross-Section Fibers

Prediction of the cross-sectional shape of fibers through the analysis of polymer flow extruded from the noncircular cross-section spinning nozzle is an important subject. In general, cross-sectional shape of fibers is dull in comparison with the sharp cross-sectional shape of spinning nozzle because of the viscoelastic effect during extrusion of polymer flow through spinning nozzles and also the effect of surface tension after the extrusion. For the latter case, Takarada et al. analyzed the variation of the cross-sectional shape of flat fiber and hollow fiber along the spin-line through the online measurement and numerical analysis [31]. More recently, Zhou et al. reported the good agreement of the experimental results and results of numerical analyses based on the finite element method with the application of optimized equation of surface tension [32]. In ordinary melt spinning processes, surface tension dominates the change in cross-sectional shape because of relatively low viscosity. This tendency can be clarified through the analysis of the cross-sectional shape of extrudates prepared using a noncircular cross-section nozzle under various extrusion conditions as shown in Fig. 3.3. In this figure, a nondimensional number "complexity" is defined for the cross-sectional shape as the square of circumference divided by cross-sectional area. It can be seen that the complexity increases with the increases of polymer viscosity and extrusion rate.

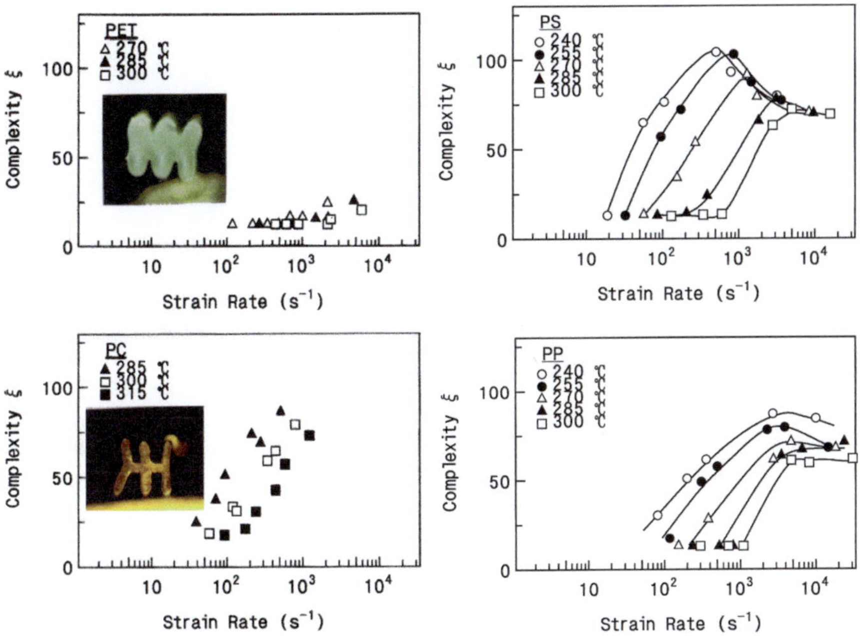

Fig. 3.3 Variation of cross-sectional shape of extrudates of four polymers prepared using noncircular spinning nozzle at various extrusion conditions

3.3.2 *Non-steady-state Spinning*

From the viewpoint of producing fibers in stable condition, researches on the time-course variations of melt spinning behavior caused by disturbances is crucial. It is astonishing to realize that Kase and Matsuo dealt with this problem when their first report appeared in 1965. In addition, there were intensive researchers on the intrinsic instability of melt spinning process known as "draw resonance," which tends to occur under the condition of isothermal spinning with high drawdown ratio. More recently, Ishihara et al. conducted the experimental and theoretical investigations on the non-steady-state melt spinning behavior by introducing mechanical disturbance to the melt spinning process equipped with a water-quenching bath. Conditions for the occurrence of draw resonance were also investigated in detail applying the power-law fluid model [33, 34].

In the theoretical prediction of draw resonance, a narrow stable region was found for the condition of extremely short air gap. Validity of this theory was confirmed through the online measurement of spin-line diameter fluctuation in the melt spinning process equipped with a water-quenching bath [35]. Result of online diameter measurement is shown in Fig. 3.4. When the position of the water-quenching bath is shifted toward the spinneret, amplitude and frequency of diameter fluctuation increase. However, if the position of water-quenching bath becomes close enough to the spinneret, spin-line suddenly stabilizes. On the other hand, with

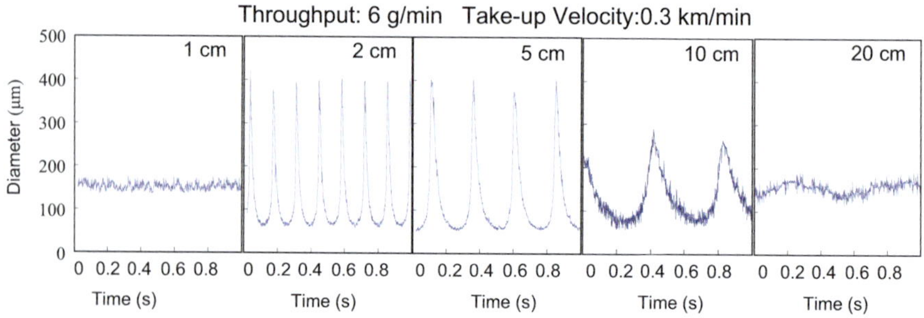

Fig. 3.4 Time-course variation of spin-line diameter for melt spinning with a water-quenching bath of various distances from the spinneret

the aim of introducing designed disturbances to the spin-line, effect of the periodic modulation of take-up velocity on spinning behavior as well as the structure and properties of resultant fibers was investigated by Kazama [36].

3.3.3 Development of Higher-Order Structure

Based on the phenomenological analysis of the melt spinning process, researches on the development of higher-order structure such as molecular orientation and crystallization in the melt spinning process have been attempted rigorously [37–40]. Such analysis is important for the prediction of the properties of fibers prepared under certain spinning conditions because the characteristics of fibers are strongly dependent on their higher-order structure. Nevertheless, we need to say that the theoretical backgrounds for the prediction of higher-order structure from the spinning conditions and also for the prediction of the characteristics of fibers from the higher-order structure still are not matured.

Prediction of higher-order structure development accompanied by crystallization is one of the most crucial problems in this field. It is well known that crystallization rate can be increased significantly if molecular chains are oriented. Once crystallization starts, however, crystallization rate may be a function of the degree of orientation in the region remaining as a melt. In this context, understanding of the crystallization kinetic is a difficult problem since understanding of crystallization behavior from the view of molecular level is necessary.

Effect of flow history on the flowability of polymer melt itself may be introduced applying the integral-type constitutive equation, while the effect of flow field on the higher structure development may be introduced applying the stress-optical rule and crystallization kinetics for non-isothermal crystallization under the effect of stress. However, difficulty for the understanding of structure formation behavior in the melt spinning process also arose from the viewpoint of the effect of structure development on spin-line dynamics. In other words, flow history of the polymer melt as well as the structure development affect the processing behavior itself through the change in the flowability or deformability of the material in the process.

On the other hand, regarding the prediction of the molecular orientation in the final products, effect of orientation-induced crystallization on the change in the state of molecular orientation needs to be considered. In other words, under the effect of flow, crystallization itself would promote the further increase of molecular orientation because crystallization can be considered as the spontaneous ordering process in nature.

3.3.4 *Effect of Molecular Entanglement*

As described in the previous sections, along with the analysis of the melt spinning process from a macroscopic view utilizing fundamental mass, momentum, and energy balance equations based on the transfer phenomena, microscopic analysis of the process is indispensable from the viewpoint of higher-order structure development in the fibers. For this purpose, combining of numerical simulation programs for the analyses of melt spinning dynamics and molecular dynamics has been proposed. The concept of this analysis is the same as that of "seamless zooming" in the research project on "Platform for designing high functional materials (NEDO)" conducted by Doi [41] that is an attempt for understanding the behavior of polymeric materials through continuous views from microscopic/short-term level to macroscopic/long-term level with the connection in between the two extremes by applying theoretical analyses of several mesoscopic levels. In the project for the development of high-strength polyester fibers, which was one of the subthemes in the program called "Nanostructured Polymeric Materials" sponsored by NEDO, we have discussed the change in the state of the entanglement of molecular chains and the effect of such change on the mechanical properties of resultant fibers by introducing the concept for the control of melt structure in the melt spinning process [42, also see Chap. 8 in this book]. For this purpose, numerical analyses on the change in the state of molecular entanglement were conducted [43]. Similar attempts have been made more recently by some other researchers [44, 45].

3.4 Development of Technology for Online Measurement of Spinning Process

For the analysis of melt spinning process, online measurements of the macroscopic parameters such as temperature, velocity, diameter, and tension of the spin-line as well as the microscopic analyses on the higher-order structure development through the measurements of birefringence, X-ray scattering, etc. are necessary.

Online measurement for the crystallization of polyolefins in the melt spinning process utilizing X-ray scattering analyses has been conducted for many years starting from 1970 by Katayama, Koyama, and Zachmann [46–48]. Online X-ray measurement of condensation polymers such as polyamides and polyesters is more difficult since the orientation-induced crystallization starts to occur at much higher take-up velocities. Attempts for the analysis of the melt spinning process of polyamides were made firstly by White in 1970 [49] and then by Haberkorn of BASF [50]; the latter successfully detected the crystallization of polyamide 6 in the spinning line at the take-up velocity of 5.5 km/min. Online measurement of the orientation-induced crystallization of poly(ethylene terephthalate) was successful only by using the synchrotron radiation facilities [51, 52]. In these cases, suppression of air scattering by introducing helium gas as a cooling medium was indispensable. Experimental setup and examples of wide-angle X-ray diffraction patterns for the high-speed melt spinning of high molecular weight poly(ethylene terephthalate) obtained at the synchrotron radiation facility, SPring-8, are shown in Figs. 3.5 and 3.6. On the other hand, fiber structure development in the drawing process of undrawn fibers accompanied by necking deformation was conducted by Ohkoshi utilizing the carbon dioxide laser irradiation for the contactless fixation of the position of necking [53].

There have been development of the technology for the online measurements of spin-line diameter, velocity, temperature, and tension [54]. Through these analyses, it was found that so-called necklike deformation occurs in the high-speed spinning line when the take-up velocity is high enough for the occurrence of orientation-induced crystallization [38, 40]. Detailed analysis on the exact shape of necklike deformation in the high-speed melt spinning process and in-line drawing process was also carried out [40, 55–57]. Typical example regarding the change in the detailed shape of necklike deformation with take-up velocity for high-speed melt spinning of poly(butylene terephthalate) is shown in Fig. 3.7. Maximum strain rate evaluated through this analysis exceeded 20,000 s^{-1}.

From the viewpoint of the analysis of orientation development, online measurement of birefringence development was also carried out utilizing the optical system for the analysis of optical retardation [58, 59]. Typical example of the optical system for the online birefringence measurement and the result of online birefringence development analysis for the high-speed spinning of high molecular weight PET are shown in Figs. 3.8 and 3.9.

As described in the previous section, control of the state of molecular entanglement in the melt spinning process attracts much attention these days; however, there is no method for the in situ evaluation of the state of molecular entanglement. Considering the relatively long relaxation time of such structure, however, classical method for analysis of the spin-line through the measurement of samples captured from the spin-line may provide a certain level of information. True stress versus true strain curves of samples captured at various distances from the spinneret for the

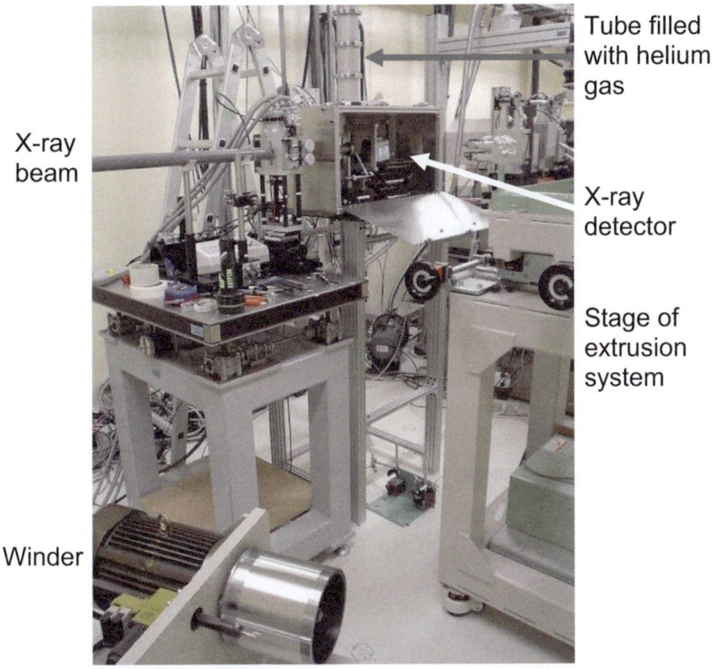

Fig. 3.5 Apparatus for online wide-angle X-ray diffraction measurement of melt spinning line set in a beam hatch at synchrotron radiation facility SPring-8

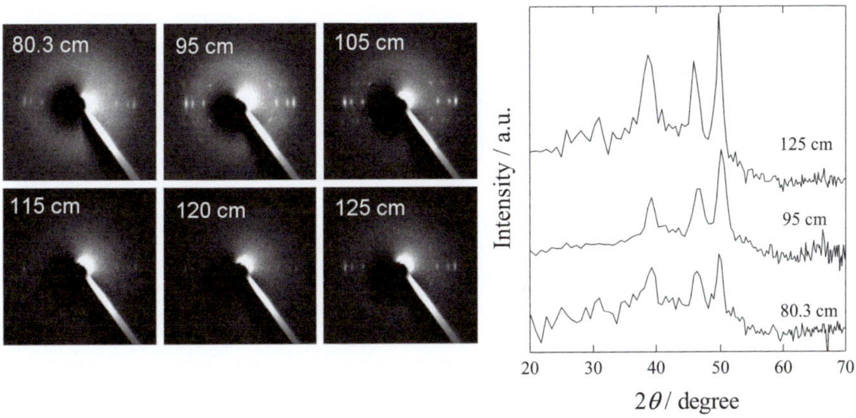

Fig. 3.6 2-D and 1-D wide-angle X-ray diffraction intensity patters obtained from the online measurement of PET spinning line at various distances from the spinneret

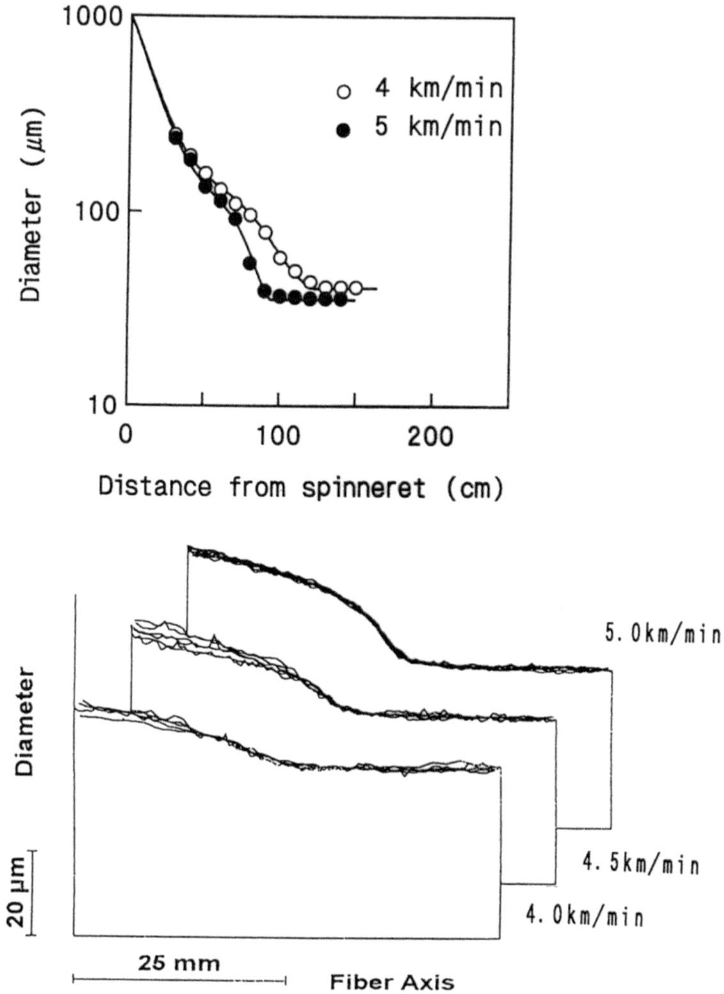

Fig. 3.7 Variation of the detailed shape of necklike deformation with take-up velocity for high-speed melt spinning of poly(butylene terephthalate). Overall thinning behavior of the spin-line is also shown at the top

high-speed spin-line of PET of two different take-up velocities are shown in Fig. 3.10 [60]. More recently, Murase applied the spin-line capturing method for the analysis of the gel-spinning process of high molecular weight polyethylene. Mechanism for the development of shish-kebab structure was discussed through the analysis of the captured samples by the transmission electron microscope [61].

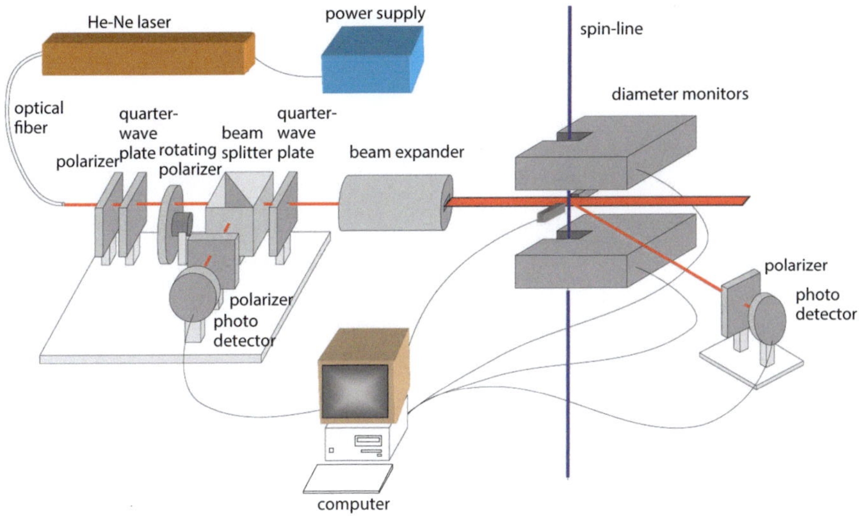

Fig. 3.8 Optical system for the online measurement of birefringence development along the spin-line

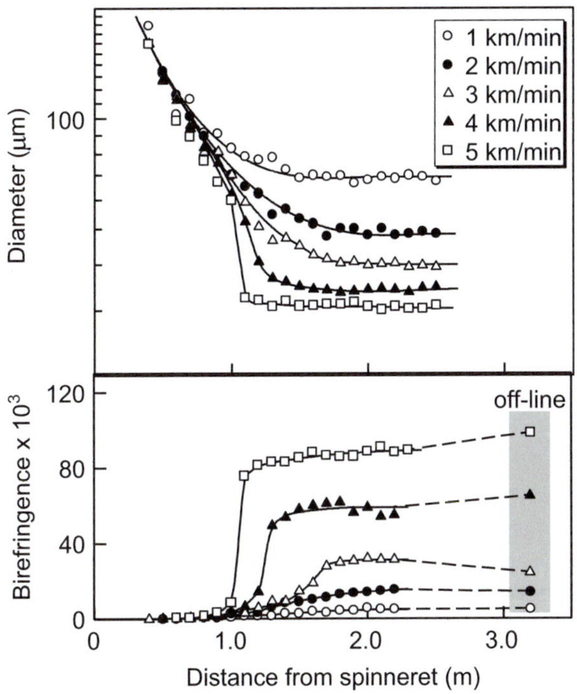

Fig. 3.9 Variations of diameter and birefringence along the high-speed spin-line of high molecular weight PET. Steep birefringence increase accompanied by necklike deformation was observed at the take-up velocities of 4 and 5 km/min

Fig. 3.10 True stress versus true strain curves of the samples captured at various distances from the spinneret for the high-speed spin-line of PET

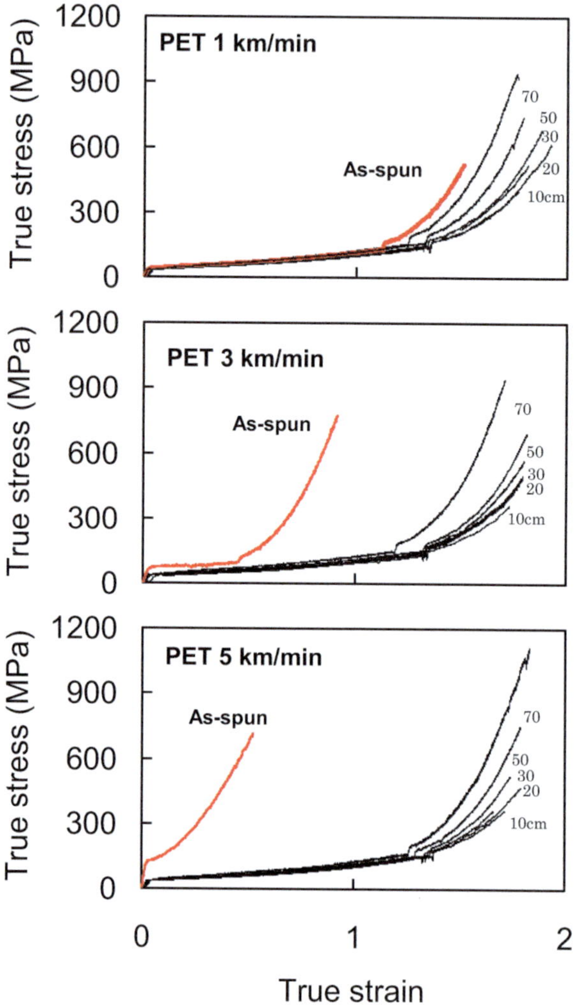

3.5 Distinctive Fiber Spinning Technologies

For the development of functionalized fibers or for fulfilling the necessity of preparing fibers from polymers with difficulty for fiber formation, various distinctive fiber spinning technologies have been developed.

Among those, some technologies utilize the extrusion at low temperatures for the production of unique fibers. Molten polymer can be extruded at a temperature below its melting temperature if residence time of the polymer in such a low-temperature region is short enough in comparison with the crystallization rate of the polymer because crystallization rate is relatively low at a temperature close to its melting point. In these cases, effect of polymer flow would enhance the formation of crystalline structure to introduce unique characteristics to the resultant fibers. Cao has reported that the extrusion of high molecular weight PET is possible at 265 °C, the temperature very close to its melting point, which leads to the starting of orientation-induced crystallization even at extremely low speed of 2 km/min [62]. On the other hand, extrusion of polymer powders to form fibers at a temperature close to its melting temperature leads to the development of unique structure in that microfibrils are connecting polymer powders. Such structure is known to exhibit a unique characteristic of negative Poisson's ratio when tensile strain is applied. The fibers with such characteristics are called auxetic fibers. Alderson et al. reported that thin auxetic fiber of 300 μm diameter could be obtained through the extrusion of polypropylene at a temperature close to its melting temperature [63]. On the other hand, Shirakashi [64] and Xu [65, 66] reported that the fibers with rough surface can be prepared through the melt extrusion of PA6/PET blend with PET being the minor component at a temperature below the melting temperature of PET. They discussed the mechanism of rough surface development through the crystallization of PET and also the applicability of such fibers for production of artificial hair. Variation of surface characteristics of PA6/PET blend fibers extruded at various temperatures is shown in Fig. 3.11.

There also are noteworthy technologies for the extrusion of polymers. In recent years, production of spider silk through the combination of molecular design, gene synthesis, microbial fermentation, and spinning technologies attracts much attention [67], whereas technology for mimicking the extrusion system of silkworm for formation of silk fibers utilizing the microfluidic channel technology has been studied. Luo et al. reported the concentration of regenerated silk fibroin aqueous solution using microfluidic devices based on regenerated cellulose membrane [68]. Renberga et al. also reported the spinning of spider silk fibers utilizing the microchip technology [69].

Various attempts also have been made for the introduction of unique characteristics to the extrudates in the melt spinning process. Researches on the numerical analysis of polymer flow in the static mixer provided deep insight for the designing of die for bicomponent extrusion [70]. Oh et al. reported that shape of boundary in side-by-side bicomponent spinning can be manipulated through the control of flow channel, which leads to the control of crimp behavior in the bicomponent fibers [71]. Tseng et al. reported the micro-manufacturing of complicated spinning die utilizing the LIGA and microinjection molding technologies [72]. Kamiyama et al. reported the production of nano-fibers through the bicomponent spinning of islands-in-the-sea-type fibers with about 1000 islands [73]. Spinning die for bicomponent spinning of polymers with a different thermal resistance was designed introducing temperature control system to independently maintain separate molten polymer streams at selected temperatures [74]. Multicomponent extrusion introducing a logo design consisting of 100 × 100 matrix into the fiber cross section was also developed [75]. More recently, development of technology for introducing arbitrary nanoscale structure in the cross section of bicomponent fibers was disclosed [76].

In addition to the high controllability of cross-sectional configuration of fibers, utilization of bicomponent spinning for the control of the structure development of

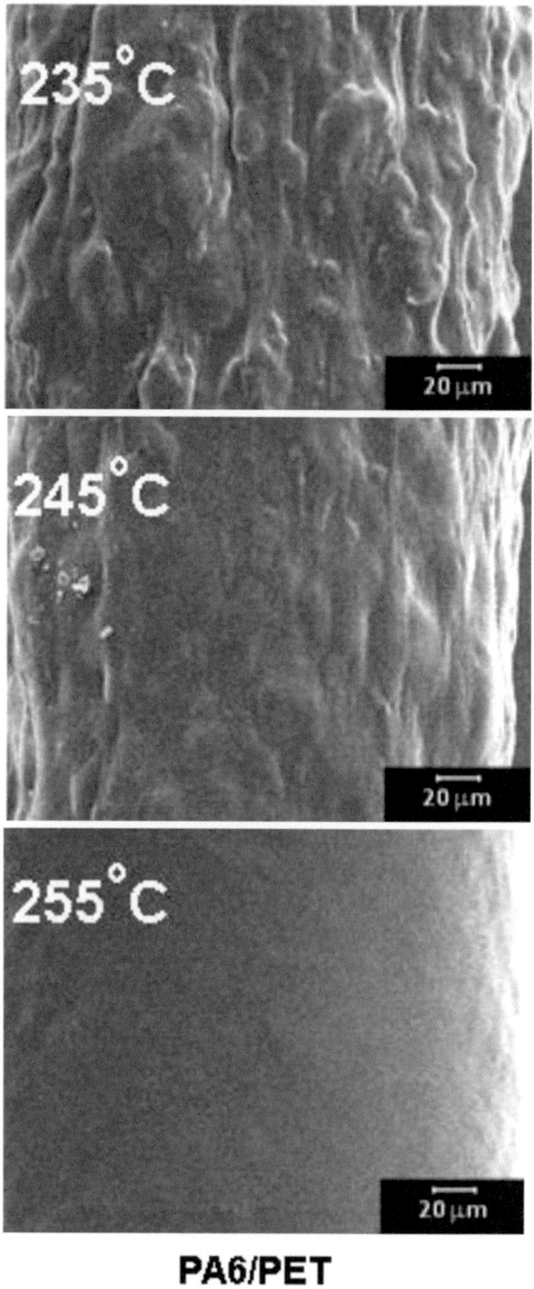

Fig. 3.11 Variation of the surface characteristics of PA6/PET blend fibers extruded at various temperatures

individual components utilizing the kinematic mutual interaction of two components in the melt spinning process also attracts much attention [77, 78]. A typical example is shown in Fig. 3.12. In the bicomponent spinning of PET/polystyrene

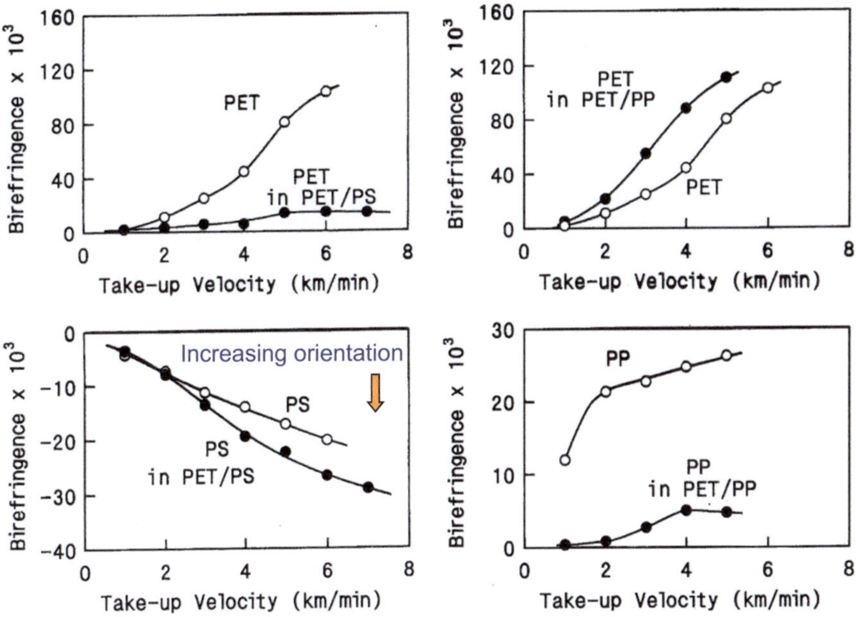

Fig. 3.12 Effect of kinematic mutual interaction in spin-line for birefringence development of individual components in high-speed bicomponent spinning process

(PS), orientation development of PET was significantly suppressed in comparison with its single-component spinning, while that of PS was enhanced. On the other hand, in the bicomponent spinning of PET/PP, orientation development of PET was enhanced, while that of PP was suppressed in comparison with the single-component spinning of individual component. Such mutual interaction was found to be governed by the temperature dependence of viscosity and solidification temperature of individual components.

3.6 Concluding Remark

In this chapter, new fiber spinning technology for the melt spinning of polymeric materials was mainly introduced. Because of the limitation of the length of this article, technologies for production of nonwoven fabrics through the electrospinning, spunbonding, and melt-blowing processes were not discussed. Structure formation behavior of individual components in the bicomponent spinning also could not be described n detail even though this is considered to be another important research field. Nevertheless, it should be emphasized that the understanding of the basic principle which governs the rather complicated melt spinning behavior is the most important issue.

References

1. H. Maeda, K. Kanamaru, Sen'i. Gakkaishi. **19**, 230–242 (1963)
2. A. Takizawa, S. Ueda, Sen'i. Gakkaishi. **19**(11), S35–S41 (1963)
3. S. Ueda, Sen'i. Gakkaishi. **21**(5), S105–S106 (1965)
4. A. Ziabicki, *Fundamentals of fiber formation* (Wiley, London, 1976)
5. O. Ishizuka, Sen'i. Gakkaishi. **30**(3), P-172–P-177 (1974)
6. J. Shimizu, Sen'i. Gakkaishi. **40**(4–5), P-173–P-179 (1984)
7. H. Yamada, Sen'i. Gakkaishi. **45**(12), P-529–P-534 (1984)
8. J. Shimizu, Sen'i. Gakkaishi. **50**(6), P-229–P-234 (1994)
9. M. Matsui, Sen'i. Gakkaishi. **60**(6), P-166–P-171 (1994)
10. H. Ito, Seikei. Kakou. **16**(7), 414–419 (2004)
11. H. Ito, Seikei. Kakou. **17**(7), 437–443 (2005)
12. K. Miyata, Seikei. Kakou. **18**(7), 460–463 (2006)
13. K. Miyata, Seikei. Kakou. **19**(7), 403–405 (2007)
14. K. Kim, Y. Ohkoshi, Seikei. Kakou. **20**(7), 395–399 (2008)
15. K. Kim, Y. Ohkoshi, Seikei. Kakou. **21**(7), 376–380 (2009)
16. W. Takarada, Seikei. Kakou. **22**(7), 333–338 (2010)
17. W. Takarada, Seikei. Kakou. **23**(7), 387–392 (2011)
18. M. Takasaki, Seikei. Kakou. **24**(7), 360–363 (2012)
19. M. Takasaki, Seikei. Kakou. **25**(7), 302–305 (2013)
20. T. Kikutani, Seikei. Kakou. **20**(8), 509–513 (2008)
21. M. Fukuhara, Research report on technological systematization (Gizyutsu no Taikeika). Natl. Mus. Nature. Sci. **7**, 123–178 (2007)
22. J. Shimizu, T. Kikutani, J. Appl. Polym. Sci. **83**, 539–558 (2002)
23. T. Kikutani, Sen'i. Gakkaishi. **63**(12), P-417–P-422 (2007)
24. A. Ziabicki, H. Kawai (eds.), *High-Speed Fiber Spinning* (Wiley-Interscience, New York, 1985)
25. T. Nakajima (ed.), *Advanced Fiber Spinning Technology* (Woodhead Publishing Lmited, Cambridge, 1994)
26. D.R. Salem (ed.), *Structure Formation in Polymeric Fibers* (Hanser, Munich, 2000)
27. T. Kikutani, in *Handbook of Textile Fibre Structure, Vol. 1: Fundamentals and Manufactured Polymer Fibres*, ed. by S.J. Eichhorn, J.W.S. Hearle, M. Jaffe, T. Kikutani (Woodhead publishing, Cambridge, 2009), pp. 157–180
28. R. Beyreuther, H. Brunig, *Dynamics of Fiber Formation and Processing* (Springer, Berlin, 2010)
29. S. Kase, T. Matsuo, J. Polym. Sci. Part A **3**, 2541–2554 (1965)
30. S. Kase, T. Matsuo, Sen-i. Kikai. Gakkaishi. (J. Text. Mach. Soc. Jpn.) **18**, 188 (1965)
31. W. Takarada, H. Ito, T. Kikutani, N. Okui, J. Appl. Polym. Sci. **80**, 1575–1581 (2001); **80**, 1581–1588 (2001); **80**, 1589–11600 (2001)
32. J. Zhou, J. Li, W. Yu, X. Lin, X. Zhou, Polym. Eng. Sci. **50**(10), 1935–1944 (2010)
33. K. Ikeda, M. Shibaya, H. Ishihara, Seikei. Kakou. **17**(12), 849–854 (2005)
34. H. Ishihara, M. Shibaya, K. Ikeda, Seikei. Kakou. **17**(3), 204–212 (2005)
35. T. Enya, H. Ito, T. Kikutani, *Fiber Society Annual Meeting* (University of California, Davis, 2007)
36. W. Takarada, K. Kazama, H. Ito, T. Kikutani, Intern. Polym. Process. **19**(4), 380–387 (2004)
37. K. Katayama, M. Yoon, Sen'i. Gakkaishi. **38**(9), P-434–P441 (1982)
38. T. Kikutani, PhD Thesis (Tokyo Institute of Technology) 1982
39. A.K. Doufas, A.J. McHugh, C. Miller, J. Non-Newtonian Fluid Mech. **92**, 27–66 (2000)
40. J. Shimizu, N. Okui, T. Kikutani, in *High-Speed Fiber Spinning*, ed. by A. Ziabicki, H. Kawai (Wiley, New York, 1985), pp. 173–201
41. OCTA Home Page. http://octa.jp/
42. M. Masuda, W. Takarada, T. Kikutani, Intern. Polym. Process. **25**(2), 159–169 (2010)

43. Y. Masubuchi, in *Fundamental and Practical Technologies for Nano-structured Polymeric Materials*, ed. by S. Nakahama (CMC Shuppan, 2008), Tokyo, pp. 66–74
44. K. Takahashi, T. Taniguchi, Seikei. Kakou. **13**, 102 (2013)
45. T. Ishikawa, T. Aoyagi, K. Toyoda, J. Kojima, W. Takarada, T. Kikutani, *14th Symposium on Computational Science on Polymers*, 13 March 2014
46. K. Katayama, in *Sen-i no Keisei to Kozo no Hatsugen* (*Formation and Structure Development of Fibers*), ed. by Sen'i Gakkai (Kagaku-Dojin, 1969), Tokyo, pp. 159–176
47. O. Ishizuka, K. Koyama, Polymer **18**, 913–918 (1977)
48. R. Kolb, S. Seifert, N. Stribeck, H.G. Zachmann, Polymer **41**, 1497–1505 (2000)
49. V.G. Bankar, J.E. Spruiell, J.L. White, J. Appl, Polym. Sci. **21**, 2341–2358 (1977)
50. H. Haberkorn, K. Hahn, H. Breuer, H.-D. Dorrer, P. Matthies, J. Appl. Polym. Sci. **47**, 1551–1579 (1993)
51. H. Hirahata, S. Seifert, H.G. Zachmann, K. Yabuki, Polymer **37**, 5131–5137 (1996)
52. H. Nishimura, M. Sei, K. Chizuka, M. Masuda, H. Yamazaki, J. Kojima, K. Yamamoto, S. Okamoto, K. Nakamae, K. Kotera, T. Kikutani, S. Sakurai, Zairyo **62**(1), 8–12 (2013)
53. K.H. Kim, T. Yamaguchi, Y. Ohkoshi, Y. Gotoh, M. Nagura, H. Urakawa, M. Kotera, T. Kikutani, J. Polym. Sci. Pt-B. Polym. Phys. **47**, 1653–1665 (2009)
54. T. Kikutani, Seikei. Kakou. **5**(12), 831–838 (1993)
55. T. Kikutani, T. Matsui, A. Takaku, J. Shimizu, Sen'i. Gakkaishi. **45**(11), 441–446 (1989)
56. T. Kikutani, Y. Kawahara, N. Ogawa, N. Okui, Sen'i. Gakkaishi. **50**(12), 561–566 (1994)
57. W.-G. Hahm, H. Ito, T. Kikutani, Intern. Polym. Process **21**, 536–553 (2006)
58. T. Kikutani, K. Nakao, W. Takarada, H. Ito, Polym. Eng. Sci. **39**(12), 2349–2357 (1999)
59. W.-G. Hahm, W. Takarada, H. Ito, T. Kikutani, Intern. Polym. Process. **24**, 281–287 (2009)
60. K. Fukutome, H. Ito, T. Kikutani, *Fiber Preprints, Japan*, (2001) p. 89
61. H. Murase, Y. Ohta, T. Hashimoto, Macromolecules **44**, 7335–7350 (2011)
62. J. Cao, J. Appl, Polym. Sci. **102**, 3078–3082 (2006)
63. K.L. Alderson, A. Alderson, G. Smart, V.R. Simkins, P.J. Davies, Plast. Rubber. Composties. **31**(8), 344–349 (2002)
64. Y. Shirakashi, O. Asakura, S. Ito, J. Ishibashi, X. Xu, W. Ttakarada, T. Kikutani, Seikei. Kakou. **23**(6), 358–364 (2011)
65. X. Xu, Y. Shirakashi, J. Ishibashi, W. Takarada, T. Kikutani, Text. Res. J. **82**(13), 1382–1389 (2012)
66. X. Xu, Y. Shirakashi, J. Ishibashi, W. Takarada, T. Kikutani, Text. Res. J. **83**(19), 2042–2050 (2013)
67. J. Sugawara, Fiber Preprints, Japan, pp.15–18 (2014); http://www.spiber.jp/
68. Jie Luo, Yaopeng Zhang∗, Yan Huang, Huili Shao, Xuechao Hu, *Sensors and Actuators*, **B162**, 435–440 (2012)
69. B. Renberga, H. Andersson-Svahna, My Hedhammarb. Sensors Actuators **B195**, 404–408 (2014)
70. O.S. Galaktinov, P.D. Anderson, G.W. Peters, H.E.H. Meijer, Intern. Polym. Process. **18**, 138–150 (2003)
71. T.H. Oh, Intern. Polym. Process. **21**, 374–378 (2006)
72. S.-C. Tseng, C.-T. Shi, Intern. Polym. Process. **23**, 277–285 (2008)
73. M. Kamiyama, T. Soeda, S. Nagajima, K. Tanaka, Polym. J. **44**, 987–994 (2012)
74. Hills, Inc., *USP* 7,252,493 (2007)
75. http://www.hillsinc.net/
76. Toray, http://www.toray.co.jp/news/fiber/detail.html
77. T. Kikutani, J. Radhakrishnan, S. Arikawa, A. Takaku, N. Okui, X. Jin, F. Niwa, Y. Kudo, J. Appl, Polym. Sci. **62**, 1913–1924 (1996)
78. T. Kikutani, H. Ito, Sen-i. Kikai. Gakkaishi. (J. Text. Mach. Soc. Jpn.) **55**(5), 38–44 (2002)
79. T. Kikutani, Sen'i. Gakkaishi. **70**(9), P-388–P-391 (2014)

Chapter 4
History of Polyester Resin Development for Synthetic Fibers and Its Forefront

Masatoshi Aoyama and Yoichiro Tanaka

Abstract Polyethylene terephthalate (PET) was invented about 50 years ago, and now it is widely used for synthetic fibers, films, bottles, and so on. In a half century, the polymerization process has been improved, for example, from DMT method to direct polymerization method and from batch process to continuous one. Various new polymerization catalysts also have been developed. For production of fibers, a variety of copolymerization composition has been developed to control and improve the properties, such as flame retardancy, dyeability, solubility, and shrinkage. In addition to PET, polyethylene naphthalate, polybutylene terephthalate, and polytrimethylene terephthalate are now used for various usages. Nowadays, bio-based polyesters polymerized from bio-based ethylene glycol, propylene glycol, terephthalic acid, and so on are also developed and some are already commercialized.

Keywords Polyester • Polycondensation process • Flame retardant • Polyethylene terephthalate • Bio-based polyester

4.1 Introduction [7]

Due to the good balance between performance and cost, polyester fiber has addressed rapid growth in global scale among synthetic fibers and has the largest production quantity today among different types of fibers including natural fibers.

During the dawn of synthetic fibers, the first great topic is the invention of Nylon 66 by Carothers at the US DuPont. However, the types of polyester that had been examined at the same time were mainly aliphatic polyesters. As the polyester is appropriate for synthetic fibers, later invention of the aromatic polyester introducing the terephthalic acid unit, namely, polyethylene terephthalate (PET) by Whinfield, Dickson, and so forth at British Calico Printers, had to be waited. Production of the fiber using this PET began by ICI and DuPont during the

M. Aoyama (✉) • Y. Tanaka
Toray Industries, Inc. Fibers & Textiles Research Laboratories, Tokyo, Japan
e-mail: Masatoshi_Aoyama@nts.toray.co.jp

Society of F.S. Technology, Japan (ed.), *High-Performance and Specialty Fibers*,
DOI 10.1007/978-4-431-55203-1_4

mid-1950s. In Japan, Teikoku Jinzo Kenshi Co., Ltd. and Toyo Rayon Co., Ltd. introduced the technology jointly in 1957, starting production in 1958.

This chapter takes the general view on the history of development and current situation of polyester resins for synthetic fibers with main focus on PET.

4.2 Methods of Synthesis [6] [8] [9]

4.2.1 DMT Method and Direct Polymerization Method

Typical manufacture methods for PET include the DMT method which uses dimethyl terephthalate (DMT) and ethylene glycol (EG) as the starting materials and the direct polymerization method (TPA method) to start from terephthalic acid (TPA) and EG, and in the beginning DMT method was adopted. DMT was popularly used because of its simple refining processes such as distillation and recrystallization compared to TPA which had no boiling point, poor solubility, and high difficulty in refining.

However, it can also be said that DMT method contains an extra process to replace the methanol added to TPA, which is added to facilitate purification, with EG and collect by transesterification during the PET production processes. Because of this, direct polymerization method became more popular and has become the mainstream today as a technique to manufacture TPA with high purity (called PTA) was developed later.

Manufacture of PET mainly comprises of the transesterification (EI reaction) to form the intermediate bishydroxyethyl terephthalate (BHT or BHET) from DMT and EG or the direct esterification reaction (DE reaction) to form BHT from TPA and EG in the first part and the polycondensation reaction to form PET polymer from BHT in the second part. In addition, processes to facilitate the reactions efficiently are usually adopted in PET manufacture systems by dividing the system into a reaction tank for EI or DE reaction process and a reaction tank for polycondensation reaction.

In EI reaction, DMT and EG are heated and made into a uniform phase in a reaction tank, and the reaction is facilitated under the existence of an appropriate catalyst and atmosphere of normal pressure while removing methanol, which is a reaction by-product. Since DMT has the melting point approximately 140° C, reaction uniformity and production efficiency can be improved by melting it in a separate dissolver and supplying it into the EI reaction tank as necessary, although it can also be injected into the reaction tank as powder or flakes.

The factors that affect the EI reaction rate include the reaction temperature, the ratio of charged quantities of DMT and EG (molar ratio), and EI reaction catalysts, and catalysts are especially important as they also have great influence on the quality of the polymer that is formed in the end.

As EI catalysts, various compounds have been tested after those seen in initial patent by Whinfield and Dickson, and there have been many reports on the reaction mechanism and reaction kinetics. For example, in Japan, there was a series of examinations by Ida et al. [1] which organized the EI reaction rate by its relationship to the electronegativity of the catalyst compound.

While DE reaction is classified as an esterification reaction between carboxylic acid and alcohol which are generally well known, it is rather difficult to handle in studies on reaction kinetics as TPA has poor solubility to EG, and thus creative measures are required in industrial processes. As DE reaction is an autocatalytic reaction in which proton from the acid functions as a catalyst, reaction continues by mixing and heating TPA and EG together and removing water, which is the reaction by-product. This reaction starts in a heterogeneous system (slurry state) from the beginning until the middle of the process since TPA is poorly soluble in EG as discussed above. Therefore, how this TPA/EG slurry properties are controlled is one of the most important points in controlling the DE reaction. Examples of various creative measures in industrial processes to improve the handleability and fluidity of slurry and ES reactivity include optimization of the size and shape of TPA particles as well as mixture ratio with EG and use of BHT instead of EG.

4.2.2 Polycondensation Reaction

BHT which is formed by transesterification or esterification reaction is normally transferred to the polycondensation reaction tank, which is a separate reaction tank, to go through the polycondensation reaction under high temperatures and high vacuum to form PET.

Unlike polycondensation of polyamides, the equilibrium constant for PET polycondensation is close to 1. Therefore, how fast EG forming as a by-product in polycondensation in order to facilitate the reaction in product system is removed from the system becomes the important point in reaction process. That is, the creative measures taken include setting the reaction tank shape with shallow liquid depth and large surface area for reacting substances and promotion of driving the by-product EG out of the system by making contrivances on mixing impeller shape.

It is also necessary to set the reaction temperature to extremely high around 280° C in order to obtain sufficient polycondensation reaction rate as the melting point for the polymer formed is high. However, coloration due to thermal decomposition and so forth are also facilitated under high temperatures, indicating that control and optimization of polycondensation temperature and good design on polycondensation catalyst are important. Polycondensation catalysts will be discussed in more detail later.

4.2.3 Molecular Weight

The molecular weight of the PET manufactured through polycondensation reaction needs to be optimized depending on the purpose. Molecular weight is an important basic property which can influence the mechanical properties as well as forming processability when used as fiber.

The average molecular weight for PET used in fiber for clothes is approximately 20,000 (g/mol) with degree of polymerization approximately 100 and intrinsic viscosity about 0.62–0.65. This value of molecular weight is similar to that of biaxially oriented polyester film used for packaging or general industrial material applications.

Compared to this, PET for tire cords and seatbelts which are required to have high strengths usually has molecular weight of 30,000 (degree of polymerization 150 and intrinsic viscosity 0.80) or higher.

Furthermore, the PETs used in fiber for clothes usually have the melt viscosity of around 2000 poises (200 Pa s) at 290° C, which can be stirred, and thus PET can be manufactured with the specified molecular weight by only melt polymerization. However, PETs of higher molecular weights have extremely high viscosity, which prevents uniform stirring of melt and continuation of melt polymerization. Therefore, solid state polymerization method is used to further increase the molecular weight after melt polymerization.

4.2.4 Polymerization System

PET polymerization processes started from batch method, and continuous polymerization method became more widely used as the production quantity increased.

While the two-tank structure comprising of the reaction tank to conduct EI reaction or DE reaction process and the reaction tank to conduct polycondensation reaction is typical in general in batch polymerization, multiple reaction tanks are often adopted in continuous polymerization method, due to the necessity to vary the temperature and degree of vacuum in stages to increase the degree of polymerization in polycondensation reaction. For example, a system may adopt a five-tank system with two EI (or DE) reaction tanks and three polycondensation reaction tanks.

The continuous polymerization process is suited to mass production of only a few species or favorably one species, and cost reduction is addressed by the merits of efficiency improvement through continuous production and scale increase with large equipment. To produce multiple species of copolyesters and so forth in relatively small quantities, on the other hand, batch method is better suited.

4.2.5 Polycondensation Catalysts

Although the polycondensation reaction for PET is a transesterization reaction in a similar fashion to the reaction to form BHT from DMT and EG, it has a high reaction temperature compared to the EI reaction between DMT and EG. Due to the restrictions including consideration of competing reaction called thermal decomposition, metallic compounds different from EI catalysts are selected in many cases.

Antimony compound catalyst which is most generally used at present was already indicated in the initial patent by DuPont [2], indicating that they had thoroughly searched for the catalyst at the time.

As the catalyst added/contained in polymer by several tens to hundreds of ppm can affect the various properties of fiber products such as thermal stability in addition to the polymerization reactivity, catalyst design is positioned as one of the most important elemental technology in PET polymerization and has motivated various research and development activities. Here, this section mentions some of the attempts to further activate and improve the various properties of polymer by catalytic compounds other than antimony from the 1990s to 2000s to exceed the antimony compound catalyst which has been used most generally.

It has been long known that a titanium compound has catalyst activity 10–20 times that of antimony compound. However, use of it in PET polymerization had been limited because coloration of the polymer cannot be avoided as it also has high decomposition activity. To address this problem, new titanium catalysts have been developed by several domestic and international companies since the 1990s. Although many of the details in catalyst design have not been disclosed, various methods have been tried to control the catalyst activity of titanium such as a composite oxide with elements other than Ti such as Si and a titanium compound with modification using organic complexing agents based on the disclosed information.

In addition, DuPont [3] and Toyobo Co., Ltd. [4] proposed attempts to use aluminum compounds, which they had thought to have little catalytic activity against polycondensation, as polycondensation catalysts. Especially the latter case is interesting in that it claims that the activity is improved by combining the aluminum compound with no catalyst activity by itself with a specific compound.

4.3 Copolymerization Polyester

4.3.1 Copolymerization Components and Manufacture Method

An example of copolymer components used in the production of copolyesters is shown in Tables 4.1 and 4.2. By copolymerizing these components, it is possible to make polymer design to suit the application by adding mechanical properties,

Table 4.1 Monomer adopted for copolymerization of polyester. (1) Polyvalent carboxylic acids

Terephthalic acid	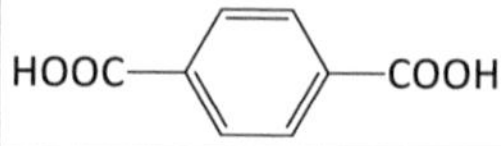
Isophthalic acid	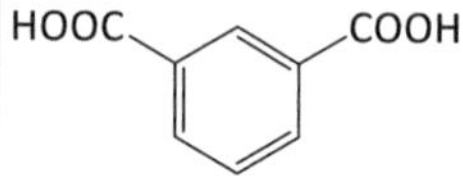
Orthophthalic acid	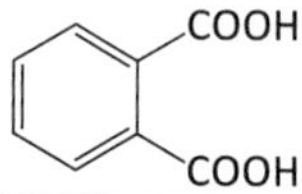
2,6-Naphthalene dicalboxylic acid	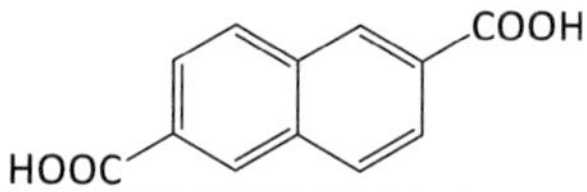
Paraphenylene dicalboxylic acid	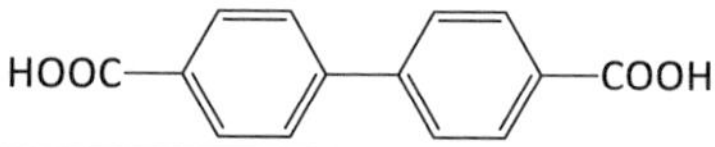
1,4-Cyclohexane dicalboxylic acid	
Succinic acid	$HOOC(CH_2)_2COOH$
Glutaric acid	$HOOC(CH_2)_3COOH$
Adipic acid	$HOOC(CH_2)_4COOH$
Suberic acid	$HOOC(CH_2)_4COOH$
Azelaic acid	$HOOC(CH_2)_7COOH$
Sebacic acid	$HOOC(CH_2)_8COOH$
Dodecanedioic acid	$HOOC(CH_2)_{10}COOH$
1,2,4-Benzenetricarboxylic acid	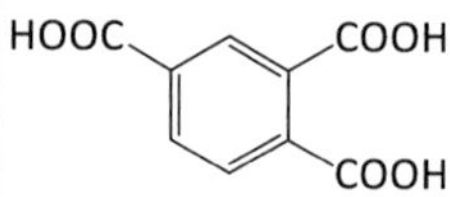
1,2,4,5-Benzenetetracarboxylic acid	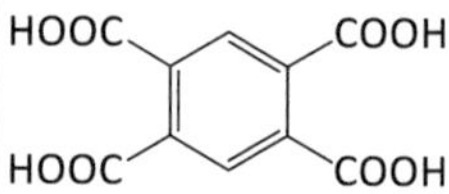
5-Sulfoisophthalic acid monosodium salt	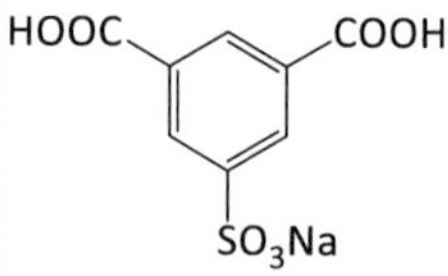

thermal properties, chemical properties, and so forth to polyester that cannot be delivered by homogeneous PET.

The basic manufacture method is similar to that of homogeneous PET, and it is manufactured by mixing the copolymer components with TPA and EG in

Table 4.2 Monomer adopted for copolymerization of polyester. (2) Polyhydric alcohol

Ethylene glycol	$HO(CH_2)_2OH$
Propylene glycol	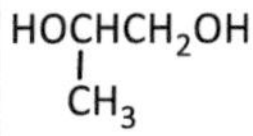
1,4-Butanediol	$HO(CH_2)_4OH$
1,5-Pentanediol	$HO(CH_2)_5OH$
1,6-Hexanediol	$HO(CH_2)_6OH$
Neopentyl glycol	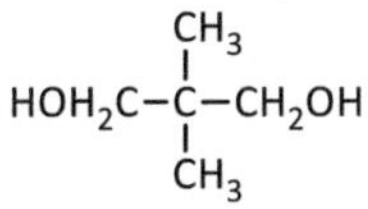
Diethylene glycol	$H(OCH_2CH_2)_2OH$
Triethylene glycol	$H(OCH_2CH_2)_3OH$
Polyethylene glycol	$H(OCH_2CH_2)_nOH$
Polytetramethylene ether glycol	$H[O(CH_2)_4]_nOH$
1,4-Cyclohexane dimethanol	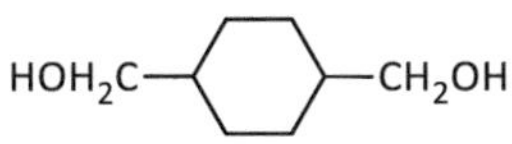
Bisphenol A ethylene oxide	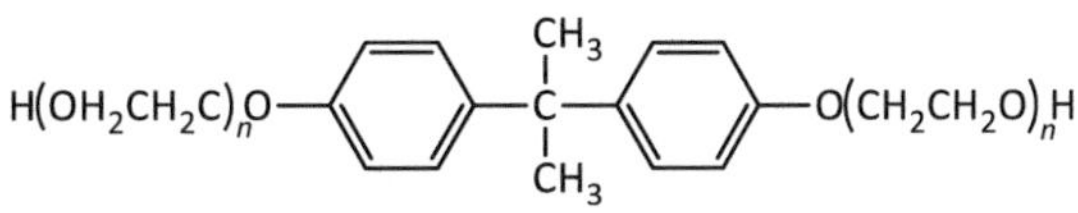
Trimethylol propane	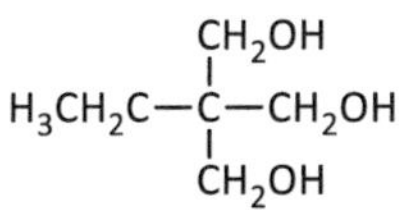
Pentaerythritol	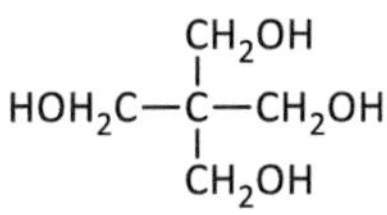

polymerization process to go through the EI or DE reaction followed by polycondensation reaction under high vacuum.

While the melting point for polyester deteriorates due to the molar fraction on copolymer components, thermal properties also vary depending on the structure of copolymer components. The glass transition temperature (Tg) for a random copolymer varies depending on the weight fraction of the copolymer component. Tg is lowered as the quantity of softening components such as adipic acid and sebacic acid is larger and increased as the quantity of rigid molecular structures such as paraphenylene dicarboxylic acid and naphthalene dicarboxylic acid is larger. Crystallinity also deteriorates due to copolymerization in general.

Copolymerization is also used as a measure to add various properties such as flame retardant introducing phosphorus element into molecular chains, improving affinity to dyestuff introducing ionic group, in addition to addressing control on melt viscosity and control of affinity to solvents or alkaline solutions.

4.3.2 Dyeability

One of the most important copolymerization technologies regarding fiber polyester would be the technology to improve the dye affinity. There is a problem with homogeneous PET that it is difficult to dye with its glass transition temperature around 70° C, which is close to the boiling point of water under normal pressure.

In the design concept to promote dye diffusion by copolymerizing a soft 3rd component and lowering Tg and crystallinity, polyethylene glycol (PEG) is an effective compound as a copolymerization component which can reduce Tg while suppressing the drop in melting point. However, there is an issue of PEG polymerization of certain amount being necessary to address sufficient coloring in dyeing under atmospheric pressure, and obtaining balance with other physical properties of fiber being difficult, and thus industrial application seems to be limited.

In contrast to this, a method to address brighter dyeing by using a cationic (basic) dye has also been examined. To address this method, a copolymerization component with an acidic functional group that can react with the dye would be required for PET. However, this copolymerization technique has already been applied for patent [5] by DuPont in 1957, and the sulfoisophthalic acid sodium salt indicated here is still used as the most typical cationic dyeable component. The issue about this copolymerization component is that it leads to increased melt viscosity of the polymer as the intermolecular interactions are strong, leading to lower fiber strength as the final polymerization degree for the polymer needs to be lowered due to the restrictions of the polymerization system. As methods to compensate for this flaw, the use of sulfoisophthalate quaternary phosphonium salt with lower intermolecular interactions, a method to further copolymerize a soft component with viscosity reduction effects such as PEG and so forth, has been proposed.

4.3.3 Hydrolyzability

While increasing hydrolysis resistance is demanded from the viewpoint of addressing durability of the fiber products, polyester with high rate of hydrolysis is required when shaping it into fiber with special cross section shapes by alkaline hydrolysis and removal of one polymer component during the fiber product manufacture process.

The sulfoisophthalic acid sodium salt copolymerization polyester described in the previous section has also been used as an easily soluble component in manufacture of modified cross section yarns and extremely thin yarns as its hydrolysis rate is extremely high under the existence of alkali.

4.3.4 Flame Retardant Property

In fiber products, flame retardant property is also considered an important function other than the stainability. Synthetic fibers such as polyester, nylon, and acryl have low LOI (limiting oxygen index) values in general and are flammable, and flame resistance is improved by polymer modification and post-processing.

Although various methods have been adopted for polyester including kneading or copolymerizing flame retardants into the polymer, copolymerization of flame retardants into polymer before spinning has an advantage that it causes relatively low deterioration in mechanical properties and that its flame retardant property persists as excelling in durability.

In copolymerization, the method to increase phosphorus content as polyester by compolymerizing a compound containing phosphorus atom which introduces flame retardant property to polymer is most typical. The major copolymerization components are shown in Fig. 4.1. Figure 4.1a has phosphorus atom on the main chain of polyester, while Fig. 4.1b is believed to be resistant to decomposition of the main chain as phosphorus atom exists on the side chain.

4.3.5 Mixed Filaments of Different Shrinkage

Introduction of the 3rd component into PET by copolymerization disturbs the regularity of polymer chain, and rigid segments especially deteriorate the crystallinity. The fiber made of this amorphous polyester has large rate of shrinkage, and the filament yarn combined with homogenous PET gives difference of yarn lengths due to the difference in shrinkage properties by adding thermal history during the post-processing and thus gives bulky touch to the fabric.

Copolymerization technique is an effective method for the polyester properties, and it is expected to be examined further in various ways to be utilized as a method to clear the paths to new applications of fibers in the future.

4.4 Bio-based PET

4.4.1 Background

Petroleum, which is a fossil resource, is a major raw material in chemical industry. However, it is a limited resource which is feared to be depleted in the future, and it causes a global environmental problem as it emits a large amount of carbon dioxide during the manufacture process as well as disposal by incineration. Under such circumstances, attempts to replace the petroleum chemical products which are currently mass consumed with recyclable products made from biomass resources

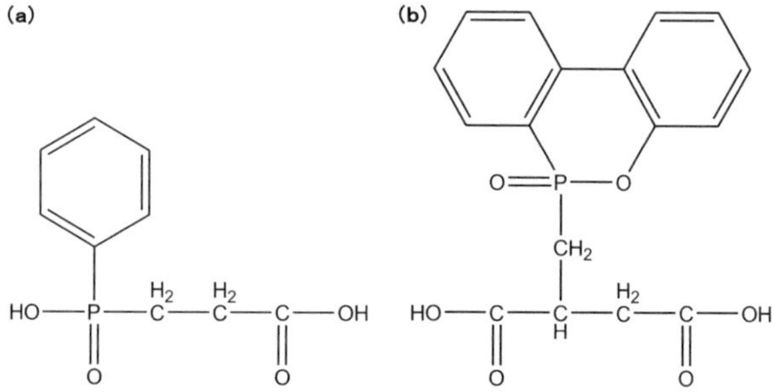

Fig. 4.1 Major copolymerization components for flame retardant property with phosphorus atom on the main chain (**a**) and on the side chain (**b**)

or products with lower environmental loads have attracted great attention. Consumption of Polyethylene terephthalate (PET) is especially large as various mold goods including fiber, films, and bottles. Movements to develop bio-based versions of such polymers with large consumption levels have become active these few years. It was Coca Cola which announced the adoption of bio-based PET bottles and released their sales ahead of the world in 2009 [10] that triggered these movements. These bio-based PET bottles (PlantBottle™) were manufactured by developing bio-based ethylene glycol as PET raw material, and their biomass content (^{14}C concentration measurement method: described later) is approximately 20 %.

It is possible to produce bio-based PET by using bio-based ethylene glycol and terephthalic acid as raw materials in existing polymerization facilities. Therefore, raw materials made from biomass resources would be the key point.

4.4.2 Bio-based Ethylene Glycol

India Glycols has manufactured bio-based ethylene glycol since 1989 by using exhausted molasses from sugarcane, and its manufacture method comprises of forming bioethanol from the sugar by fermenting, dehydrating it into ethylene, oxidizing into ethylene oxides, and hydrolyzing into ethylene glycol. At present, Toyota Tsusho and China Man-Made Fiber in Taiwan founded Greencol Taiwan Corp. (GTC) by joint venture with matching investments [11] and established a technique to manufacture bio-based ethylene glycol industrially through a similar route. The bio-based ethylene glycol manufactured by this method is high in purity and can be used to produce PET with equivalent performance as that of PET manufactured from petroleum-based ethylene glycol. However, they still have the problem of high costs compared to petroleum-based PET.

This tendency to develop businesses for PET manufacture with bio-based ethylene glycol is seen in active market expansion promotion for bottles and automotive interior materials by Toyota Tsusho, films by Iwatani Corporation and Dai Nippon Printing [12], bottles by Kirin Beverage Company [13], thermal shrinking films and long fiber nonwoven fabric by Toyobo [14], and fibers for clothes by Teijin [15] and Toray in addition to the bottles by Coca Cola mentioned above [16].

4.4.3 Bio-based Terephthalic Acid

Various examinations have also been made regarding development of bio-based terephthalic acid (TPA), which is the other raw material for PET. If both ethylene glycol and terephthalic acid can be bio-based, biomass content of PET made with those raw materials will be 100 %. PepsiCo made a press release in March 2011 that it had addressed the completely bio-based bottles [17], and Toray succeeded in trial manufacture of completely bio-based PET film and fiber for the first time in the world in a joint effort with Gevo in June and November 2011 [18, 19]. Gevo formed bio-based paraxylene by forming isobutylene through dehydration after forming isobutanol with fermentation method using sugar obtained from sweet corn and so forth as the raw material, followed by dimerization by radical reaction and cyclization. Completely bio-based PET fiber and film were delivered eventually as Toray converted this bio-paraxylene into terephthalic acid with its proprietary technique. Virent, Anellotech, and so forth have also been working on examination of industrialization of bio-based paraxylene production besides Gevo above.

In the future, both bio-ethylene glycol and bio-terephthalic acid will secure their markets once their relatively high costs compared to petroleum-based products are mitigated and supply stability is ensured. It is also an important matter to use nonedible resources with no competition for food as the starting materials in order to form larger markets in the future. Since PET is popularized in wide applications due to its excellent properties, bio-based PET is superior to other bio-based polymers in versatility, demand scale, and replaceability. It is surely expected to become popularized widely.

4.4.4 Biomass Content of PET

This section mentions how to measure the biomass content (^{14}C concentration measurement method) for reference.

Biopolymers can be analyzed for how much carbon originating from biomass resources is contained in the total carbon atoms comprising the polymer. Specifically, it is calculated by measuring the ^{14}C concentration (ratio between ^{14}C and ^{12}C detection values in substance) using Accelerator Mass Spectrometry (AMS). ^{14}C is an isotope of ^{12}C, and it is a radioactive element (half-life: approx. 5730 years). While ^{14}C is supplied into the atmosphere at a stable quantity every year as it is formed by cosmic rays falling onto the earth constantly, it also vanishes by radioactive decay in a certain quantity every year. The quantity of ^{14}C in atmosphere is therefore nearly constant in equilibrium. As ^{14}C supplied into the atmosphere is distributed widely into plants by photosynthesis and then into animals through the consequent food chain, the concentration of ^{14}C (ratio between ^{14}C and ^{12}C detection values) is nearly constant for all organisms on earth.

Meanwhile, the most convincing theory for the origin of petroleum is that it was made from organisms from 100 to 200 million years ago, and the ^{14}C concentration in petroleum is nearly zero as its half-life is long past. Since biomass products contain certain level of ^{14}C and petroleum-based products do not as explained above, it is possible to calculate the ratio of carbon derived from biomass or the rate of bio-based production for the sample by measuring the ^{14}C concentration in the sample [20].

4.5 Polyesters Other Than PET (3GT, PBT, and PEN)

4.5.1 Major Aromatic Polyesters Other Than PET

While polyester represented by PET is synthesized by condensation polymerization between diol and dicarboxylic acid, polyester fibers other than PET include 3GT (PTT) fiber using propanediol instead of ethylene glycol as diol, PBT fiber using butanediol, and PEN fiber using naphthalenedicarboxylic acid instead of terephthalic acid as the dicarboxylic acid. An outline is provided for each of them below.

4.5.2 3GT Fiber [21]

3GT fiber (polytrimethylene terephthalate fiber) delivers both morphological stability of polyester and softness of nylon. It also excels in stain resistance and has no yellowing as nylon does. It also has large molecular deflection and excels in elasticity.

A major development target is the bulk textured yarn mainly for carpets. Side-by-side bimetal composite yarns in which the stretch recovery property is maximized have also been manufactured. In this purpose, high crimping is addressed by utilizing the thermal shrinkage property of the polymer comprising the fiber to form crimps during the heating stage in high-order processing so that excellent soft stretch property and stretch recovery property unlike any other fibers are addressed. It is also applied in other daily commodities such as clothes to be worn in the middle layer with characteristics of small thickness, lightweight, heat retention, moisture permeability, and flexibility; padding with excellent shape recovery, no dust, and moisture permeability; and toothbrush with little opening of tips and high plaque removal capability.

Meanwhile, it also attracts attention for biomass conversion of raw materials. Although the ethylene oxide method by Shell and acrolein hydrating method by Degussa (both using petroleum-based resources) were available as the industrial production methods for propanediol, DuPont developed a technique to convert sugar derived from plant materials into glycerin and further convert it into propanediol using microorganisms. By producing 3GT with this propanediol, 3GT made from biomass resources has been addressed with approximately 27 % biomass content (^{14}C concentration measurement method as described earlier).

4.5.3 PBT Fiber

Polybutylene terephthalate is a typical engineering plastic, and it is widely used in electric and electronic parts, automotive parts, and so forth as it excels in thermal stability, dimensional precision, and electric properties while being easy to process with high crystallization rate although it is inferior to PET in melting point. However, use of it as fiber is small in quantity. The production quantity remains small due to the restriction in price, although it is easily dyed and excels in elasticity during the false twist process (suited to stretching materials).

4.5.4 PEN Fiber [22]

Characteristics of polyethylene naphthalate fiber include high melting point (265° C), high Tg value, high strength, high modulus, low heat shrinkability, high chemical resistance, and high hydrolysis resistance.

It is often favorably used in industrial fiber applications utilizing these characteristics, such as fiber to reinforce the automobile tires (tire cords), reinforcing materials for hoses and transmission belts for automobiles, and heat-resistant canvas. In Japan, Teijin succeeded in industrialization in 1989 and sales was begun in 2001.

4.6 Concluding Remark

Now, we reviewed some current topics of polymer for synthetic fiber. Development of novel polymer materials has brought new high-performance fibers or enlarged the field of the fibers to be applied until now and also in the future.

References

1. K. Tomita, H. Ida, Polymer **16**, 185 (1975)
2. USP2, 578, 660 (DuPont)
3. USP5, 512, 340 (DuPont)
4. S. Gyobu, POLYESTER2005 (10th World Congress) Proceeding, Maack Business Services (2005)
5. USP3, 018, 272, Patent Gazette Showa No.34-10497 (DuPont)
6. *Handbook on Saturated Polyester Resins*, ed. Yugi, Nikkan Kogyo Shimbun (1989)
7. Mototada Fukuhara, *National Museum of Nature and Science, Investigation Report on Systemization of Technologies, 7th Series (History of Polyester Fiber Technologies Used for Making Textiles)* (National Museum of Nature and Science, Tokyo 2007), pp. 125–178
8. Shigeji Konagaya, *Lecture Note of Textile College*, (The Textile Machinery Society of Japan, Tokyo 2012)
9. T. Suzuki, Sen'i. Gakkaishi. **59**, 220–227 (2003)
10. *Kagaku Kogyo Nippou*, 12.18 (2009)
11. *Nihon Keizai Shimbun*, 10.8 (2010)
12. *Kagaku Kogyo Nippou*, 2.1 (2012)
13. *Kagaku Kogyo Nippou*, 1.6 (2012)
14. *Kagaku Kogyo Nippou*, 7.1.(2011)
15. *Kagaku Kogyo Nippou*, 3.1.(2011)
16. *Market Analysis on PET Resins and Products*, Osaka Chemical Marketing Center, 2012, pp. 31–38
17. *Nikkei Sangyo Shimbun*, 3.18 (2011)
18. *Kagaku Kogyo Nippou*, 6.28 (2011)
19. *Nikkei Sangyo Shimbun*, 11.18 (2011)
20. M. Kunioka, Eng. Mater. **56**(2), 27–31 (2008)
21. *Analysis on Polyester Resins at Internationalization Age* (CMC Research, 2009), pp. 187–214, 249
22. *High Function Fibers*, (Investigation Division, Toray Research Center, 2013), pp. 26–28

Part II
High-Strength High-Modulus Organic Fibers

Chapter 5
History of Super Fibers: Adventures in Quest of the Strongest Fiber

Hiroki Murase and Kazuyuki Yabuki

Abstract The high-strength and high-modulus fibers have been widely spread into many industrial applications since the last quarter of the twentieth century. Key technologies leading to the industrial success of the fibers are the molecular designs of rigid polymers giving rise to a spontaneous molecular organization and the gel spinning of flexible polymers. The continuous efforts of a great number of researchers in this field over 40 years are summarized in this chapter.

Keywords High-strength and high-modulus fibers • Super fiber • History • Liquid crystalline • Aramid fiber • Polyarylate fiber • Ultrahigh molecular weight polyethylene fiber • Gel spinning

5.1 Introduction

Our modern, safe, and comfortable lives are supported with advanced fibers. The application of the fibers has spread across various fields, for example, sports gears, personal protective textiles (e.g., body armor, protective gloves, firefighter garment, etc.), slings, ropes, tension members for optical cables, automotive applications, and aerospace fields. "Super fiber," which is defined as a fiber showing tensile strength higher than 20 g/day (or 2 GPa), is a typical advanced fiber and has contributed to the prosperous growth of the market since the 1970s. In this chapter, the history of super fibers will be summarized with focusing on the great efforts of researchers having led to the innovation.

The invention of Nylon® fiber in the late 1930s can be doubtlessly regarded as a great innovation in the field of man-made fibers. After this invention, the market of synthetic fibers rapidly grew and its application widely spread not only to clothing

H. Murase (✉)
Research Center, TOYOBO Co., Ltd., 2-1-1 Katata, Ohtsu, Shiga 520-0292, Japan

Department of Textile and Clothing, Faculty of Home Economics, Kyoritsu Women's University, 2-2-1 Hitotsubashi, Chiyoda-ku, Tokyo 101-8437, Japan
e-mail: hmurase@kyoritsu-wu.ac.jp

K. Yabuki
Research Center, TOYOBO Co., Ltd., 2-1-1 Katata, Ohtsu, Shiga 520-0292, Japan

Society of F.S. Technology, Japan (ed.), *High-Performance and Specialty Fibers*,
DOI 10.1007/978-4-431-55203-1_5

textiles but also to industrial uses. One example of the application in the industrial uses is reinforcement of tires. The tire cord was firstly industrialized using cotton fiber and later replaced with rayon fibers. The higher-strength rayon fibers had been intensively developed in the 1940s–1950s. However, the rayon fibers for tire reinforcement were driven out and replaced by Nylon® fiber after its invention. This is the typical example of remarkable progresses in synthetic fibers. The same situations were able to be observed in various industrial applications. The success of Nylon® fiber proved the tremendous usefulness of synthetic fibers and realized us the promising future of the high-strength synthetic fibers. The industrial background behind the inventions of the super fibers, which can be regarded as the second innovation of man-made fibers, would be constructed through the growth of the market. In the 1960s, the fiber-producer's eyes were on the growing tire code market, and their anticipation expecting a higher-strength fiber in this market would probably pump up the motivation to develop the super fibers [1].

Moreover, the scientific insights on polymer solids were concurrently and rapidly deepened in parallel to the growth of fiber industry. The strenuous efforts of a great number of researchers in the field of polymer science elucidated the complicated fiber structures. Their significant insights shed light on the complicated but essential nature of polymer and implied the limitation of the conventional melt spinning technology as a method making high-strength fibers. An innovative and novel manufacturing technology to obtain the strongest fiber had been explored by many researchers not only in academy but also in industry under such scientific and industrial circumstances. The researchers in American and European companies attained great discoveries concerning the innovative molecular organizations giving rise to high-strength fibers in the 1970s to early 1980s.

A brief survey of the history would be helpful to the readers in getting the overview of this field before we move into more details of each fiber's history. In these days, we can categorize the technologies for producing the super fibers into two streams, i.e., one is the fiber made from a rigid polymer and the other a flexible one. The two epoch-making events should be worthy to note, i.e., the discovery of liquid crystalline nature of solutions of para-aramid polymer founded by Kwolek, a researcher in DuPont company [2], and the invention of "gel spinning" attained by Smith and Lemstra, the researchers in the Netherlands company DSM [3].

In advance of the para-aramid technology, meta-type aramid fiber had already been commercialized as a heat-resistant fiber in the middle 1960s by DuPont. However, the insight of researchers into the potential of rigid polymers was less general than today. Therefore, Kwolek and her colleague's works, which showed the self-organization of para-aramid molecules into nematic liquid crystalline state in its solution and the commercialization of Kevlar® fiber utilizing such a liquid crystalline nature, should be regarded as an epoch of the high-strength fiber industry. The commercial success of Kevlar® fiber stimulated the development of various high-strength fibers along this line. An aramid copolymer, which contains a less rigid unit in its molecular moiety, was commercialized as "Technora®" from Teijin company. The wholly aromatic polyester fiber made by thermotropic liquid crystalline spinning process was also developed and commercialized. Moreover, the researchers in the US Air Force pursued an approach to improve molecular rigidity in the 1980s and developed the fibers made from polymers including heterocyclic rings, for example, poly-*p*-phenylene benzobisoxazoles (PBO), which showed extremely high strength and high modulus. This extremely rigid polymer fiber was researched furthermore and developed to PBO fiber (Zylon®) by Japanese company Toyobo in 1998.

The other stream leading to the production of super fiber, which is a totally different approach from rigid polymers, was originated from an epoch made in the 1960s. The discovery [4, 5] of the intriguing flow-induced super structure "shish-kebab" in the middle 1960s is a noteworthy event as a prologue of the next innovation concerning to the super fiber made from polyethylene. Pennings, who is a discoverer of the flow-induced structure, ingeniously utilized the flow field for molecular manipulations and developed high-strength and high-modulus fibers. The method, called as "surface growth" method [6], was never industrialized due to its low productivity. His works indicated the promising potential of polyethylene to be a high-strength fiber, and it should be regarded as a precursor of the gel spinning. Smith and Lemstra developed a new spinning technology based on an optimum control of number density of entanglements in the solution of ultrahigh molecular weight polyethylene (UHMWPE). The average number of entanglements was adjusted with the concentration of polymer in the solution to be an optimum state leading to good drawability. The optimum state was fixed by the crystallization of polyethylene in the subsequent cooling process with extensional deformation after the throughput of the solution from the spinneret [7]. This crystallized solid fiber contained much solvent and looked like a swollen-gel, which was able to be drawn by several tens times the original length. The gel spinning exhibited higher productivity than the surface growth method, and the commercial availability led to its success in the super fiber market.

A brief survey of the history of the super fibers has been shown in this section. Let us move into the detailed stories behind the discovery of each super fiber in the following sections with keeping the above-mentioned two technological streams in mind.

5.2 Rigid Polymers

5.2.1 Aramid Fibers

Developments of rigid polymers containing aromatic rings in their molecular moiety were featured as a new approach to obtain high-strength fibers in the 1960s. However, the researchers encountered the problem of lower processability due to its rigidity giving rise to higher melting temperature than its decomposition temperature and lower solubility into conventional organic solvents. A researcher in DuPont, Dr. Kwolek, synthesized a great number of rigid polymers, and finally she discovered the liquid crystallinity of para-aramid polymer in sulfuric acid solution [2, 9, 10]. The list of her synthesized polymers can be found in a literature [11]. The potential of "para"-type aramid polymer being high-strength fiber had not been previously recognized even at the time. In 1968, they got a fiber made from poly-para-phenylene terephthalamide (PPTA), which is the same polymer as current Kevlar®. However, its tensile strength was still low. Air-gap wet spinning, which was invented by Blades in 1970, brought about breakthrough [12]. The molecules in the liquid crystalline solution are spontaneously oriented parallel to each other due to the exclude-volume interaction between those rigid molecules. Nevertheless, the orientation usually does not propagate to whole volume of the solution and its long range consistency is kept no more than several tens or hundreds micrometer in length. The region, which keeps the consistent molecular orientation, is called a "domain." The neighboring domains are usually independent, and therefore the macroscopic molecular orientation of the solution is random in the static state. It is the most crucial point in the liquid crystalline spinning that a shear and/or extensional flow field applied on the fiber after throughput from a spinneret aligns the molecules into the mean direction in each domain, resulting in the molecular orientation parallel to the fiber axis. The high deformation rate can be applied on the fiber under spinning in the narrow space between the spinneret and coagulation bath, and this is the crucial point of engineering success of the air-gap wet spinning. The pilot plant of Kevlar® fiber was constructed in DuPont at Richmond (VA), and the semicommercial production was started in 1971. Moreover, the US trade committee named the aromatic polyamide "aramid" in 1974, which was followed by ISO (International Organization for Standardization) to distinguish the polymer from the aliphatic polyamides (e.g., nylons).

The growth of the market of high-strength fibers, initiated by Kevlar®, stimulated the technological evolution along this line utilizing rigid polymers and enhancing molecular orientation along the fiber axis via air-gap wet spinning. Akzo, a Dutch company, followed DuPont and started in 1976 the manufacturing and selling of PPTA fiber which has the same chemical components with Kevlar®. They fell behind DuPont due to the economically worse situations through the first oil crisis although they started the research of the fiber as early as DuPont in the 1960s. They constructed a pilot plant in 1976 and started commercial production at Emen with a capacity of 5,000 ton/year in 1986. They applied a special technology for preparing PPTA solutions: polymer powder is mixed with frozen sulfuric acid and the temperature of the mixture is elevated to a given melting temperature to obtain the condensed PPTA solutions [13]. The business unit of the PPTA fiber, which is firstly named "Arenka®" and later renamed "Towaron®," was separated from Akzo to Acordis company in 1999 and finally took over with Teijin. Teijin Aramid, which is a group company of Teijin, is manufacturing and selling Towaron® at present. In 1987, Japan Aramid Co. was founded as a joint company of Enka and Sumitomo Chemical, and they imported Twaron® but the company itself was now dissolved.

The different ways to make high-strength aramid fibers through isotropic solution (not liquid crystalline solution) were developed at roughly the same time as Kevlar® along the approach to make a polymer with high solubility into organic solvents. Monsanto company developed a wholly aromatic polymer, i.e., polyamide-hydrazides, which can be dissolved into organic solvents via the introduction of flexible groups in the molecule and made a high-strength fiber from this polymer [8]. The fiber was called "X-500." They also developed an improved version "X-702," but they got out from the business at the beginning of the 1970s. On the other hand, Teijin developed a copolymer-type aramid fiber in 1974, named firstly "HM-50" and later "Technora®." The polymer contains 3,4-diaminodiphenylether as a less rigid component in the molecule and can be dissolved into organic solvents. N-Methylpyrrolidone (NMP) was a polymerization solvent for the polymer, which can be used for spinning solution directly. This nature gave a great advantage for manufacturing, whereas Kevlar® has to choose sulfuric acid as a solvent. The less rigid nature also gives isotropy in the solution and the as-spun fiber shows 2 g/day in strength at most [14]. The as-spun fiber can be drawn at draw ratio much larger than ten times at elevated temperatures around 500 °C and results in higher tensile strength than Kevlar®.

5.2.2 *Polyarylate Fiber*

The successes of aramid fibers opened the new market of super fibers and called researcher's attentions to the promising potential of rigid polymers as an advanced material. This movement activated the development of rigid wholly aromatic polyesters in the 1970s. Molecular design of the polymers lowered the melting temperature to a temperature being below the degradation temperature of the polymer and allowed the cost-conscious melt spinning with keeping the liquid crystalline state. This is a big advantage over aramid polymers, which can be manufactured only through the solution spinning utilizing strong acids as a solvent. Poly-para-hydroxybenzoate (PHB, Ekonol®), which was developed in Carborundum company, has a melting temperature beyond 600 °C due to its simple chemical structure compared to the following variations of the wholly aromatic polyesters, and therefore many companies modified molecular components based on Ekonol® and succeeded to lower the melting temperature to below 300 °C. For example, Eastman Kodak company developed "X7G" being a copolymer with PHB and polyethylene terephthalate (PET). Later Carborundum released a melt-processable polymer "Ekkcel®," which is the modified version of Ekonol®. The molecular modifications applied on the aromatic polyesters can be categorized as follows: (1) introduction of large functional groups (e.g., PHQT, DuPont [15]); (2) introduction of less-linear units in place of para-phenylene group, for example, naphthalene unit (e.g., Vectra®, Celanese company[16]) or isophthalic acid (e.g., Xydar®, Dartcore company); and (3) introduction of a flexible alkyl chain (e.g., X7G, Eastman Kodak). As described above, a great number of variations of the thermotropic liquid crystalline polyesters were developed [17]. However, the as-spun fibers made from those polymers showed disappointing tensile properties. Post-polymerization of the as-spun fiber under an elevated temperature led to the high strength deserved to be a super fiber, but this process required extremely long elapsed time and pushed up its cost. Celanese developed a copolymer of p-acetoxybenzoic acid and hydroxynaphthoic acid, which was named "Vectra®," and explored its marketability with the name "F117." Kuraray company succeeded its commercialization as "Vectran®" fiber under the license from Celanese. The high performance of Vectran® fiber is derived via liquid crystalline spinning and post-polymerization after spinning at high temperature. The extremely high molecular orientation induced by the liquid crystalline spinning gives the as-spun fiber very high tensile modulus of 500–600 g/day (62–75 GPa). The post-polymerization results in higher tensile strength with increasing molecular weight, and the strength finally exceeds that of Kevlar® [17]. The economic advantage of melt spinning would be balanced with the additional thermal treatment after spinning, and therefore the price advantage against Kevlar® was probably lost. However, the fiber is accepted from market with its outstanding properties, i.e., low moisture uptake and low creep. A great number of polymers were put into market on trial; nevertheless, only Vectran® fiber survives in the super fiber market to the present. The improvement of spinnability utilizing low molecular weight oligomers and the post-polymerization worked well to those polyesters. However, Vectran® exhibits higher spinnability with relatively high molecular weight oligomers and higher rate of post-polymerization comparing to the other polyesters [18]. Such advantage of Vectran® fiber would preferentially work to survive in the super fiber market.

5.2.3 Heterocyclic Polymer

Challenges to obtain the strongest fiber using the more rigid polymer, which should be higher than aramid fibers, were started from the early 1980s. The polymer including heterocyclic rings in their molecule moiety, for example, poly-*p*-phenylenebenzobisthiazole (PBZT) or poly-*p*-phenylene benzobisoxazole (PBO), was found to be rigid enough to show the longer persistence length, a measure of molecular rigidity, than PPTA. These polymers were firstly studied by the researchers in the US Air Force [19, 20]. Helminiak presented that the PBZT fiber showed as high tensile strength as para-aramid fiber and twice higher tensile modulus than the latter [21]. Later the research was inherited by the Stanford Research Institute (SRI), who filed the patents concerning the synthesis and production of poly-benzazoles [22]. Dow Chemical Company obtained the patent licenses from SRI and developed an industrially available synthesis route of PBO independently [23]. Firstly, they tried to develop the PBO fiber solely, but Toyobo company joined the project as a collaborator in 1991 and Dow Chemical Company quitted the project due to the change in their business strategy in 1994. Toyobo solely continued the development and finally accomplished the commercialization of PBO fiber with the trade name "Zylon®" in 1998. Zylon® is the strongest organic fiber in the world even today among the commercialized fibers. AkzoNobel company invented the fiber [24] made from special heterocyclic polymer, poly (2,6-diimidazo(4,5-b:4′5′-e) pyridinylene-1,4(2,5-dihydroxy)phenylene) (PIPD), which forms the intermolecular hydrogen bonds and gives the highest compressive strength (1–1.7 GPa), whereas PBO shows lower compressive strength due to its weaker inter- and intramolecular interactions. The fiber was called "M5."

AkzoNobel lost an interest in this fiber and transferred it to Acordis. Finally, Magellan Systems International obtained the patents and intellectual properties of the fiber. DuPont Advanced Fiber Systems signed a definitive agreement to acquire a majority stake in Magellan Systems International in 2005, and they constructed a pilot plant at Richmond. The fiber still does not come onto the market today.

5.3 Semirigid Polymers

Super fibers are not only obtained from rigid polymers but also from flexible polymers. However, semirigid polymer, for example, polyethylene terephthalate (PET), is difficult to get the super fiber. Owing to the high economical consistency with properties, PET fiber is one of most manufactured fibers in the world. Its application field has been spreading not only to clothing textiles but also industrial fibers, for example, tire cord. The industrial impact must be tremendous if we could produce a PET fiber with the property close or superior to the super fibers. The organization of Japanese government, New Energy and Industrial Technology Development Organization (NEDO), launched a national project with the aim to double the tensile strength of PET fiber within 1.5 times increase of its cost. Seven-year intensive efforts with leading researchers in academic (Tokyo Institute of Technology, Shinshu University, Kyushu University, Kyoto Institute of Technology, and Kyoto University) and researchers from five companies (Asahi Kasei, Teijin, Toray, Toyobo, and Unitika) accomplished a great result giving the PET fiber with 1.72 times higher strength (1.72 GPa) [25].

5.4 Flexible Polymers

It must deserve attention that we can obtain super fiber using flexible polymers such as polyethylene. The industrially prosperous spinning method called "gel spinning" was invented by Dutch researchers, Smith and Lemstra, in DSM in the early 1980s [3]. Before focusing on the gel spinning, we need to pay an attention on the previous researches as a prologue. Polyethylene, the polymer produced in a largest scale in the world, has attracted many attentions by a great number of researchers. The discovery of the characteristic superstructure "shish-kebab," which was independently found in 1963–1965 by Mitsuhashi [4] and Pennings [5] in the stirred polyethylene solutions, deserves to be picked up here as an event inducing the later epoch. The structure was named "shish-kebab" because of its characteristic structure, which comprised with a long center core crystal and many overgrown lamellar crystals, due to its resemblance with the Turkish cuisine. We must note Pennings's keen insight into the potential of the structure as a high-strength material. He applied a Couette flow on a solution of ultrahigh molecular weight polyethylene filled in a space between two cylinders and set a seed crystal in the flowing solution [6]. A fiber was continuously grown from the seed crystal and the fiber showed tremendous properties of 3 GPa in strength and 100 GPa in modulus [26]. This method was called "surface growth" method and attracted researcher's attentions in the context of a discovery suggesting the potential of polyethylene as a high-strength fiber, which surpasses the strength of the aramid fiber. However, the growth rate of the fiber was very low [6], several centimeters an hour; therefore, this method fade from the mainstream without commercialization. Since the early 1970s, many researchers tried to make high-strength polyethylene fibers with the various approaches, for example, solid-state co-extrusion [27, 28], zone drawing [29], gel-press drawing [30], hot drawing [31, 32], etc. The gel spinning emerged in the early 1980s as a method being consistent with productivity and performance [3]. The flexible polyethylene chains are mutually entangled to form a random coil, or the so-called Gaussian chain, in the highly condensed solution or in the melt. The highly entangled state interferes with continuous elongation of the melt-spun fiber even under elevated temperatures. The optimization of number density of the chain entanglements is a key technology of the gel-spinning method. Semi-dilute solution [33], which gives a partially entangled state, was extruded from a spinneret, and the state was fixed via crystallization in subsequent cooling process and the gel-like as-spun fiber attained very high draw ratio, more than several tens times, by the hot drawing at a temperature higher than the nominal melting temperature of polyethylene. The drawn fiber exhibited the strength much higher than 3 GPa. The essences of the gel-spinning method are summarized as follows: (1) the reduction of number density of entanglements by diluting the solution with a solvent, (2) the larger molecular weight between entanglements using an ultrahigh molecular weight polymer, (3) the fix of less entangled state in the solution via the crystallization after the throughput from a spinneret, and (4) the high draw ratio at an elevated temperature. The challenge to industrialize the gel spinning of ultrahigh molecular weight polyethylene was started in 1984 under the collaboration between DSM and Toyobo. The first commercial plant was constructed in Japan in 1989, followed by the second plant in Heerlen, Netherlands, in 1990. Now they produce and sell the fiber named "Dyneema®" and "IZANAS". In the USA, AlliedSignal Inc. (now Honeywell) independently developed the high-strength polyethylene fiber following DSM, and finally they were granted a license from DSM to manufacture "Spectra®" fiber at Petersburg (VA) [34]. Mitsui Petrochemical company also developed a high-strength polyethylene fiber utilizing the same process as the gel spinning using a relatively lower molecular weight polyethylene and a higher polymer concentration of the spinning solution. However, they quitted from the market at the present day. The concept of gel spinning was applied to poly(vinyl alcohol) (PVA), and the high-strength PVA fiber was commercialized as "Kuralon K-II™" from Kuraray company in 1998 [35].

5.5 Concluding Remarks

A graphical summary of the super fiber's history mimicking a genealogical tree of biologic evolution is shown in Fig. 5.1. To simplify the figure, the fibers avoided from market are not featured here. Moreover, the historical changes of tensile strength of each fiber are also shown in Fig. 5.2. These figures would be helpful for the readers to obtain the perspectives on the history. The innovative events being worthy to be emphasized are shown in Fig. 5.1 with red stars. White open circles denote the year when commercial or semicommercial businesses were started for each fiber. It is worthy to note here that many super fibers (Technora®, Vectran®, Zylon®, Dyneema®, and Kuralon K-II™) are commercialized by Japanese companies. The contribution of the Japanese should not be overlooked, whereas the originality should be attributed to the American and the European companies, and we have to pay our highest respect to them.

From the bottom to the top in Fig. 5.1, the position roughly represents polymer rigidity. The most rigid polymer, PBO, gives the highest strength and modulus. The approach along pursuing the polymer rigidity had worked well in the twentieth century; however, it seems to be encountered a limitation in these days. In the case of another approach utilizing the gel spinning, its availability is also limited. A novel approach must be desired for further advance of super fibers; however, a sign

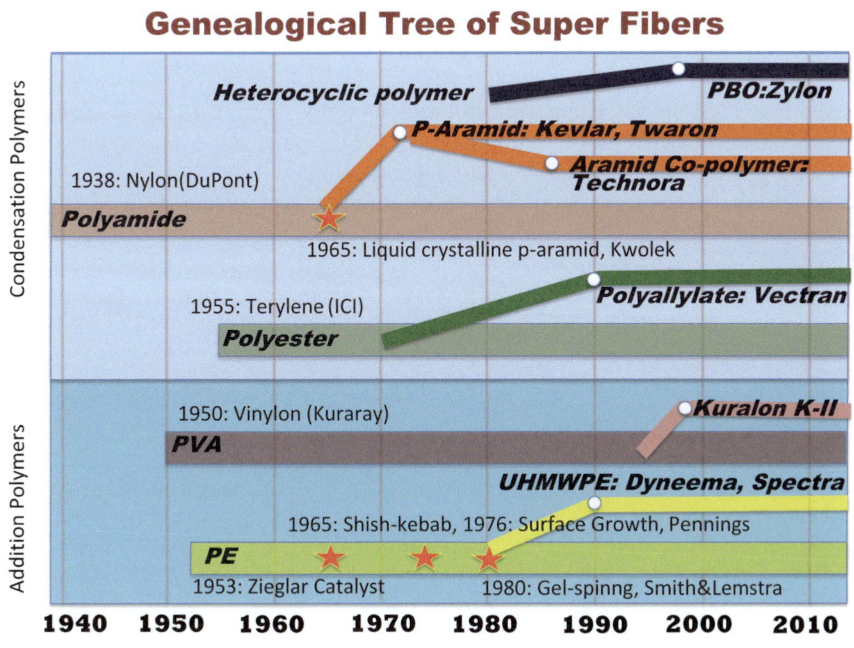

Fig. 5.1 Graphical image mimicking a genealogical tree of biology represents history of super fibers. *Red stars* denote an epoch-making event which deserves to be noted. *White circles* denote a year when commercial or semicommercial business was started for each super fiber

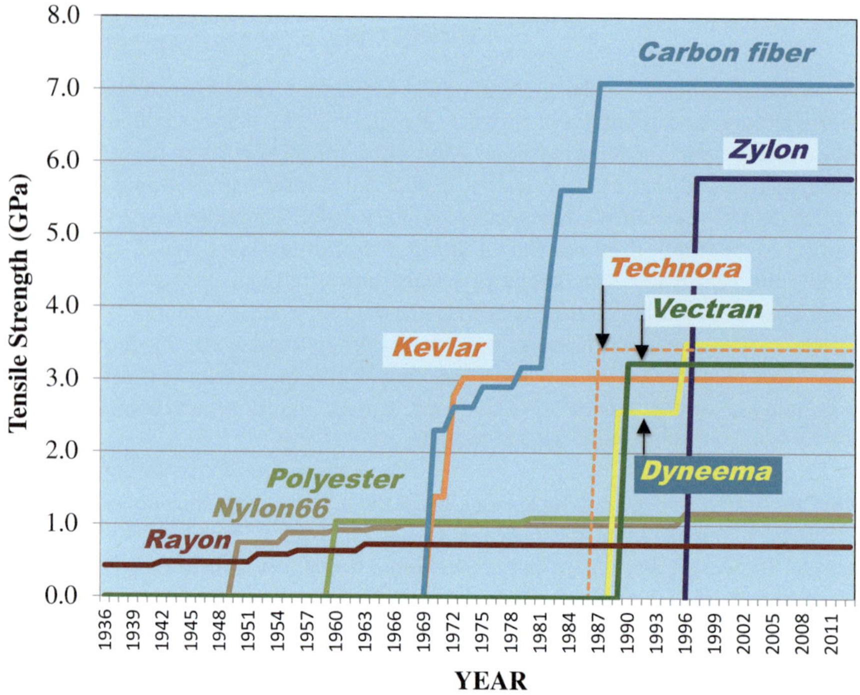

Fig. 5.2 Historical change of tensile strength of super fibers

of the innovation has not appeared in these 15 years. We need a game-changing idea. The scientific trend of bio-mimicry or bio-inspired materials might give a possible solution for overcoming the stagnation. Plant cells generate extended-chain crystals of cellulose on its cell membrane via continuous polymerization of glucoses and concurrent crystallization through the complex of enzymes. If we could mimic the bottom-up process of native cellulose generation with economically available methods, a new frontier of the super fiber would be opened.

References

1. J.W.S. Hearle, in *High-Performance Fibres*, ed. by J.W.S. Hearle (Woodhead Publishing Ltd, Cambridge, 2001), p. 16
2. H.W. Hill, S.L. Kwolek, P.W. Morgan, US patent 3,006,899 (1961); S.L. Kwolek, P.W. Morgan, W.R. Sorenson, US patent 3,063,966 (1972); S.L. Kwolek, US patent 3,671,542; US patent 3,819,587 (1974)
3. P. Smith, P.J. Lemstra, J. Mater. Sci., 15, 505–514 (1980); P. Smith et al., European Patent 0,077,590 (1982); P. Smith et al., US patent 4,436,689 (1982)
4. S. Mitsuhashi, Bull. Text. Res. Inst. J **66**, 1–9 (1964)
5. A.J. Pennings, A.M. Kiel, Kolloid Z. Z. Polym. **205**, 160–162 (1965)
6. A.J. Pennings, Coll. Polym. Sci. **254**, 868–881 (1976)

7. H. Yasuda, K. Ban, Y. Ohta, in *Advanced Fiber Spinning Technology*, ed. by T. Nakajima, K. Kajiwara, Mclntyre (Woodhead Publishing Ltd, Cambridge, 1994), pp. 172–186
8. W.B. Black, in *High-Modulus Wholly Aromatic Fibers*, ed. by W.B. Black, J. Preston (Marcel Dekker, New York, 1973), p. 7
9. S.L. Kwolek, in *Liquid Crystalline Polyamides*. 179th American Chemical Society Meeting, Houston (1980)
10. H. Sakai, Sen'i Gakkaishi J. **43**(4), 125–129 (1987)
11. H.H. Yang, Composite material series volume 2, in *Fiber Reinforcements for Composite Materials*, ed. by A.R. Bunsell (Elsevier, Amsterdam, 1988), pp. 251–253
12. H. Blades, US patent 3,767,756 (1973); US patent 3,869,429 (1975)
13. H.T. Lammers, US patent 4,320,081 (1980)
14. S. Kato, Sen'i Gakkaishi J. **43**(4), 130–134 (1987)
15. C.R. Payet, US patent 4,159,365 (1979)
16. G.W. Calundann, US patent 4,067,852 (1978); US patent 4,161,470 (1979); US patent 4,256,624 (1981)
17. J. Nakagawa, in *Advanced Fiber Spinning Technology*, ed. by T. Nakajima, K. Kajiwara, J.E. Mclntyre (Woodhead publishing, Cambridge, 1994), pp. 160–171
18. W.J. Jackson Jr., in *Contemporary Topics in Polymer Sc., Vol.5*, ed. by E.J. Vandenborg (Plenum Publishing Corporation, New York, 1984), p. 177
19. J.F. Wolfe, F.E. Arnold, Macromolecules **14**, 909 (1981)
20. J.F. Wolfe, B.H. Loo, F.E. Arnold, Macromolecules **14**, 915 (1981)
21. T.E. Helminiak, ACS Org. Coast Plat. Chem. **40**, 475 (1979)
22. J.F. Wolfe et al, US patent 4,225,700 (1980); US patent 4,533,724 (1985); US patent 4,533,692 (1985); US patent 4,533,693 (1985)
23. Z. Lysenko, US patent 4,766,244 (1988)
24. D. Sikkema, Polymer **39**, 5981 (1998)
25. M. Masuda, W. Takarada, T. Kikutani, International polymer processing. Hanser **25**(2), 159–169 (2010)
26. A. Zwijnenburg, A.J. Pennings, J. Polym. Sci. Lett. **14**, 339 (1976)
27. S.H. Southern, R.S. Porter, J. Macromol. Sci. Phys. **B4**, 541 (1970)
28. R.S. Porter et al, Polym. Eng. Sci., **16**, 200 (1976); J. Polym. Sci. Polym. Phys. Ed., **18**, 575 (1980)
29. T. Takahashi, T. Tanaka, R. Kamei, N. Okui, M. Takahiro, S. Umemoto, T. Sakai, Japan. J. Polym. Sci. Tech. **45**, 201 (1988)
30. Y. Sakai, K. Miyasaka, Polymer **29**, 1608 (1988)
31. G. Capaccio, I.M. Ward, Polymer **15**, 233–238 (1976)
32. G. Capaccio, T.A. Crompton, I.M. Ward, J. Polym. Sci. Part B Polym. Phys. **18**, 301–309 (1980)
33. P.G. deGennes, *Scaling Concepts in Polymer Physics* (Cornel University Press, New York, 1979)
34. J.L.J. Van Dingenen, in *High-Performance Fibres*, ed. by J.W.S. Hearle (Woodhead Publishing Ltd, Cambridge, 2001), p. 64
35. E. Sasagawa, Sen'i Gakkaishi J. **54**, 21 (1998)

Chapter 6
Microscopically Viewed Relationship Between Structure and Mechanical Property of Crystalline Polymers: An Important Guiding Principle for the Development of Super Fibers

Kohji Tashiro

Abstract The crystalline phase of polymer substance may be assumed as an idel state for the study of ultimate mechanical property of this polymer. The historical development in the study of structure-property relationship of polymer crystals has been reviewed by focusing on the crystallite modulus along the chain axis, the anisotropy in the lateral plane perpendicular to the chain axis, and the strength. This information is quite important as a guiding principle of the development of fibers with ultrahigh modulus and ultrahigh strength. In other words, the history of the development of these super fibers has been a challenge for approaching the goal of ultimate mechanical property. The experimental and theoretical methods to estimate the ultimate mechanical property of polymer crystals have been reviewed in Sects. 6.2, 6.3, 6.4, and 6.5. In the experimental approach, the X-ray diffraction peak shift and Raman band shift caused by the application of tensile stress are measured to detect the mechanical deformation of crystalline region, but the data analysis must be performed by taking the heterogeneous distribution into account. The quantum mechanical prediction of the ultimate strength of polymer chain has been reviewed in Sect. 6.6. The factors governing the strength of bulk polymer sample have been discussed concretely.

Keywords Young's modulus • Strength • Chain conformation • Crystal structure

K. Tashiro (✉)
Department of Future Industry-Oriented Basic Science and Materials, Toyota Technological Institute, Nagoya, Japan
e-mail: ktashiro@toyota-ti.ac.jp

Society of F.S. Technology, Japan (ed.), *High-Performance and Specialty Fibers*,
DOI 10.1007/978-4-431-55203-1_6

6.1 Introduction

Historically viewed, the recognition of the possibility that the synthetic polymers possess the elastic modulus beyond the steel was made much earlier than the actual development of fibers with high Young's modulus and high strength. Let us see the case of polyethylene (PE). In 1949, Shimanouchi and Mizushima discovered the so-called accordion vibrational bands for a series of *n*-alkanes in the low-frequency region of Raman spectra [1]. The peak positions were found to shift inversely in proportion to the chain length. They interpreted these bands in terms of longitudinal acoustic vibration of a continuous body. This mode was found to correspond to the propagation of acoustic phonons in the polyethylene crystal along the chain axis [2]. In almost the same years, Lyons (1959) [3] and Treloar [4] calculated the Young's modulus of polyethylene zigzag chain on the basis of the theory about the balance of forces between the internal coordinates such as bond lengths and bond angles. Although the calculation was relatively rough, the theoretical evaluation of the Young's modulus of PE chain was made for the first time. In 1966, Odajima [5] calculated the elastic constants of orthorhombic polyethylene crystal on the basis of lattice dynamical theory developed by Born [6]. The vibrational frequency-phase angle dispersion curve of polyethylene crystal was calculated by Tasumi and Shimanouchi [7] and Krimm [8]. From the slope of the dispersion curve of the acoustic branch along the chain axis, the Young's modulus of polyethylene was estimated. In 1971, the neutron scattering measurement gave the dispersion curves experimentally and evaluated the Young's modulus as about 329 GPa [9]. On the other hand, in 1966, Sakurada et al. performed the X-ray diffraction measurement for the oriented polyethylene crystal under the constant tensile force, and the Young's modulus was estimated as 235 GPa, where the homogeneous stress distribution was assumed, i.e., the stress working on the crystal region was equal to the tensile stress in the bulk sample [10]. In the 1970s–1980s, the theoretical and experimental techniques to estimate the ultimate mechanical property of the various types of polymer crystals were developed remarkably [11]. In this way, it should be noticed that the prediction of the ultimate mechanical property of polymers was made in much earlier years before the development of actual polyethylene fiber with ultrahigh modulus in the 1980s.

In the present article, the development of the molecular theoretical study on the relationship between structure and mechanical property of polymer materials in this half century is reviewed briefly, which has been an important guiding principle for the development of fibers of ultrahigh modulus and ultrahigh strength.

6.2 Experimental Evaluation of Ultimate Elastic Constants of Polymers

6.2.1 X-Ray Diffraction Method

Important mechanical property in the discussion of high-modulus fibers is the Young's modulus along the chain axis, which is named here the crystallite modulus E_c. Of course, the most ideal method for this purpose is to use a giant single crystal of polymer. The strain along the chain axis is estimated directly by applying a tensile force along the chain axis [12, 13]. Unfortunately, the usage of such a giant single crystal is limited at present to the polymers obtained by the photo-induced solid-state polymerization reaction, e.g., polydiacetylenes [12, 14] and poly (muconic ester) [13]. In the case of partially crystalline polymers, the X-ray diffraction method is one of the most excellent techniques [10]. The signal from the small crystallites is detected by measuring the X-ray diffractions. This point can be evaluated well in such a meaning that the mechanical property of small crystallites can be extracted even in the bulk sample consisting of the crystalline and amorphous regions. As pointed out in the introduction, this X-ray diffraction technique was accomplished by Sakurada et al. in the 1960s [10]. The small strain of the crystal is estimated by tracing the small shift of the X-ray diffraction peak corresponding to the lattice planes along the chain axis by applying a tensile force to the oriented sample. The shift of X-ray diffraction peak can be converted into the strain ε_c. Unfortunately, the stress σ_c working on the crystalline region cannot be known explicitly, and so it *must* be assumed to be equal to the macroscopic stress calculated from the actually applied tensile force and the cross-sectional area of the sample. This was named the assumption of homogeneous stress distribution. The E_c can be estimated as the initial slope in the $\sigma_c - \varepsilon_c$ curve, which is now termed as $E_c^{\text{X-ray}}$. For example, $E_c^{\text{X-ray}}$ is 235 GPa for polyethylene (PE) crystal, 40 GPa for *isotactic* polypropylene (*it*-PP) [15, 16], 53 GPa for polyoxymethylene (POM) [17], 100 GPa for poly(ethylene terephthalate) (PET) [18, 19], and so on.

As pointed above, these X-ray values were obtained under the assumption of homogeneous stress distribution. This assumption causes a serious problem in the evaluation of *true* Young's modulus of the crystalline region (E_c^{true}) in the semi-crystalline polymer samples. In fact, in many cases, the E_c^{true} and $E_c^{\text{X-ray}}$ gave serious gap between them. For example, a giant single crystal of poly(muconic ester) of several centimeter lengths gives the E_c of 47 GPa, which was measured accurately by reading out the change in sample length under the application of tensile forces [13]. This value might be assumed as E_c^{true} in a good approximation. Luckily, this polymer can be melted and stretched to give the oriented semicrystalline samples. The crystallite modulus was estimated by the X-ray method under the assumption of homogeneous stress distribution. However, the $E_c^{\text{X-ray}}$ was only 17–35 GPa depending on the sample preparation conditions, far lower than the E_c^{true}. In this way, one of the most significant points about the X-ray diffraction method is that the $E_c^{\text{X-ray}}$ is not always common to all the samples, but it depends on the preparation condition of the sample. This situation was found to occur also in the Raman scattering experiment, as will be mentioned in a later section. For the solution of this problem, Ward et al. [20] and Tashiro et al. [13, 21–23] utilized a complex mechanical model to show the heterogeneous stress distribution, which was proposed by Takayanagi et al. in 1963 [24]. The Young's modulus of the bulk sample (E_{bulk}) and $E_c^{X\text{-ray}}$ (and Raman shift factor) were expressed in an explicit form on the basis of this mechanical model. The E_c^{true} was determined so that the observed data of E_{bulk} and $E_c^{X\text{-ray}}$ (and Raman shift factor) were reproduced consistently. The above-mentioned relation of $E_c^{true} = 47$ GPa and $E_c^{X\text{-ray}} = 17–35$-GPa observed for poly(muconic ester) samples could be interpreted reasonably using this complex mechanical model [13]. This method was applied to the observed $E_c^{X\text{-ray}}$ values of the various polymers, giving E_c^{true} of 75 GPa for POM ($E_c^{X\text{-ray}}$ was 53–59 GPa for normally drawn samples and 72 GPa for ultra-drawn sample) [21], 260 ± 10 GPa for PE [21], and so on.

6.2.2 *Vibrational Spectroscopic Method*

In this way the correction of heterogeneous stress distribution gives in general a higher Young's modulus compared with $E_c^{X\text{-ray}}$. Such higher E_c values are detected in the vibrational spectroscopic experiments. As mentioned in the introduction, Shimanouchi and Mizushima discovered the accordion bands in the low-frequency region of Raman spectra of *n*-alkanes, which were interpreted reasonably using the longitudinal acoustic dispersion curve of polyethylene crystal along the chain axis [1, 2]. The evaluated Young's modulus was quite high, ca. 358 GPa. Strobl et al. [25] and Kobayashi et al. [26] corrected an effect of end-to-end interactions between the *n*-alkane molecules in the neighboring layers on the LAM (longitudinal acoustic mode) frequency and derived the E_c^{LAM} as 280–290 GPa, which is nearer to the E_c^{true}. The dispersion curve measured using coherent neutron scattering from the fully deuterated polyethylene crystal gave 329 GPa [9]. The theoretical calculation based on the lattice dynamical method gives 316 GPa [27]. For reference, the Young's modulus of ultra-drawn polyethylene fiber at low temperature was 288 GPa [28], which is closer to these spectroscopic values and far beyond the $E_c^{X\text{-ray}}$.

The Raman bands corresponding to the vibrational modes of the skeletal chains were shifted by applying the tensile force [29, 30]. In a small strain region, the shift is expressed as a linear function of stress ($\nu = -\alpha \cdot \sigma$). Similarly to the case of $E_c^{X\text{-ray}}$, the shift factor α is also affected sensitively by the morphology of the sample [21]. This fact is combined with the $E_c^{X\text{-ray}}$ data to derive the heterogeneous stress distribution as already mentioned in the previous sections [13, 21–23].

6.3 Theoretical Evaluation of Ultimate Elastic Constants of Polymers

As mentioned in the introductory section, the theoretical estimation of E_c was performed already in the 1960s. The principle of the calculation was based on the static balance of forces between the internal coordinates such as bond lengths, bond angles, and torsional angles [3, 4, 31–34]. This method was developed by deriving the equation of E_c on the basis of lattice dynamical theory [5, 11, 31–36]. In this calculation the atomic displacements were expressed in terms of the accurate atomic coordinates and the reliable force constants, from which the three-dimensional elastic constants matrix can be derived. Odajima et al. calculated the elastic constant matrix of orthorhombic polyethylene crystal by the lattice dynamical theory for the first time [5]. Miyazawa et al. rewrote the equation by using matrices corresponding to the force constants and the atomic coordinates [35]. Tashiro et al. introduced the space group symmetry to the lattice dynamical equation and reduced the dimension of the giant matrices remarkably so as to accelerate the calculation rate of elastic constants, and they made it possible to apply the equation to any type of polymer crystal of complicated aggregation structure of atoms [36]. Another method is based on the molecular mechanics (MM), in which the potential energy of polymer chain (or polymer crystal) is numerically calculated by changing the repeating period along the chain axis step by step by considering the interactions between the atoms, from which the potential energy is expressed as a function of strain ($V = (1/2)E_c\varepsilon^2$) and so the Young's modulus E_c is obtained as the second derivatives of potential energy V with respect to the strain ε [37]. The ab initio molecular orbital (MO) method is more sophisticated and calculates the potential energy based on Hamiltonian and a basis function set. The elastic constant matrix is obtained by the calculation of second derivatives of the thus calculated potential energy with respect to the Cartesian coordinates [38, 39]. These MM and MO methods give in general the overestimated values than the lattice dynamical method. In the latter case, the force constants between the internal displacement coordinates such as bond stretching, bond angle deformation, etc. were refined so that the calculated vibrational frequencies are in good agreement with the actually observed infrared and/or Raman spectra. The MM and MO give the relatively higher values to the vibrational frequencies. Sometimes the "adjustable parameter" was introduced in the MO calculation so that the thus-calculated frequencies were matched to the observed data.

The temperature dependence of the elastic constants can be estimated theoretically on the basis of molecular dynamics (MD) method [40–42]. The time-dependent fluctuation of potential energy of a crystal is calculated at the various temperatures, from which the elastic constant is obtained as a function of temperature.

6.4 Relationship Between Chain Conformation and Young's Modulus

As easily imagined, the Young's modulus E_c is sensitively dependent on the geometry or conformation of polymer chain and the interactions between the internal coordinates. More concretely, the E_c is expressed in an approximation as follows [11]:

$$1/E_c \propto \sum_{\mathrm{i}} (\partial d/\partial R_{\mathrm{i}})^2/F_{\mathrm{ii}} \propto \sum_{\mathrm{i}} (\mathrm{PED})_{\mathrm{i}}$$

where $(\partial d/\partial R_{\mathrm{i}})$ is the first derivative of the pitch of one monomeric unit with respect to the internal coordinate R_{i}, F_{ii} is the diagonal term of force constant corresponding to R_{i}, and $(\mathrm{PED})_{\mathrm{i}}$ is the potential energy distribution to R_{i}. When a tensile force is applied along the chain axis, the potential energy is increased as a result of chain deformation, which is distributed to the various changes of the internal coordinates. This equation indicates that the E_c is related to the easiness of chain deformation $(\partial d/\partial R_{\mathrm{i}})$ and the rigidity of the change in R_{i} (F_{ii}) and that the internal coordinates having larger PED contribute to the determination of E_c. For example, Fig. 6.1 shows the deformation of molecular chains with the various PED values calculated for PE, PET, POM, PEOB (poly(ethylene oxybenzoate) α form), and PBO (poly-*p*-phenylene benzobisoxazole) [11, 42]. PE shows a high E_c because of its slim form and the PED to the skeletal C-C bond stretching and C-C-C angle deformation in almost equal percentage. PET chain is deformed by changing mainly the bond angles of ester carbon atoms (O-C(O)-benzene) so that the virtual bond passing through the benzene ring stands up toward the chain axis. Although PET chain possesses rigid benzene rings and ester groups, the E_c is not determined by the rigidity of these groups but by the easiness of deformation of zigzag angle. POM chain takes a helical form and the deformation occurs mainly through the change of COCO torsional angles as well as the change in bond angles (COC and OCO). POM shows relatively high E_c value because of the large contribution of bond angle deformation in addition to the torsional angle changes. PEOB α form takes a large zigzag conformation consisting of long arms and ethylene segmental parts. The chain deformation occurs mainly through the torsional motion around the ether bonds, giving quite low E_c of about 2 GPa. In the case of PBO, the whole of chain is so rigid, and the deformation occurs mainly through the stretching of linkages connecting the benzene and oxazole rings as well as the deformations of these rigid rings. The E_c value of 460 GPa is in the highest level among the various polymers.

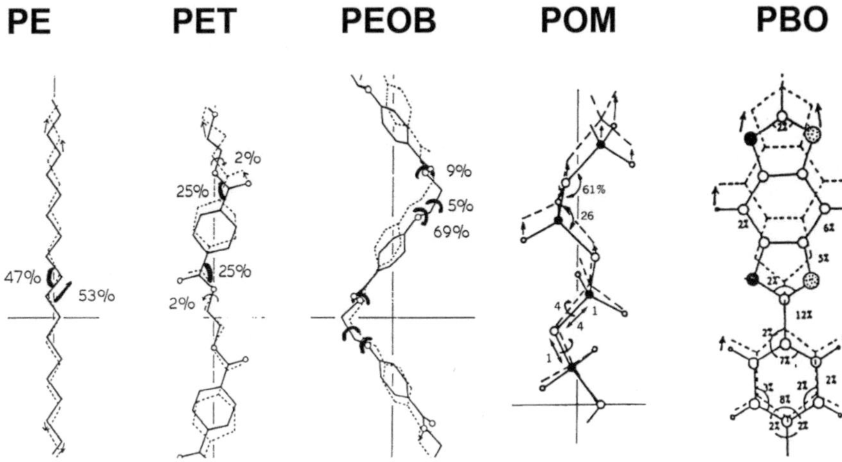

Fig. 6.1 Chain deformation and potential energy distribution calculated for the various polymers

6.5 Crystal Structure and Anisotropic Mechanical Property

The Young's modulus along the chain axis is one of the mechanical properties. The anisotropy in Young's modulus is also important in the discussion of the relationship between crystal structure and mechanical property [11]. The H...H intermolecular interactions in the crystal lattice govern mainly the anisotropy of the mechanical property. Therefore the Young's modulus in the lateral directions is one to two order lower compared with the modulus along the chain axis. Figure 6.2 shows the anisotropic curves of Young's modulus calculated for the various types of polymer crystals [11, 43]. In the case of orthorhombic PE crystal, the anisotropy is not very high, but the modulus is almost isotropic in the lateral directions because of the weak van der Waals forces between the hydrogen atoms. In the case of nylon 6, the sheet planes built up by hydrogen-bonded zigzag chains are stacked by weaker van der Waals interactions, resulting in the highly anisotropic mechanical property. These intermolecular interactions do not affect the Young's modulus along the chain axis very much. Poly-*p*-phenylene terephthalamide (PPTA) forms also the sheet structure similar to that of nylon 6 [44]. Therefore this polymer shows appreciably highly anisotropic mechanical property in the lateral direction as seen in Fig. 6.2. In the case of poly-*m*-phenylene isophthalamide (PMPI), the intermolecular hydrogen bonds are formed between the neighboring chains in a similar way as those of poly-*p*-phenylene terephthalamide, but when the direction of hydrogen bonds is viewed along the chain axis, it changes alternately perpendicularly along the *a* and *b* axes. As a result, the mechanical property is isotropically strong in the lateral directions [45]. POM chain takes a compact

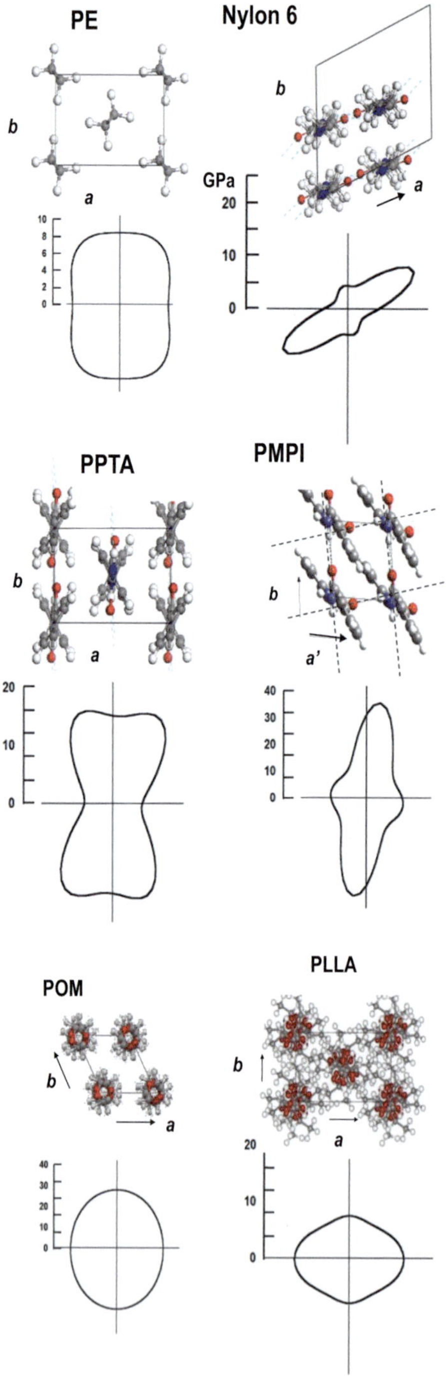

Fig. 6.2 Anisotropic Young's modulus in the plane perpendicular to the chain axis calculated for the various types of polymer crystals (*PE* polyethylene, *PPTA* poly-*p*-phenylene terephthalamide, *PMPI* poly-*m*-phenylene isophthalamide), *POM* polyoxymethylene, and *PLLA* poly(L-lactic acid)

helical form, and these chains are closely packed in a hexagonal-type cell, resulting in the appreciably strong but isotropic Young's modulus in the lateral plane perpendicular to the chain axis [43]. Another typical example of helical chain conformation is observed for poly(L-lactic acid) [46]. The helices of large radius are packed in the crystal lattice in relatively loose manner, giving appreciably small and isotropic curve of the Young's modulus [43].

6.6 Strength of Polymer Chains

So far we have discussed the Young's modulus of polymer crystals or the mechanical property in the infinitesimally small deformation region. When the crystal is deformed in higher strain region, the crystal will be broken. The stress at this point is called the (fracture) strength. The fracture occurs by cutting the covalent bonds of the skeletal chain. If the potential energy of bond stretching mode is harmonic ($V = (1/2)E_c\varepsilon^2$), the chain recovers to the original form necessarily even when the chain is stretched to a quite high strain. The breakdown of the covalent bond can be expressed using the anharmonic potential function, typical form of which is a so-called Morse potential function $\left(V = V_o\left(1 - e^{-k\Delta r}\right)^2\right)$. As the deformation Δr ($= r - r_o$) becomes larger and reaches a certain value, then the potential energy reaches the flat value, where the recovery force becomes zero. This stress corresponds to the breakage stress. Since the breakage of covalent bond occurs by changing the electronic structure, more sophisticated method is based on the quantum mechanics. The repeating period along the chain axis is changed step by step, and the potential energy is calculated using a Hamiltonian operator. The recovery force is calculated as the first derivative of the potential energy. The recovery force reaches maximum and reduces sharply to zero at a certain strain, as shown in Fig. 6.3. The strength of PE chain was calculated to be about 45 GPa by the several researchers [47–49]. The sp^3 orbital of carbon atom changes to the sp^2 form when the radicals are generated at the ruptured C-C bond (see Fig. 6.3). However, the thus-estimated stress is too high compared with the actually observed strength of ultra-drawn PE sample, at most 6 GPa. The similar situation can be seen also for such rigid polymers as PBO [38].

These calculations did not consider the kinetic factor. As shown in Fig. 6.4, the scission of covalent bond is a chemical reaction. The system must cross the energy barrier E^* in the transition process from the bonded C-C bond to the fractured C atoms. When the stress σ is applied to the system, the barrier is reduced, and the probability to cross the barrier, expressed as $\exp[-(U - \beta\sigma)/kT]$, becomes higher, where k is the Boltzmann constant, T is an absolute temperature, and β is a proportional coefficient. Using this idea, the fracture strength was calculated by a Monte Carlo method, giving a few GPa, comparative to the realistic value [50–52].

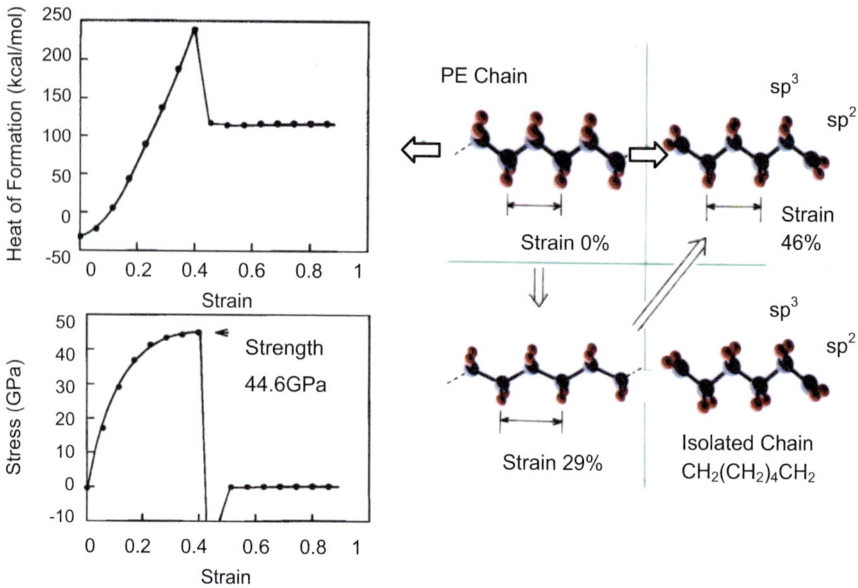

Fig. 6.3 Fracture of polyethylene zigzag chain calculated by quantum mechanical method: (**a**) the potential energy plotted against the strain, (**b**) the corresponding force, and (**c**) the chain deformation procedure

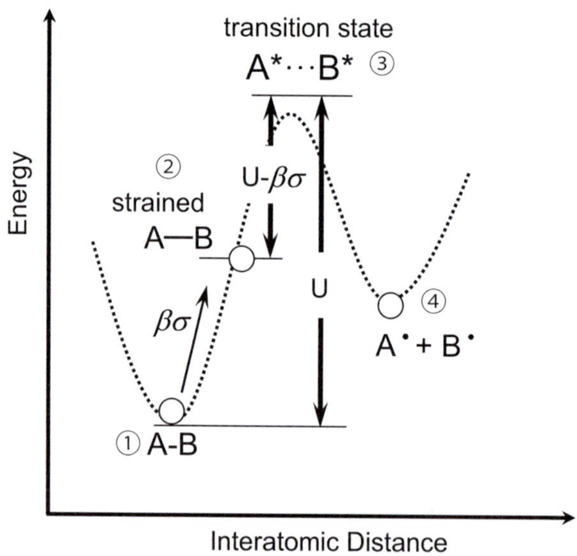

Fig. 6.4 Potential energy barrier in the A-B bond breakage process. (*1*) The original A-B bond, (*2*) the stretched A-B bond. The potential energy is increased by strain and (*3*) the barrier to cross the transient state A*. . .B* becomes lower. (*4*) The broken bond

In the actual case, the fracture of fibers does not only occur by the scission of C-C covalent bonds in the fully extended chains but also it occurs at the various structural defects through the separation of chains in the vicinity of chain ends, the slippage between the microfibrils, and so on. Because of these additional factors, the strength of a bulk sample becomes lower.

In this way the fracture of fibers must be interpreted by taking these various factors into consideration. Sometimes the phase transition might occur before the fracture, and the strength predicted must be changed correspondingly. Another factor important for the discussion of strength is the heterogeneous stress distribution. The local parts with structural defects might be damaged most easily if the stress concentration occurs there. One example is seen in carbon fiber (CF) [23, 53, 54]. The inner structure of a CF monofilament is not homogeneous but consists of the complicated aggregation of crystalline and amorphous regions. Roughly speaking, the CF monofilament takes a skin-core structure as revealed by the wide-angle X-ray scattering measurement using a synchrotron X-ray beam of micrometer size [53, 54]. The $E_c^{\text{X-ray}}$ and Raman shift factor (α) were found to be different depending on the CF monofilaments prepared at the different temperatures [53, 54]. The analysis was performed in a similar way using a complex mechanical model, from which the heterogeneous stress distribution was estimated in a quantitative manner. As shown in Fig. 6.5, the stress at the skin part is much higher than the core part. The skin part or the surface of CF monofilament is annealed at a higher temperature than the core part. As a result the Young's modulus of the surface part is higher and owes the stress at a higher level. The structure defect is said to generate relatively easily on the surface of CF monofilament, causing easier fracture of the CF starting from the surface part. In other words, the homogeneous heat treatment is needed for the production of the CF monofilament with homogeneous stress distribution. One idea is to use a microwave technique by which the PAN precursor is heated homogeneously to be a CF monofilament with higher modulus and higher strength than before [55].

Tie chain is also an important concept in the discussion of strength [56]. The tie chains pass through the neighboring lamellae and protect the stacked lamellae from the fracture. So far the tie chain had been assumed as only a hypothetical object. But one example is seen to show the actual existence of tie chains. This is a study of structural change in the isothermal crystallization of POM from the melt [57]. As well known, the infrared spectra of POM change sensitively depending on the morphology or the aggregation state of helical chains [58, 59]. The ECC (extended chain crystal) and FCC (folded chain crystal) can be distinguished by measuring the IR band characteristic of these morphologies. The time-dependent measurement of FTIR spectra during the isothermal crystallization from the melt revealed that the FCC bands appeared at first in parallel to the formation of the stacked lamellae as known from the small-angle X-ray scattering measurement. After that, the ECC bands started to appear, just when the daughter lamellae were generated between the mother lamellae, as shown in Fig. 6.6. The good correspondence between the appearance of ECC bands and daughter lamellae make it possible to imagine that the fully extended tie chain segments pass through the neighboring lamellae. When

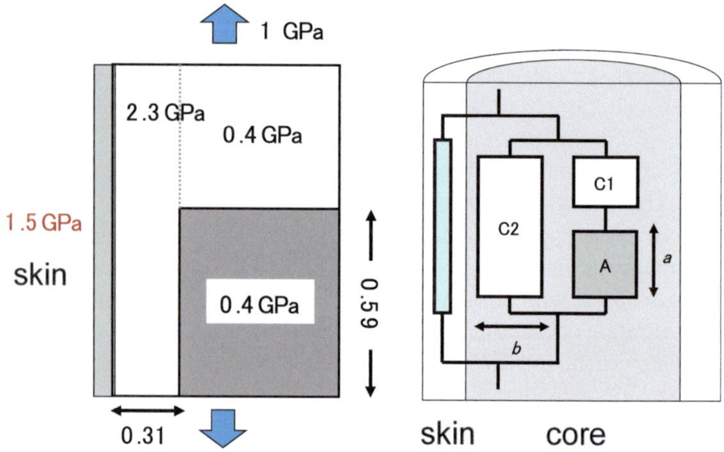

Fig. 6.5 The heterogeneous stress distributions calculated for the carbon fiber monofilament subjected to a tensile stress 1 GPa. The stress in the skin part is 1.5 times higher than the average stress

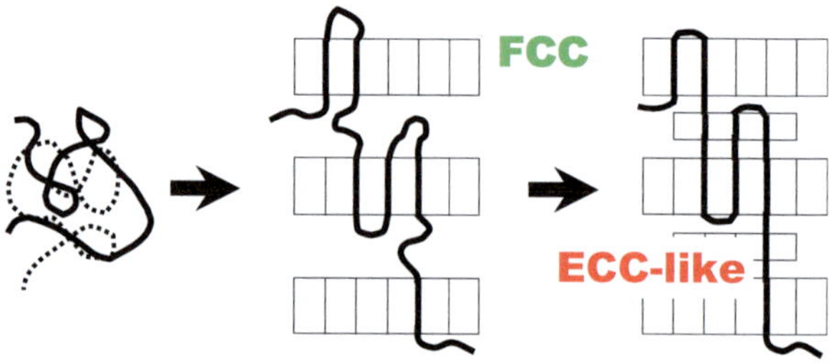

Fig. 6.6 Formation of tie chain segments in the melt-isothermal crystallization of polyoxymethylene

the POM sample is stretched, these tie chain segments may be tensioned and protect the fracture of stacked lamellae up to the critical stress causing the covalent bond breakage.

In this way the strength of bulk polymer material is governed by the various factors including the ultimate strength of fully extended chain, the kinetic effect of transient energy barrier, the structural defects, the slippage of microfibrils, the heterogeneous stress distribution, and the existence of taut tie chains. The theoretical estimation is still difficult at present for understanding the microscopically viewed correlation between the mechanical breakage and the chain aggregation state in the bulk polymer material by taking all of these factors into account.

References

1. S.I. Mizushima, T. Shimanouchi, J. Am. Chem. Soc. **71**, 1320 (1949)
2. R.F. Schaufele, T. Shimanouchi, J. Chem. Phys. **47**, 3605 (1967)
3. W.J. Lyons, J. Appl. Phys. **29**, 1429 (1958)
4. L.R.G. Treloar, Polymer **1**, 95 (1960)
5. A. Odajima, T. Maeda, J. Polym. Sci. Part C **15**, 55 (1966)
6. M. Born, K. Huang, *Dynamical Theory of Crystal Lattice* (Oxford University Press, London, 1954)
7. M. Tasumi, T. Shimanouchi, J. Chem. Phys. **43**, 1245 (1965)
8. M. Tasumi, S. Krimm, J. Chem. Phys. **46**, 755 (1967)
9. L. Holliday, J.W. White, Pure Appl. Chem. **26**, 545 (1971)
10. I. Sakurada, T. Ito, K. Nakamae, J. Polym. Sci. Part C **15**, 75 (1966)
11. K. Tashiro, Prog. Polym. Sci. **18**, 377–435 (1993)
12. G. Wu, K. Tashiro, M. Kobayashi, Macromolecules **22**, 188–196 (1989)
13. S. Nakamoto, K. Tashiro, A. Matsumoto, J. Polym. Sci. Part B Polym. Phys. **41**, 444–453 (2003)
14. G. Wegner, Pure Appl. Chem. **49**, 433 (1977)
15. K. Tashiro, M. Kobayashi, H. Tadokoro, Polym. J. **24**, 899–916 (1992)
16. C. Sawatari, M. Matsuo, Macromolecules **19**, 2653 (1986)
17. K. Nakamae, T. Nishino, K. Hata, T. Matsumoto, Kobunshi Ronbunshu **42**, 241 (1985)
18. M. Matsuo, C. Sawatari, Polym. J. **22**, 518 (1990)
19. T. Thistlethwaite, R. Jakeways, I.M. Ward, Polymer **29**, 61 (1988)
20. S. Jungnitz, R. Jakeways, I.M. Ward, Polymer **27**, 1651 (1986)
21. K. Tashiro, G. Wu, M. Kobayashi, Polymer **29**, 1768–1778 (1988)
22. T. Kitagawa, K. Tashiro, K. Yabuki, J. Polym. Sci. Part B Polym. Phys. **40**, 1281–1287 (2002)
23. T. Kobayashi, K. Sumiya, Y. Fujii, M. Fujie, T. Takahagi, K. Tashiro, Carbon **50**, 1163–1169 (2012)
24. M. Takayanagi, K. Imada, T. kajiyama, J. Polym. Sci. Part C Polym. Symp. **15**, 263 (1967)
25. G.R. Strobl, R. Eckel, J. Polym. Sci. Polym. Phys. Ed. **14**, 913 (1976)
26. M. Kobayashi, K. Sakagami, H. Tadokoro, J. Chem. Phys. **78**, 6391 (1983)
27. K. Tashiro, M. Kobayashi, H. Tadokoro, Macromolecules **11**, 914–918 (1978)
28. P.J. Barham, A. Keller, J. Polym. Sci. Polym. Lett. Ed. **17**, 591 (1979)
29. K. Tashiro, S. Minami, G. Wu, M. Kobayashi, J. Polym. Sci. Part B Polym. Phys. **30**, 1143–1155 (1992)
30. R.J. Young, R. Young, C. Ang, Strain Measurement in Fibres and Composites Using Raman Spectroscopy, Development in the Science and Technology of Composite Materials, Springer Netherlands, pp. 685–690 (1990).
31. T. Shimanouchi, M. Asahina, S. Enomoto, J. Polym. Sci. **59**, 93 (1962)
32. M. Asahina, S. Enomoto, J. Polym. Sci. **59**, 113 (1962)
33. H. Sugetam, T. Miyazawa, Polym. J. **1**, 226 (1970)
34. K. Tashiro, M. Kobayashi, H. Tadokoro, Macromolecules **10**, 731–736 (1977)
35. Y. Shiro, T. Miyazawa, Bull. Chem. Soc. Jpn. **44**, 2371 (1971)
36. K. Tashiro, M. Kobayashi, H. Tadokoro, Macromolecules **11**, 908–913 (1978)
37. G.C. Rutledge, U.W. Suter, Polymer **32**, 2179 (1991)
38. S.G. Wierschke, Mater. Res. Soc. Symp. **134**, 313 (1989)
39. J.R. Shoemaker, T. Horn, P.D. Haaland, R. Pachter, W.A. Adams, Polymer **33**, 3351 (1992)
40. J.R. Ray, Comp. Phys. Rep. **8**, 109 (1988)
41. K. Tashiro, Comp. Theor. Polym. Sci. **11**, 357–374 (2001)
42. K. Tashiro, M. Kobayashi, Macromolecules **24**, 306–3708 (1991)
43. K. Wasanasuk, K. Tashiro, Macromolecules **45**, 7019–7026 (2012)
44. K. Tashiro, M. Kobayashi, H. Tadokoro, Macromolecules **10**, 413–420 (1977)
45. K. Tashiro, M. Kobayashi, Polymer **32**, 1516–1526 (1991)

46. K. Wasanasuk, K. Tashiro, M. Hanesaka, T. Ohhara, K. Kurihara, R. Kuroki, T. Tamada, T. Ozeki, T. Kanamoto, Macromolecules **44**, 6441–6452 (2011)
47. D.S. Boudreaux, J. Polym. Sci. Polym. Phys. Ed. **11**, 1285 (1973)
48. B. Crist, M.A. Ratner, A.L. Brower, J.R. Sabin, J. Appl. Phys. **50**, 6047–6051 (2009)
49. K. Tashiro, J. Text. Mach. Soc. Jpn. **48**, 425 (1995)
50. Y. Termonia, P. Meakin, P. Smith, Macromolecules **19**, 154 (1986)
51. Y. Termonia, P. Smith, Macromolecules **20**, 835 (1987)
52. Y. Termonia, P. Smith, Macromolecules **21**, 2184 (1988)
53. T. Kobayashi, K. Sumiya, Y. Fukuba, M. Fujie, T. Takahagi, K. Tashiro, Carbon **49**, 1646–1652 (2011)
54. T. Kobayashi, K. Sumiya, Y. Fujii, M. Fujie, T. Takahagi, T. K., Carbon **53**, 29–37 (2013)
55. B.B. Balzer, J. McNabb, J. Ind. Tech. **24**, 1 (2008). http://atmae.org/jit/Articles/balzer061608.pdf
56. S.N. Zhurkov, V.E. Korsukov, J. Polym. Sci. Polym. Phys. Ed. **12**, 385 (1974)
57. H. Hama, K. Tashiro, Polymer **44**, 6973–6988 (2003)
58. M. Shimomura, M. Iguchi, Polymer **23**, 509 (1982)
59. M. Kobayashi, M. Sakashita, J. Chem. Phys. **96**, 748–760 (1992)

Chapter 7
Dyneema®: Super Fiber Produced by the Gel Spinning of a Flexible Polymer

Yasunori Fukushima, Hiroki Murase, and Yasuo Ohta

Abstract Dyneema® is the ultrahigh molecular weight polyethylene (UHMWPE) fiber showing the higher tensile strength than that of aramid fibers. In addition to the extremely high tensile properties, the fiber shows other unique characteristics, for example, lightweight, chemical stability, shock absorption, negative thermal expansion, etc., and its application area has been widely spread by the long-continued efforts of TOYOBO company and DSM company from the late 1980s.

The UHMWPE fibers are manufactured by a unique spinning technology, i.e., gel-spinning method, invented by researchers in DSM in the early 1980s, and this method can be defined as an innovation leading to the industrialization of the super fiber made of flexible chain polymers. In this chapter, the essence of the gel-spinning technology, the structure evolution in the spinning and the drawing process, fiber properties, and applications of Dyneema® are presented. Especially, the reasons, why we can achieve the superior mechanical properties of the UHMWPE fibers with the gel-spinning method, are described by focusing on the microstructures of the fibers.

Keywords Ultrahigh molecular weight polyethylene fiber • Dyneema® • Super fiber • Gel-spinning • Flexible polymer • Semi-dilute solution • Drawability • Shish-kebab structure

7.1 Introduction

Ultrahigh molecular weight polyethylene (UHMWPE) fibers—one of the typical super fibers—are produced by a special method known as gel spinning. The gel-spinning method was invented by Smith and Lemstra of DSM, a Dutch company, in the late 1970s [1]. Thereafter, TOYOBO and DSM jointly developed the

Y. Fukushima (✉) • Y. Ohta
Research Center, TOYOBO Co., Ltd., 2-1-1 Katata, Otsu, Shiga 520-0292, Japan
e-mail: yasunori_fukushima@toyobo.jp

H. Murase
Faculty of Home Economics, Department of Textile and Clothing, Kyoritsu Women's University, 2-2-1 Hitotsubashi, Chiyoda-ku, Tokyo 101-8437, Japan

Society of F.S. Technology, Japan (ed.), *High-Performance and Specialty Fibers*,
DOI 10.1007/978-4-431-55203-1_7

Table 7.1 Mechanical properties of UHMWPE fibers and other super fibers

		Tensile strength		Tensile modulus		Density	Moisture regain	Limiting oxygen index	Melting or decomposition temperature
		cN/dtex	GPa	cN/dtex	GPa	g/cm³	wt%	–	°C
Aramid	Kevlar®49 (DuPont)	20.8	3.0	780	112	1.44	3.5	29	560
	Towaron®HM (Teijin)	20.7	3.0	720	105	1.45	2.5		
	Technora® (Teijin)	24.7	3.4	520	72	1.39	2.0	25	500
Polyarylate	Vectran® HT (Kuraray)	22.9	3.2	530	75	1.41	<1	28	450
PBO	Zylon® AS (Toyobo)	37.0	5.7	1,150	180	1.54	2.0	68	650
	Zylon® HM (Toyobo)	37.0	5.8	1,720	270	1.56	0.6		
UHMWPE	Dyneema® SK60 (Toyobo)	27	2.6	810	79	0.97	0	16.5	150
	Dyneema® SK71 (Toyobo)	36	3.5	1,270	123	0.97	0		
	Spectra®2000 (Honeywell)	30–36	2.9–3.5	810–1,280	79–124	0.97	0		

fibers based on the dominant patent acquired by DSM. In the late 1980s, TOYOBO and DSM succeeded in the industrialization of these fibers (trade name Dyneema®) for the first time in the world [2]. Subsequently, Spectra® by Honeywell company (formerly Allied Signal company) and Tekmilon® by Mitsui Chemicals company (formerly Mitsui Petrochemical Industries company) entered the market; these fibers were manufactured using the same gel-spinning method (Tekmilon® is no longer produced). Table 7.1 lists the mechanical properties of the UHMWPE fibers manufactured by the abovementioned companies and other commercial super fibers. The table indicates that UHMWPE fibers have the highest tensile strength and highest tensile modulus among the super fibers except for PBO fiber [3]. UHMWPE fibers are used in industrial materials, impact resistance materials, and cut resistance materials such as high-performance ropes and nets that are required to have high tenacity and high tensile modulus. Since UHMWPE fibers comprise polyethylene and do not degrade by water absorption, they are widely used for outdoor and marine applications such as ropes, fishing lines, fishing nets, and nets.

This chapter provides an overview of the gel-spinning technology. Although there is often a trade-off between good mechanical properties and high productivity, achieving both is essential for the industrialization of the fibers. Furthermore, technological developments to improve physical properties could lead to improving the productivity without degrading the mechanical properties. This concept is applicable to the technological development of gel spinning as well as melt spinning. This chapter also explains the concept of the gel-spinning method and the technical key elements to achieve successful industrialization. Moreover, the reasons why the superior mechanical properties of UHMWPE fibers were attainable with the gel-spinning method are described by focusing on the microstructures of the fibers. Then, various physical properties and principal uses of UHMWPE fibers are discussed. Finally, the future outlook for the gel-spinning method is presented.

7.2 Essence of the Gel-Spinning Technology

7.2.1 Important Points for Increasing the Tenacity of Polyethylene Fibers

Many previous academic investigations have reported polyethylene fibers with a tensile strength of over 6 GPa [4]. However, the experimental tenacity was less than a quarter of the theoretical value (31 GPa [5], 25 GPa [6]), calculated using the energy of the C-C covalent bond, as shown in Table 7.2. On the other hand, the value of the experimental tensile modulus was almost equal to that of the crystal modulus (240 GPa) of polyethylene. A crystal modulus of 240 GPa indicates that the tensile modulus of the polyethylene orthorhombic crystal is oriented along the *c* axis (molecular axis). The similar values of the macro tensile modulus of fibers and the crystal modulus suggest that polyethylene molecules in the fibers are almost completely oriented in the fiber axis direction. Despite the ideal state of molecular

Table 7.2 Theoretical and experimental tensile properties of UHMWPE fibers

Tensile strength	Theoretical strength [5, 6]	Commercial strength	Experimental strength [4]
	25–31 GPa	3–4 GPa	6.9 GPa
Tensile modulus	Theoretical modulus [5]	Commercial modulus	Experimental modulus [4]
	240 GPa	103–137 GPa	238 GPa

arrangement, the tensile strength remains less than a quarter of the ideal value because various defective structures exist in the fibers. Reducing as many defects as possible in the fibers is essential to increase their tensile strength.

This paragraph elaborates on the defects in fibers. UHMWPE fibers contain hierarchical structures over a wide spatial scale ranging from nanometer scale, micrometer scale, up to millimeter scale. The minimum structural unit in these fibers is a polyethylene molecular chain. Furthermore, the fibers have microfibril structures with diameters ranging from several nanometers to several tens of nanometers as a bundle of polyethylene molecules. Figure 7.1 shows the microscopic higher-order structure model proposed by Hosemann for stretched linear polyethylene [7]. The figure depicts the different types of macrolattices found in individual filaments. In addition, this model shows that the fibers contain many kinds of microdefects such as molecular ends, branches, and entanglement of molecular chains at the interface and inside the crystalline and amorphous structures. It is presumed that UHMWPE fibers essentially have similar defect structures, leading to a reduction in their tensile strength. If we imaged an ideal fiber structure, all extended molecules should perfectly align parallel to the fiber axis and the defect of a chain end should homogeneously dispersed in the fiber without forming aggregations of the defects. The ideal structure will give a maximum tensile strength and modulus when the fiber is stretched in the direction parallel to the fiber axis. Moreover, the denser packing of the molecules and the lesser density of the defects would give the higher strength. As described above, the actual fiber structure shows a large discrepancy from the ideal molecular organization, but we can recognize that the gel-spinning technology is one of promising methods to achieve the ideal nanostructure, even though the obtained state still includes many defects, because the extremely high tensile properties indicate the current UHMWPE fiber attained the higher molecular orientation and the lesser number of defects comparing to the conventional melt-spun fibers giving the imperfect structure as modeled by Hosemann.

On the other hand, structural defects in the structure ranging from the micrometer to millimeter scale can be considered to have been caused by the inconsistency in the diameter of individual filaments or the fluctuations in the mechanical properties of fibers in the longitudinal direction. Since a bundle of numerous single filaments with a diameter of several tens of micrometers is practically used, the presence of these defects or unevenness causes reduction in the tenacity of the aggregated fibers. From an industrial perspective, it is necessary to reduce various defects over the spatial scale ranging from nanometers to millimeters.

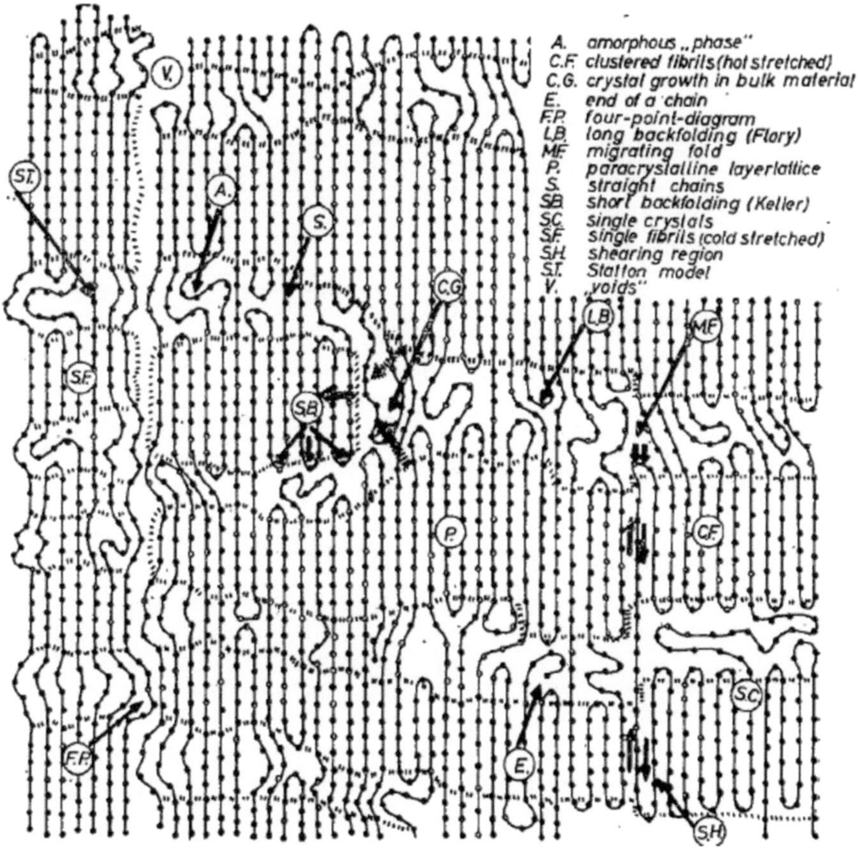

Fig. 7.1 Structure model of the stretched linear polyethylene [7]

This chapter describes in detail the importance of reducing structural defects in nanometer-scale structures such as crystalline and amorphous structures. In recent years, the shish-kebab structure, described at Chap. 5 in this part, has attracted significant attention as it is considered an important nanometer-scale high-order structure. The formation process of the shish-kebab structure and its effect on the physical properties are described in detail in the second half of this chapter.

7.2.2 Evolution of the Fundamental Concepts of the Gel-Spinning and Industrial Efforts on Its Commercialization

The gel-spinning process of UHMWPE fibers is schematically shown in Fig. 7.2 [8]. The crucial points of the technologies are shown as follows: (1) polymers being

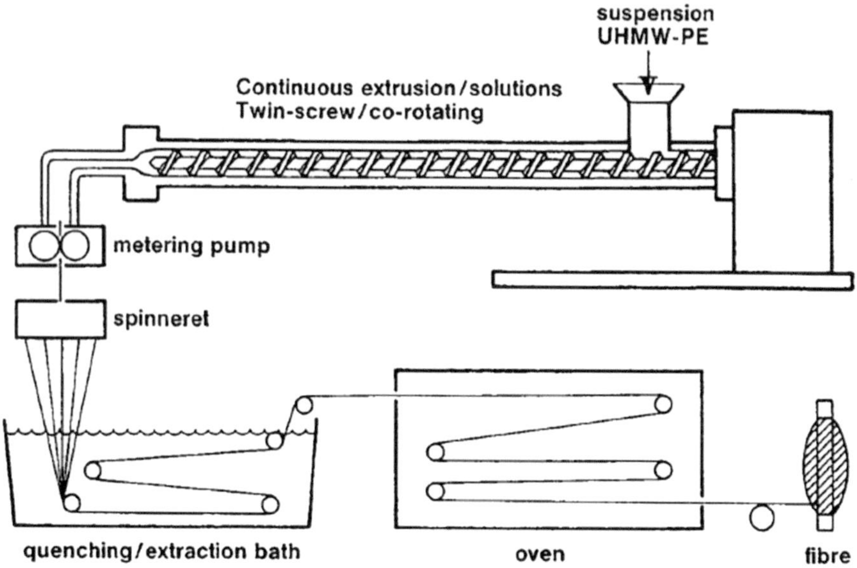

Fig. 7.2 Schematic image presenting the gel-spinning process of UHMWPE fiber [8]

ultrahigh molecular weight are used; the weight average molecular weight M_w is at least higher than 600,000 g/mol and typically higher than 1,000,000 g/mol; (2) the number average of the entanglements in a polyethylene chain is decreased using a solution system with a solvent, i.e., decalin or paraffin, as the spinning dope; (3) the less-entangled state is fixed by crystallization in the spinning line; and (4) the as-spun fiber can be drawn at an ultrahigh draw ratio higher than several tens times to the initial length at the elevated temperature higher than the nominal melting temperature of polyethylene. The drawing process induces the large-scale molecular rearrangements from the lower ordered state including the chain folding to the highly ordered state consisting of the extended chains. Hence the higher draw ratio gives the lesser density of the defects in the obtained fiber and results in higher tensile strength and modulus. The controlled state of the entanglement of polymer chains leading the high drawability as abovementioned is the essence of the gel-spinning technology, but we may note here that the drawable nature of polyethylene crystal due to the less molecular interaction between the polymer chains should not be overlooked as a crucial point leading the success of UHMWPE fiber. The gel-spinning technology was applied on PAN and PVA [9]; however, the achieved tensile strength of the fibers was limited. This might be caused by the less drawability due to the higher molecular interactions, for example, intra- and/or intermolecular hydrogen bonds, of the polymers than polyethylene [10]. The detailed explanations of the technical elements (1) to (4) will be shown in the following sections.

7.2.2.1 Controlled State of Entanglements Using Ultrahigh Molecular Weight Polymers and Semi-dilute Solution Systems

The degree of polymerization of polyethylene is one of the most important factors for the gel-spinning technology. UHMWPEs having the weight average molecular weight M_w more than 600,000 g/mol are generally used. One possible reason, why such ultrahigh molecular weight polymers are used, can be simply considered that the higher molecular weight polymer leads to a lesser density of defects due to less number of chain ends in the solid state. These defects will act as the starting point of crack propagations, which causes lower tensile strength. But the chain-end effect is not an exclusive reason for the utilization of the ultrahigh molecular weight polymer. We can understand the high tensile properties of the gel-spun fibers on the basis of the dependency of the maximum draw ratio as a function of the molecular weight as shown in the following paragraph.

UHMWPE molecules in the melt state possess extremely high number of chain entanglements per one molecule comparing to moderate molecular weight polymers. The extremely high viscosity due to the high density of entanglements makes it difficult to deform the polymer via the conventional melt process. Hence, the dilution with solvent is one of key technologies of the gel spinning. The various types of solvents have been reported, i.e., tetralin, decalin, naphthalene, mineral oil, paraffin oil, and paraffin wax [11].

The degrees of the number average of entanglements on a single molecule can be expressed by the overlap concentration (C^*) where the random coil macromolecules begin to overlap each other. C^* can be formulated as the function of the average molecular weight (M) as follows [12]:

$$C^* \propto M_e/M \tag{7.1}$$

where M_e is the molecular weight between entanglement points.

As expressed by this equation, C^* becomes lower with increasing molecular weight M. It is worthy to note here that the optimum concentration of the polymer in the spinning solution depends on the entanglements of polyethylene molecules. If the concentration is lower than C^*, the polyethylene molecules are isolated in the solution. Hence there is no entanglement between the polyethylene molecules. At the concentration, the molecules do not connect to each other, and the less-entangled state is also maintained in the solid state after the spinning process. The less-entangled state allows less drawability. We will also find a less drawability state at the strongly entangled solution as well as the dilute solution. When the concentration is much larger than C^*, the highly entangled state of polyethylene chains also leads to less drawability as mentioned in the previous section. Therefore, there must be an optimum concentration and it might be the concentration close to or slightly larger than C^*. Such partly entangled solution is defined as "semi-dilute" solution [13], and the maximum draw ratio increases with the increasing molecular weight for the almost isolated but partially entangled polymer chains as follows [12]:

$$\lambda_{max} \propto M^{0.5} \tag{7.2}$$

On the actual industrialization of the technology, we should consider the productivity of the fiber. In the case of the UHMWPE having $M_w = 2.0 \times 10^6$, C^* is approximately 0.5 %. Such a low concentration cannot be accepted due to its low productivity. Hence we have to make efforts to increase the concentration and we will find a more complicated situation requiring the optimization of concentration and molecular weight. In the case of the concentration region much higher than C^*, Bastiaansen experimentally indicated the maximum draw ratio of gel-spun fibers as a function of concentration C and molecular weight M with following equation [14].

$$\lambda_{max} \propto C^{-0.5} M^{-0.5} \tag{7.3}$$

The maximum draw ratio decreases with increase of M and C, which can be explained with an increasing of the entanglement number. We should explore the best set of M and C to balance the tensile properties and the productivity. Moreover, we should take care of the process velocity as well as the polymer concentration. The higher spinning and drawing velocity will improve the productivity and here we should consider the dependency of the molecular weight of the polymer on the velocity. It will be discussed in the section of 7.2.2.3.

7.2.2.2 Spinning Process as a Process for Controlling the Crystalline Morphology Leading to High Drawability

The optimized entangled state of the spinning solution is fixed through the crystallization after throughput of the solution from a spinneret and the subsequent cooling process under uniaxial deformation in the spinning line. This is the third crucial point of the gel-spinning technology. If the crystallization is slower than the solvent evaporation from the spinning fiber, the entanglements in the obtained fiber would be increased into the highly entangled state close to the original UHMWPE polymer. Therefore, the quick and deep quench in the spinning process is important, and hence the water bath is generally applied for the quenching process of the gel spinning [8].

We should consider the rheological behavior of the UHMWPE solution to operate the practical spinning process steadily. The semi-dilute solution of UHMWPE shows a non-Newtonian behavior [15]. The shear viscosity of the solution drastically decreases with increasing shear rate. The solution shows elastic behavior at a high shear rate region due to its long relaxation time whereas the shear viscosity drops to a very low value at the shear rate. Such the strong non-Newtonian behavior generally induces unstable flow in the spinneret and results in unstable spinning condition leading frequent fiber brakes. Therefore, the careful design of the spin-line, especially spinneret, will be necessary to stabilize the spinning condition.

7.2.2.3 Drawing Process

As mentioned in the previous sections, the controlled state of entanglements via the solution system and subsequent crystallization leads to high drawability. Here we would like to discuss the drawing velocity as well as the draw ratio. We will achieve the maximum draw ratio according to the Eq. (7.3) using the as-spun fiber if we can apply a slow drawing velocity. However, we should speed up the drawing velocity on the practical industry to improve productivity. The Eq. (7.4) derived by Ward [16] gives the relation of the maximum draw ratio and the deformation rate based on the molecular dynamics:

$$\lambda_{\max} \propto \nu \sim 1/\tau \tag{7.4}$$

where ν is the macroscopic deformation rate of the given process and τ is the characteristic relaxation time of the system. Namely, if we apply a higher deformation rate than the inverse of τ ($\nu > 1/\tau$), the molecular chains cannot relax an excess stress generated at a local point of the molecules, and then the molecular chain will be broken at the local point. On the other hand, in the condition of $\nu < 1/\tau$, the molecular chains can relax the excess stress via disentanglement. Here we assume that the as-spun fiber still contains much solvent and the drawing is conducted at the temperature higher than the nominal melting temperature of polyethylene. On such condition, the fiber would be deformed on a state like a solution and we can understand the drawing process on the basis of molecular dynamics derived for the entangled polymer solutions. From the molecular theories which deal with the molecular weight and concentration dependence on the relaxation time, τ can be generally expressed as Eq. (7.5). And in Graessley model [17], $\alpha = 1.5$ and $\beta = 3.5$ and, in Doi-Edwards model [18], $\alpha = 1.0$ and $\beta = 3.0$ were reported. These values are much higher than those of Eq. (7.2), and actually we also have experienced a similar relation to these theoretical power laws with high speed drawing experiments.

$$\tau \propto C^{\alpha} M^{\beta} \tag{7.5}$$

$$\lambda_{\max} \propto C^{-\alpha} M^{-\beta} \tag{7.6}$$

The Eq. (7.6) and large β suggest that the small increase of molecular weight gives a large effect on the drawing velocity and the achieved draw ratio. We should carefully choose a molecular weight giving the best balance of the higher draw ratio and the faster drawing velocity.

7.3 Structure Evolution in Spinning and Drawing Processes

The structures evolved in the gel-spun fibers will be summarized in this section. The UHMWPE fiber consists of a highly extended chain assembly, and the structure is evolved through large-scale molecular rearrangements from an initial state in an as-spun fiber to extended chain states in the drawing process. The less-entangled molecular state is the crucial molecular organization leading high drawability of the as-spun fiber. It is a key factor of the gel-spinning process as described in the previous sections. In this section, we aim to present an important role of nanostructures developed in the as-spun fiber as well as the controlled entanglements. Here we pay an attention on the intriguing superstructure "shish-kebab" which is already suggested as a major structure comprising the as-spun fibers in the previous section. First, we present the brief introduction of the discovery of shish-kebab and next the formation of the structure in the gel-spinning process will be shown. Finally the morphological changes of the shish-kebab will be presented in the drawing process under a given elevated temperature.

7.3.1 Discovery of Shish-Kebab Structure in Dilute UHMWPE Solutions

Shish-kebab structure was firstly found in 1963–1965 in stirred dilute UHMWPE solutions as already described in Chap. 5, [19, 20]. Crystalline polymers generally show a characteristic crystalline organization, which is called "spherulite," on the crystallization under a static state. Nevertheless, the structure developed in stirred polymer solutions has a totally different crystalline arrangement with the spherulite as shown in Fig. 7.3, and the structure is usually called as "shish-kebab." Shish-kebab structure is constituted with two characteristic parts, i.e., "shish" including extended chain crystals and "kebab" consisting of folded chain crystals, respectively. The structure has attracted attentions of a great number of researchers due to its intriguing morphology and high tensile properties. Moreover, the discovery stimulated intensive works on the basis of flow-induced crystallization and led the general insight that shish-kebab structures were universally observed in the flowed melt polymer systems as well as the solutions [21, 22]. Several types of mechanism of shish-kebab formation in the dilute solutions were proposed by many researchers, for example, a shish formation via molecular extension induced by the extensional flow between Taylor vortices [23] and pulling out extended chain bundles from an absorbed gel-like layer on the flow cell [24].

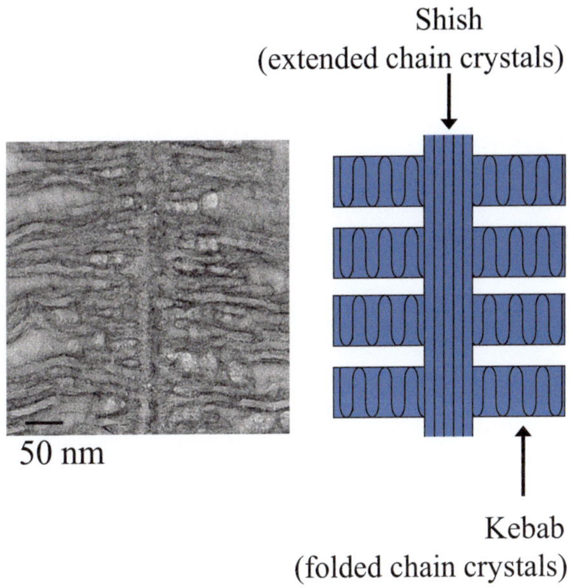

Fig. 7.3 TEM micrograph and schematic representation of shish-kebab structure

7.3.2 Structural Development of Shish-Kebab Structure in Entangled Semi-dilute Solutions

A great number of researches have been investigated after the discovery of shish-kebab to elucidate its structure and the mechanism of its formation. The models for the mechanism have frequently been proposed by the leading scientists. The effect on chain extension via extensional flow was dominantly discussed for shish-kebab formation in dilute solutions as described in the previous section and the deformation of molecular networks was focused in the case of melt polymer systems [25]. Here we should notice that all those models mainly focused on the flow effects on chain extension as a dominant driving force of the shish formation. Recently we found that development of a dissipative structure precedes crystallization in the process of the flow-induced crystallization of semi-dilute solutions of UHMWPE [26–28]. When a semi-dilute polymer solution is subjected to a shear flow, the shear flow enhances the concentration fluctuations and finally induces liquid-liquid phase separation. The phenomenon is called as "shear-induced phase separation" and has been studied mainly on the noncrystalline polystyrene solutions from the middle 1970s. The new insights drawn from the in situ observation call our reconsideration on the shish-kebab formation at least for the entangled solution system.

Murase et al. elucidated the transient structures in the gel-spinning process via TEM observations of gel-fibers, which were sampled at the position close to the spinneret with a quick quench [29]. The UHMWPE solution used for the gel spinning was a homogeneous solution in the static state and the as-spun fiber showed shish-kebab structures. Therefore the samples obtained at the position close to the spinneret might include the transient structures from the homogeneous solution to the shish-kebab. This is the motivation of the experiment (Fig. 7.4). Figure 7.5 shows the TEM micrograph of the fiber obtained at the most close to the spinneret (30 mm from the surface of the spinneret). Many ellipsoidal structures were observed. The structures are extended perpendicular to the fiber direction, suggesting an oblate-ellipsoidal shape with the axis of revolution oriented parallel to the flow direction. Their mass centers are randomly aligned in space. These ellipsoidal structures are the demixed domains developed via the flow-induced phase separation. We might emphasize here that the structures were originally liquid-liquid phase separated structure and the structures "in situ" developed in the spinning process were solidified with the crystallization when the fiber was sampled and quickly quenched. The intriguing flow-induced phenomenon is

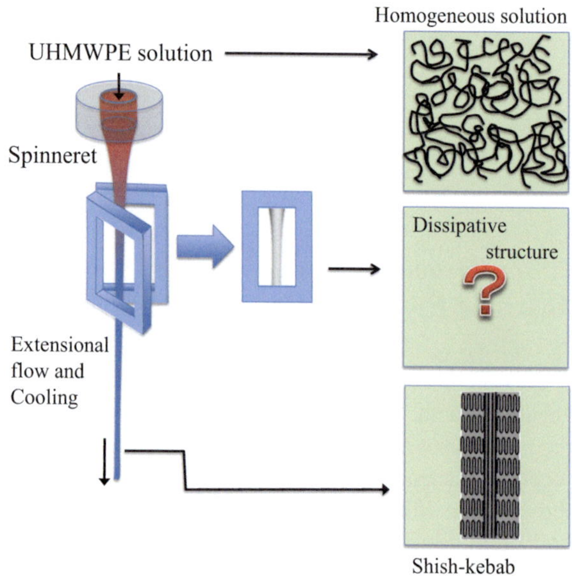

Fig. 7.4 Schematic representations of the fiber spinning process and the method for sampling of the fibers

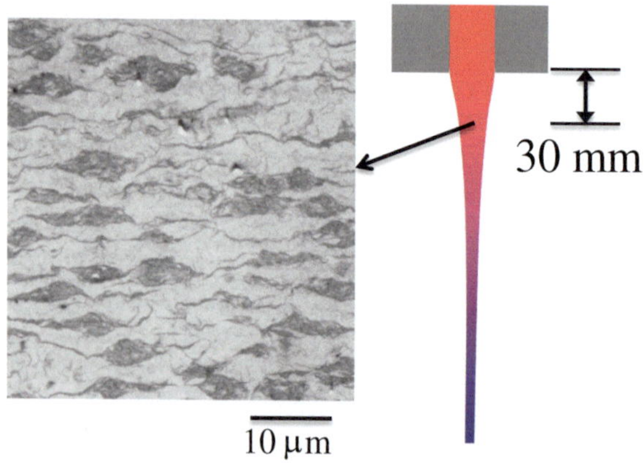

Fig. 7.5 TEM micrograph of the fiber obtained at the most close to the spinneret (30 mm from the surface of the spinneret)

theoretically explained with a concept coupling stress and diffusion [30–33]. In a quiescent state, the thermal concentration fluctuations give rise to the regions having more entanglement points and less entanglement points in a semi-dilute solution. When a shear flow is imposed on the solution, the stress tends to become higher in the more entangled regions than the less-entangled regions. The local stress variation or spatial variation of the elastic free energy built up by the deformation of the entangled polymer chains is relaxed by disentanglement when the shear rate is less than the maximum relaxation rate being the longest relaxation time. Under this condition the semi-dilute solution remains homogeneous solution even under flow. However when the shear rate is larger than the maximum relaxation rate, the excess elastic free energy stored is released only by squeezing solvent from the more entangled regions. This solvent squeezing results in enhancement of concentration fluctuations under shear flow against osmotic pressure. This occurs when the external mechanical force outweighs the thermodynamic driving force to maintain a homogeneous solution.

Figure 7.6 is a TEM micrograph of the gel-fiber obtained at a position further downstream in the spinning line. We can observe the transient structure in this

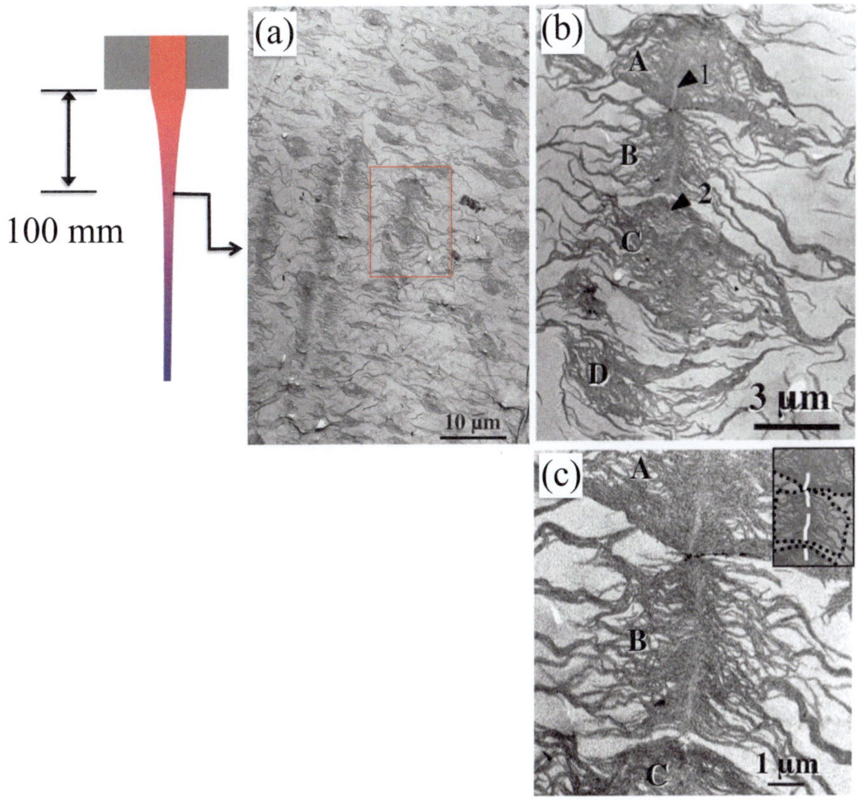

Fig. 7.6 (**a**) TEM micrograph of the fiber obtained at 100 mm downstream from the spinneret, (**b**) zoom-in image of the assembled demixed domains indicated by the square in (**a**), and (**c**) zoom-in image of shishs developed in the impinged demixed domains. The figure inserted in the *upper right* of (**c**) shows a traced image to emphasize the shishs and the outline of the demixed domains

micrograph. The demixed domains aligned parallel to the flow direction (denoted by the characters A, B, C, and D in Figs. 7.6b, c) and shishs were developed between the demixed domains in the string-like assembly as indicated by the arrows (1 and 2) in Fig. 7.6b. The parallel assembly of the demixed domains might be arranged via hydrodynamic interactions. Hydrodynamic random motions of the demixed domains along the long axis of the assembly induce a collision of neighboring domains and molecular exchanges at the interface of the domains. Moreover, the impinged demixed domains create bundles of stretched chains within the bridging domains formed at the boundary of the domains where the local elongational strain rate imposed on polymer chains may be effectively enhanced and become larger than the bulk strain rate, then promote a growth of the bundles from the boundaries toward the interiors of the demixed domains. This growth process of shish within the string-like domain assembly was proved by a TEM micrograph of the fiber obtained more downstream in the spinning line [29]. These ex situ observations call for us to propose the new scenario of kinetic pathway of shish-kebab formation. Figure 7.7 shows a schematic model of the kinetic pathway [27], and the scenario is summarized as follows: (a) the demixed domains aligned parallel to the flow, (b) formation of the bridging domains or bridging chains between the domains which are densified in the string, (c) formation of bundles of stretched chains, (d) nucleation and growth of shishs within the bundles of stretched chains in the bridging domains at the boundaries of demixed domains, (e) growth of the bundles of stretched chains and shishs from the boundaries to the

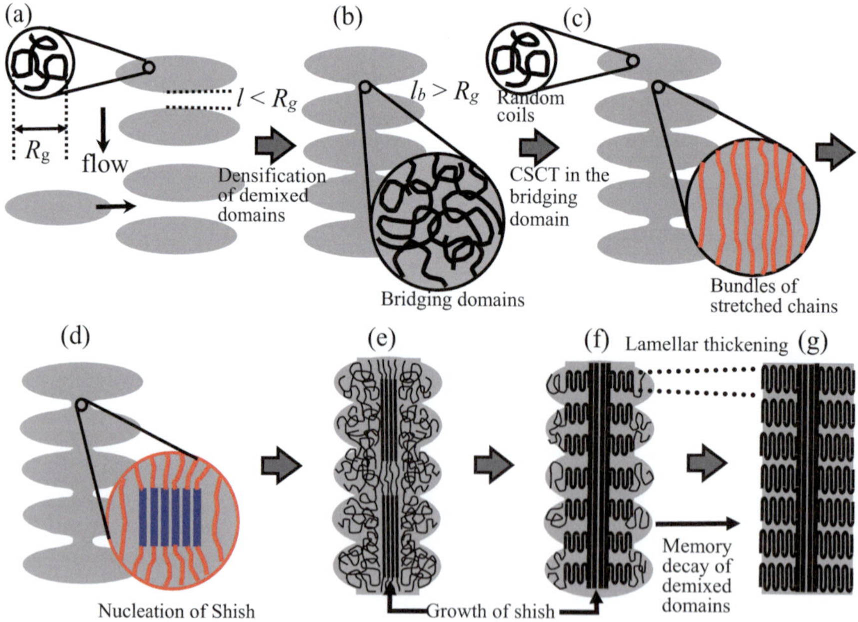

Fig. 7.7 Schematic model of the kinetic pathway and the scenario of shish-kebab formation [29]

interiors of the domains, (f) the epitaxial crystallization of the random coils in the demixed domains into kebab lamellae from shishs, and (g) the long-range rearrangement of chain molecules within the as-grown shish-kebab, which causes the memory of the demixed domains within the shish-kebab decay to form a more ordered shish-kebab.

7.3.3 Structural Changes in the Drawing Process: Transformation of Shish-Kebab into Microfibrous Structure

The structural changes in the drawing process are also important as well as the spinning process. We elucidated the transformation forms shish-kebab into highly extended chain state through the ex situ TEM observations. Figure 7.8 shows the TEM micrographs obtained from fibers which were drawn at elevated temperatures higher than the nominal melting temperature of polyethylene at various draw ratios [34, 35]. Shish-kebabs were deformed and the kebabs seem to be transformed into a shish. It is worthy to note here that the dimensional analysis of the characteristic parts of shish-kebab as a function of the draw ratio gives an intriguing insight into the morphological dependence on the strength of the drawn fibers [34, 35]. Figure 7.9 shows the diameters of shish and kebab as a function of the draw ratio. Diameter of the shish keeps a constant value whereas that of the kebab decreases with the increasing draw ratio. This suggests that the continuous transformation from kebabs to a shish. The molecules in the kebabs are supplied to the stretched shish and hence the shish keeps its diameter during the drawing. Here we can assume a model of the structure-dependent physical property. The breaking force (F_b) of the single drawn

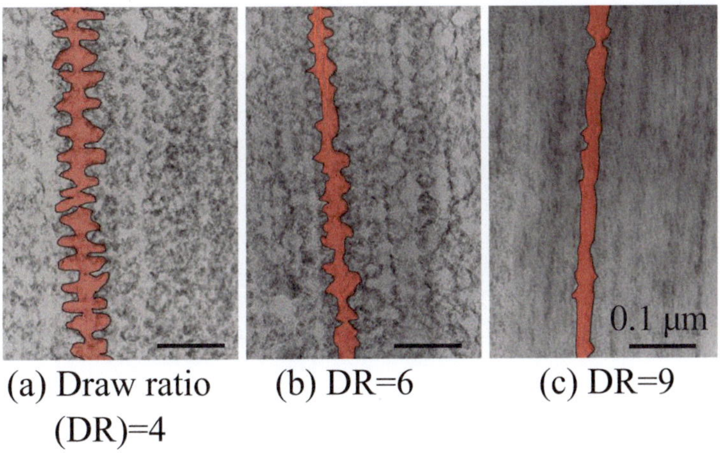

Fig. 7.8 TEM micrographs obtained from drawn fibers at various draw ratios

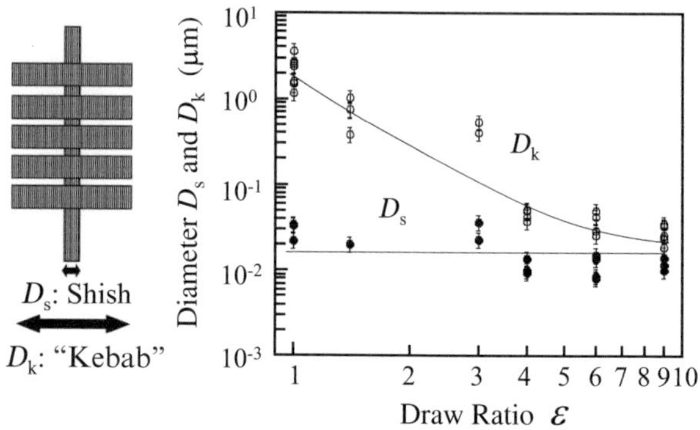

Fig. 7.9 Diameters of shish and kebab as a function of draw ratio [35]

fiber may be expressed in an approximation with the averaged breaking force (f_b) and the number (n) of the shish-kebabs contained in the single fiber as follows,

$$F_b = nf_b \tag{7.7}$$

and the strength (σ_b) of the fiber at the breaking point can be calculated as follows,

$$\sigma_b = F_b/\left[\pi(D/2)^2\right] \tag{7.8}$$

where D is the diameter of the single fiber. If the fiber is deformed according to an affine deformation with a constant volume, the diameter must be reciprocally proportional to square root of the draw ratio ε.

$$D \propto 1/\varepsilon^{1/2} \tag{7.9}$$

Then

$$\sigma_b \propto F_b\varepsilon \tag{7.10}$$

Therefore, if the total number (n) of the shish-kebabs per single fiber and breaking force f_b per single shish-kebab are constant during the drawing process, the breaking force of the fiber as a whole (F_b) should be independent of the draw ratio. If this is the case, the fiber strength should be proportional to draw ratio ε and reciprocally proportional to square of the fiber diameter. Figure 7.10 shows the force F_b at the breaking point of the drawn fibers as a function of the draw ratio. As mentioned above, F_b is independent of the draw ratio. This strongly suggests that the assumption is plausible and the initial structure generated in the spinning process has a crucial role on the strength of the drawn fiber. Those results presented in this section are summaries in our previous works [35] on the structure formation and changes from the spinning to the drawing processes. The plausibility of our hypothesis

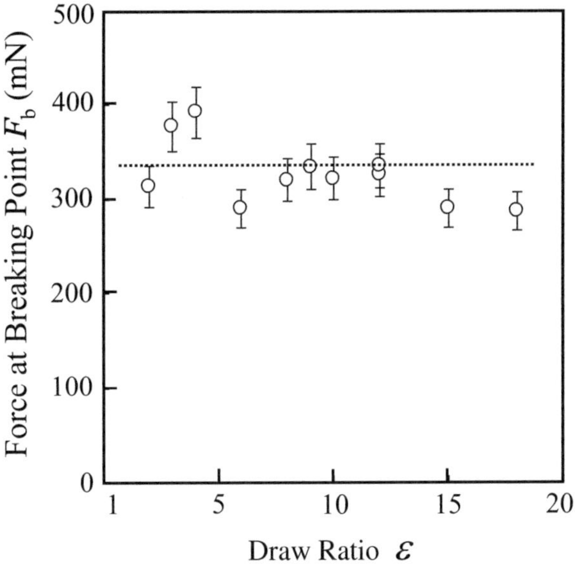

Fig. 7.10 Force at breaking point of UHMWPE fiber as a function of draw ratio [35]

should be checked and confirmed by a further research and it would deserve to be future works.

7.4 Fiber Properties and Applications

Basic properties of Dyneema® are shown in Table 7.1. Tensile strength of Dyneema® is 40 % higher than that of the *p*-aramid fibers. And the specific gravity is 0.97, this means floatable nature of the fiber on fresh water. Other representative advantages of Dyneema® are easy fabrication, no moisture regain, weather resistance, flexural durability, abrasion resistance, chemical resistance, cut resistance, and incineration without noxious-gas [36].

Major limitations of Dyneema® on the practical applications are its low thermal resistance and low creep resistance. Compared to the heat-resistant fibers, mechanical properties are sensitive to the given temperature in the case of the polyethylene fiber due to its relatively lower melting temperature. In the case of SK-60 which is the standard grade of Dyneema®, the tensile strength at 40 °C and 70 °C is 89 % and 62 % of the strength at 24 °C, respectively. Young's modulus also shows the same tendency. The modulus at 40 °C and 70 °C is 88 % and 59 % of the modulus at 24 °C, respectively [37]. Dyneema® can be processed at the temperature lower than 120 °C without decreasing strength. Moreover, Dyneema® fabrics can pass the FMVSS 302 test (Fig. 7.11) although the fiber is made from polyethylene. Rapid thermal dissipation due to the high thermal conductivity of the fiber [38] and the high molecular weight might cause the flame resistance. Dyneema® shows a high

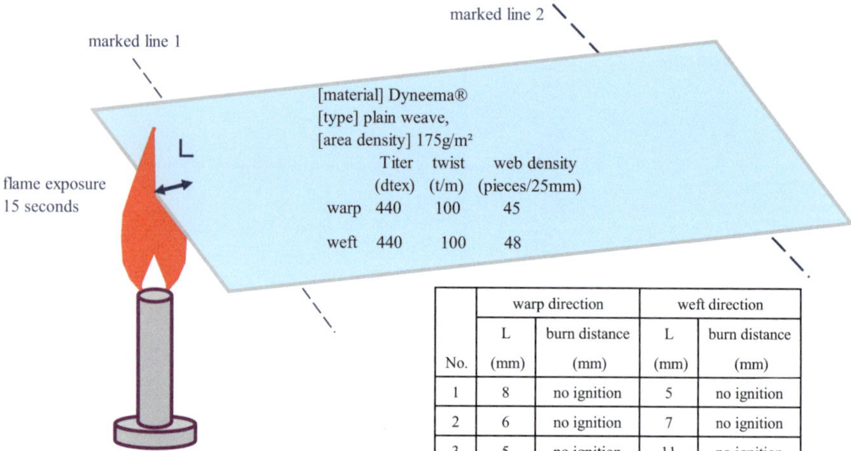

No.	warp direction L (mm)	warp direction burn distance (mm)	weft direction L (mm)	weft direction burn distance (mm)
1	8	no ignition	5	no ignition
2	6	no ignition	7	no ignition
3	5	no ignition	11	no ignition
4	6	no ignition	6	no ignition
5	5	no ignition	6	no ignition

Fig. 7.11 Schematic image of the burn test (FMVSS 302) of Dyneema® woven fabrics

Table 7.3 Dyneema® applications

Primary feature	Secondary feature	Application
High-strength, high-modulus, and low density	Flexional durability, abrasion resistance	High-performance ropes and cables
	Weather resistance	Land nets, fishery ropes, and nets
	Impact resistance	Helmets, ballistic panels, explosion protection fabric
	Low elongation	Sports shoes, fishing line
	Cut resistance	Working gloves, protection clothes
	Chemical resistance	Concrete reinforcement, medical thread
Highly oriented structure	Thermal conductivity	Comfortable textiles (ICEMAX®)
	Negative coefficient of thermal expansion	Cryogenic engineering parts
High molecular weight polyethylene	Low friction coefficient	Lead rope for infrastructure construction
	Low dielectric constant	Radar dome
	Low electric conductivity	Spacer for super conductor

creep rate at a high load [39, 40]. Creep resistance can be improved by modification of polymer structure, for example, copolymerization with short branches into the molecular moiety [41, 42].

Applications of Dyneema® are summarized in Table 7.3. Dyneema® shows unique thermal properties. We would like to introduce the applications using the unique thermal properties. One is a negative thermal expansion in the direction parallel to the fiber axis [43] and the combination of Dyneema® with the material showing a positive thermal expansion gives a thermally neutral material. Such a thermally stable material is very useful in the field of superconducting applications. Coil bobbins made of Dyneema® and thermoplastic resins can reduce the mechanical losses of superconducting magnets [44]. Other intriguing thermal property is its high thermal conductivity. Thermal conductivity of Dyneema® is comparable to steel in spite of its nature of electronic insulation [38]. One unique application is the comfortable fabric "ICEMAX®" [45]. The fabric gives a cool touch due to the high thermal conductivity and this fabric is good for bedclothes giving rise to comfortable sleeping even at a hot summer night in Japan.

7.5 Future Perspectives

The gel-spinning technology opened a gate for us to produce the super fiber using flexible chain polymers. The extremely high strength and the characteristic properties, for example, lightweight and high chemical stability, have led broad applications of the fiber. In these two decades after starting the industrialization of the UHMWPE fiber, we have continued the research not only to improve the tensile properties but also to accommodate the market needs via improving the other properties required by our customers. Under the current market situation, the demand of the customers requiring the higher strength is not strong. However, we dare say here the research improving the basic tensile strength and modulus still keeps important meaning for our business because the technology leading the higher tensile properties also gives a useful insight into the cost-improving technologies. Therefore we should continue the effort on the development of a next generation technology of the gel-spinning method. The recent trends of the development are introduced in this section and the outlook for the future is summarized.

7.5.1 Recent Trends for Polymer Development

The invention of the new catalyst, i.e., metallocene catalyst, is a promising technology to open up a new frontier of the polyolefin applications. One of the characteristic features of the linier polyethylene polymerized using the metallocene catalyst is the low polydispersity of molecular weight comparing to that of the conventional linear polyethylene synthesized by the Ziegler-Natta catalyst. The effects of the lower polydispersity on the process condition and the fiber properties are still not clarified. But the new polymer is expected to bring an evolution into the UHMWPE fiber. Moreover, there is a possibility that the new catalyst improves the productivity of the fiber. Phenoxy-imine catalyst, which is so-called post-metallocene catalyst, shows extremely high activity [46]. This high activity leads the lower concentration of the catalyst on the polymerization and results in a less impurities in the obtained polymer. The lower concentration of impurities in the polymer gives stable spinning condition due to less frequent fiber breaks.

As frequently mentioned in this chapter, the lesser molecular entanglements lead the higher draw ratio and result in the higher tensile properties. The approach from the polymerization to reduce the entanglements was reported. Rotzinger and coworkers proposed a method to reduce the entanglements through a low temperature polymerization and indicated the good drawability of the as-received polymers [47]. This method is called as "virgin polymer method." Kanamoto and his coworkers also obtained the high-strength and high-modulus films via the solid-state extrusion of the as-polymerized polymers [48]. Aida and coworkers developed a unique polymerization method [49], and Zhibin et al. obtained a polyethylene fiber showing a high strength of 1.3 GPa via the approach along the Aida's method [50]. They fixed catalysts on the inner surface of meso-porous silica and injected ethylene gas into the meso-pores. The ethylene gas was polymerized into polyethylene at the inner surface of the meso-porous silica. Hence polyethylene fibers were continuously grown and extruded from the meso-pores. This unique bottom-up process indicates a promising approach to obtain the fiber giving the ideal nanostructure. The simultaneous crystallization after the polymerization in the confined space would be able to induce the highly extended molecular arrangement and less defect state in the obtained fiber.

7.5.2 *New Spinning and Drawing Technologies*

We would like to introduce several works concerning to new spinning and drawing approaches reported in the literature. Fujimatsu and coworkers reported a unique spinning method that a fiber was pulled up from a UHMWPE solution. They obtained a high-strength fiber corresponding to a super fiber [51].

Nanotechnology has shown tremendous progresses in these days and has been also applied on this super fiber field. We can find several reports concerning to the blending technology of carbon nanotubes into the solution of UHMWPE [52]. The research group of Stony Brook University attained more than 500 % improvement of the elongation at break of UHMWPE films using the blend system of MCNF (surface-modified carbon nanofibers) [53]. Nanofibers produced through the electrospinning method have been studied by a great number of researchers using almost every polymer, which can be dissolved into organic solvents. Polyethylene is no exception and Yoshioka reported the shish-kebab formation in the nanofiber produced via the electrospinning method [54]. Their work suggested that the shish-kebab can generate in the confined nano-space and this would give an important insight into the mechanism of shish-kebab formation. The advanced nanotechnology will be actively applied on the gel-spinning technology and it is expected to open up a frontier of the field.

We should also continue to make efforts on the improvement of the drawability. We revealed the continuous transformation from shish-kebab to the fibrous extended chain molecular organization in the conventional drawing process. The basic morphological insight of the drawing process is established but the more detailed molecular order mechanism of the transformation is not still clarified. The innovative drawing process is required not only to improve the tensile strength giving rise to the less defects structure but also to deepen the insight of the molecular order structure development in the drawing process. The new approach to improve the drawability can be found in the paper reported by Bartczak and coworkers [55]. They tried to reduce the entanglement in the solid polyethylene using a phenomenon of the thickening of lamellar crystal through the transition from orthorhombic crystal to hexagonal crystal under high-temperature and high-pressure condition. Garcia and coworkers showed another approach. They investigated the drawing of UHMWPE fibers in the presence of supercritical carbon dioxide [56]. They attained the improvement of drawability with high drawing stress at a high-temperature condition in the presence of supercritical carbon dioxide in comparison with the conventional atmosphere.

Finally we would like to make a brief introduction of other technologies that are recently developed. TOYOBO developed a melt-spun, high-strength, and high-modulus polyethylene fiber and recently started its commercialization as the trade name of "Tsunooga®" [57]. Teijin Aramid company announced that they developed a high-strength polyethylene tape using a non-solvent process [58]. The environment-friendly processes suggested by these materials will be one of most important research targets of the super fibers.

We introduced the new technologies concerning to polymerization, spinning, and drawing in this section. These technologies seem to be promising, but are still exploratory researches. We should verify practical effectiveness of these technologies on the actual process and the market of the UHMWPE fiber. An overwhelming advantage of the achieved properties would be necessary to drive out the existing technology and it must be a tough work to overcome the highly sophisticated modern gel-spinning technology. Truly innovative technology must be required. We believe the more effective drawing technology without cost increment would deserve to be challenged as a future research target. Moreover, it must be meaningful to examine the combination of the new technologies abovementioned and the old proven technology, for example, the spin-draw method which was evolved for melt spinning.

7.6 Conclusions

The basic concepts of the gel-spinning technology were summarized in this chapter with shedding light on the nanostructures evolved in the spinning process and the large-scale molecular rearrangement of the super structures during the drawing process. We paid a great attention on a characteristic super structure, i.e., "shish-kebab," and introduced the recent progress of the work revealing the kinetic pathway of its formation in the spinning line. The intriguing role of the dissipative structure, so-called the flow-induced phase separation, evolved in the flow field was presented in this chapter in the context of a structure which has been overlooked on the studies of the flow-induced crystallization. The new insight of the phenomenon would allow us to propose a new scenario of the shish-kebab formation which is not still clarified even nowadays. We also presented the essential effects of the shish-kebab structure on the tensile properties of the finally obtained fiber after the drawing through the careful observation of the structures in the drawing process using TEM. The proposed hypothesis of the structure-dependent fiber strength can explain the actual change of the strength as a function of the draw ratio. We basically explained the effects of molecular weight and concentration of the polymer on the spinning process and the drawability under a simple assumption on the basis of the dynamical equations derived for entangled solutions in this chapter. As mentioned above, the shish-kebab structure has a crucial effect on the strength of the gel-spun fiber. The molecular weight and the concentration of the solution must have an essential effect on the shish-kebab formation through the dissipative phenomenon. Therefore the total understanding of the shish-kebab formation with respect to the dependence with molecular weight, polymer concentration, and temperature etc., will become more important to brush up the gel-spinning technology and deserves to be a future work.

In these days, we are encountering the complicated global economic issues. Moreover the competition between the manufacturers becomes harder and harder globally. The continuous effort to improve the basic properties of the fiber is still important to keep our advantage in this field. In addition to the economical aspect of the motivation for the development, we should pay more attention on the environmental issues surrounding us in these days. We believe the effort to improve the tensile strength and the productivity of the fiber will be able to contribute to reduce CO_2 emission not only through the reduction of energy for manufacturing itself but also through the reduction of weight of the products using the fiber. We would like to continue the research to make an innovation on this field and try to attain this challenge on the collaboration with Japanese and worldwide academia.

The trade name of UHMWPE fiber for TOYOBO was changed from Dyneema® to IZANAS®.

References

1. P. Smith, P.J. Lemstra, A.J. Pennings, N.L. Patent, 177,840, 1979
2. P.J. Lemstra, N.L. Patent, 8,500,428, 1985
3. Y. Teramoto, WEB J. **89**, 31 (2007)
4. M. Matsuo, K. Inoue, N. Abumiya, Sen'i Gakkaishi(J) **40**, 275 (1984)
5. T. Itoh, *Kouseinou Koubunnshikei Fukugouzairyou*, (Maruzen, 1990), p. 7
6. H.F. Mark, *Polymer Science and Materials* (Wiley-Interscience, New York, 1971), p. 236
7. R. Hosemann, Polymer **3**, 349 (1962)

8. P.J. Lemstra, N.A.J.M. van Aerle, C.W. Bastiaansen, Polym. J. **19**, 85 (1987)
9. K.D. Yangu, K. Shierudon, P.S. Dasan, J.P. Patent 1,904,029, 1984. M. Mizuno, H. Tanaka, F. Ueda, J.P. patent 2,113,964, 1986. H. Sano, J.P. Patent 1,973,830, 1987
10. H. Narukawa, H. Noguchi, Sen'i Gakkaishi(J) **38**, 466 (1990)
11. F. Okada, T. Ohta, J.P. Patent 1,617,837, 1989
12. P. Smith, P.J. Lemstra, H.C. Booij, J. Polym. Sci. Phys. Educ. **19**, 877 (1981)
13. P.G. de Gennes, *Scaling Concepts in Polymer Physics* (Cornel University Press, New York, 1979)
14. C.W.M. Bastiaansen, J. Polym. Sci. B, Polym. Phys. **28**, 1475 (1990)
15. Y. Ohta, H. Murase, H. Sugiyama, H. Yasuda, Polym. Eng. Sci. **40**, 2414 (2000)
16. I.M. Ward (ed.), *Development in Oriented Polymer-2* (Elsevier Applied Science Publishers, New York, 1987), p. 39
17. W.W. Graessley, J. Chem. Phys. **54**, 5143 (1971)
18. M. Doi, S.F. Edwards, *The Theory of Polymer Dynamics* (Oxford University Press, Clarendon, 1986)
19. S. Mitsuhashi, Bull. Text. Res. Inst. (J) **66**, 1–9 (1964)
20. A.J. Pennings, A.M. Kiel, Kolloid Z. Z. Polym. **205**, 160 (1965)
21. M.J. Hill, A.J. Keller, J. Macromol. Sci. Phys. **B3**, 1531 (1969)
22. A.J. Keller, M.J. Machin, J. Macromol. Sci. Phys. **B1**, 41 (1967)
23. A. Zwijnenburg, A.J. Pennings, Coll. Polym. Sci. **254**, 686 (1976)
24. J.C.M. Torfs, A.J. Pennings, J. Appl. Polym. Sci. **26**, 303 (1981)
25. M. Seki, D.W. Thurmana, J.P. Pberhauser, J. Kornfield, Macromolecules **35**, 2583 (2002)
26. H. Murase, T. Kume, T. Hashimoto, Y. Ohta, Macromolecules **38**, 6656 (2005)
27. H. Murase, T. Kume, T. Hashimoto, Y. Ohta, Macromolecules **38**, 8719 (2005)
28. T. Hashimoto, H. Murase, Y. Ohta, Macromolecules **43**, 542 (2010)
29. H. Murase, Y. Ohta, T. Hashimoto, Macromolecules **44**, 7335 (2011)
30. E. Helfand, G.H. Fredrickson, Phys. Rev. Lett. **62**, 2468 (1989)
31. A. Onuki, Phys. Rev. Lett. **62**, 2472 (1989). J. Phys., Condens. Matter, **9**, 6119(1997)
32. M. Doi, A. Onuki, J. Phys. II France **2**, 1631 (1992)
33. S.T. Milner, Phys. Rev. E **48**, 3674 (1993)
34. Y. Ohta, H. Murase, T. Hashimoto, J. Polym. Sci. Polym. Phys. Educ. **43**, 2639 (2005)
35. Y. Ohta, H. Murase, T. Hashimoto, J. Polym. Sci. Polym. Phys. Educ. **48**, 1861 (2010)
36. http://www.toyobo-global.com/seihin/dn/dyneema
37. B. Dessain, O. Moulaert, R. Keunings, A.R. Bunsell, J. Mater. Sci. **27**, 4515 (1992)
38. A. Yamanaka, T. Takao, Sen'i Gakkaishi(J) **68**, 111 (2012)
39. M.A. Wilding, I.M. Ward, Polymer **19**, 969 (1978)
40. M.A. Wilding, I.M. Ward, Polymer **22**, 870 (1981)
41. Y. Ohta, H. Sugiyama, H. Yasuda, J. Polym. Sci. Polym. Phys. Ed. **32**, 261 (1994)
42. I.M. Ward, M.A. Wilding, J. Polym. Sci. Polym. Phys. Ed. **22**, 561 (1984)
43. N. Sekine, S. Tada, T. Higuchi, T. Takao, A. Yamanaka, S. Fukui, IEEE Trans. On Appl. Super. **13**, 1161 (2004)
44. T. Maeda, Sen'i Gakkaishi(J) **68**, 184 (2012)
45. http://www.toyobo-global.com/seihin/dn/dyneema/youto/textile.htm
46. S. Nishijima, J. Saito, Y. Nakayama, Y. Ogo, J.P. Patent 4,139,341, 2005
47. B.P. Rotzinger, H.D. Chanzy, P. Smith, Polymer **60**, 1814 (1989)
48. T. Kanamoto, A. Tsuruta, K. Tanaka, M. Takeda, R.S. Porter, Macromolecules **21**, 470 (1988)
49. K. Kageyama, J. Tamazawa, T. Aida, Science **285**, 2113 (1999)
50. Y. Zhibin, F. Wei, Z. Shiping, Y. Qiang, Macromol. Rapid Commun. **27**, 1217 (2006)
51. H. Fujimatsu, J.P. Patent 2007-056388, 2007
52. S. Ruan, P. Gao, T.X. Yu, Polymer **47**, 1604 (2006)
53. X. Chen, K. Yoon, C. Burger, I. Sics, D. Fang, B.S. Hsiao, B. Chu, Macromolecules **38**, 3883 (2005)
54. T. Yoshioka, R. Dersch, M. Tsuji, A.K. Schaper, J. Polym. Sci. **48**, 1574 (2010)

55. Z. Bartczak, Macromolecules **38**, 7702 (2005)
56. M. Garcia-Leiner, J. Song, A.J. Lesser, J. Polym. Sci. **41**, 1375 (2003)
57. http://www.toyobo-global.com/news/2008/release_406.html
58. http://www.teijinendumax.com/products/products/

Chapter 8
Development of High-Strength Poly(ethylene terephthalate) Fibers: An Attempt from Semiflexible Chain Polymer

Takeshi Kikutani

Abstract Poly(ethylene terephthalate) (PET) dominates the market of synthetic fibers because of its good performance and low cost; however, maximum strength of PET fibers available in the market remains at a low level. Based on such background, a project for the development of high-strength PET fibers was started in 2001 gathering researchers from major fiber companies and universities. Concept of the melt structure control was introduced and intense effort was made for the preparation of undrawn fibers which possess high potential for becoming high mechanical performance fibers after applying the drawing and annealing processes. It was found that the reduction of Deborah number in the melt spinning process by irradiating carbon dioxide laser to the spin-line and also by the use of a spinning nozzle of small diameter was found to be effective for the improvement of mechanical properties. To clarify the mechanism for such behavior, theoretical analysis was applied combining the numerical simulations for the spin-line dynamics and the coarse molecular dynamics to estimate the change in the state of molecular entanglement. Reduction of Deborah number in the melt spinning process was found to yield homogeneous entanglement structure with narrow distribution of molecular weight between adjacent entanglement points.

Keywords Molecular entanglement • High strength • High toughness • Poly (ethylene terephthalate)

8.1 Introduction

Research project on the "development of high-strength fibers" was initiated in 2001 as one of the subprojects in the project called "Nano-structured Polymeric Materials" [1, 2]. This project was one of the eight main projects in Nanotechnology

T. Kikutani (✉)
Department of Organic and Polymeric Materials, Graduate School of Science and Engineering, Tokyo Institute of Technology, Tokyo, Japan
e-mail: kikutani.t.aa@m.titech.ac.jp

Society of F.S. Technology, Japan (ed.), *High-Performance and Specialty Fibers*, DOI 10.1007/978-4-431-55203-1_8

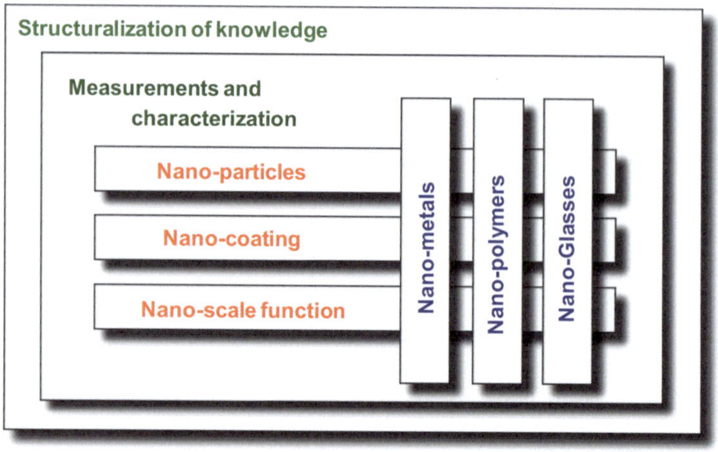

Fig. 8.1 Subthemes of "Nanotechnology Program (2001–2007)" sponsored by NEDO

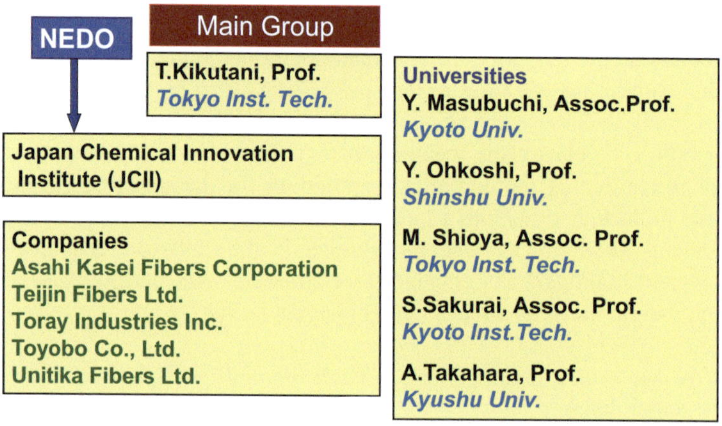

Fig. 8.2 Research organization for the project "development of high-strength fibers"

Program sponsored by NEDO (New Energy and Industrial Technology Development Organization) of Japan (Fig. 8.1).

Final goal of the project "development of high-strength fibers" was set to improve the strength of conventional high-strength fibers consisting of semiflexible chain polymers such as polyesters and nylons. The target is to achieve fiber strength of 2 GPa, which is two times higher than the strength of conventional high-strength fibers in the market, at the production cost less than two times of that of conventional high-strength fibers.

Five major fiber-producing companies and five university professors were involved in this project. Structure of the research group is shown in Fig. 8.2. Main part of research was carried out in Tokyo Institute of Technology, where researchers from different companies worked in the same office, conducted experiments together, and shared the results of experiments. Patents based on the achievement of researches conducted by those researchers were applied with the names of five companies together. Uniqueness of this research organization is the expected collaboration of researchers from companies competing with each other in this research field. Researchers from universities helped the project from the view point of the fundamental analyses on the fiber formation processes as well as the structure and properties of resultant fibers. In this article, essence of the achievements accomplished in this project will be introduced.

8.2 Background of Research

In the year of 2001, when the project was started, about 75 % of the annual production of synthetic fibers in the world was occupied by poly(ethylene terephthalate) (PET). Tendency of the domination of PET in the market has been continuing. As of 2012, more than 86 % of synthetic fiber in the world is occupied by PET [3].

On the other hand, with regard to the high-strength fibers, various types of high-strength fibers have been commercialized from 1970' to 1990' while those with excellent mechanical properties with their strengths higher than 2 GPa were developed either from flexible-chain polymers such as polyethylene (PE) and poly(vinyl alcohol) (PVA) or rigid-chain polymers such as poly(*p*-phenylene terephthalamide) (PPTA) and poly(*p*-phenylenebenzobisoxazole) (PBO) (Fig. 8.3) [4]. Even though high-strength fibers for common uses such as tire

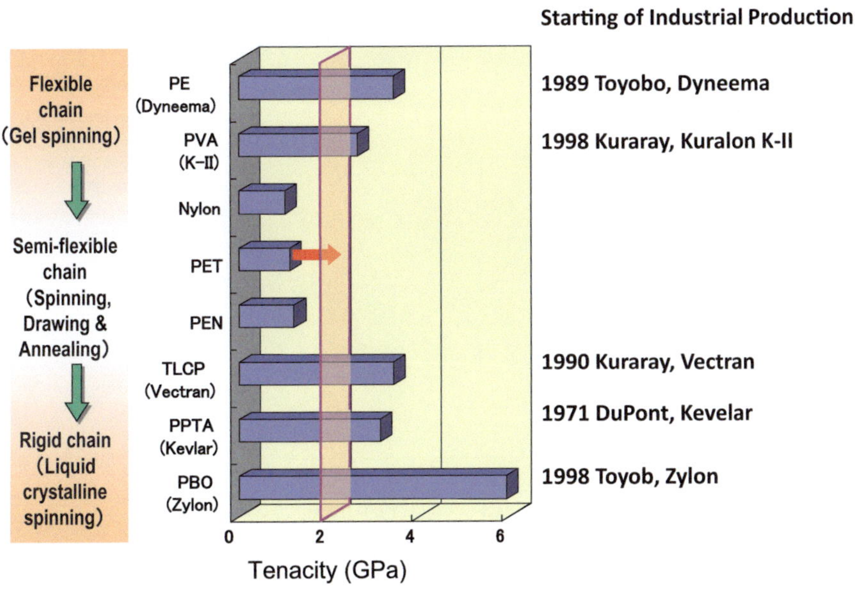

Fig. 8.3 Tensile strength of various fibers in the market

cords have been produced from semiflexible chain polymers such as PET and polyamide (PA), their strength remained at a level of around 1 GPa. In this regard, improvement of the strength of PET fibers had been highly anticipated.

8.3 Strategy for Development of High-Strength PET Fibers

Development of high-strength PET fibers had been attempted in the fiber industry in the world for decades since the starting of the production of PET fibers in 1965; however, no significant progress was made. Considering such background, basic strategy for the development of high-strength PET fibers was ascertained in this project as shown in Fig. 8.4. In addition to rather common three research subjects, i.e., "molecular weight control," "improvement of drawing and annealing processes," and "characterization and evaluation of the structure and properties of fibers," a new concept "melt structure control" was introduced as the main objective of the project.

This new concept can be depicted as follows: Control of the stress and temperature histories of spinning line to produce undrawn fibers of unique structure which essentially has high potential to be high-strength and high-toughness fibers after

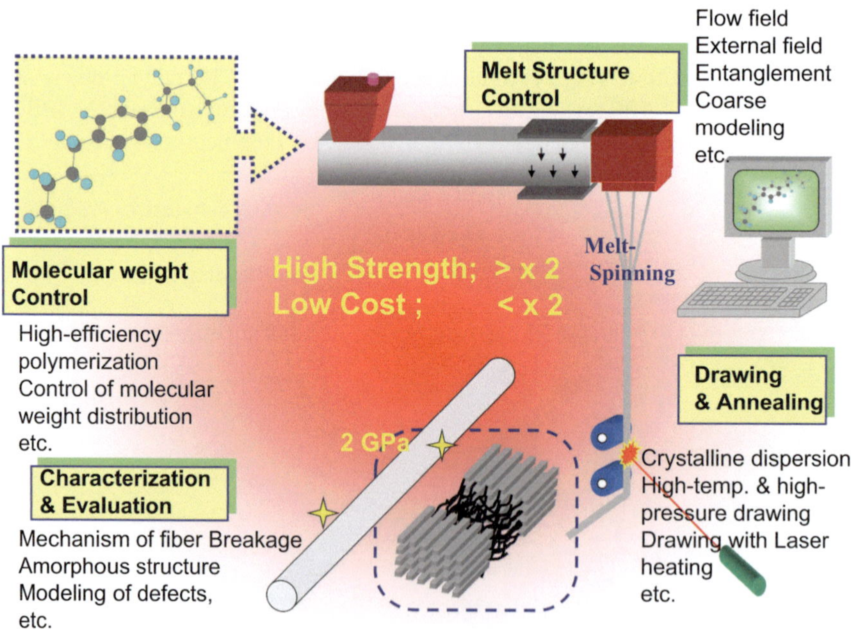

Fig. 8.4 Strategy for development of high-strength PET fibers

applying the drawing and annealing processes. More concrete image of this strategy is explained in Fig. 8.5. In the ordinary fiber production process, firstly, so-called undrawn yarn (UDY) is produced through the melt spinning process. The UDYs generally have low strength and high elongation at break, and the characteristics of the stress–strain behavior of UDY are the necking drawing. After applying the drawing process to UDY, which includes drawing at a temperature near the glass transition temperature and further annealing at elevated temperatures, so-called drawn yarn (DY) can be produced. Mechanical properties of the DY can be represented by high tensile strength with relatively low elongation at break. The mechanical properties can be manipulated mainly by changing the draw ratio. When draw ratio is increased, tensile strength generally increases whereas elongation at break decreases. Eventually, toughness of the fiber remains almost constant. Nevertheless, there always is limitation of maximum draw ratio and tensile strength. Strategy of this project for the development of high-strength PET fiber was to introduce modification in the melt spinning process to produce the UDY of different characteristics, i.e., the fiber which has higher potential for becoming high-strength and high-toughness fiber after applying appropriate drawing process. Shift of the toughness line, the line connecting the elongation at break versus strength relation of fibers of various draw ratios, to the directions of higher

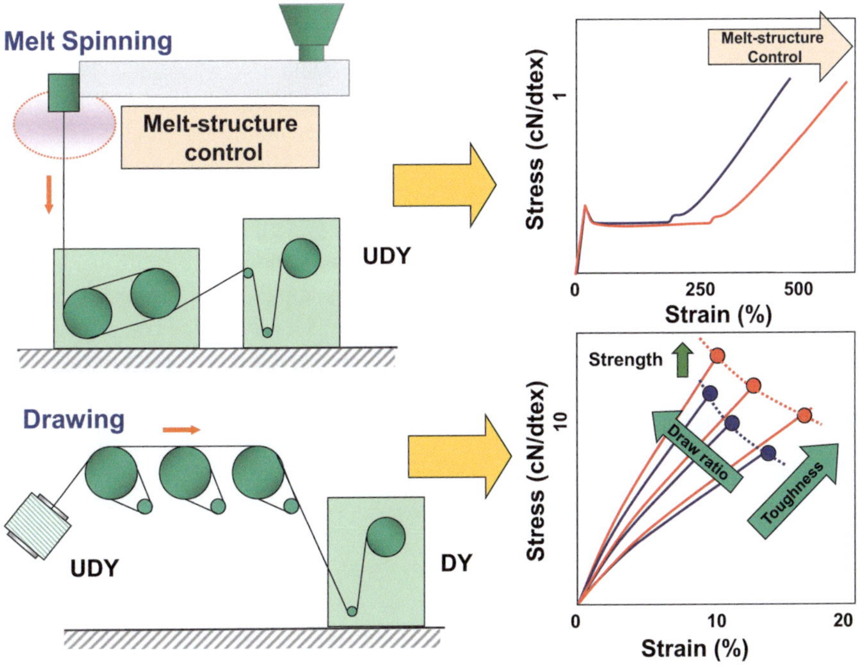

Fig. 8.5 Concept of melt structure control for development of high mechanical performance fibers

elongation and higher strength implies that the UDY essentially has higher potential for becoming high-strength and high-toughness fibers.

8.4 Various Technologies Applied for the Modification of Spinning Process

To strengthen fibers of conventional polymers of semiflexible molecular chain such as PET and PA, main target was set to establish the technology for the control of melt structure as stated previously. Emphasis was given to the combination of the melt structure control technology with the molecular weight control technology [5]. Various technologies for the modification of the spinning process attempted in this project will be briefly introduced below.

8.4.1 Modification of Spinning Process Through Addition of Modifier

With the aim of controlling the characteristics of extrudates and spinning behavior of PET, effect of the addition of a small amount of nucleating agent was investigated [6]. Chemical structure of the nucleating agent (NCA) used is shown in Fig. 8.6. It should be noted that the applied nucleating agent also has a function of plasticizing the molten PET. Blending of a small amount thermoplastic liquid crystalline polymer (TLCP) was also attempted [7]. Significant improvement of drawability was found both for PET modified with a nucleating agent and TLCP. As a result, fibers with tensile strength of 11.2 cN/dtex (1.5 GPa) were obtained. When high-speed spinning process was applied, orientation-induced crystallization was suppressed in both cases, indicating that the stress applied to the spin-line in the region where orientation-induced crystallization was expected to occur was suppressed because of the modification of thinning behavior of the spin-line.

Fig. 8.6 Chemical structure of the nucleating agent/plasticizer

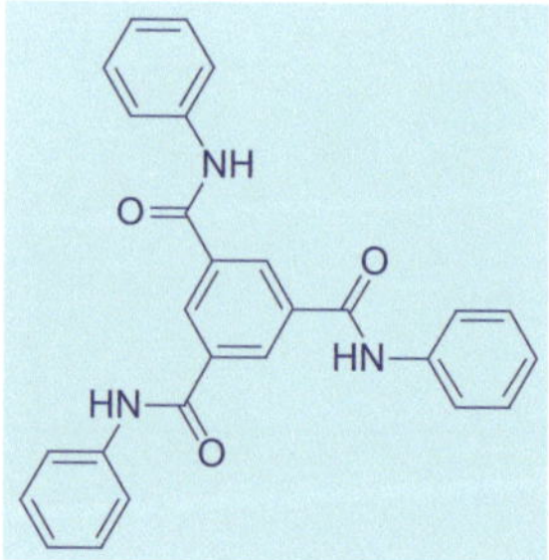

Trimesic acid trianilide, Tm=321 °C

8.4.2 *Utilization of Pressurized Medium*

For the improvement of the flowability of high molecular weight PET, plasticizing of molten PET through the incorporation of pressurized carbon dioxide was investigated. As shown in Fig. 8.7, a certain degree of viscosity reduction was confirmed through the measurement of viscosity using a capillary viscometer equipped in the spinning head of the extrusion system for fiber spinning. When the molted polymer containing carbon dioxide is extruded through a spinning nozzle, foaming of the material took place because of the pressure drop. To overcome such problem, further attempt was made for the modification of the spinning process by constructing an equipment for extruding the fiber into the medium of high-pressure carbon dioxide [8]. A technique for producing fibers with this equipment attached to the spinning head was established. A unique temperature and stress history could be applied to the spin-line, which passes through not only the high-pressure medium but also the jet-flow of gas experiencing adiabatic inflation. Drawing and annealing of the as-spun fibers yielded fibers exhibiting a unique stress–strain behavior.

8.4.3 *Heating of Spin-Line Immediately Below the Spinneret Irradiating Carbon Dioxide Laser*

Carbon dioxide laser beam was irradiated to the spin-line of PET at the position immediately below the spinneret to modify the spinning process as shown in Fig. 8.8 [9, 10]. By utilizing the laser heating, high temperature spinning of PET could be mimicked preventing the thermal degradation of the polymer, which is unavoidable if the polymer stays in the spinning die of high temperature for a long time. After the drawing and annealing of as-spun fibers prepared with the laser irradiation heating, high-strength and high-toughness fibers with tensile strength of 10 cN/dtex (1.4 GPa) and elongation at break of 15 % were obtained.

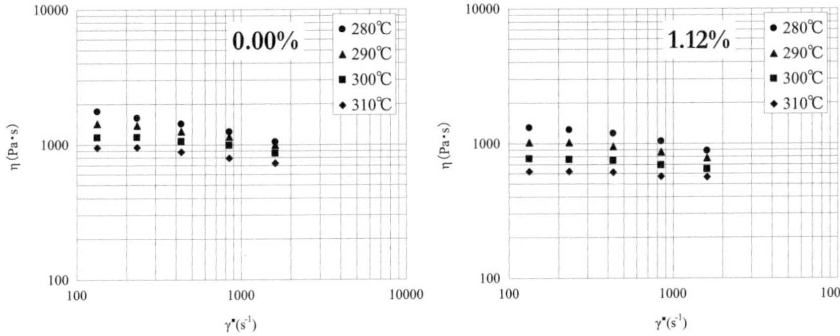

Fig. 8.7 Variation of shear viscosity of PET through the absorption of carbon dioxide

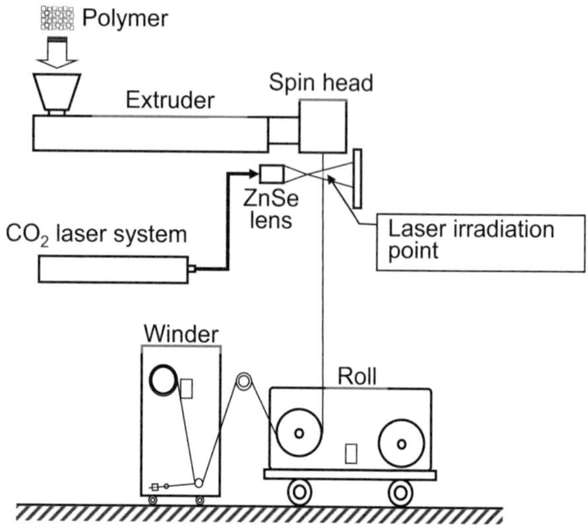

Fig. 8.8 Schematic of spinning system with laser irradiation

8.4.4 *Modification of Spin-Line Introducing the Concept of Direct Spin-Drawing*

Direct spin-drawing of PET was also conducted by changing the position and temperature of the heated first godet roll to modify the diameter profile in the spin-line, and also to apply the drawing process to the fiber without lowering its temperature below the glass transition temperature. In other words, a hot roll was placed at a position where the spin-line is still in a molten state. Schematic diagram of the spinning apparatus is shown in Fig. 8.9. Controllability of the fiber structure via the melt spinning process was also investigated by installing two liquid isothermal baths of different temperatures in the spin-line as shown in Fig. 8.10 [11]. In the direct spin-draw process, an increase in the fiber strength of drawn and annealed fibers was observed using the as-spun fibers prepared by gradually decreasing the distance from the spinneret to the first godet roll. On the other hand, by installing two LIBs and also by setting a lowered liquid temperature for the second LIB, fibers with highly developed molecular orientation and with relatively high strength were obtained. Mechanism for the enhancement of fiber structure development using two LIBs was clarified through the numerical simulation of the melt spinning process equipped with LIBs.

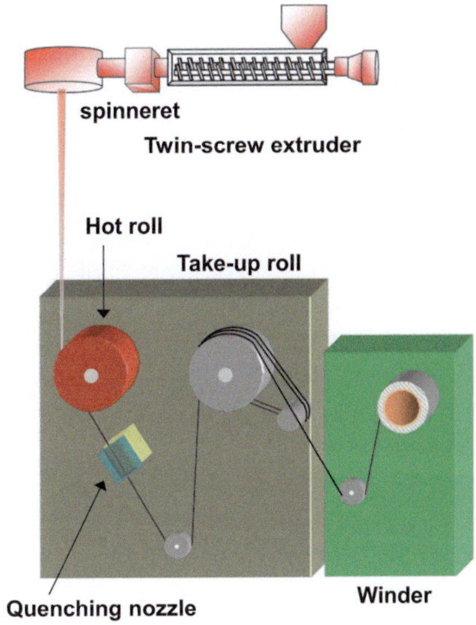

Fig. 8.9 Schematic diagram of in-line drawing system equipped with a hot roll placed at a position where the spin-line is still in a molten state

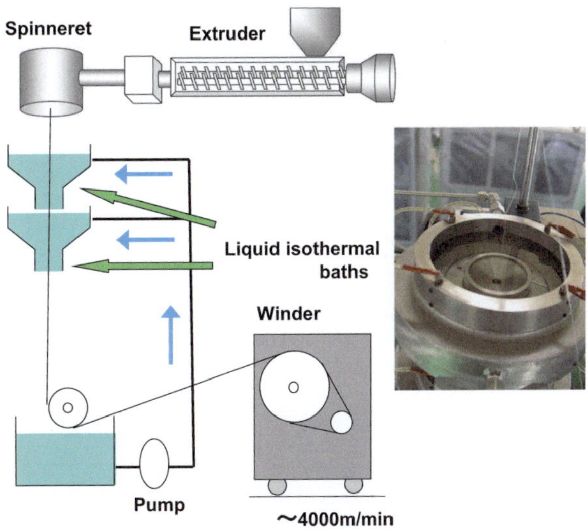

Fig. 8.10 Schematic diagram of spin-line equipped with two liquid isothermal baths (LIBs)

8.5 Concept for Strengthening of PET Fibers

Through the various attempts for improving the mechanical properties of PET fibers stated above, a hypothesis of keeping the melt structure of molten polymer in the melt spinning process through the reduction of Deborah number, i.e., nondimensional number defined as the product of strain rate and relaxation time, came to arise. To confirm such concept, effect of the combination of two spin-line modification methods, i.e., laser irradiation heating and application of small-diameter spinning nozzle, on mechanical performance of resultant fibers was investigated [12–14].

The relation between tensile strength and elongation at break for the drawn fibers of various draw ratios prepared from the as-spun fibers of four different spinning conditions, i.e., standard (SD), laser irradiation heating (LIS), use of small-diameter nozzle (SC), and combination of LIS and SC (LIS+SC), is shown in Fig. 8.11. Shift of the toughness curve toward upper-right direction in the order of SD → LIS ≅ SC → LIS+SC indicated that the toughness of drawn fibers were improved in this order. Stress–strain curves of the drawn fibers of various draw ratios prepared from the as-spun fibers of LIS+SC condition shown in Fig. 8.12 indicated that the highest tensile strength of 12.1 cN/dtex (1.68 GPa) with the elongation at break of 9.1 % was achieved at the draw ratio of 7.0. This strength is significantly higher in comparison with the tensile strength of ordinary PET tire cord of about 1 GPa.

Variations of velocity gradient of the spin-line with spin-line temperature for the four different spinning conditions were estimated utilizing the numerical simulation program based on the spin-line dynamics. The results are shown in Fig. 8.13. Near the spinneret where the temperature is around 300 °C, velocity gradient for the LIS +SC showed the highest value whereas that for the SD showed the lowest value. On the contrary, when spin-line is cooled down to the temperature below around 250 °C, velocity gradient for the SC+LIS showed the lowest value whereas that for the SD showed the highest value. The order of SD > LIS ≅ SC > SC+LIS corresponds to the order of the strength and toughness curve of the drawn fibers prepared from the undrawn as-spun fibers of four different spinning conditions. In other words, lowering of Deborah number yielded the as-spun fibers which

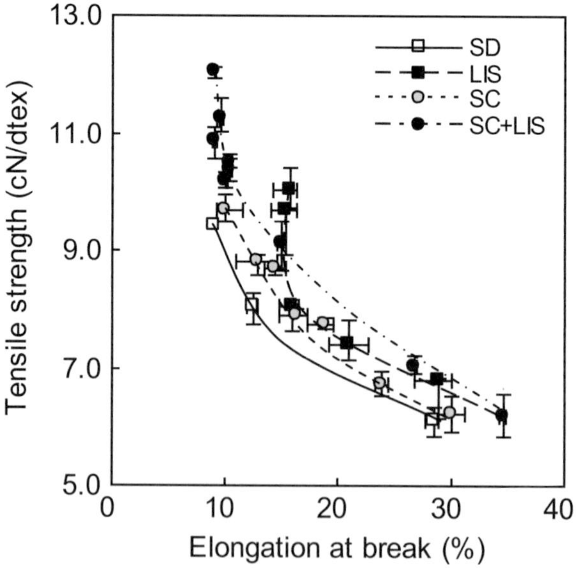

Fig. 8.11 Relation between tensile strength and elongation at break for drawn fibers of various draw ratios prepared from as-spun fibers of four different spinning conditions. *SD* standard, *LIS* with laser irradiation, *SC* with use of small-diameter nozzle and SC+LIS: with both laser irradiation and use of small-diameter nozzle

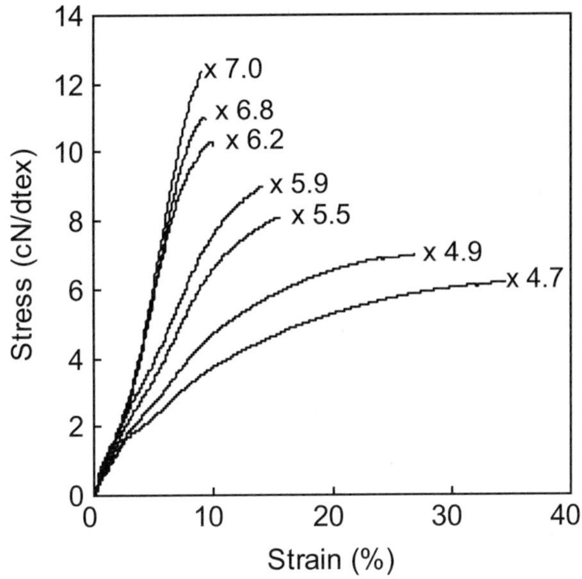

Fig. 8.12 Stress–strain curves of drawn fibers of various draw ratios prepared from as-spun fiber of laser irradiation spinning with small-diameter nozzle (SC +LIS) at 0.5 km/min

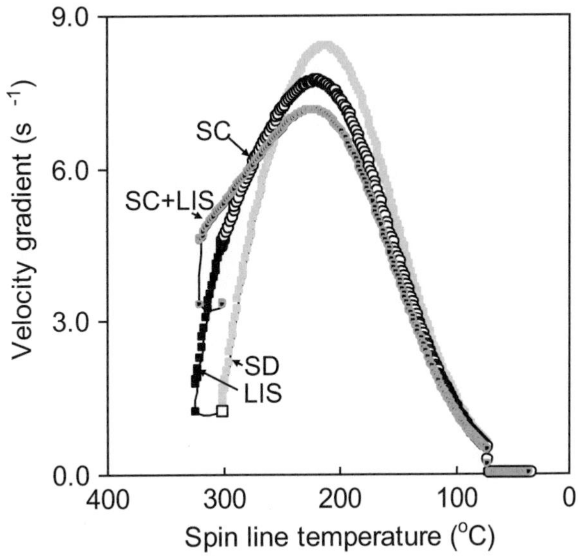

Fig. 8.13 Calculated spin-line velocity gradient (tensile strain rate) profiles for four different spinning conditions

possesses the characteristics of becoming high-strength and high-toughness fibers after applying drawing process. It is important to note that the additivity of the effect of modifying the spinning conditions for the lowering of Deborah number was also confirmed.

8.6 Estimation for the Change in the State of Molecular Entanglement

Combination of numerical simulations of spin-line dynamics and coarse molecular dynamics was utilized to estimate the change in the state of molecular entanglement in the spinning line [15]. In this analysis, macroscopic spin-line dynamics was analyzed applying the conventional mass, momentum, and energy balance equations (see Chap. 3 in this book), whereas a software called "NAPLES," which is based on the primitive chain network model, was applied for the prediction of the changes in the state of molecular entanglement in the flow [16, 17]. Result of calculation comparing the melt spinning of ordinary spinning condition and that of high extrusion temperature, which corresponds to the spinning with carbon dioxide laser irradiation, are shown in Fig. 8.14. In the case of ordinary melt spinning process, Deborah number in the intermediate region of the spinning line is higher. Accordingly there is a more distinct reduction of entanglement density. In the melt spinning with laser irradiation, Deborah number is lower and the reduction of entanglement density is milder.

At this point, it should be noted that this result contradicts to the general concept for the production of high-strength fibers. In the production of high-strength PE fibers, the gel spinning technology is applied for the reduction of the entanglement density. In the case of PPTA and PBO, designing of rigid-chain molecules itself implies the concept of reducing entanglement density. In the case of the production of high-strength and high-toughness PET fibers, experimental results suggested the

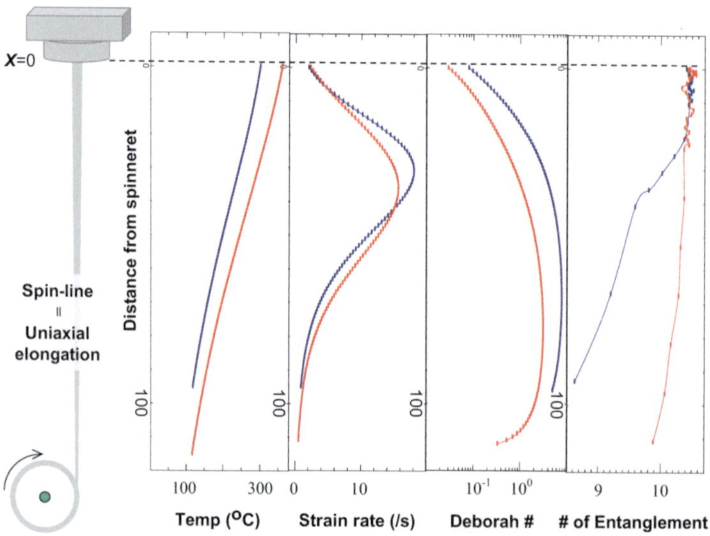

Fig. 8.14 Calculated variations of temperature, strain rate, Deborah number, and number of entanglement points per chain along the spin-line for the conditions of standard spinning (*blue line*) and spinning with laser irradiation heating (*red line*)

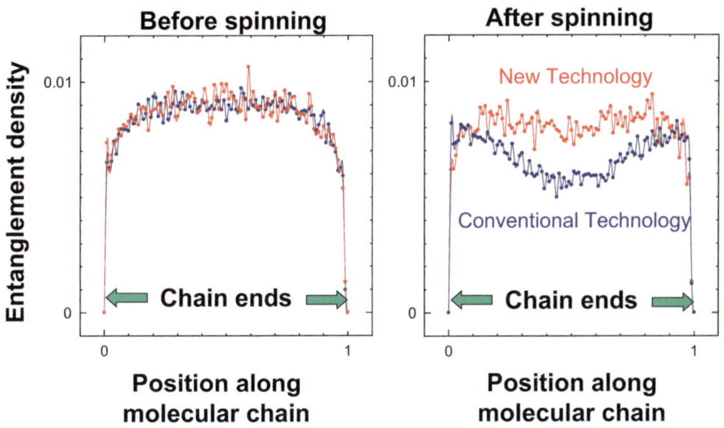

Fig. 8.15 Entanglement density distribution along individual molecular chain before and after application of elongational flow in the melt spinning process for the conditions of standard spinning (*blue line*) and spinning with laser irradiation heating (*red line*)

opposite concept, i.e., prevention of the reduction of entanglement density is necessary for the improvement of mechanical properties. Therefore, results of coarse molecular dynamics simulation was investigated more in detail. Figure 8.15 shows the variation of entanglement density distribution along individual molecular chain. In the polymer melt without flow, entanglement density is slightly lower near the both chain ends because of higher mobility of molecular chain. When the polymer melt is subjected to elongational flow in the spin-line, overall entanglement density reduces, in that reduction of entanglement density occurs exclusively in the center region of molecular chain. It is known that disappearance and generation of entanglement points occur continuously in the polymer melt. In the center part of molecules, generation of new entanglement points is rather difficult because of the lower mobility of molecular chain. With this mechanism, distribution of the molecular weight between two adjacent entanglement points tends to be wider.

Mechanism for the improvement of tensile strength and toughness is speculated based on the abovementioned results of numerical simulation as follows. It can be considered that the mechanical properties of fibers can be improved if the state of all the molecules is similar. This means that similar level of stress is applied to all the molecules when the tensile stress is applied to the fibers. If the narrower distribution of the molecular weight between two adjacent entanglement points can be incorporated in the as-spun fibers, stress concentration to certain molecules can be avoided during the drawing process, which should lead to the higher drawability with less possibility of fiber breakage. In other words, if all the molecules bear similar level of stress during either drawing process or tensile test, drawability as well as strength and toughness of the resultant fibers can be improved. The concept described above is schematically depicted in Fig. 8.16.

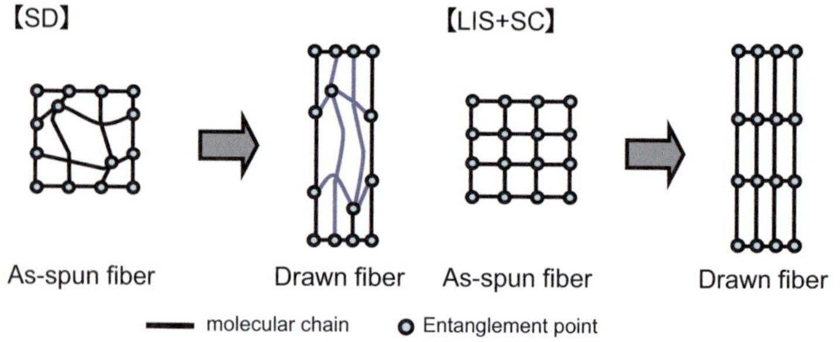

Fig. 8.16 Schematic of speculated structural difference between standard spinning (SD) and spinning modified with laser irradiation heating and small-diameter nozzle (LIS+SC)

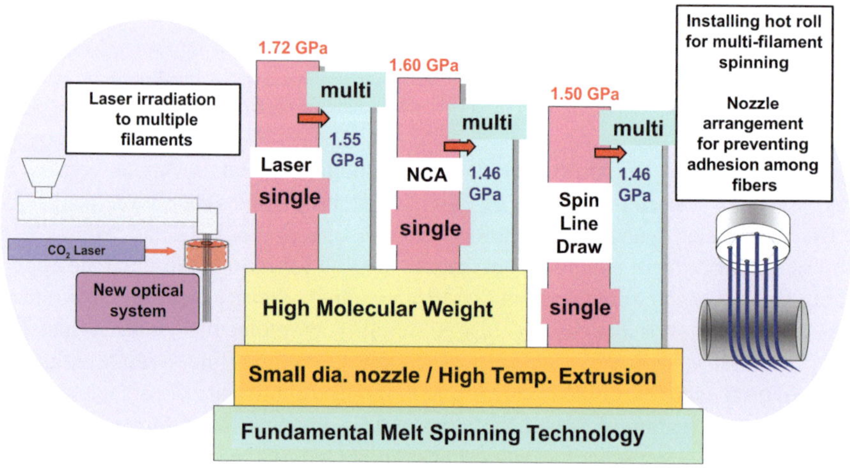

Fig. 8.17 Variation of fiber strength from single filament to multifilament fiber formation. Key technologies for multifilament processing are also indicated

8.7 Concluding Remark

In this research, melt spinning of single fiber was mainly conducted. Application of these technologies for multiple fiber spinning process was also attempted. As expected, strength of multiple fiber bundle was about 10 % lower in comparison with the single fiber spinning as shown in Fig. 8.17; however, basic trend for the preparation of good mechanical performance PET fibers was reconfirmed. As the future research subject, experimental verification of the basic concept for the improvement of the mechanical properties of PET fibers acquired in this project, that is the homogenization of the state of molecular entanglement in the system so as to evenly distribute the tensile load to individual molecular chains in the system, needs to be pursued.

References

1. *Nanostructured Polymeric Materials and Technologies*, ed. S. Nakahama, (CMC Shuppan, 2004)
2. *Fundamental and Practical Technologies for Nano-structured Polymeric Materials*, ed. S. Nakahama, (CMC Shuppan, 2008)
3. Japan Chemical Fibers Association, http://www.jcfa.gr.jp/data/japan/2_2.html
4. T. Kikutani, J. Appl, Polym. Sci. **83**, 559–571 (2002)
5. K. Chizuka, T. Kikutani, Sen'i Gakkaishi **60**, 331–337 (2004)
6. Jap. Pat. (Laid Open), 2006-336132
7. J. Kojima, T. Kikutani, Sen'i Gakkaishi **61**, 29–37 (2006)
8. Jap. Pat., 4,588,538, Jap. Pat. (Laid Open), 2006-328613
9. Jap. Pat., 4,224,813
10. M. Masuda, Y. Funatsu, K. Kazama, T. Kikutani, Sen'i Gakkaishi **60**, 338–345 (2004)
11. Jap. Pat. (Laid Open), 2006-328550, Jap. Pat. (Laid Open), 2008-308786
12. Jap. Pat., 4,791,844, Jap. Pat., 4,757,655, Jap. Pat., 4,792,410, Jap. Pat. (Laid Open), 2006-350324, Jap. Pat., 4,958,108, Jap. Pat., 5,134,417
13. M. Masuda, W. Takarada, T. Kikutani, Sen'i Gakkaishi **65**, 118–126 (2009)
14. M. Masuda, W. Takarada, T. Kikutani, Intern. Polym. Process. **25**, 159–169 (2010)
15. Y. Masubuchi, in "*Fundamental and Practical Technologies for Nano-structured Polymeric Materials*", ed. S. Nakahama, (CMC Shuppan, pp. 66–74), 2008
16. Y. Masubuchi, J. Takimoto, K. Koyama, G. Ianniruberto, G. Marrucci, F. Greco, J. Chem. Phys. **115**, 4387–4394 (2001)
17. http://masubuchi.jp/NAPLESweb/

Chapter 9
Technora® Fiber: Super Fiber from the Isotropic Solution of Rigid-Rod Polymer

Shigeru Hayashida

Abstract After the launch of Kevlar®, a new para-aramid fiber was researched on its molecular design concept in Teijin Ltd. As a result, Technora®, which is a para-aramid copolymer, was developed. Typically Technora® yarn has tenacity of 24.5 cN/dtex, tensile modulus of 530 cN/dtex, and 2 % equilibrium moisture regain under the standard condition. The super fiber was generated from the isotropic solution unlike Kevlar®. An overview of Technora® is given with emphasis on its molecular design, process, and its physical and chemical properties, comparing to PPTA.

Keywords Technora® • Twaron® • Kevlar® • Aramid fiber • PPTA • 3,4′-Diaminodiphenyl ether • Polymer sequence • Dry-jet wet spinning

9.1 Introduction

In 1971, Teijin America Incorporated, which was a subsidiary company of Teijin Ltd. in New York, sent the information about Fiber-B (code name of Kevlar® in developing stage) of DuPont to Central Research Laboratories, Teijin Limited. According to the information, the followings were informed:

- In the 1960s DuPont started the research of fibers with high tenacity and high modulus for automotive tires to avoid the risk that existing man-made fibers could be replaced by steel cord in tire industries.
- Aramid (whole aromatic polyamide) and arylate (whole aromatic polyester) were the candidates for the fibers.
- In 1965, Stephanie Kwolek developed the first liquid crystal polymer which provided the basis for Kevlar brand fiber.

At that time, the high performance fiber division of the Central Research Laboratories, Teijin Limited, dedicated to build up a semicommercial plant of Teijinconex® which is meta-aramid fiber. But the division had strong interest in

S. Hayashida (✉)
Teijin Limited, Osaka, Japan
e-mail: ICD01151@nifty.com

Society of F.S. Technology, Japan (ed.), *High-Performance and Specialty Fibers*,
DOI 10.1007/978-4-431-55203-1_9

the Fiber-B, and Ozawa, who was a top manager of the division, decided to start the research.

The research was started with tracing patents of Kevlar®. After some experiments, the researchers of the division recognized that a para-aramid fiber could become a high performance fiber which they had not seen before in terms of high tenacity and high modulus.

After the research phase, the development was executed in the Manufacturing Technology Institute located in Iwakuni City. In 1979 there was the second oil crisis, and the project was broken up for 2 years. However, thanks to persistent efforts of researchers, it was resumed in 1981 and finally Technora® [1–3] was commercialized in 1987.

9.2 Polymer Research

By tracing Kevlar® patents, it was found that Kevlar® fibers had excellent tenacity and modulus. However, they noticed that there was an issue that Kevlar® needed sulfuric acid as its spinning solvent. Sulfuric acid has a strong corrosive property to the process facility and is harmful for human health. So Teijin researchers set the first target that a new polymer should be soluble in an organic solvent instead of strong acids. They thought that the excellent properties were generated from the para-aramid polymer which had rigid polymer chains with parallel orientation of the molecules. Para-aramid copolymer could also have excellent properties as well as Kevlar® (poly-paraphenylene terephthalamide) which is a homopolymer and is produced by polymerizing para-phenylenediamine and terephthaloyl dichloride. Then the direction of the research was decided to seek the third monomer to be added to poly-paraphenylene terephthalamide (PPTA).

After starting the research, many efforts were made for a new polymer research containing monomer syntheses, polymerizations, measuring reactivity and solubility in a solvent, spinning and drawing, and evaluation of the fiber properties. It took approximately 3 years to find the third monomer which makes the polymer possible to dissolve in an organic solvent and the properties of its fibers comparable to Kevlar®. Tables 9.1, 9.2, and 9.3 show main chemical substances for the third monomers, which were tested during the research phase.

9.3 Technora Polymer

After approximately 3 years of experiments, it was found that 3,4′-diaminodiphenyl ether (3,4′DAPE) was the best third monomer which could give superior properties to existing man-made fibers. The chemical formulation is shown in Fig. 9.1. Ozawa has reported that the physical properties of Technora® fiber such as tenacity, elongation at break, and modulus change depending on the molar ratio of two

Table 9.1 The list of third component for copolymer (diamine component)

	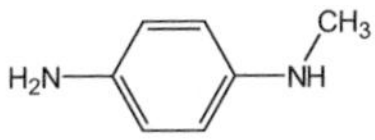
CH_3 CH_3 CH_2NH-CH_2NH	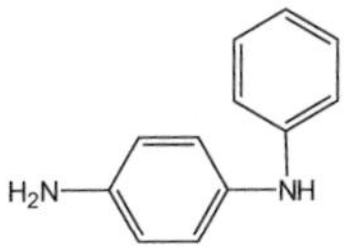
$CH_2NH_2-CH_2NH_2$	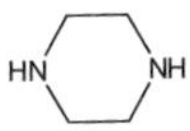
	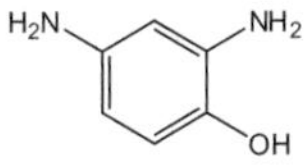
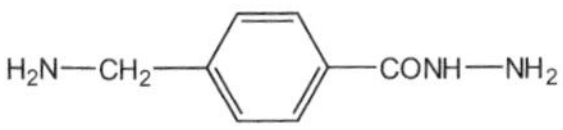	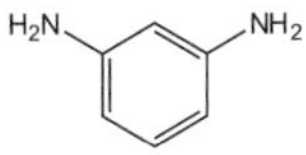
	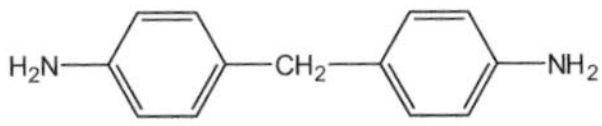
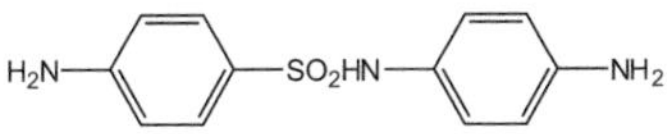	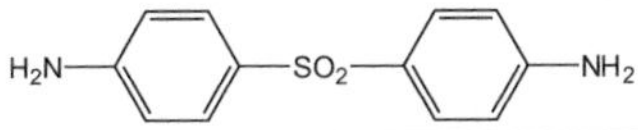
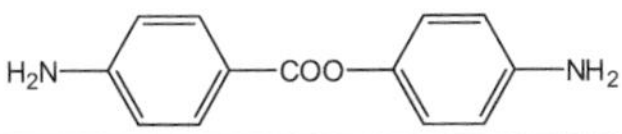	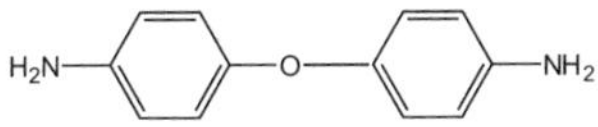

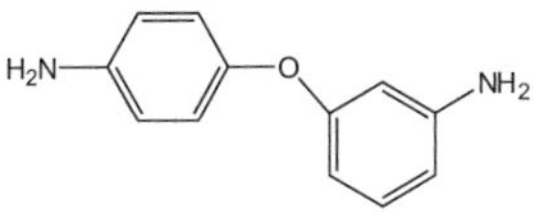	

diamine components, and there is an optimum composition to achieve excellent physical properties [3].

The physical properties will be discussed later.

9.4 Polymer Preparation

Kevlar® (poly-paraphenylene terephthalamide) (PPTA) can be prepared by the low temperature polycondensation of p-phenylene diamine (PPD) and terephthaloyl dichloride (TPC) in an amide solvent. The amide solvents for this synthesis include hexamethylphosphoramide, N-methyl-2-pyrrolidone, and dimethylacetamide.

Table 9.2 The list of the third component for copolymer (acid component)

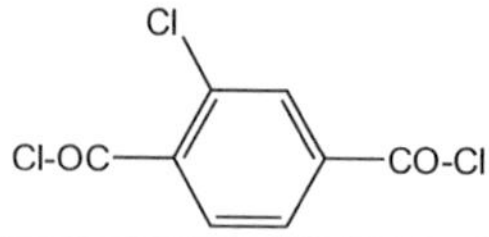	ClCO-COC1
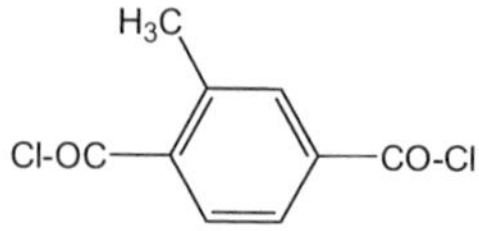	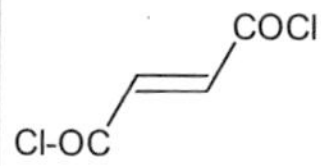
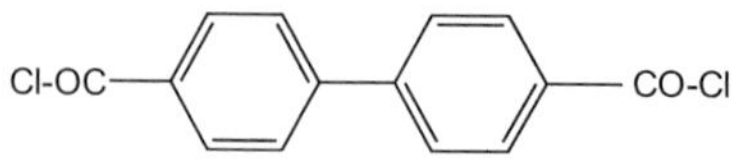	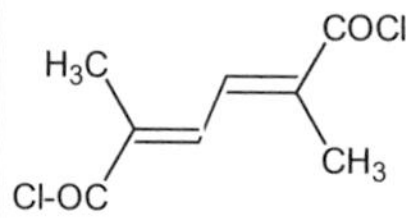

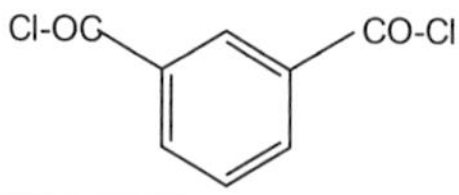	

Table 9.3 The list of the third component for copolymer (heterocyclic amine component)

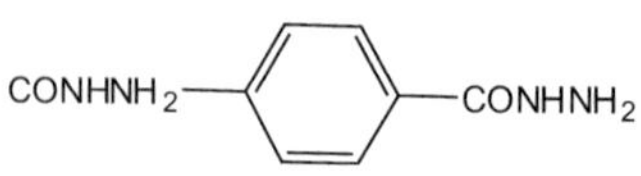	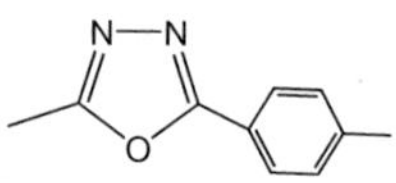
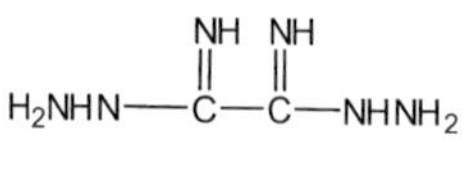	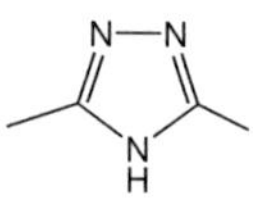
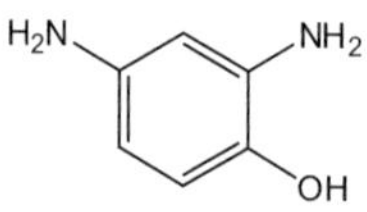	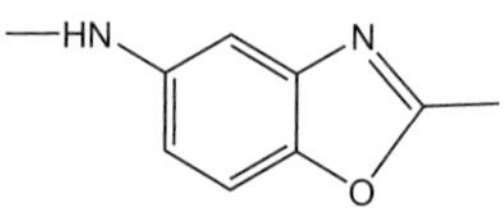
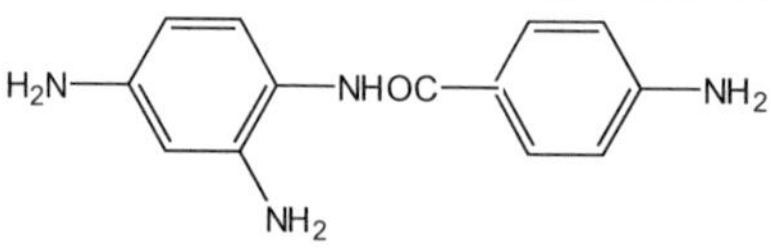	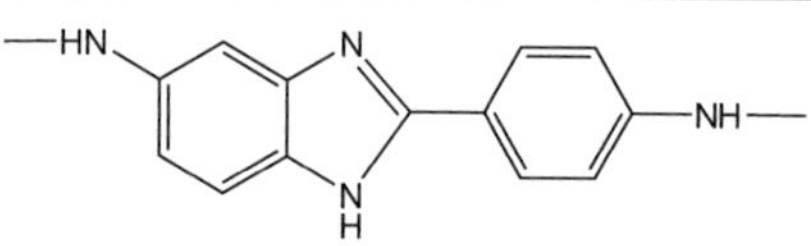

However, hexamethylphosphoramide is not used because it was suspected to be carcinogenic by the results of animal tests [4]. In the manufacturing process, N-methyl-2-pyrrolidone is used in the presence of an inorganic salt like lithium chloride or calcium chloride. The polymer is isolated by coagulation with water and

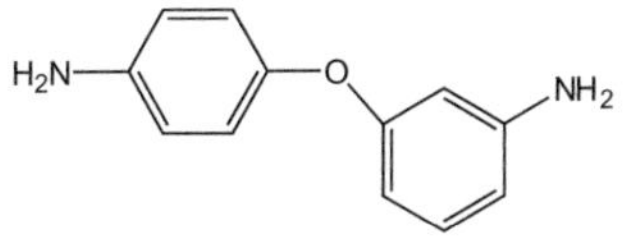

3,4'- diaminodiphenyl ether

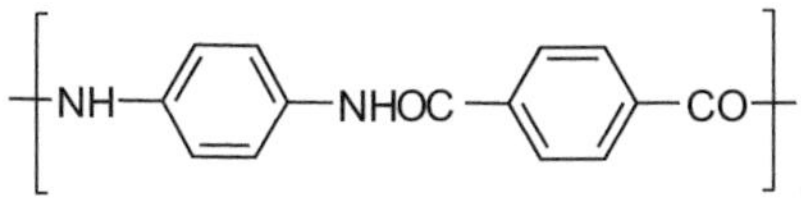

Kevlar® (poly-paraphenylene terephthalamide) (PPTA)

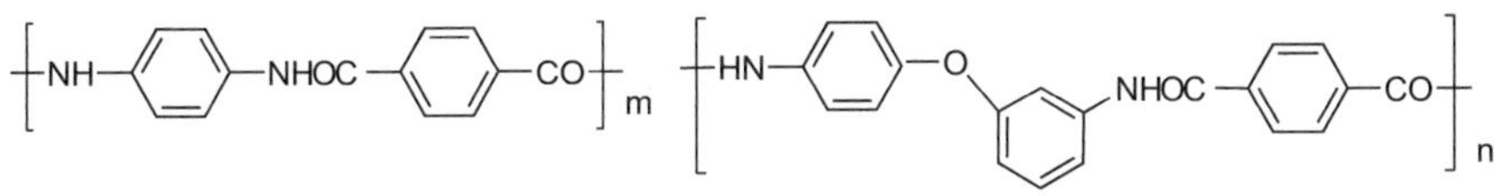

Technora® Co-poly-(paraphenylene/3,4'-diphenylether terephthalamide)

Fig. 9.1 PPTA and Technora® Polymer

neutralized by alkaline substance like calcium oxide or calcium hydroxide. After that the polymer is washed by water and dried.

Technora®(copoly-paraphenylene/3,4′-diphenyletherterephthalamide) also can be prepared by the low temperature polycondensation of 40–60 mol% of PPD, 60–40 mol% of 3,4′DAPE, and 100 mol% of TPC in an amide solvent. As for the amide solvent in the manufacturing process, normally N-methyl-2-pyrrolidone (NMP) is selected. An inorganic salt is not necessary for Technora® polymer solution. The polymer solution is neutralized by adding calcium hydroxide/NMP suspension. And the polymer solution can be used for the spinning solution without a polymer powdering process like PPTA.

9.5 Spinning Solutions

PPTA polymer cannot be dissolved in organic solvents. It can be dissolved in strong acid such as 100 % sulfuric acid, chloro- and fluorosulfonic acid, nitric acid, and hydrogen chloride. As it is well known, Kwolek [5] discovered that PPTA/H_2SO_4 solution at 10 % polymer concentration was anisotropic. In comparison, the solution of Technora® polymer does not show such an anisotropic behavior, but the solution shows isotropic characteristic.

Figure 9.2 presents a correlation of viscosity of PPTA/ H_2SO_4 solution vs. PPTA concentration [6, 8].

In the case of PPTA, the concentration of polymer for the spinning solution might be about 20 % at 90 °C. Above this concentration, the solution becomes biphasic with the appearance of a PPTA solid phase.

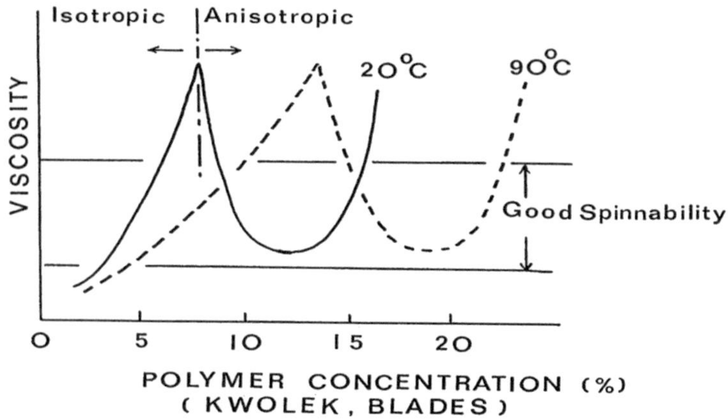

Fig. 9.2 Anisotropic behavior of PPTA/H_2SO_4 solution [6, 8]

On the other hand, in the case of Technora®, as the solution viscosity increases sharply with increasing polymer concentration, the concentration was chosen in the range with good spinnability. It is 6.0 wt.% concentration of the polymer in the solution.

9.6 Fiber Spinning

The dry-jet wet spinning process was invented by Blades [6] in order to produce PPTA fibers efficiently from an anisotropic solution of PPTA. The key feature of this process is that an anisotropic solution is extruded from spinneret holes through an air gap into a coagulation bath as shown in Fig. 9.3. The coagulated filaments are washed, neutralized by caustic soda, and finally dried.

As shown in Fig. 9.4, PPTA fiber is structurally developed through a dry-jet wet spinning system. A liquid crystalline solution of PPTA is extruded through a spinneret capillary, and the liquid crystalline domains near the wall may orient along the direction of flow due to the capillary shear. At the capillary exit, some de-orientation of liquid crystalline domains occurs because of the Barus effect. However, the de-orientation is quickly stopped when the polymer flow gets thin in the air gap by the elongational tension. When they are coagulated, filaments keep the highly oriented molecular structure.

In case of Technora®, the conventional wet spinning process, where the spinning nozzle is immersed in the coagulant, was tested to be applied. However, it was difficult to harmonize the temperature of the spinneret and the temperature of the coagulation bath, because the temperature of the spinneret should be raised as high as possible to decrease the viscosity of the polymer solution. On the other hand, the temperature of the coagulation bath should be kept moderately to handle the wet

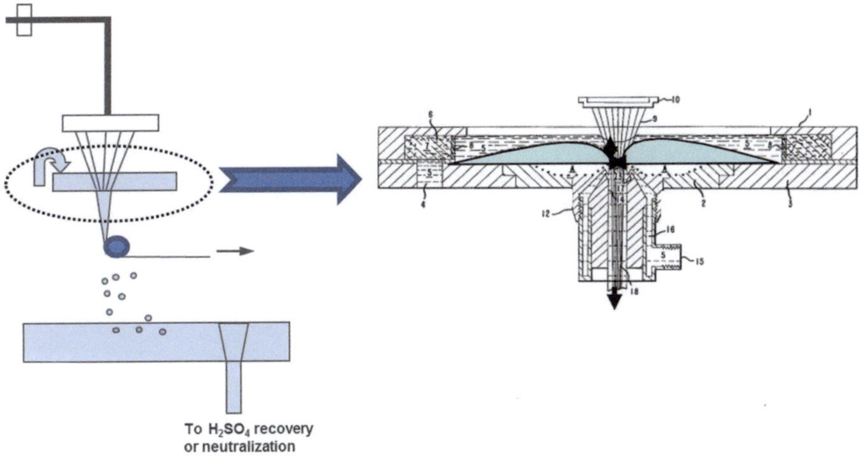

Fig. 9.3 PPTA dry-jet wet spinning system [7]

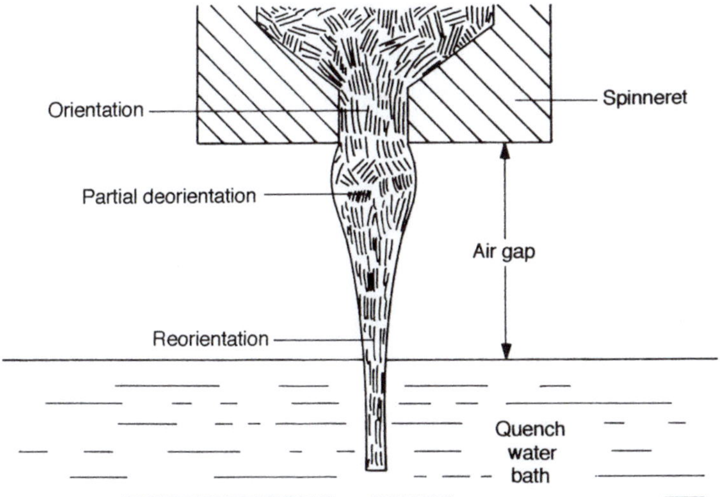

Fig. 9.4 Fiber structure development in dry-jet wet spinning process [8]

yarn in the bath. So the dry-jet wet spinning process is suitable for Technora® spinning as well as PPTA as shown in Fig. 9.5. The process has the merit which is allowed to have different temperatures between the spinneret and the coagulation bath. Because Technora® polymer solution is isotropic, the fiber structural development in a dry-jet wet spinning cannot be expected. Actually the coagulated Technora® filaments have no oriented structure in as-spun yarns.

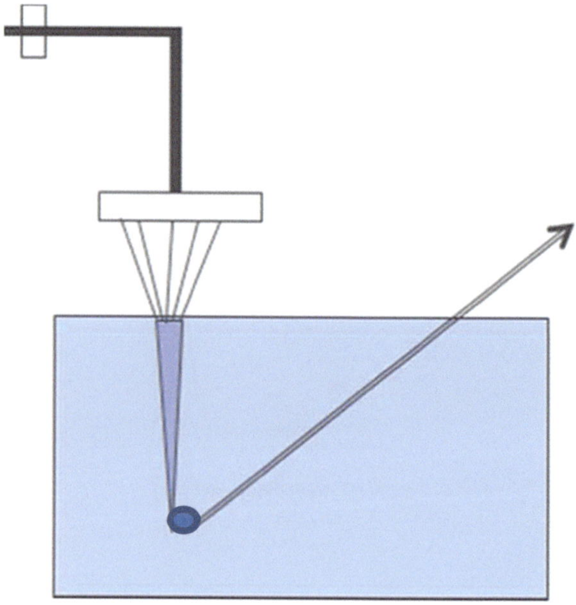

Fig. 9.5 Technora® dry-jet wet spinning system

Comparing two types of coagulation processes, while PPTA uses the flow tube type coagulation system, Technora® uses the bath type coagulation system. Because the concentration of Technora® polymer in the solution is 6 %, it takes time to diffuse most of the solvent from the filament to the coagulant. It can be concluded that the bath type is suitable for Technora®.

9.7 Fiber Drawing

In the case of PPTA fiber, no drawings are necessary for a high degree of molecular orientation and a high degree of crystallinity.

Technora® fiber needs drawings for a high degree of molecular orientation and a high degree of crystallinity like commodity man-made fibers such as polyester, nylon, and so on. However, the drawing temperature and the ratio of draw should be extremely high compared to common fibers as Fig. 9.6 shows the drawing characteristics of Technora® fibers. To be more precise, the draw ratio is more than ten times at the temperature of about 500 °C to achieve the excellent properties. It is particularly worth noting that above the drawing temperature of 470 °C, the tenacity of the drawn fiber is increased enormously. Above the temperature the molecular has the thermal mobility to some extent, so that the molecular can be oriented along the fiber axis. As a result of this, Technora® has the tenacity more than 24 N/dtex.

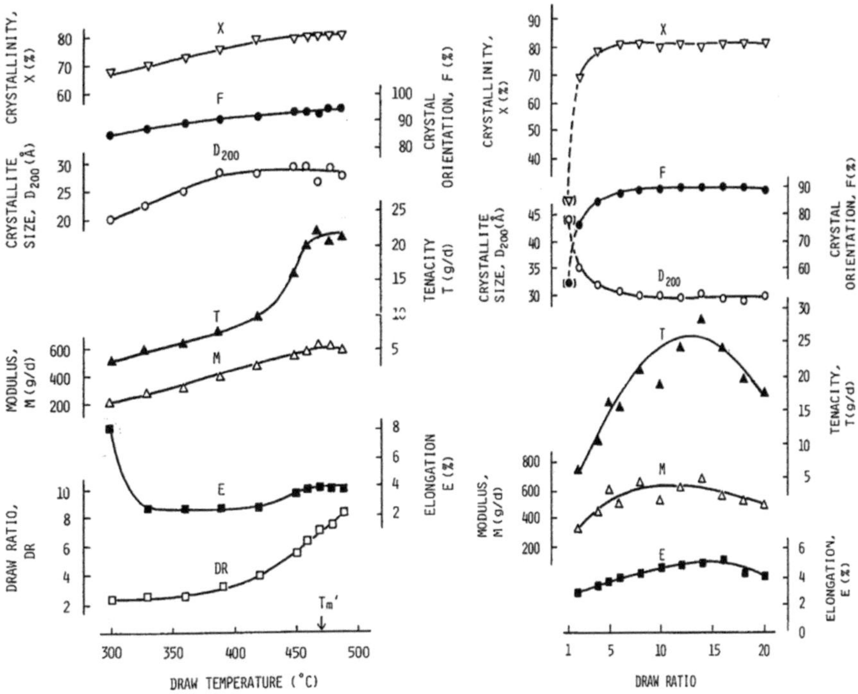

Fig. 9.6 Drawing characteristics of Technora® fiber

Technora® polymer contains 3,4′-DAPE component in its polymer chain so that the polymer chains cannot be rigid mainly due to the flexible ether bond of 3,4′-DAPE. However, to have the high tenacity and the high modulus, the linear orientation of the polymer chains is essential. The reason why Technora® needs the heat drawing can be thought that the polymer chains should be extended as linear as possible. As the result of extended polymer chains arrangement, the crystallinity is increased. Figure 9.7 shows the linearity of 3,4′-diaminodiphenyl ether (3,4′-DAPE) and 4,4′-diaminodiphenyl ether (4,4′-DAPE) as a reference. Although it seems that 4,4′-DAPE has intuitively more linearity than 3,4′-DAPE, actually 3,4′-DAPE has linearity after heat drawing, while 4,4′-DAPE has a bend structure.

Figure 9.8 shows the thermal decomposition of Technora® fiber.

It is interesting to note that Technora® fiber can be drawn at 500 °C without melting down which is starting temperature for the polymer decomposition. In case of common man-made fibers like polyester, it is impossible to draw the fibers at the temperature close to its decomposition temperature. Judging from this phenomenon, it might be possible to think Technora polymer has semirigid chains unlike polyester or nylon.

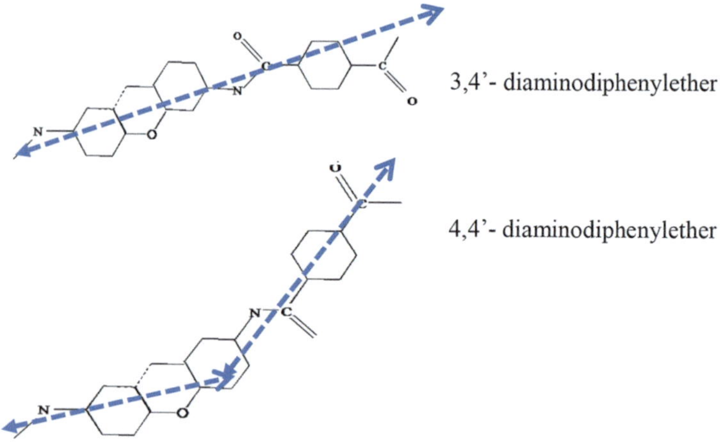

Fig. 9.7 Linearity of 3,4′-diaminodiphenyl ether and 4,4′-diaminodiphenyl ether

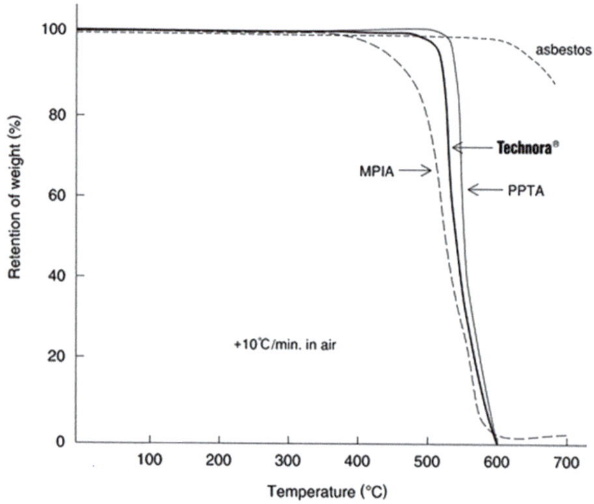

Fig. 9.8 Thermal decomposition of Technora®

9.8 Manufacturing Process of PPTA and Technora®

Figure 9.9 shows the manufacturing process of Technora® compared to the process of PPTA.

Compared to the manufacturing process of PPTA, the manufacturing process of Technora® is successful in terms of the following:

- N-methyl-2-pyrrolidone (NMP) is the solvent for the polymerization, and the polymer solution can be used for the spinning process without polymer powdering process which includes coagulation, neutralization, washing, and drying processes.

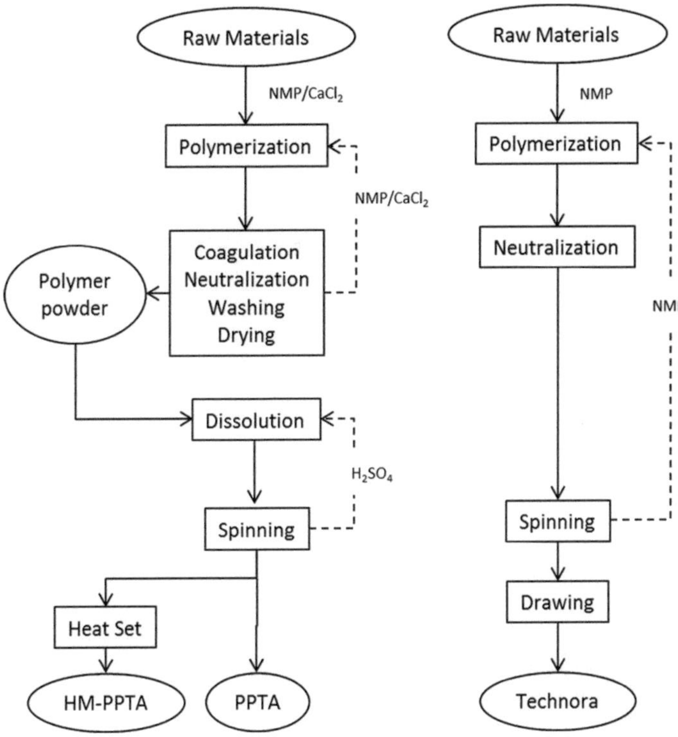

Fig. 9.9 Manufacturing process of PPTA and Technora®

- No sulfuric acid is needed for the spinning solution. So yarn handling can be done safely at the coagulation and the washing processes.
- One recovery process for NMP is necessary, while PPTA needs two recovery processes for NMP and sulfuric acid.
- The recovery process of NMP can be simpler than the acid recovery process in terms of the system and the materials of facility.

However, the Technora® process is inferior to PPTA process in points of the following:

- The polymer concentration in the solution is 6.0 wt.% compared to 20 wt.% of PPTA in the polymerization solution, which means the size of Technora® facility should be about three times larger than PPTA facility at the same capacity of production.
- In spinning process, the drawing at the high temperature is necessary, because Technora® does not make a liquid crystal solution, and to achieve the excellent properties, the heat drawing is inevitable.

Table 9.4 Types of Technora®

Available products	Type	Dtex/cut length	Applications
Filament	T-200	800dtex	Rubber reinforcement, rope, cord, general industrial materials
		1110, 1670dtex	
	T-200M	2500dtex	
	T-202	1670dtex	Rubber reinforcement
Yarn	T-221	1100, 1670dtex	Rope, cord, general industrial materials
	T-240	110dtex	Woven and knitted fabrics, composites
		220, 440dtex	
		1110, 1670dtex	
	T-240B	440, 830, 1100dtex	
	(Dope dyed black)	1670dtex	
Chopped	T-320	1.7dtex X 1, 3, 6, 12 mm	Resin and cement, reinforcement, asbestos replacement
Fiber	T-321	1.7dtex X 30 mm	Cement reinforcement
	T-322EH	1.7dtex X 1, 3, 6 mm	Resin reinforcement (FRTP)
	T-323SB	1.7dtex X 1, 3, 6 mm	Rubber reinforcement
	T-330		
Staple	T-330B		Spun yarn, nonwoven fabrics
Fiber	(Dope dyed black)	1.7dtex X 51 mm	
	T-330G		
	(Dope dyed gray)		
Spunnized	T-360	30/–, 20/–, 10/–	Protective clothing
Yarn	T-400	110, 220, 330, 440dtex	General industrial materials
			Rubber reinforcement

9.9 Technora® Aramid Products

As Table 9.4 shows, Technora® has continuous filament yarns, chopped fibers, staple fibers, and spunnized yarns. The type of Technora® can be chosen according to applications.

Table 9.5 presents the typical properties of Technora® and PPTA aramid yarn. Note that the yarn has typically 24.5 cN/dtex tenacity, 530 cN/dtex tensile modulus, and 2 % equilibrium moisture regain under the standard condition.

Figure 9.10 shows the stress-strain curves of several fibers. As shown in the figure, para-aramid fibers are quite different from glass fiber, steel fiber, polyester fiber, and nylon fiber in terms of the high tenacity, high modulus, and low elongation at break. To be more precise, the tenacity of Technora® is more than five to six

Table 9.5 Standard performance of Technora® and PPTA

Property	Unit	Technora®	PPTA
Color	–	Gold	Yellow
Density	g/cm^3	1.39	1.44
Filament diameter	μΦ	12	12
Tensile strength	Kg/mm^2	24.5	20.3
Tensile modulus	cN/dtex	530	490
Tensile modulus	Kg/mm^2	7500	6900
Tensile modulus	GPa	74	70
Elongation at break	%	4.5	3.6
Thermal decomposing temperature	°C	>500	>500
	F	>932	>932
Heat of combustion	cal/g	6800	8400
Specific heat (25 °C)	Cal/g·°C	0.26	0.34
Limited oxygen index	–	25	29
Equilibrium moisture regain	%	2	6

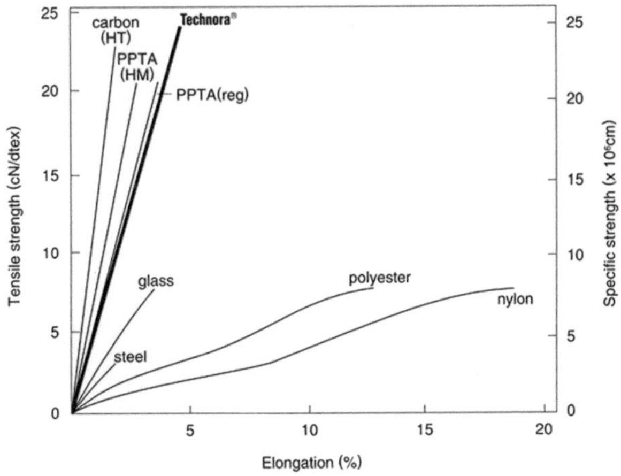

Fig. 9.10 Stress-strain curves (per weight)

times of steel wire and twice of polyester, nylon, and glass fibers. Technora yarn has high initial modulus of 530 cN/dtex which is almost twice that of steel wire or fiber glass. It may be worth to note that Technora has higher tenacity than that of PPTA.

9.10 Structure and Morphology

There are some proposed crystalline models about PPTA fibers like a skin-core model [10], a paracrystalline structure model [11], and a pleat structure model [12]. Although there are some differences in the model, characteristics common to all are the highly oriented crystalline and the presence of periodical pleat structure

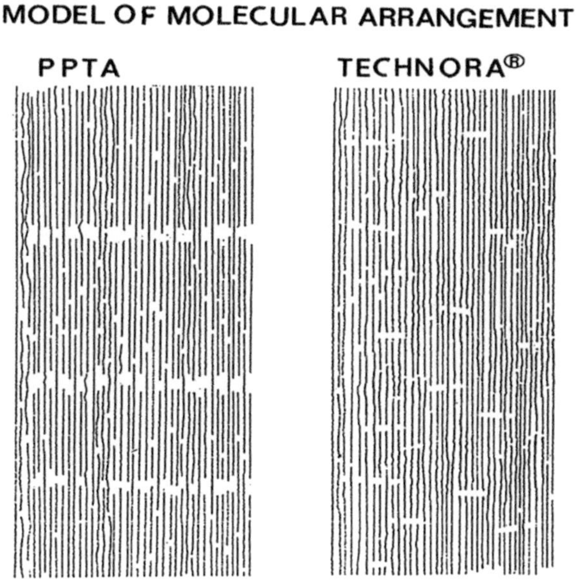

Fig. 9.11 Model of molecular arrangement of PPTA and Technora® [9]

Table 9.6 Crystalline structure parameter

Item	Unit	Technora®	PPTA
Density	g/cm^3	1.39	1.44
Crystallinity	%	66	66
Orientation	%	93	91
Crystalline size (D200)	A	32	46

along the fiber axis. As for Technora®, there are no proposed crystalline models. However, Tashiro et al. proposed the model of molecular arrangement in Fig. 9.11 as the results of X-ray analysis [9]. Technora® may have highly ordered molecular along the fiber axis, and no periodical defects, while PPTA has periodical defects along the fiber axis. It may be related with the randomness of copolymer [13]. It will be discussed later (Table 9.6).

9.11 Chemical Resistance

The chemical resistance is one of most important characteristics of Technora®, which is related with crystalline structure. As explained in 5.10 Structure and Morphology, PPTA has periodical defects in the structure. The periodical defects mean the amorphous region may have an effect on its chemical resistance in the sense to be easily attacked by chemicals. On the other hand Technora has relatively dense layers in its surface and homogenous molecular arrangement without clear periodical defects.

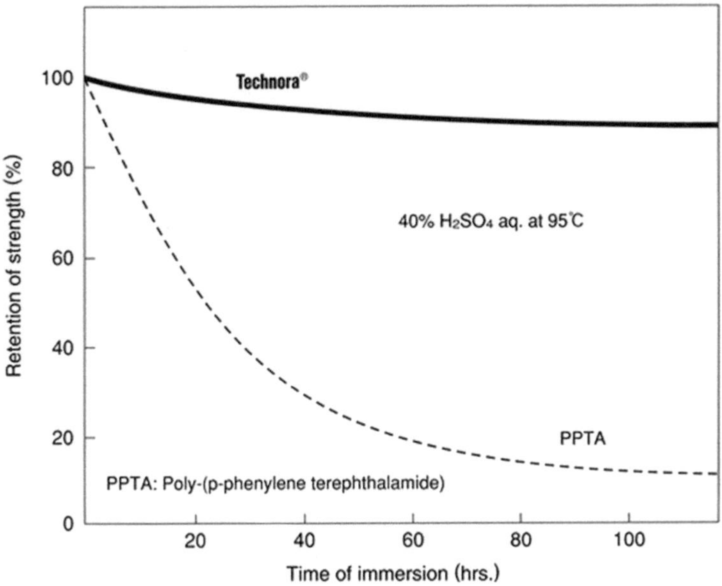

Fig. 9.12 Acid resistance

It is well known that PPTA fiber is vulnerable to hydrolysis under acidic conditions or alkaline conditions, and the amide group in the PPTA chain is set off the bond cleavage according the reaction below:

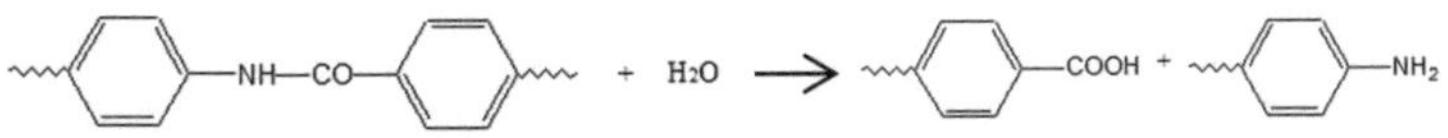

The hydrolytic degradation causes the reduction of the tenacity of PPTA fiber.

Characteristically Technora® possesses the strong chemical resistance against strong acids such as sulfuric acid, hydrochloric acid, and phosphoric acid. Figure 9.12 shows the acid resistance of Technora® and PPTA. Under the severe condition of 40 % sulfuric acid at 95 °C, Technora® keeps the retention of strength more than 90 % after 100 h, where PPTA has only 10 % of the retention.

Similarly Technora® possesses the strong chemical resistance against alkali substances such as sodium hydroxide, Portland cement, and sodium hypochlorite. Figure 9.13 exhibits the alkali resistance of Technora® and PPTA. Under the serious condition of 10 % caustic soda at 95 °C, Technora® keeps the retention strength of about 80 % after 100 h, where PPTA has only 20 % of the retention.

Also Technora® possesses the strong steam resistance. Figure 9.14 exhibits the steam resistance in 120 °C saturated steam. Technora® fiber does not change the tenacity even after 400 h. However, PPTA fiber loses the strength till 20 % as well as polyester fiber. Figure 9.15 shows Technora® has higher temperature resistance than PPTA. It might be possible to think that salts in PPTA fibers act as a catalyst.

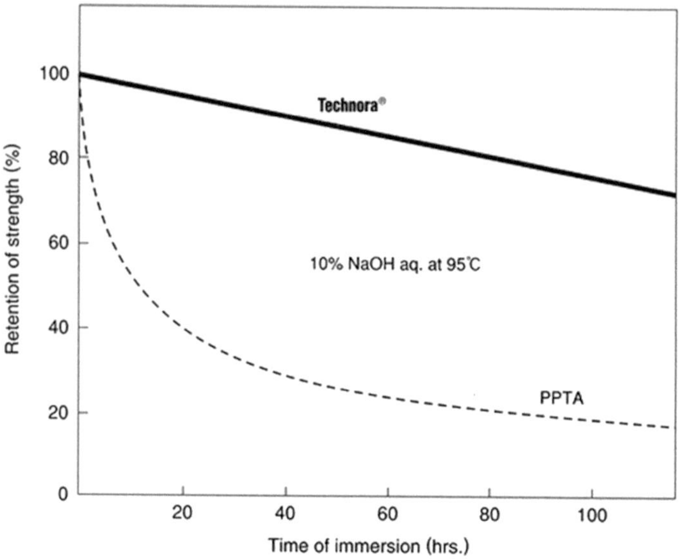

Fig. 9.13 Alkali resistance

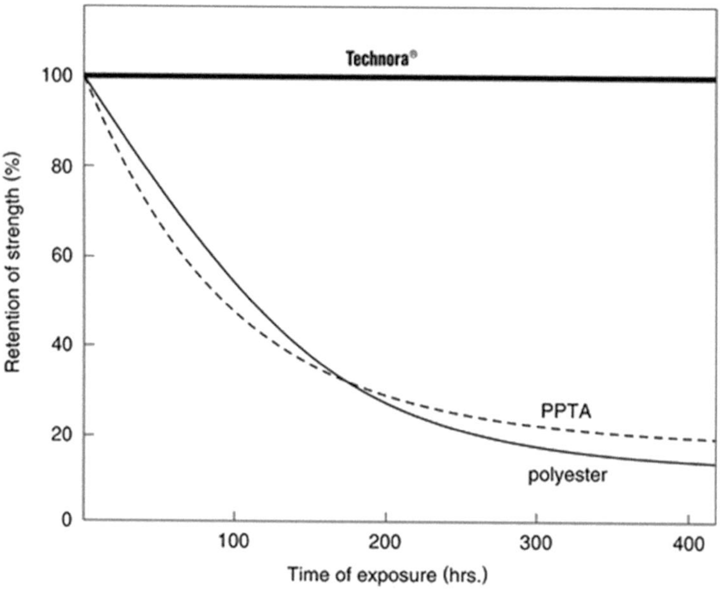

Fig. 9.14 Steam resistance (in 120 °C saturated steam)

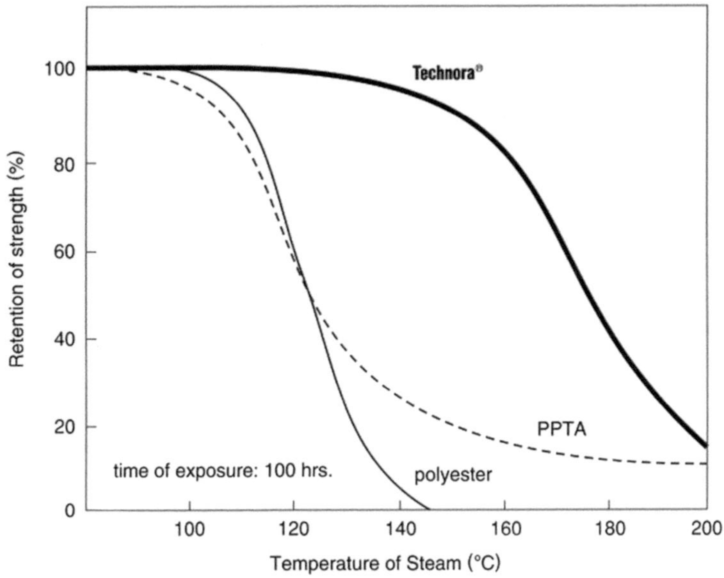

Fig. 9.15 Steam resistance (in 100–200 °C saturated steam)

Table 9.7 Chemical resistance (1)

Chemicals	Conc. (wt.%)	Temp. (°C)	Time (h)	Technora®	PPTA reg	PPTA HM	MPIA (FY)	PET
Sulfuric acid	20	95	20	99	13	50	65	–
	20	95	100	93	2	29	20	–
	40	95	100	89	1	20	2	–
Hydrochloric acid	20	20	100	98	42	87	74	–
Nitric acid	10	20	100	99	52	90	77	–
	20	70	366	14	–	–	–	–
Phosphoric acid	10	20	100	96	11	57	84	–
Sodium hydroxide	10	95	20	93	15	38	20	–
	10	95	100	75	4	18	0	–
Portland cement	Saturated	95	100	98	23	87	60	2
	Saturated	180	15	70	13	25	50	0
Sodium hypochlorite	10	95	20	95	8	–	–	–
Hydrofluoric acid	100 ppm	20	336	100	99	–	96	–
Seawater	3	95	1000	98	85	89	98	64
Water	100	100	100	100	90	–	–	96
	100	160	100	84	17	–	–	0
	100	180	100	40	10	–	–	0

*Figures show retention of strength (%)

Table 9.8 Chemical resistance (2)

Chemicals	Conc. (wt.%)	Temp. (°C)	Time (h)	Technora®	PPTA reg	PPTA HM	MPIA (FY)	PET
Formic acid	90	95	100	82	10	32	62	–
Acetic acid	40	95	100	97	10	70	90	–
Gasoline	100	20	784	>95	>95	>95	>95	–
Benzene	100	20	784	>95	>95	>95	>95	–
Para-xylene	100	95	100	97	94	92	100	99
Methyl ethyl ketone	100	20	1000	97	94	92	100	100
Ethyl acetate	100	20	1000	96	96	95	97	100
Ethylene chloride	100	20	1000	100	94	96	100	100
Ethylene glycol	100	95	300	94	96	92	90	63
Phenol	100	95	300	95	97	95	95	0
N-Methyl pyrrolidone	100	95	100	31	96	93	16	–

*Figures show retention of strength (%)

Tables 9.7 and 9.8 show the chemical resistance of Technora® fibers and PPTA fibers compared to MPIA (meta-phenylene isophthalamide) fibers and PET fibers.

The chemical resistance against Portland cement is an important characteristic for Technora®, because the cement reinforcement is one of the potential markets for Technora® fibers. Actually, under the serious condition of saturated Portland cement at 95 °C, Technora® keeps the retention strength of about 98 % after 100 h, where PPTA has only 23 % of the retention.

Technora® fiber has enough chemical resistance against organic solvents except NMP.

9.12 Fibrillar Structure and Fatigue Resistance

As shown in Table 9.9, the abrasion durability life and tube durability life of aramid are much shorter than PET.

Generally aramid fiber fibrillates easily by abrasion because it contains a highly ordered fibrillar structure. The tendency for fibrillation can be caused by the lack of lateral forces, except for the van der Waals and the hydrogen bond forces between polymer chains.

Although Technora® fiber is weaker than polyester fiber in the abrasion durability and tube durability life, Technora® is stronger than PPTA in the abrasion durability and tube durability life. Because it can be thought that Technora® has smaller crystallines than PPTA and does not have highly oriented skin layer in the surface of the fibers as PPTA has, Technora is suited for the rubber reinforcement,

Table 9.9 Fatigue and durability of Technora® [14,15]

Fatigue and durability		Technora®	PPTA	PET	Test method and conditions
Abrasion durability	Cycles	360	130	1600	Method: Cross abrasion
Life to breakage					Cord: 1670 dtex/1 × 2
Fiber to fiber		790	200	4000	Weight: 600 g
Fiber to metal					Load: 0.18 cN/dtex
Flexural fatigue	%	52	36	99	Method: Fatigue on roll
Retention of strength					Cord: 1670 dtex/1 × 2
					Diameter: Roll/Cord = 11/0.7 mm
					Load: 0.53 cN/dtex
					Number of cycles: 2000
Disc fatigue	%	85	85	90	Method: Rotary disc
Retention of strength					Cord: 1670 dtex/2, 40 × 40 turns/10 cm
					Strain: Stretch/compression = 6/12 %
					Temperature: 20 °C
					Time: 72 h
Tube durability	Cycles	30×10^5	10×10^5	80×10^5	Method: Mallory tube
Life to breakage					Cord: 1670 dtex/2, 40 × 40 turns/10 cm
					Angle of bending: 80°
					No. of ends: 32 ends/in.
					No. of rotation: 850 rpm

like reinforce cords of rubber timing belts for automobiles which need high tenacity, high heat resistance, and tube durability.

9.13 Polymer Sequence Distribution Analysis

Because Technora® polymer is the condensation co-polyamide prepared irreversibly in nonequilibrium state in N-methyl-2-pyrrolidone solution, the sequence is not necessarily random. It is important whether the sequence of the Technora® polymer is random or blocky in connection with the physical properties such as solubility or hydrolytic stability. As shown in Fig. 9.16, the dyad sequence analysis of Technora® was reported using ^{13}C NMR in trifluoromethanesulfonic acid at 100 °C. For the detailed peak assignment, some model copolymers were synthesized and the ^{13}C NMR spectra were observed. Well-resolved carbonyl carbon peaks of the terephthalic units of Technora® were observed, which made it possible to analyze

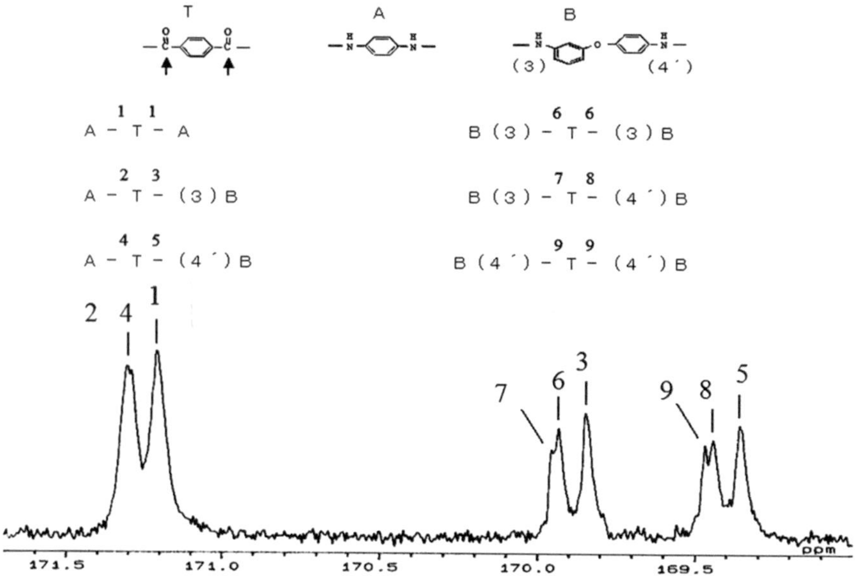

Fig. 9.16 ^{13}C NMR spectra of Technora® polymer and the model polymers in CF_3SO_3H at 100 °C

Table 9.10 Results of the sequence distribution analysis of the Technora® polymer by ^{13}C NMR[a]

Dyad sequences	Fractions[b]		Randomness	Average sequence lengths
A-T-A	f_{AA}	0.248(0.250)	R_{AB} 1.03(1.00)	L_A 1.96 (2.00)
A-T-B	f_{AB}	0.514(0.500)	R_{BB} 0.99(1.00)	L_B 1.93 (2.00)
B-T-B	f_{BB}	0.238(0.250)		
A-T-(3)B	f_{A3B}	0.499(0.500)		
A-T-(4')B	$f_{A4'B}$	0.501(0.500)		
B(3)-T-(3)B	f_{3B3B}	0.251(0.250)		
B(3)-T-(4')B	$f_{3B4'B}$	0.496(0.500)		
B(4')-T-(4')B	$f_{4'B4'B}$	0.253(0.250)		

[a]The values in parentheses were theoretical values on the basis of a Bernoullian dyad distribution
[b]The molar fractions were calculated as $f_{AA}+f_{AB}+f_{BB}=1$, and $f_{3B3B}+f_{3B4'B}+f_{4'B4'B}=1$.

the p-phenylenediamine units and 3,4′-DAPE units at dyad level. The information on the head-to-head, tail-to-tail, and head-to-tail sequence of the 3,4′-DAPE units was also obtained. It was concluded that the sequence of Technora® was completely random [13] (Table 9.10).

9.14 Conclusion

The history and fundamental principles of Technora® have been explained. The process, product types, chemical and physical properties, structures and morphologies were discussed comparing to PPTA fibers.

In December 2000, Teijin merged the Twaron® business commercialized by Akzo BV of the Netherlands in 1987. Twaron® is made from PPTA and has the same composition as Kevlar®. As the results of that, Teijin has two types of para-aramid fibers. The main markets of Twaron® are pulp, ballistics, and optical fiber cable, and the main markets of Technora® are mechanical rubber goods, hose reinforcement for deep sea oil, and reinforcement of concrete. The markets seem to be segregated for them, based on their characteristics.

Because para-aramid fits the trend for maintenance free, light weight, and energy saving, the demand of para-aramid fibers including Technora® and Twaron® (PPTA) fibers has increased exponentially in the last 25 years. In 2020 the demand is anticipated to be increased by 5–7 % per year and the size of the para-aramid market will be 60,000 tons a year. In the future new products and new end uses of para-aramid can be expected for a sustainable society. The technological development will be proceeded to be suited for the new applications. Para-aramid fibers will not only be high performance fibers but also grow in importance in our society.

References

1. S. Ozawa, Y. Nakagwa, K. Matsuda, T. Nishihara, H. Yunoki, U.S. Pat. 4,075,172, (1978)
2. K. Matsuda, Polym. Prepr. Am. Chem. Soc. Div. Polym. Chem. **20**, 122 (1979)
3. S. Ozawa, Polym. J. **19**, 119 (1978)
4. J.A. Zapp Jr., Science **190**, 422 (1975)
5. S.L. Kwolek, US Pat. 3,671,542 (1972); US Pat. 3,819,587 (1974)
6. H. Blades, US Pat. 3,767,756 (1973); US Pat. 3,869,429 (1975)
7. H.H. Yang, US pat. 4,340,559 (1980)
8. H.H. Yang, 'Aramid fibers', in *Fiber Reinforcement for Composite Materials*, ed. by A.R. Bunsell (Elsevier, Amsterdam, 1988), pp. 249–329; 'Aromatic High-Strength Fibers', (Wiley-Interscience, New York, 1989)
9. K. Tashiro, Y. Nakatsu, T. Ii, M. Kobayashi, Y. Chatani, H. Tadokoro, Sen-I Gakkaishi **43**, 627 (1987)
10. R.J. Morgan, C.O. Pruneda, W.J. Steele, J. Polym. Sci. Polym. Phys. Ed. **21**, 1757 (1983)
11. R. Barton, J. Macromol. Sci. Phys. B. **24**, 119 (1985)
12. M.G. Dobb, D.J. Johnson, B.P. Saville, J. Polym. Sci. Polym. Phys. Ed. **15**, 2201 (1977)
13. H. Matsuda, T. Asakura, Y. Nakagawa, Macromolecules **36**, 6160 (2003)
14. JIS L 1017 (2002)
15. ASTM D6588/D6588M-11

Chapter 10
Vectran®: Super Fiber from the Thermotropic Crystals of Rigid-Rod Polymer

Hideki Hoshiro, Ryokei Endo, and Forrest E. Sloan

Abstract Vectran® is a high-performance multifilament yarn spun from liquid crystal polymer (LCP) which is the only commercially available melt spun LCP fiber in the world. Vectran® exhibits several unique properties such as excellent mechanical properties and high abrasion resistance and so on. In this article, we firstly show Vectran® and its several properties, and then we introduce recent research as for the crystal structural change on thermal treatment. Finally, we introduce composite applications of Vectran®.

Keywords Thermotropic liquid crystal • Vectran • Crystal structure • Composite materials

10.1 General Introduction

Recently there is much interest in liquid crystal polymers (LCPs) as useful polymeric materials. Because LCPs show high modulus and high strength, superior heat resistance and excellent chemical stability were compared to other conventional polymers [1–3]. Therefore, LCPs have been widely used in a broad range of industrial application such as automotive, electric, and electronic fields [4]. Industrial materials have recently required high performance, and hence there has been increasing demand for LCPs. As a result, LCPs also have attracted much attention as future materials.

It is well known that LCPs are classified as either lyotropic or thermotropic liquid crystals. The former shows liquid crystalline nature in the solvent at a certain temperature and concentration. In contrast, the latter utilizes thermal change under cooling and heating. The temperature range at which liquid crystalline nature is observed depends on the chemical structure of the thermotropic LCPs [5].

H. Hoshiro (✉) • R. Endo • F.E. Sloan
Fibers and Textiles Company, Kuraray Co., Ltd., Kurashiki, Japan
e-mail: hideki_hoshiro@kuraray.co.jp

Society of F.S. Technology, Japan (ed.), *High-Performance and Specialty Fibers*,
DOI 10.1007/978-4-431-55203-1_10

In general, wholly aromatic polyamides like a Kevlar belong to the lyotropic LCPs. Since these polymers have a high melting temperature which is close to their decomposition temperature, they are generally spun from a solution in a special solvent. Thermotropic LCPs include wholly aromatic polyesters, and they can be processed in a variety of ways, such as injection molding, extrusion molding, and melt spinning. The spinning of wholly aromatic polyester is quite simple, although wholly aromatic polyamide must be spun under several complicated steps from the original polymer to the final fiber product. Consequently, thermotropic LCP fibers have attracted much attention from an economic and environment standpoint.

Kuraray is the world's only one supplier of Vectran® which is wholly aromatic polyester fiber. Its capacity is 1000 tons per year. Kuraray has also completed an expansion project at its manufacturing plant in Saijo, Japan, and also brought additional capacity online at its Fort Mill, SC.

Vectran® is suitable for performance and safety applications that require strength, durability, and dimensional stability. Kuraray supplies the fiber to a wide variety of manufacturing segments including aerospace, military, industrial rope and cable, and composites for such end uses as NASA space systems [6], inflatable wind turbine generators, and helicopter slings.

In this article, first of all, we explain the characterization of Vectran®. Secondly, we introduce the latest research of crystal structure of Vectran® and its composite application.

10.2 Characterization of Vectran®

10.2.1 Fiber Chemistry

Vectran® offers a balance of properties unmatched by other high-performance fibers. This unique fiber's history spans 30 years of research and development in thermotropic LCPs [5, 7–9]. The molecules of thermotropic LCPs are stiff, and its rodlike structure makes the self-assembled order that shows an anisotropic behavior even in the melt state. Figure 10.1 shows structural formation of Vectran® during the spinning process. Vectran® is formed by melt extrusion of the LCP through fine-diameter capillaries. Then the molecular domains orient parallel to the fiber axis. After the melt spinning, solid-state polymerization is conducted at temperature range from 240 to 320 °C in order to develop the fiber strength [11].

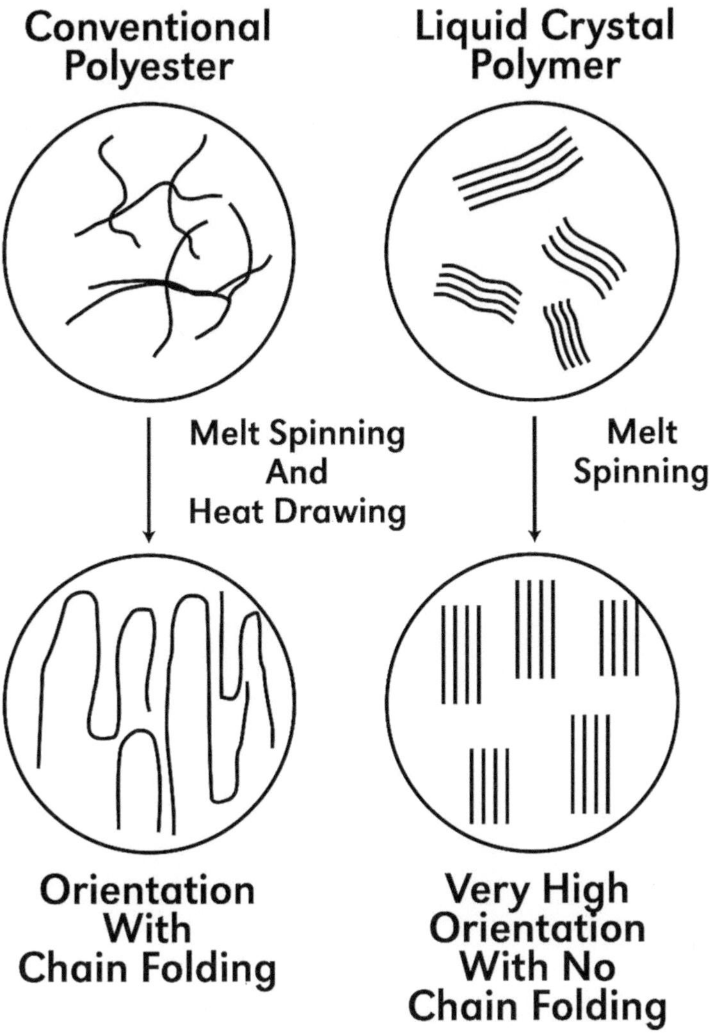

Fig. 10.1 Schematic illustration of deformation process of molecular chain during melt spinning for conventional polyester and liquid crystal polymer [10]

10.2.2 Molecular Structure

Vectran® is thermotropic liquid crystalline copolyesters composed of 4-hydroxybenzoic acid (HBA) and 2-hydroxy-6-naphthoic acid (HNA) monomers in a molar mass ratio of HBA/HNA = 73/27 as shown in Fig. 10.2.

With conventional polyethylene terephthalate (PET), the molecular chains are flexible. The as-spun fiber from such polymer shows low molecular orientation and its strength is not good enough for the industrial uses. In order to increase the

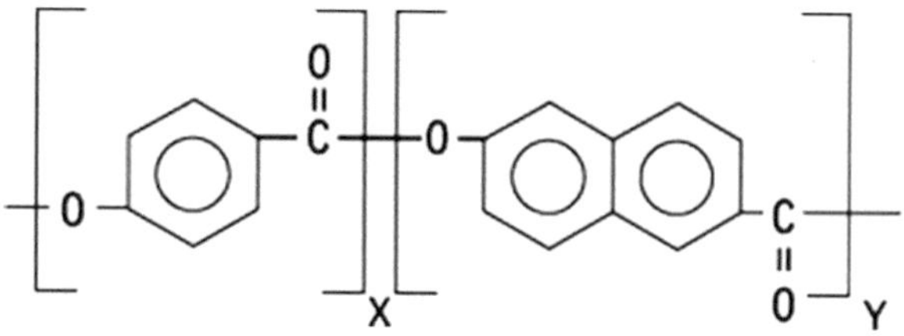

Fig. 10.2 Chemical structure of Vectran, X/Y = 73/27

strength of that fiber, the subsequent drawing is necessary. However, Vectran® having liquid crystalline nature is locked in directly during the melt-spinning process, thanks to the molecular structure and liquid crystalline nature of the starting polymer.

Vectran® is different from other high-performance fibers such as aramid and ultrahigh molecular weight polyethylene (UHMW-PE). Vectran® is melt spun above its melting temperature. On the other hand, aramid like Kevlar is solvent spun due to its lyotropic nature. UHMW-PE fiber is generally gel spun in order to control the density of chain entanglements.

10.2.3 Mechanical Properties

Vectran® offers a distinct advantage over traditional metals in terms of strength-to-weight ratios as shown in Table 10.1 due to the light weight of an organic material. Table 10.2 also shows the mechanical properties of Vectran® compared with the properties of other industrial fibers.

10.2.4 Thermal Properties

The excellent heat resistance of Vectran® is summarized below and in Table 10.3:

- Good limiting oxygen index (LOI) and low-smoke generation
- Low thermal shrinkage in hot air, boiling water, and laundry process

Table 10.1 Comparison of mechanical properties of various engineering materials

Materials	Density (g/cm^3)	Tensile strength (GPa)	Specific strength (km[a])	Tensile modulus (GPa)	Specific modulus (km[b])
Vectran®	1.4	3.2	229	75	5300
Titanium	4.5	1.3	29	110	2500
Steel	7.9	2.0	26	210	2700
Aluminum	2.8	0.6	22	70	2600
E-glass	2.6	3.4	130	72	2800

[a]Specific strength = strength/density (also divided by force of gravity for SI units). Also known as breaking length, the length of fiber that could be held in a vertical direction without breaking
[b]Specific modulus = modulus/density (also divided by force of gravity for SI units). This measure increases with increasing stiffness and decreasing density

Table 10.2 Comparison of mechanical properties of various industrial fibers

Materials	Density (g/cm^3)	Tensile strength (GPa)	Tensile modulus (GPa)	Elongation at break (%)	Moisture regain (%)
Vectran®	1.40	3.4	70	4	<0.1
PET	1.38	1.1	14	15	<0.5
Nylon	1.14	1.0	10	20	6–8
Aramid	1.44	3.0	65	4	4–6
UHMW-PE	0.95	3.4	110	4	<0.1

Table 10.3 The thermal properties of Vectran® compared with aramid

	Vectran®	Aramid
LOI	28	30
Melting temperature, °C	320	None
Heat shrinkage at 180 °C for 30 min, %	<0.2	<0.2
50 % strength retention temp., °C	145	400
TGA (20 % weight loss), °C	>450	>450

– No dripping in vertical flammability tests
– Good strength retention after hot air and radiant energy exposure
– Small negative coefficient of thermal expansion
– Excellent property retention in a broad temperature range

Retention of mechanical properties during or after thermal exposure is a key concern in many applications. Most commonly, high temperatures are encountered during a downstream processing step, such as coating or laminating. Care must be taken to minimize line tensions or other mechanical loads during the high-temperature step. That shown in Fig. 10.3, which describes the tensile strength loss at the elevated temperatures, should be used as a reference in selecting process conditions. For high-temperature processing at low mechanical load, Fig. 10.4 shows that Vectran® will have excellent strength after processing. In fact, it is superior to aramid.

Mechanical properties of Vectran® at low temperature were evaluated by ILC Dover company during the design of the airbag system for the 1997 Mars Pathfinder mission. ILC reported that Vectran actually increased in strength in tests at −62 °C, leading to its selection for the airbag fabric and external assembly tendons [12]. This distinguishing characteristic of Vectran® is shown in Fig. 10.5.

10.2.5 Creep Property

Creep property is very important for the use of long-term static loading. Resistance to creep (or its static-strain complement, stress relaxation) is a critical design

Fig. 10.3 Temperature dependence of relative strength of Vectran® (●). The data of aramid (■) are also included for comparison

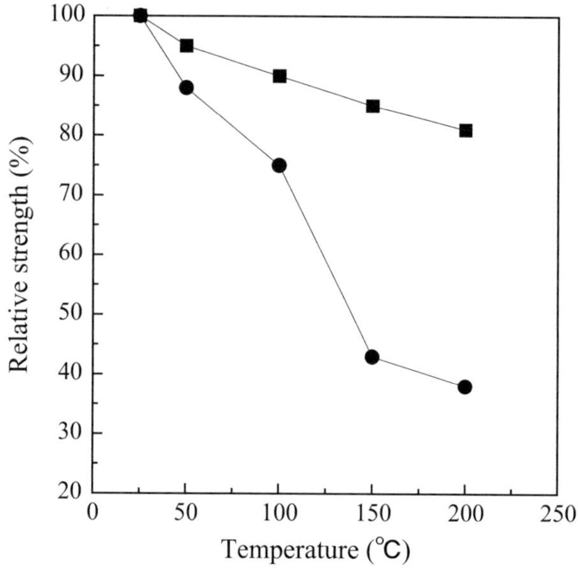

Fig. 10.4 The behavior of relative strength of Vectran ® (●) after thermal exposure for 4 days at each temperature. The data of aramid (■) are also included for comparison

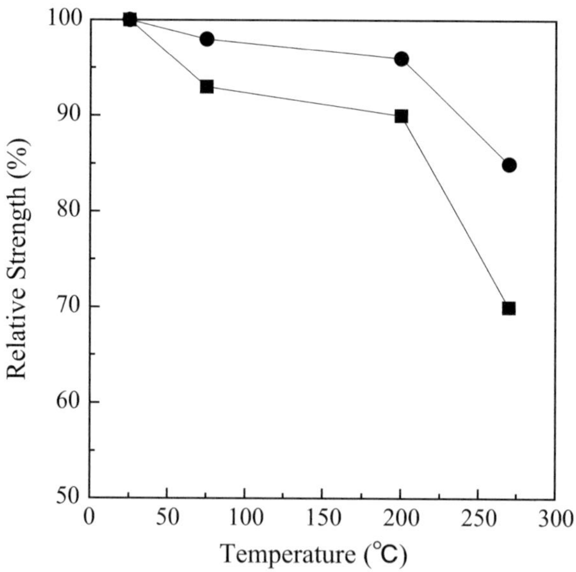

specification for the material selection of many applications requiring long-term dimensional stability.

Figure 10.6 shows the definitely small creep rate of Vectran® as compared with aramid. These tests ran for as long as 10,000 h at ambient temperatures.

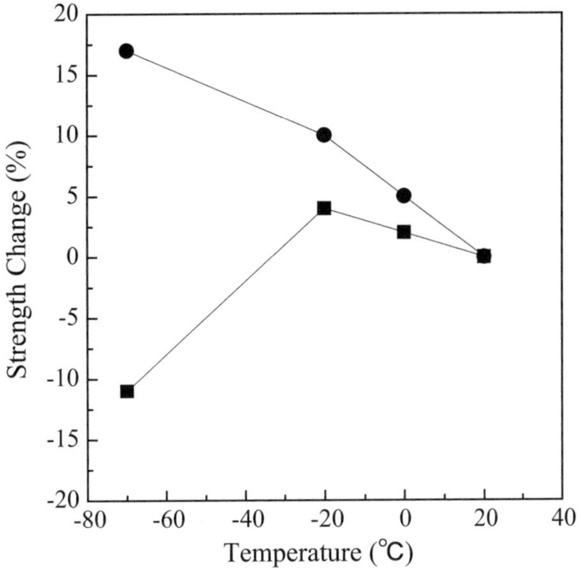

Fig. 10.5 The strength change of Vectran® (●) at low temperature. The data of aramid (■) are also included for comparison

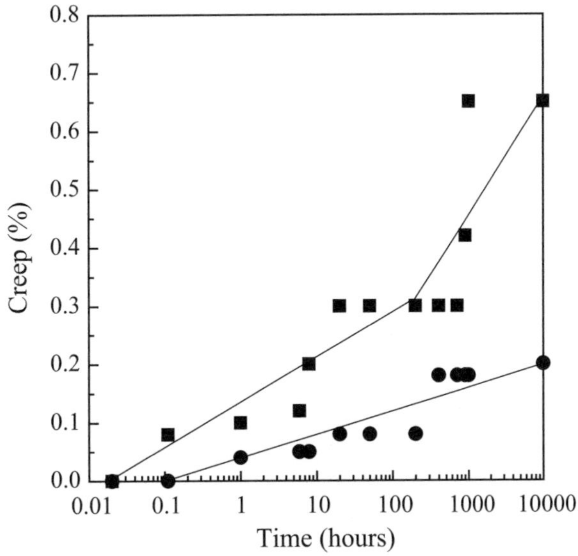

Fig. 10.6 Creep behavior of Vectran® (●) at ambient temperature (30 % of break load). The data of aramid (■) are also included for comparison

10.2.6 Vibration Damping

The comparison of vibration damping characteristics for glass fiber, carbon fiber, aramid, and Vectran® is shown in Fig. 10.7. The differences are apparent and demonstrate that Vectran® is ideal for vibration damping in sporting goods.

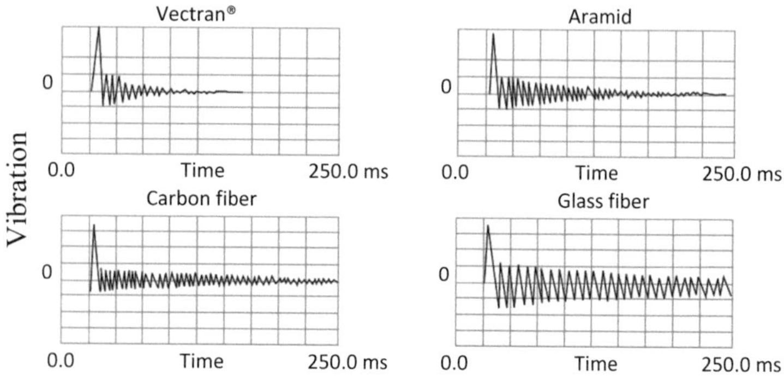

Fig. 10.7 Vibration damping of various fibers

Table 10.4 Audio engineering data for various metals and composites

Materials	Density (g/cm^3)	Speed of sound (m/s)	Elastic modulus (GPa)	Modulus rigidity (%)	Internal loss (%)
Vectran®	1.40	4288	28	8.2	0.070
Carbon fiber	1.42	6902	68	15	0.035
Glass fiber	2.00	3216	21	20	N/A
Titanium	4.54	4773	103	1.1	0.002
Steel	7.90	5125	207	0.4	0.002
Magnesium	1.74	5000	44	8.3	0.004

Table 10.4 presents a list of performance characteristics of various metals and composite materials used by a manufacturer of audio components.

10.2.7 Cut Resistance

Many and various cut-resistant methods were prepared. The uniformity of test sample and the cutting edge are critical in all tests. For in-house cut resistance comparisons, a Sintech tensile testing machine attached with the cutting blade and the knitted hose leg holder was used as shown in Fig. 10.8.

Inspection of the round blade to assure a clean cutting edge is critical. Table 10.5 compares the cut resistance of various fibers.

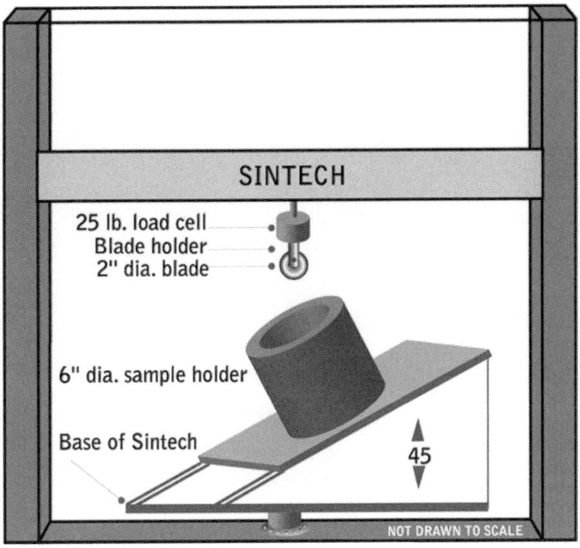

Fig. 10.8 Schematic illustration of the cut resistance test apparatus using a tensile testing machine

Table 10.5 Sintech cut resistance of Vectran®. The data of other materials are also included for comparison

Materials	Denier	Relative road
Vectran®	1500	3.4
Aramid	1500	1.1
UHMW-PE	1500	1.0

10.3 Crystal Structure of Vectran®

10.3.1 Introduction

In the case of Vectran®, it is well known that mechanical properties should be developed by annealing at high temperature. However, little is known about the mechanism of formation of crystal structure during the annealing process. Recently we have analyzed correlation between annealing temperature and the crystal structure using WAXD method. Therefore we introduce our latest study in this section.

10.3.2 Crystal Structural Change on Annealing Process

Figure 10.9 shows the WAXD photographs for the as-spun fiber and the annealed fibers. All WAXD shows well-oriented fiber patterns. However additional diffraction spots are observed on WAXD pattern of the sample annealed at 260 °C.

Figure 10.10 shows the WAXD equatorial profiles for samples annealed at indicated temperatures for 12 h. In case of samples annealed beyond 240 °C, the (200) reflection was observed. Figure 10.11 shows the lattice parameters (a, b) as a

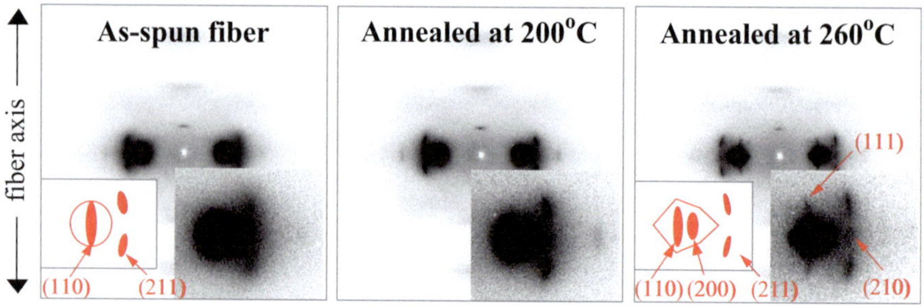

Fig. 10.9 WAXD photographs for as-spun fiber and annealed samples

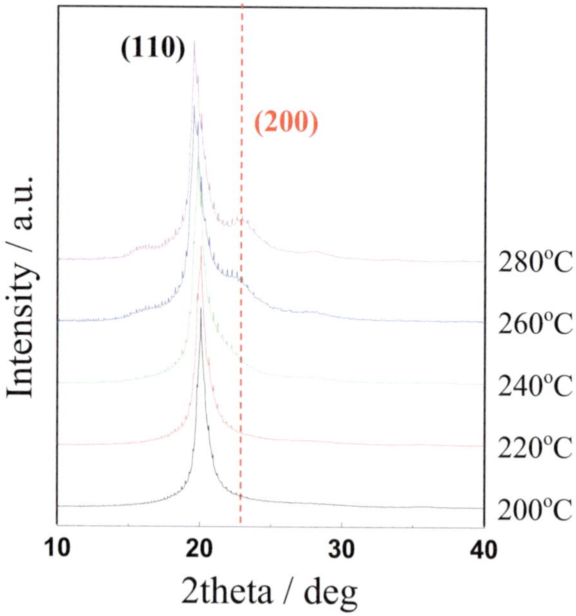

Fig. 10.10 WAXD profiles for samples annealed at several temperatures

function of annealing temperature. The crystal structures for as-spun fiber and the samples annealed below 240 °C are defined as the hexagonal. In contrast, the crystal structure for the sample annealed beyond 240 °C is defined as the orthorhombic.

Kaito et al. also reported that, in the case of Vectran having random orientation, crystal structures changed from hexagonal to orthorhombic on annealing process [13]. According to their study, this change is attributed to the formation of HBA-rich crystal caused by chain-sliding diffusion upon annealing process. Thus it is expected that the lattice parameter of chin axis also change. Figure 10.12 shows lattice parameter, *c*, as a function of annealing temperature.

Lattice parameters (*a*, *b*, and *c*) of HBA crystal were also indicated in the figure. It was found that the lattice parameter, *c*, of Vectran® having high orientation does

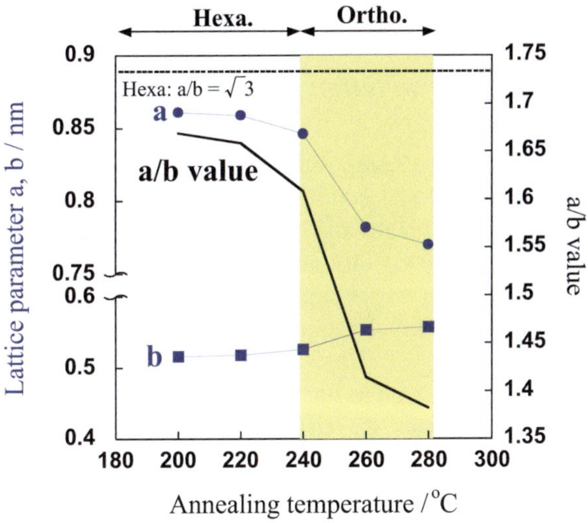

Fig. 10.11 Lattice parameters (a, b) as a function of annealing temperature

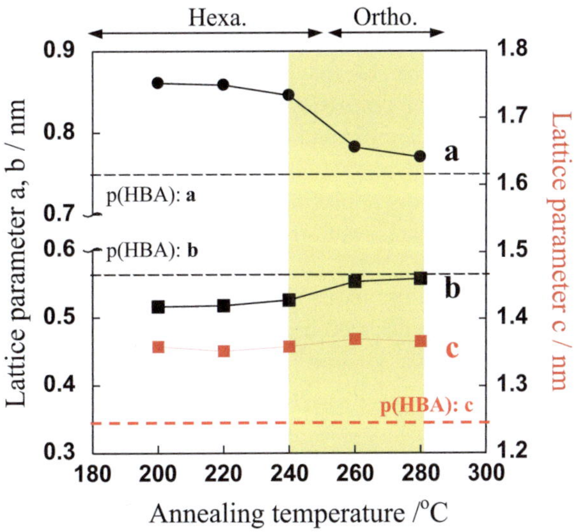

Fig. 10.12 Lattice parameter, c, as a function of annealing temperature

not change irrespective of annealing temperature. Though the reason for the difference in our study and that of Kaito et al. is not yet clear, but one possible explanation may be that the increase of dipole-dipole interaction which is caused by orientation suppresses the chain-sliding diffusion [14].

10.4 Composite Application of Vectran®

10.4.1 Introduction

Fiber-reinforced composites are routinely used for structures ranging from civil engineering to spacecraft. Fibers for these structural composites should ideally exhibit high strength and stiffness in both the tensile and compressive directions, to provide high rigidity and to resist failure in bending. However, there are also many end uses where high tensile strength is required, but flexibility is needed rather than rigidity. Examples include sailcloth, lighter-than-air craft, and inflatable structures [15–18]. These flexible composites are typically composed of one or multilayers of a plain open-weave fabric coated with a high-elongation polymer matrix, but there are also many nonwoven constructions. Because these thin composites offer little resistance to folding or buckling, high-fiber compressive strains occur during storage and deployment.

As an example, consider the airbeam structure of Fig. 10.13 [19]. The inflatable beams in this structure are roughly 90 cm in diameter, span a width of more than 26 m, and are more than 36 m in overall length. Yet they are transported as tightly packed rolls in a space less than 1 m^3. The coated fabrics that make up these airbeams must be able to withstand repeated packing/folding cycles with little to no loss of physical properties and no changes in inflated shape, even during wide swings of temperature and humidity during transport.

Vectran® is used today in the production of carbon/epoxy composites as preform stitching threads, replacing aramid whose high moisture regain can lead to microcracking and swelling [20–25]. Vectran® is also used in hybrid constructions with carbon and glass, to provide vibration damping and/or a more ductile or controllable failure mode. There are also specialty applications where Vectran® is used alone in reinforced fiber with standard epoxies or vinyl esters. And as will be discussed herein, Vectran® has been used for many years as reinforcing materials for flexible composites, because of their unique combination of critical performance parameters.

Fig. 10.13 Flexible composite pressurized airbeams and shipping container (Photos courtesy HDT Global company)

The purpose of this section is to summarize recent Vectran® test data relevant to performance requirements in flexible composites. The intended audience is composite engineers working with high-strength coated fabrics in demanding flexible applications or in end uses demanding durability and resistance to repeated bending. The paper will present material property data and describe the importance to success in many flexible composites.

The most important requirement for durability of a flexible composite is strength retention after high-cycle repeated flexing/folding. Other mission-specific needs are dimensional stability and compatibility with environmental conditions, e.g., temperature range and changing moisture conditions.

10.4.2 Textile Fibers for Flexible Composite

Many flexible composites occur in nature, e.g., leaves, bladders, skin, etc. All tend to have a fibrous nature within a polymeric or resinous membrane. And just as in most engineered composite applications, specific materials and structures evolved in response to steadily increasing requirements for loading and environmental compatibility.

Designers typically must trade off benefits and advantages of different materials in order to objectively select the most efficient composite. There are many possible combinations of fiber, films, and resins available. One example of a weighted property analysis is shown in Table 10.6, for comparison of reinforcing fiber options for the inflatable radome project shown in Fig. 10.14. Specific product performance requirements are listed and weighted, and then candidate materials are ranked on a 1-3-9 scale. A scoring system is used to help narrow down material choices.

Fibers with high tensile properties are not needed in all flexible composite applications. For example, inflatable children's jumping gyms can be made from standard polyester fibers. In fact the earliest inflatable aircraft were developed prior to the advent of "hi-tech" fibers.

High tensile strength is preferred where weight must be minimized and/or high internal pressures are anticipated. Military and aerospace end uses often have strict weight reduction targets due to transportation and deployment costs. Civil/emergency applications often must be designed for high system loads/pressures to account for temporary weather conditions, e.g., high wind or ice and snow accumulation on inflatable or tensile membrane structures.

The family of high-strength polymeric fibers includes oxazoles (PBO), aromatic polyamides (aramids), aromatic polyesters (LCP), and high molecular weight polyethylene (HMW-PE). Table 10.7 shows a list of physical properties for these fibers as well as for glass and a PAN-based carbon.

Tensile strength and modulus are very high in all these fibers, and all have found markets in primarily tensile applications. However note the much lower compressive properties in the polymeric fibers. Carbon and glass retain 60–80 % of their

Table 10.6 Example of material selection matrix

Solution specification	Wt.	Quartz			Glass			Aramid			LCP			HMPE		
			Rank	Total		Rank	Total		Rank	Total		Rank	Total		Rank	Total
Tensile strength	9	Excel	9	81	Good	3	27	Good	3	27	Good	3	27	Good	3	27
Tensile modulus	3	Good	3	9	Good	3	9	Good	3	9	Good	3	9	Good	3	9
Damage resistance	9	Poor	1	9	Poor	1	9	Good	3	27	Excel	9	81	Good	3	27
Compression flex fatigue	9	Poor	1	9	Poor	1	9	Good	3	27	Excel	9	81	Poor	1	9
Matrix compatibility	1	Good	3	3	Good	3	3	Good	3	3	Good	3	3	Poor	1	1
Moisture resistance	3	Good	3	9	Good	3	9	Poor	1	3	Good	3	9	Good	3	9
Temperature resistance	1	Excel	9	9	Excel	9	9	Good	3	3	Good	3	3	Poor	1	1
UV resistance	3	Good	3	9	Good	3	9	Poor	1	3	Poor	1	3	Good	3	9
Loss target	9	Excel	9	81	Poor	1	9	Good	3	27	Good	3	27	Excel	9	81
Dielectric constant	9	Good	3	27	Poor	1	9	Good	3	27	Excel	9	81	Excel	9	81
Total				246			102			156			324			254

Fig. 10.14 Air-supported flexible composite radome (Photo courtesy ILC Dover)

Table 10.7 Typical mechanical properties of several fibers

Fiber class	Carbon	Glass	PBO	Aramid	Vectran®	PE
Tensile properties (producer data)						
Modulus, GPa	250–500	70–90	180–270	72–106	75–105	70–110
Strength, GPa	3–4	3–5	5.8	2.8–3.2	3.0–3.2	2.4–3.3
Failure strain, %	1.5–2.5	5–6	2.5–3.5	2.6–3.6	2.8–3.8	2.9–3.4
Compressive properties						
Modulus, GPa	250–500	70–90	40–60	30–60	40–50	40–50
Strength, GPa	2–3	2–3	0.2–0.4	0.4–0.9	0.4–0.6	0.1–0.2
Failure strain, %	1–2	3–4	0.3–0.4	0.6–0.8	0.5–0.7	0.2–0.4

tensile properties in compression, compared with only 10–20 % for the polymeric fibers. Low compressive strength in technical fibers has been the subject of much study [26–29] and is the principal reason polymeric fibers have not replaced carbon and glass when designing composites for primary structural loads.

10.4.3 Flex/Fold Fatigue Resistance of Vectran®

In contrast to rigid structure, flexible composites are typically thin and thus easily deformed under a bending moment. If the bending loads, or bend radius, and thereby maximum compressive strains, can be controlled, then even brittle fibers can be used in flexible applications without damage. For example, thin-walled carbon/epoxy shells are rigid locally but can tolerate moderate bending excursions. Unfortunately thin-walled elements are easily buckled and after buckling are subject to low-radius folding which can lead to localized catastrophic failures. For this reason the durability of fibers to repeated folding/compressive strain is critical for long life in flexible composites.

Saito et al. studied compression fatigue in aramid and Vectran® single filaments [30–32]. Fibers were bonded to polycarbonate plates and observed during three-point bending of the plates. Critical compressive strains were defined by visual onset of kink-band formation. Critical strain in aramid was estimated at 1.0 % and for Vectran® 0.6 %, both strain levels much lower than typically observed during folding or buckling. Fibers were then subject to compression fatigue at strains of 50, 100, and 150 % of the critical value.

Vectran® data from this study are shown in Fig. 10.15. If strains were kept below critical levels where kink bands formed, residual tensile strength was largely unaffected by high-cycle compression fatigue. However, when applied strains exceeded critical values, a damage initiation period was observed followed by a damage propagation period. Vectran® exhibits a much longer period of damage initiation.

Similar results have also been observed in fabrics. Vectran® and aramid fabrics were folded over and creased with a 4.54 kg roller to simulate folding and tight container packing. After 100 cycles of creasing, the Vectran® fabric showed no significant strength loss. The aramid fabric lost approximately 20 % of its original tensile strength, which is consistent with the single filament data of Saito et al.

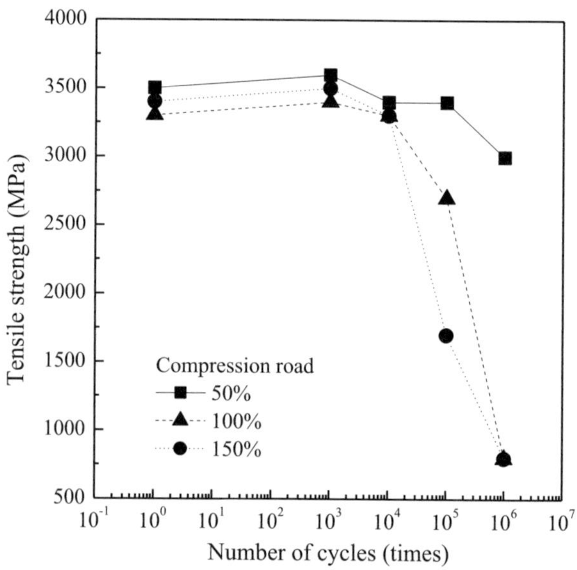

Fig. 10.15 Tensile strength of Vectran® after compression fatigue

A correlation exists between flex/fold fatigue resistance and compressive strength. Failure of material in compression is linked directly to shear strength, which for highly oriented polymeric fibers is similar to the transverse strength of the fiber. In UHMW-PE and Vectran®, there are few to no molecular bonds between polymer chains, only weak van der Waals forces. When these fibers are forced into compression, the strain is accommodated through formation of widely distributed kink bands, and little damage to load-bearing fibers result. In aramids, there are much stronger hydrogen bonds linking adjacent molecules.

The ability for Vectran® and fabrics to tolerate repeated bending/folding cycles without loss of tensile properties is key to their success in flexible composites.

10.4.4 Dimensional Stability of Vectran®

Resistance to creep elongation can be an essential property for long-term use of any composite material. In flexible composites the ability to return to a specified shape over many cycles of repeated deployments can be affected by dimensional instability. Tolerances on inflated structures can be tight, and changes in dimensions of long segments such as inflatable airbeams can lead to assembly delays and installation costs. Change of shape in inflatable airfoil sections can change lift characteristics, reducing system performance and service life.

Elastic properties of Vectran® coated and uncoated fabrics are stable and repeatable over many cycles of loading [33, 34]. Strain levels in LCP fibers and fabrics can also be monitored very accurately using Raman spectroscopy, which has useful ramifications for in-situ composite health monitoring.

Vectran® is well known to exhibit very low creep rates. Room temperature creep at 30 % loading is less than 0.04 %/decade when plotted vs log (time), one of the lowest available creep rates in a polymeric fiber. Because of the low creep rates at typical laboratory temperatures, test data must typically be accelerated by high load levels or elevated temperatures.

Developing time-to-rupture curves based on oven testing at relevant load levels is resource intensive and time consuming. For that reason Arrhenius-based acceleration methods have been developed for prediction of long-term rupture life based on short-term testing. Vectran® follows the Arrhenius temperature dependence, and master curves for long-term rupture life have been developed. Creep data from Vectran testing is shown in Fig. 10.16. Time to rupture can be limited at high loads and elevated temperatures; however LCP exhibits very long life at loads less than 50 % of the break load.

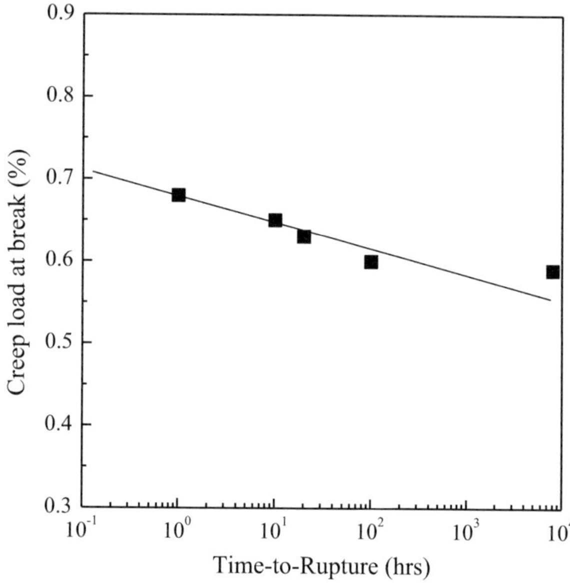

Fig. 10.16 Stress rupture of Vectran®

10.4.5 Environmental Stability of Vectran®

Like any engineering material, a flexible composite must be designed to maintain function in its service environment. Failure to account for chemical and thermal effects can lead to unexpected outcomes. However, environmental threats must be weighed against (a) their likelihood of occurrence and (b) their effect on the structure. Fibers have different base chemistries, e.g., polyester, polyamide, silica, etc., and each class of material has its own response to temperature and moisture, as shown in Table 10.8

Stability to specific chemical agents should be considered in design; however the most common environmental solvent is water. Moisture absorption/desorption by reinforcing fibers can cause the microcracking, the unanticipated weight gain, and the corrosion in adjacent metal parts; these problems often do not appear until several years after initial deployment. Weather-induced shrink/swell can also arise, e.g., flying from a dry desert environment to a wet coastal one. Dimensional changes can lead to unintended redistribution of structural loads or fatigue cracking.

Table 10.8 Basic environmental properties of composite fibers (producer data)

Fiber class	Carbon	Glass	PBO	Aramid	Vectran	PE
Density, g/cc	1.8	2.6	1.6	1.4	1.4	0.97
Moisture regain, %	<0.1	<0.1	1–2	2–6	<0.1	<0.1
½ strength temp., °C	>500	450	400	300	150	90

10.5 Conclusion

Vectran® is the only thermotropic LCP fiber which is commercially available. Vectran® shows excellent several properties such as high mechanical and thermal properties.

Regarding crystal structural change upon annealing, it was found that Vectran® shows solid-solid phase transition from hexagonal to orthorhombic on annealing at >240 °C. Lattice parameters *a* and *b* close in on that of HBA-rich crystal increasing with annealing temperature. In contrast, lattice parameter *c* does not change irrespective annealing temperature. This change may be attributed to the increase of dipole-dipole interaction caused by orientation rather than chain-sliding diffusion.

As for one of the most important applications, we showed the composite material using Vectran®. The need for robust flexible composites in inflatable and air-supported structures continues to grow rapidly across many industries, including military, aerospace, transportation, and emergency services. Many technical fibers are used in hard composite applications, but specific properties are required for flexible applications. Vectran® has high strength and is robust to long-term flex/fold cycling to maximize product life cycle. Vectran® and fabrics exhibit a unique combination of flex fatigue resistance, dimensional stability, and environmental stability ideal for use in flexible composites.

References

1. A. Pegoretti, M. Traina, in *Handbook of Tensile Properties of Textile and Technical Fibres*, ed. by A.R. Bunsell (Woodhead Publishing Ltd, Cambridge, 2009), p. 354
2. W.J. Jackson Jr., H.F. Kuhfuss, J. Polym. Sci. Polym. Chem. Ed. **14**, 2043 (1976)
3. G.W. Calundam, M. Jaffe, *Anisotropic Polymers, synthesis and properties*, Proceedings of the Robert A. Welch Conference on Chemical Research. Synthetic Polymers (1982)
4. Polyplastics Co., Ltd, LCP catalogue (2013)
5. D.Y. Yoon, N. Masciocchi, L.E. Depero, C. Viney, W. Parrish, Macromolecules **23**, 1793 (1990)
6. R.B. Fette, M.F. Soviniski, *NASA TM 2004-212773* (2004)
7. S. Hanna, A.H. Windle, Polym. Commun. **29**, 236 (1998)
8. S. Hanna, A.H. Windle, Polymer **33**, 2825 (1992)
9. A. Muhlebach, J. Lyerla, J. Economy, Macromolecules **22**, 3741 (1989)
10. A.R. Bunsell, *Handbook of Tensile Properties of Textile and Technical Fiber*, 416 (Woodhead Publishing, Oxford, 2009)

11. J. Nakagawa, in *Advanced Fibers Spinning Technologies*, ed. by T. Nakajima, K. Kajiwara, J.E. McIntyre (Woodhead Publishing Ltd, Cambridge, 1994), p. 160
12. D. Cadogen et al., *Development and Evaluation of the Mars Pathfinder Inflatable Airbag System*, ILC Dover, Inc., 49th International Astronautical Congress (1998)
13. A. Kaito, M. Kyotani, K. Nakayama, Macromolecules **23**, 1035 (1990)
14. Y. Taguchi, J. Watanabe, Macromolecules **42**, 3170 (2009)
15. R. Neglia, *Supersails for Superyachts*, Superyacht, Nautica, Spring (2007); http://www.nauticalweb.com/superyacht/539/tecnica/sails.htm
16. D. Cadogan, *Textile World*, Sept/Oct., 32 (2011)
17. P.V. Cavallaro, *2007 Yearbook of Science & Technology* (McGraw-Hill, New York, 2006)
18. A.S. Verge, Rapidly Deployable Structures in Collective Protection Systems. Research Report, US Army Natick Soldier Center (2006)
19. HDT showcases newest deployable shelter technology. *PRNewsWire*, July 15 (2009); http://www.prnewswire.com/news-releases/hdt-showcases-newest-deployable-shelter-technology-62208087.html
20. K.T. Tan, N. Watanabe, Y. Iwahori, J. Reinf. Plast. Compos. **30**, 99 (2011)
21. D.C. Jegley, NASA/TM-2009-215955 (2009)
22. K.T. Tan, N. Watanabe, M. Sano, Y. Iwahori, H. Hoshi, J. Compos. Mater. **44**, 3203 (2010)
23. M.D.K. Wood, L.L. Tong, K.A., A.R. Rispler, *J*. Reinf. Plast. Compos. **28**, 715 (2009)
24. A. Velicki, NASA/CR-2009-215932 (2009)
25. M.D.K. Wood, X. Sun, L. Tong, A. Katzos, A.R. Rispler, Y.-W. Mai, Compos. Sci. Technol. **67**, 1058 (2007)
26. A.A. Leal, J.M. Deitzel, J.W. Gillespie, Compos. Sci. Technol. **67**, 2786 (2007)
27. P. Morgan, *Carbon Fibers and their Composites* (CRC Press, 6000 Broken Sound Parkway NW, Suite 300 Boca Raton, 2005)
28. J.W.S. Hearle, *Fiber Fracture*, ed. by M. Elices, J. Llorca, (Elsevier Science Ltd., 2002)
29. I. Karacan, J. Appl. Polym. Sci. **100**, 142 (2006)
30. H. Saito, S. Inaba, M. Nakahasi, T. Katayama, I. Kimpara, *Proceedings 18th International Conference on Composite Materials*, ICCM-18, Jeju, Korea, August 21 (2011)
31. N. Inaba, M. Nakahashi, H. Saito, I. Kimpara, Y. Oomae, T. Katayama, *Proceedings 1st Japan Conference on Composite Materials*, JCCM-1, Kyoto, Japan, March 9 (2010)
32. H. Saito, S. Inaba, M. Nakahasi, I. Kimpara, Technical Report to Kuraray Co., Ltd. April 28 (2010)
33. P.V. Cavallaro, A.M. Sadegh, C.J. Quugley, Text. Res. J. **77**, 403 (2007)
34. J. Hinkle, R. Timmers, A. Dixit, J. K. H. Lin, J. Watson, 50th AIAA Structures, Structural Dynamics and Materials Conference, Palm Springs, USA, May 4 (2009)

Chapter 11
Zylon®: Super Fiber from Lyotropic Liquid Crystal of the Most Rigid Polymer

Yoshihiko Teramoto and Fuyuhiko Kubota

Abstract The research on rigid-rod polybenzazoles (PBZ) was originally intended to develop super fibers for aerospace applications. Due to the technical difficulties in synthesizing PBZ and processing it to fibers, a polybenzoxazole (PBO) fiber, Zylon® launched by Toyobo is the only fiber that has been successfully commercialized so far.

In this chapter, the history of development, manufacturing process, fiber properties, and unique applications are described for Zylon®. Chemistry of PBO and polymerization process are also presented and discussed.

Keywords PBO • Polybenzazole • Polycondensation • Liquid crystalline polymer • Dry-jet wet spinning • High-performance fiber • Advanced composite materials • Technical textile • Personal protective equipments • Heat-resistant polymer

11.1 History

PBO fiber was commercialized in Japan by Toyobo Company in 1998 [1–6]. Historically, PBO fiber was extensively studied in the USA during the 1980s and 1990s as a next-generation high-performance fiber following aramid fibers. The key synthetic method to achieve high molecular weight in polybenzazole (PBZ) chemistry was to use polyphosphoric acid (PPA) as a solvent as well as a dehydrating agent. The method had been originally developed by Iwakura and Imai in the late 1960s [7–10] and was applied to the synthetic research of ladder polymers and polybenzazoles conducted by the researchers of the US Air Force (USAF) [11]. Chemical structures of typical PBZ polymers are shown in Fig. 11.1.

Kevlar, an aramid fiber, was successfully commercialized by DuPont in 1974. It was an important landmark for high-performance fiber industry. However, even better properties in tensile strength and heat resistance than those of Kevlar were strongly desired, especially for aerospace applications that the USAF people were

Y. Teramoto (✉) • F. Kubota
Research Center, TOYOBO Co., Ltd., 2-1-1 Katata, Otsu, Shiga 520-0292, Japan
e-mail: Yoshihiko_Teramoto@toyobo.jp

Society of F.S. Technology, Japan (ed.), *High-Performance and Specialty Fibers*,
DOI 10.1007/978-4-431-55203-1_11

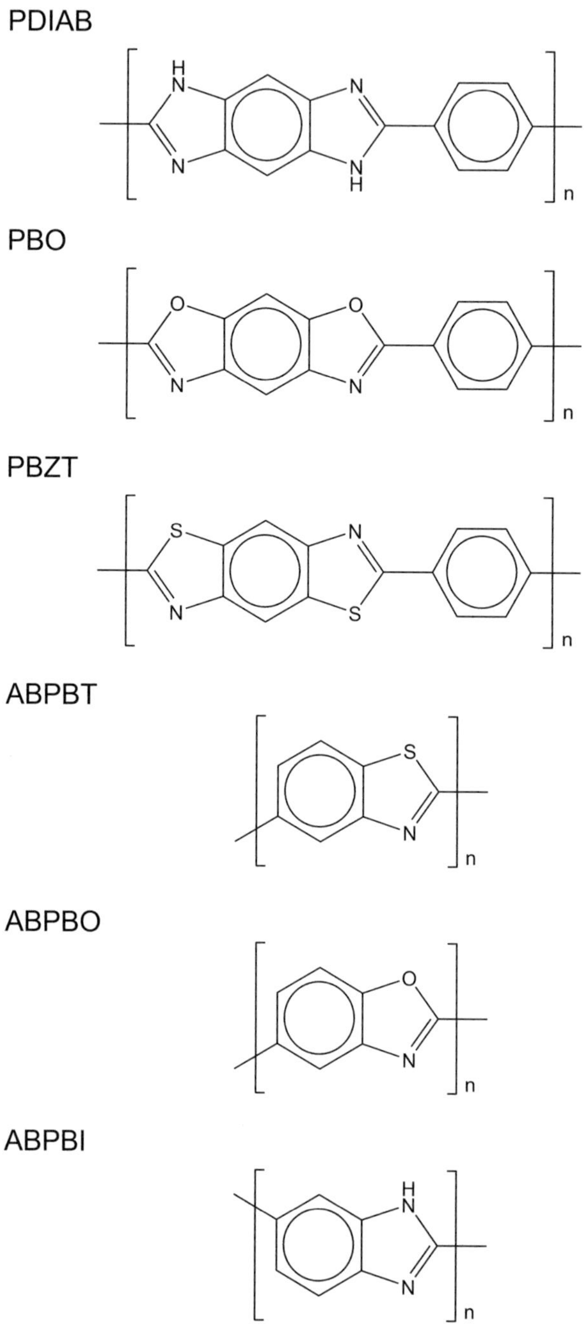

Fig. 11.1 Chemical structures of typical PBZ polymers

looking at [12]. It drove them to target active works on PBZ, especially *trans*-poly-*p*-phenylene benzobisthiazole (PBZT), which was regarded as the most promising material among PBZ family in terms of solubility of the polymer.

One of the researchers at USAF, Wolfe, moved to Stanford Research Institute (currently SRI International) and further proceeded with the development of PBZ. In PBZ polymerization, P_2O_5 content of polyphosphoric acid is the key contributor for achieving a high degree of polymerization. Wolfe et al. designed a systematic way of controlling P_2O_5 content during polymerization and successfully obtained fibers that exceed Kevlar® 49 in strength and modulus [13, 14].

In the late 1980s, Dow Chemical was trying to get into the growing market of high-performance fibers. They purchased the patents of PBZ from SRI [15] and started their project covering the whole chemistry and fiber manufacturing. From their commercial point of view, *cis*-poly-*p*-phenylene benzobisoxazole (*cis*-PBO) was more attractive than PBZT, as the monomer for *cis*-PBO, 4,6-diaminoresorcinol (DAR), was expected to be cheaper than those for PBZT. Lysenko et al. developed a new synthetic method for DAR, starting from 1,2,3-trichlorobenzene, which was a big advancement for commercialization of *cis*-PBO fiber [16].

While extensively working on the total process of *cis*-PBO fiber, Dow Chemical realized that it was more efficient to work with a fiber company to accelerate the development. Among several possible options, Dow Chemical picked up Toyobo as their partner, and the joint development started in 1991. Toyobo fiber team made a big contribution to the development of spinning technology, which was quite a big challenge even for fiber specialists, due to the difficulty in spinning resulting from the high rigidity of *cis*-PBO.

In 1994, Dow Chemical made a strategic decision to exit specialty businesses including high-performance fiber business and more focus on their commodity businesses. Under the partnership of joint development for 3 years, Dow Chemical proposed that their technologies for DAR and *cis*-PBO were transferred to Toyobo. Toyobo finally decided to commercialize *cis*-PBO fiber on their own and their polymer team took over Dow's technologies for further developing them in Japan.

In October 1998, Toyobo launched *cis*-PBO fiber, with a trade name of "Zylon®," and started the commercial production in their Tsuruga plant in Fukui, Japan. A Japanese chemical company was a new partner for Toyobo for the DAR monomer supply. It is noteworthy that PBZ chemistry which had been originally developed in Japan finally returned home and bore fruits in the form of "the world's strongest fiber" [1–6].

11.2 PBO Chemistry

11.2.1 PBZ Chemistry

PBZ polymers are typically synthesized by polycondensation of AA-type monomers and BB-type monomers. BB-type monomers for *trans*-PBZT and *cis*-PBO are 2,5-diamino-1,4-benzenedithiol dihydrochloride (DABDT) and 4,6-diaminoresorcinol dihydrochloride (DAR), respectively. The AA-type monomer for both is terephthalic acid, and its acid chloride can be used as an alternative as well. PBZ family includes AB-type polymers synthesized by self-condensation of AB-type monomers, but mechanical properties of the resulting fibers are not as good as those of the linear backbone PBZ polymers.

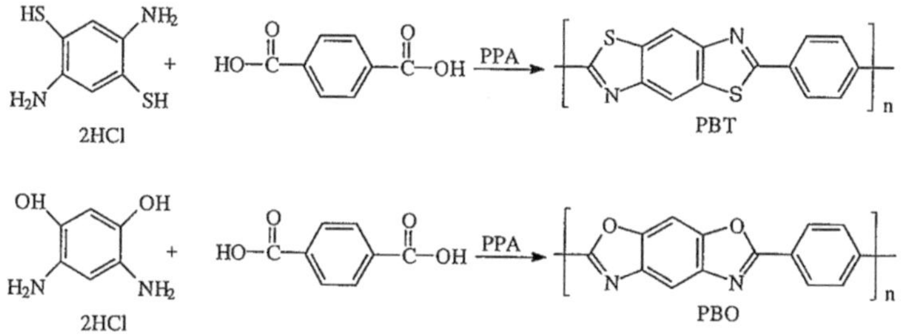

A unique PBZ polymer developed by AkzoNobel in the late 1990s was poly (2,6-diimidazo[4,5-b:4′,5′-e]pyridinylene-1,4-(2,5-dihydroxy)phenylene) (PIPD). The polymer was designed to combine the high stiffness and tenacity of the rigid-rod polymer with possible hydrogen bonding between the polymer backbones. The fiber spun from a polyphosphoric acid solution exhibited a higher compressive strength than any other organic super fiber. The fiber was called M5 in their development stage. The technology had been transferred to Magellan Systems, which DuPont acquired in 2005, to add the new fiber in their lineup of high-performance fibers. However, there has been no further information about commercialization of M5 in DuPont so far.

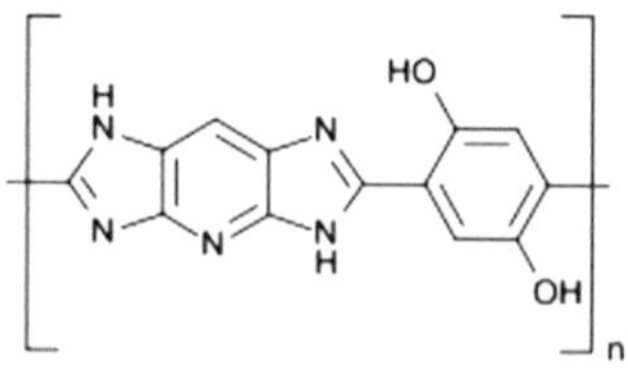

11.2.2 PBO Chemistry

11.2.2.1 Monomer Chemistry

The key monomer (BB-type) for *cis*-PBO is 4,6-diaminoresorcinol (DAR). Several starting materials, such as resorcinol and di- or tri-chlorinated benzenes, can be used for the synthesis of DAR. As high purity is required for monomers used in polycondensation, the synthetic route should be well designed to avoid formation of isomers, which may need an additional separation or purification process and lead to lower yield and higher cost.

Wolfe et al. developed a route starting with direct nitration of resorcinol, but the formation of wrong isomer and potential risk of explosive trinitro-byproduct were the issues for scale-up. A novel route starting from 1,2,3-trichlorobenzene (TCB) developed by Lysenko et al. was the first process that could be scaled up for commercialization. As the 2-position of TCB is "protected" by chlorine, there is no possible formation of the wrong isomer and explosive trinitro-byproduct. As a result, high purity can be easily achieved for DAR, although a dechlorination step is required to make the final product.

Alternatively, other routes such as diazotization of resorcinol followed by reduction and nitration of 1,3-dichlorobenzene (DCB) followed by hydrolysis and reduction are also available. However, separation of the isomer could be a challenge to achieve high purity of DAR.

DAR easily undergoes air oxidation, as it has four electron-donating groups in a single benzene ring. In order to avoid degradation through oxidation, DAR has to be handled in the form of acid salt. The most typical salt form is dihydrochloride, which evolves into hydrogen chloride (HCl) gas during the initial stage of polymerization.

Another monomer (AA-type) for *cis*-PBO is terephthalic acid (TA). As solubility of TA in PPA is very low, micronized TA is typically used to enhance the polymerization rate. Alternative AA-type monomer is terephthaloyl chloride, which is more soluble in PPA. However, evolution of HCl gas during the polymerization could be a disadvantage over the simple dehydration in the case of TA.

11.2.2.2 Polymerization

As described before, PPA is normally used as a solvent together with P_2O_5. Theoretical P_2O_5 content in the whole solvent system at the initial stage of polymerization is typically higher than 120 %. The dehydrating power of the solvent system is the driver for polymerization and the P_2O_5 content decreases, accordingly as the polymerization proceeds.

When DAR is dissolved in PPA, it readily becomes phosphoric acid salt, while releasing HCl gas. For a process on a commercial scale, evolution of toxic and corrosive gas can be a big issue. Proper care in handling HCl gas has to be taken for safe and stable operation of the process.

11.2.3 Alternative Chemistry for PBO

While handling of HCl gas is a disadvantage for a commercial process, alternative ways to avoid HCl gas evolution have been investigated. Dow developed a process of using salt of terephthalic acid and DAR (TA/DAR). When TA/DAR is used, it is quite easy to adjust the stoichiometry, for a high degree of polymerization, as the monomer is 1:1 adduct of the two monomers. However, TA/DAR is not stable under air, due to the lower acidity and stabilizing power of terephthalic acid. Therefore, extra care to avoid oxidation has to be taken in handling TA/DAR for the synthetic process and storage.

A unique monomer that can be used for PBO chemistry is a "dimer"-type monomer, whose structure is AA-BB (TA/DAR), covalently bonded DAR and TA. TA/DAR is reasonably stable under air and doesn't have to be any salt. It has the advantages of avoiding HCl gas and of achieving stoichiometry. Therefore, economical synthetic process for the monomer is highly anticipated.

11.3 Features of Zylon®

Basic properties of Zylon® (PBO fiber) are shown in Table 11.1. Strength and modulus are almost twice as high as those of poly-*p*-phenylene terephthalamide (PPTA) fiber: as a commercial fiber, Kevlar® and Twaron®. The decomposition temperature of Zylon® is 100 °C higher than that of para-aramid fiber. The limiting oxygen index is 68, which is the highest among organic super fibers. General technical information can be referred in the web page [17]. In spite of the extremely high strength and high modulus, Zylon® can be fabricated into soft fabrics and flexible cords. Zylon® behaves in a different way, unlike general synthetic fibers such as polyester or nylon, in the abrasion resistance test with the metal pins (Fig. 11.2). Abrasion resistance of Zylon® is higher than that of para-aramid fiber under the same load, but much lower than that of ultrahigh-strength polyethylene (PE) fiber.

Expositions of Zylon® have been already published by several experts [4, 18, 19]. In this section, several major properties are described and discussed.

Table 11.1 Fiber properties of Zylon (PBO fiber)

	Zylon® AS	Zylon® HM
Filament decitex	1.7	1.7
Density (g/cm^3)	1.54	1.56
Tensile strength		
(cN/dtex)	37	37
(GPa)	5.8	5.8
Tensile modulus		
(cN/dtex)	1150	1720
(GPa)	180	270
Elongation at break (%)	3.5	2.5
Moisture regain (%)	2.0	0.6
Decomposition temperature (°C)	650	650
LOI	68	68
Thermal expansion coefficient	–	-6×10^{-6}

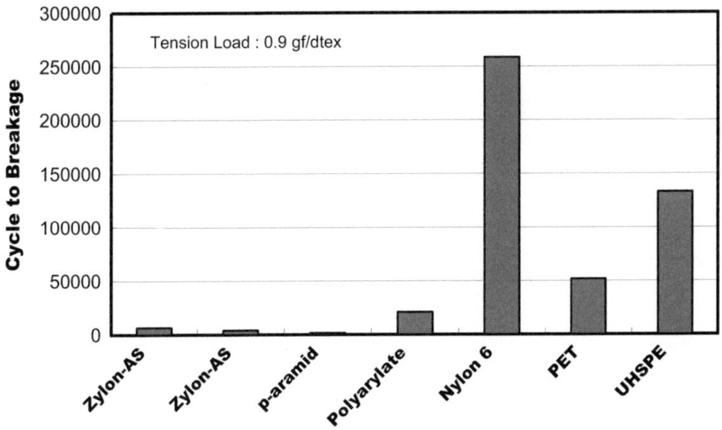

Fig. 11.2 Abrasion resistance test based on JIS L 1095:1999 9.10.2 B

11.3.1 Mechanical Properties

Stress-strain curves measured for some commercialized high-strength fibers by single filament test are shown in Fig. 11.3. Each curve shows a nearly straight line or a concave line excluding ultrahigh-strength polyethylene (UHSPE) fiber. Ultrahigh-strength PE fiber shows plastic deformation at the normal temperature with high stress in fracture test.

The theoretical strength and Young modulus are listed in Table 11.2.

In the case of high-strength fibers, the measured strength is highly dependent on the gauge length of specimen. In the case of Zylon® AS at gauge length 25 mm, the

Fig. 11.3 Stress-strain curves of commercial high-strength fibers

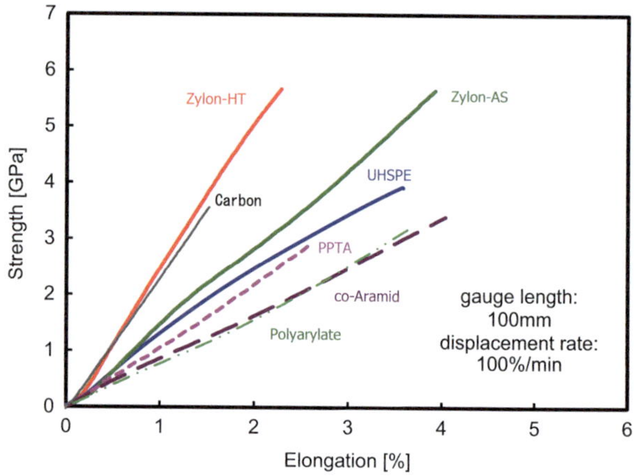

Table 11.2 Theoretical strength and Young's modulus

Fiber	Density (g/cm^3)	S^a (Å^2)	σc^b (GPa)	Ec^c (GPa)
Dyneema® [20]	0.97	18.2	31	235
Kevlar® 49 [20]	1.44	20.2	28	156
Kevlar® 149 [20]	1.47			
Technora® [20]	1.39	21.0	27	91
Vectran® [20]	1.41	19.4	29	126
Zylon® HM [21]	1.56	19.2	59	690

[a]*S*: Effective area for one C-C covalent bond
[b]σc: Theoretical strength
[c]*Ec*: Theoretical Young's modulus

mean strength is 7 GPa (Fig. 11.4 [22]). This specification of test method is commonly used for the measurement of carbon fibers. Numbers besides the circles in Fig. 11.4 are shape factor of Weibull distribution. Bigger number means the dispersion of distribution function is small. Filled plots are the mean value of strength with fiber diameter correction. Less than 10 mm gauge length occurrence of end fracture decreases the measurement of strength.

The extrapolation of the slope of −1/14.5 means that the measuring strength will reduce from 7.0 to 6.3 GPa when the gauge length increases from 25 to 250 mm. A 6.3 GPa strength is still higher than the strength measured with multifilament yarn, expecting yarn strength becomes smaller than the strength at average elongation at break of filaments composing the yarn [23]. Therefore, the average strength measured for single filaments is always higher than the strength measured for multifilament yarn.

The modulus of heat-treated Zylon® (type HM) is 270 GPa, which is 40 % of the theoretical estimation and 60 % of the crystalline modulus measured by WAXD [24]. The cause for the difference has not been elucidated directly. Two possible reasons are related to the discussions in the study of PBZ by Lee [25]. The first

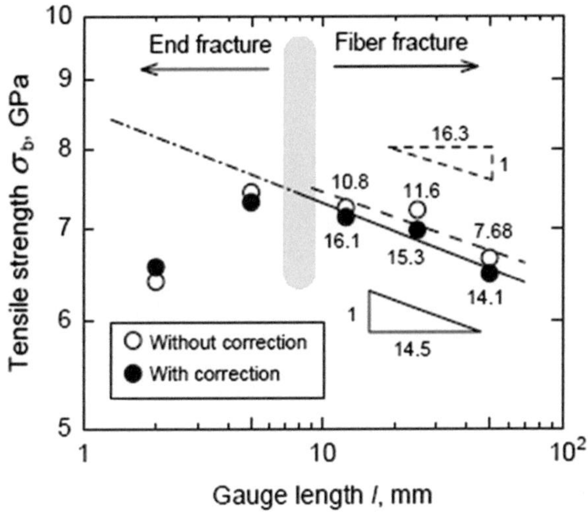

Fig. 11.4 Specimen length effect on strength for Zylon ® AS [22]

possible hypothesis is that large concentration fluctuation during coagulation process disturbs the order in the microfibril structure. The second hypothesis is caused by domain boundaries of liquid crystalline solution structure that is inherent after polymer structure development and behaves as the defect that does not transport stress. Replacement of water with an organic solvent in the coagulation process enhanced modulus of PBO fiber up to 352 GPa [26]. This supports the first mechanism. The differences in the degree of molecular orientations exist between the outer layer and inside of a fiber [27]. This is a phenomenon on several micron scales happening in the coagulation process. Kitagawa et al. applied the mechanical series model that is composed of a series of crystalline domain and low Young's modulus domain. They reported that even in high-modulus heat-treated fiber (Zylon ® HM), 14 % fiber volume remains of low modulus [28]. In the structure formation process, the evidence that domain boundary of solution structure is left at coagulation stage has not been reported. But even if domains deform with the large change of aspect ratio, the molecular chain ends tend to concentrate at the boundary. Burger et al. discussed the fine structure to account for the four-point SAXS pattern of PBO-HM [29]. Low density part might be an accumulation of chain ends [30], though. The real image of the area where a Young's modulus is low is still unknown.

11.3.2 *Compressive Strength*

The anisotropy of the elasticity constants of the PBO fiber (Fig. 11.5) is reported by Yamashita et al. [31]. As shown in Table 11.2, the compression modulus in the lateral direction is small. And the coefficient of PBO fiber is lower than that of PET

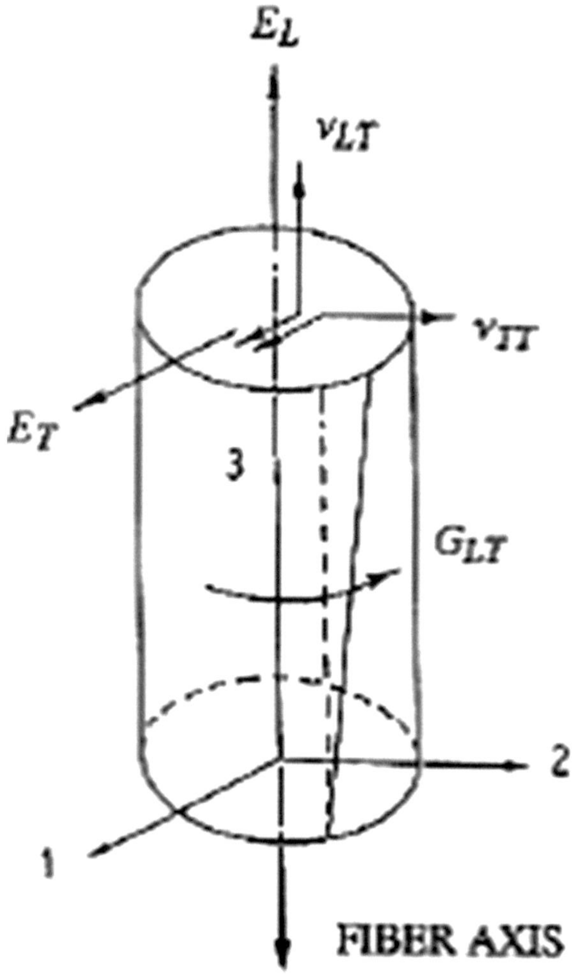

Fig. 11.5 Definition of nomenclature for elastic constants [31]

fiber. In terms of torsion rigidity, PBO fiber is comparable to PET fiber and is half as rigid as para-aramid (PPTA) fiber. Actually, codes and fabric made of PBO are softer compared with those of PPTAs.

Different methods of measuring the compressive strength of filaments are available, such as elastica loop method, mandrel bending method, bending beam method, and tensile recoil method. These are not direct measurements of stress by axial compressive strain. It is very difficult to apply axial compression for thin filament. So the value of compressive strength in each method is useful for relative comparison between materials. The meaning of compressive strength is usually the load-generating kink bands. This doesn't mean real fracture of a fiber.

Leal et al. compared compressive strength among several high-performance fibers by the elastica loop method [32]. They estimated the ratio of the compressive modulus and the tensile modulus to be 0.32 for PBO fiber. The compressive

Table 11.3 Elastic constants in Fig. 7.5 for fibers [31]

Fiber	E_L (GPa)	E_T (GPa)	G_{LT} (GPa)
PBO	315.2	0.91	1.02
PPTA (Kevlar® 29)	79.8	2.59	2.17
PPTA (Kevlar® 49)	113.4	2.49	2.01
Carbon (T-300, PAN)	234.6	6.03	18.2
Glass	77.4	67.9	42.5
PET	14.5	1.37	1.03
Nylon6	2.76	1.37	0.55

strength by this evaluation is 0.29 GPa. Assuming the compressive modulus is 73 GPa, the critical strain is about 0.4 %. This value is smaller than 0.56 GPa, which was measured by Toyobo by bending beam method. $D_e$teresa et al. found empirical linear correlation of compressive strength is 30 % of the torsion modulus (G_{LT}) [33]. Leal's data of compressive strength are close to 30 % of torsion modulus in Table 11.3.

The crystal structure of PBO was investigated by the X-ray diffraction analysis combined with the computer simulation technique. Planar shape chain forms the sheets with axial shift between the adjacent chains with $\pm c/4$ (c = fiber identity period). Those sheets stack with random shifts along the fiber direction [34, 35]. This structure suggests that PBO crystals include the plane with low interaction to sheer deformation.

Lorenzo-Villafranca et al. defined the limit of compressive strain where fibers deform to V shape. Usually, the limit of compressive strain is judged by discontinuous curvature in elastica loop method. At this compressive strain, kink band is observed, but tensile strength shows no reduction. They call V-shaped damage as "knuckle." Compressive strain where PBO fibers generate "knuckle" is 1.7 %. At this strain tensile strength decreases more than 30 %. This strain is the criteria that fiber would not return to its straight shape. The calculation of the compressive strength greatly varies with the compressive modulus to be applied. In this study, compressive strength was evaluated to be 1.24 GPa. They found plasma treatment under nitrogen improved the compressive strain approximately 40 % [36]. It is interesting that surface treatment on the shell part of a fiber improves compressive strength.

11.3.3 Fatigue

Horikawa et al. analyzed fatigue properties of Zylon® with single filament test for fiber-reinforced composite applications [37]. Under the test condition (12.5 mm gauge length, sinusoidal loading of frequency 10 Hz on three specifications of stress amplitude ratio: $R = 0.1$, 0.5, 0.7), the breaking life tended to shorten as the maximum stress increases. In addition, the influence of the stress ratio was small. HM type showed longer life than AS type (Fig. 11.6).

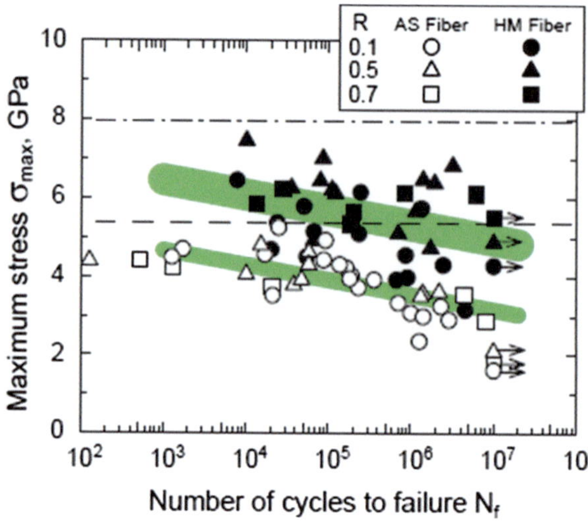

Fig. 11.6 Cycle number for filament break in single fiber fatigue test [37]. Cutoff cycle is 10^7. *Bold lines* are least mean square regression for two types of Zylon® fibers. (**a**) PPTA knitted fabric of 2/28 Nm spun yarn. (**b**) PBO knitted fabric of 2/34 Nm spun yarn

11.3.4 Flame Resistance

Having outstanding heat oxidation tolerance and incombustibility as an organic fiber, Zylon® has dramatically improved the performance for the firefighter garment and molten metal protection sheets. Excellent properties of Zylon® such as incombustibility, low smoke emission, high pyrolysis resistance, and high strength made it as the best material for heat- and flame-resistant fabrics [38]. Comparison between Zylon® and para-aramid fibers was made by cone calorimeter examination (ASTM E 1354-90) including HRR (heat release rate), THE (total heat evolved), and FIGRA (fire growth rate) index. In Fig. 11.7 remarkable differences can be observed in terms of the ignition times, the HRRs, and residue of materials [39].

11.3.5 Thermal Conductivity

Fujisiro et al. reported thermal conductivity data of PBO fiber and ultrahigh-strength PE fiber [40]. Recently, Wang reported fiber axial directional thermal conductivities measured by the high time-resolution unsteady-state heat conduction system with heating a part of cross section of single filament (Fig. 11.8). Thermal conductivities of 19 $Wm^{-1}K^{-1}$ and 23 $Wm^{-1}K^{-1}$ were reported for Zylon® HM and Zylon® AS, respectively [41]. As thermal conductivity helps release heat from the fabric, higher conductivity may give favorable results in cone calorimeter examination.

In materials, heat transfer is carried out by electrons or phonons which quantize heat vibration. In electric insulating materials, heat is transferred only by phonons, because electron transfer is restricted. Factors affecting phonon transfer reduce

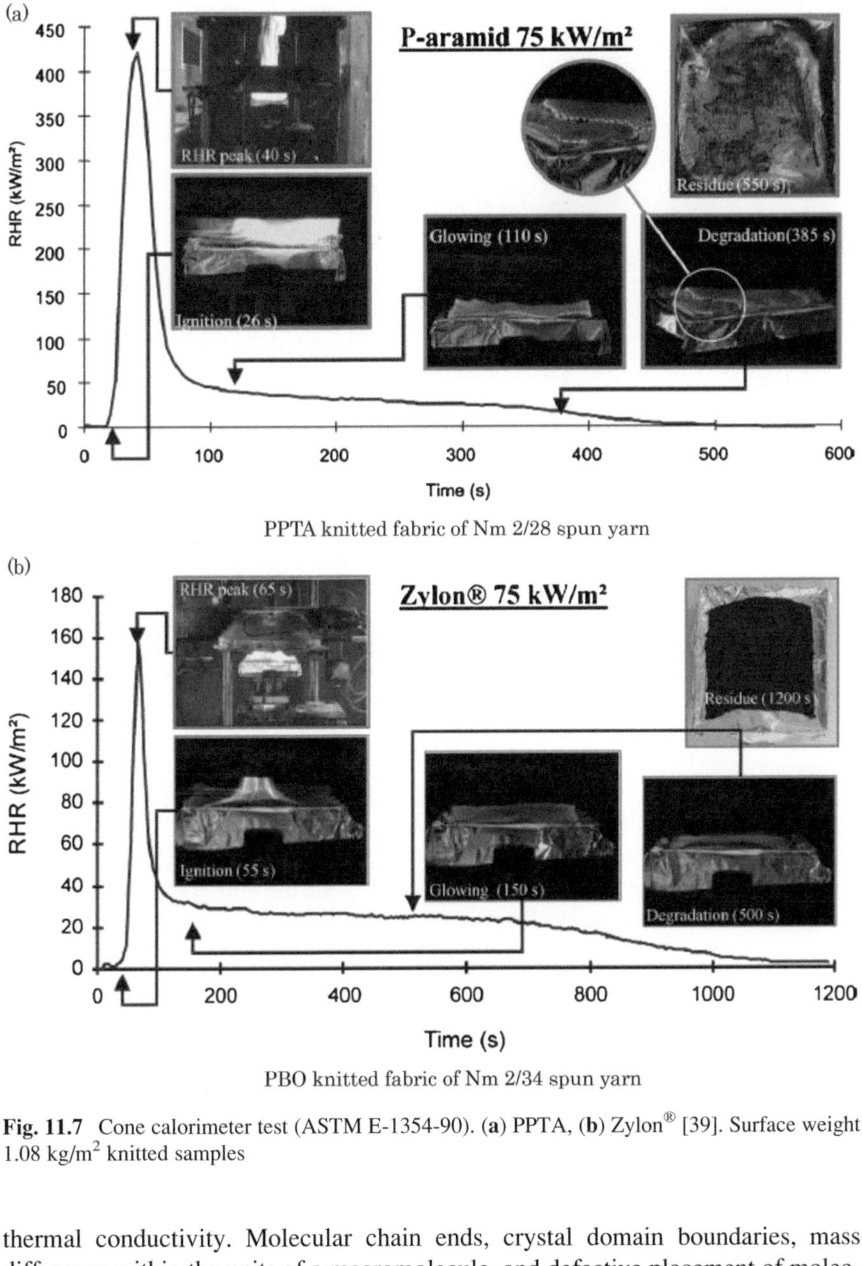

Fig. 11.7 Cone calorimeter test (ASTM E-1354-90). (**a**) PPTA, (**b**) Zylon® [39]. Surface weight 1.08 kg/m^2 knitted samples

thermal conductivity. Molecular chain ends, crystal domain boundaries, mass difference within the units of a macromolecule, and defective placement of molecular chains such as folds and dislocations can scatter phonon propagation. These factors are all similar to defects that reduce fiber strength. It is thought that Zylon® and ultrahigh-strength PE fiber show very high thermal conductivity because of two prerequisites. Firstly, their strength is very high, that is, the amount of defects, such

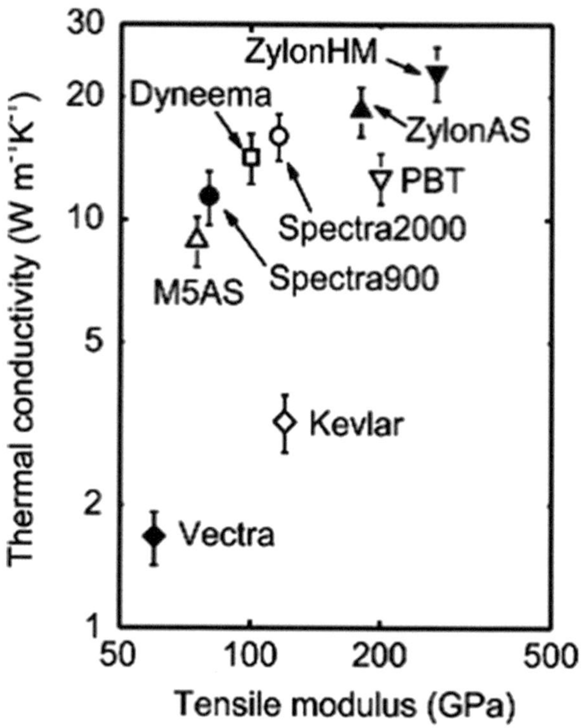

Fig. 11.8 Room temperature thermal conductivities plotted with tensile modulus [41]

as the chain fold and the chain end, in those fibers is small. Secondly, the polymer structure is suitable for phonon propagation.

Carbonized fiber made from Zylon® HM at a high temperature shows extremely high thermal conductivity. This conductivity exceeds the values of high-modulus-type pitch-based carbon fiber [42]. Kaburagi reported structure analysis of Zylon® HM carbonized graphite. The carbon fiber treated at 3000 °C has a narrow graphite interlayer spacing d002 = 0.3355 nm [43].

11.3.6 Degradation Under Hydrolytic Condition

As polybenzazoles are prepared by dehydration polycondensation, hydrolysis causes breakage of the polymer backbone. So et al. reported the study of intrinsic viscosity (IV) changes during repeated dissolution and reprecipitation of PBO polymer. Firstly, a polymer of IV = 22 was dissolved in methanesulfonic acid containing atmospheric moisture, and then, the polymer was recovered by reprecipitation in water. The IV of the recovered polymer was 15, and no carboxylic acid was detected in its IR spectrum. After repeated operation, carboxylic acid was finally detected, when the IV went down to 2. Meanwhile, when polyphosphoric acid was used as a solvent, the polymer behaved in a different way. After four times of reprecipitation, the IV dropped from 22 to 5.4 and the carboxylic acid was detected in its IR spectrum [44]. The IV difference was big until the step when amide linkages were cut after oxazole rings were hydrolyzed to amide linkage.

Fiber surface modifications were studied with strong acid [45]. A methanesulfonic acid treatment had a bigger change in surface free energy than a nitric acid treatment. It was shown that surface treatments by acid liquid produced functional group by hydrolysis. This kind of surface treatment is useful as well as plasma treatment.

Chin et al. studied strength drop in a high-temperature and high-humidity environment. Woven PBO fabrics were stored in 50 °C 37 RH% for 84 days and then in 60 °C 60 RH% for 73 days. Fiber strength dropped as the peak of oxazole ring weakened in IR spectrum [46]. Hydrolysis from benzoxazole to benzamide also affects the affinity to moisture. Strength retention data of Zylon® AS for 40 °C 80 RH% aging are disclosed in Toyobo technical information web page [17].

11.3.7 Photoaging

Photoaging is a common phenomenon that can be seen for polymeric materials. Exposure to sunlight can change their chemical, physical, and mechanical properties and such degradation may limit the scope of applications.

The strength of PBO fiber decreases with exposure to sunlight. In the case of Zylon®, photoresistance was evaluated by the use of xenon weather meter. The strength decreased sharply at the initial stage of exposure and the residual strength after 6 months exposure to daylight was about 35 % [17].

There have been several approaches to improve photoresistance of PBO fibers. Zhang et al. coated the fiber surface with zinc nanoparticles/epoxy and found that the tensile strength of coated fiber less declined than that of non-coated fiber under UV irradiation. Shielding effect of the coating layer seems to work to enhance the photostability [47]. Another approach, which is based on chemical modification of the polymer backbone, was proposed by Zhang [48]. Introduction of two hydroxyl groups on the benzene ring of PBO significantly improved UV resistance of the fiber, compared with regular PBO fiber. They suggested that the intermolecular hydrogen bonds play a primary role to improve the UV resistance, avoiding disruption of the oxazole ring, which triggers the degradation of PBO.

Although many efforts have been made so far to understand photoaging mechanism [49–51] and improve photoresistance of a PBO fiber, it is still a big challenge to make it an intrinsically photostable fiber. From a practical point of view, PBO fiber has been used in many applications where the fiber is not exposed to sunlight directly, or decline of strength is not an issue, while heat resistance is mainly appreciated.

11.4 Fiber Processing

The process in which polymerized polymer solution was directly used for fiber spinning had been developed by Celanese company [52]. The solvent for fiber spinning is polyphosphoric acid. The difficulties in using polyphosphoric acid solution as spinning dope are as follows, especially complex flow of the nematic liquid crystal solution of rigid-rod polymer ("flow instability"), pretty high viscosity of the polyphosphoric acid solution, and the corrosion of metals by polyphosphoric acid. These three drawbacks make the industrialization very difficult as well as the development of basic production technologies.

As shown in Fig. 11.9, the spinning process for Zylon® consists of the six steps:

1. Extruding the polymer dope of the nematic liquid crystalline phase through capillaries to form filaments (extrusion)
2. Stretching the filaments to make them thinner (spin-drawing)

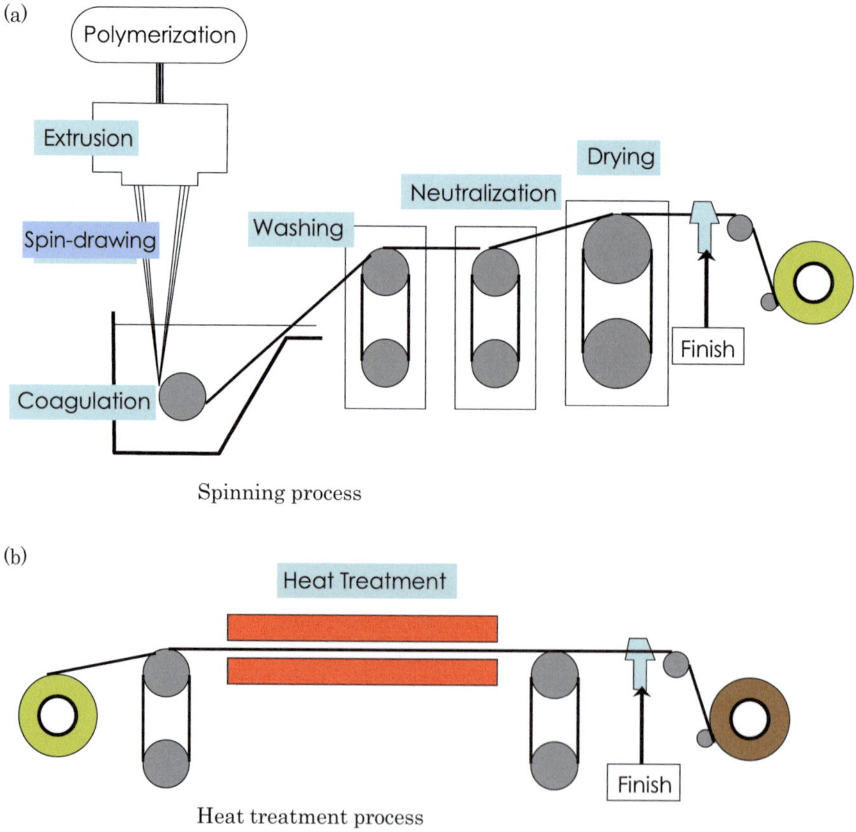

Fig. 11.9 Fiber production process. (**a**) Setup of spinning sequence consists of six steps. (**b**) Tension and temperature above 450 °C are applied on the heat treatment process

3. Immersing the small-diameter filaments in a non-solvent bath to solidify the fibers (coagulation)
4. Extracting the solvent from the fibers with water (washing)
5. Neutralizing the acid solvent in the fiber with basic solution (neutralization)
6. Drying the fibers (drying)

For the production of high modulus with small moisture regain-type HM fiber, post heat treatment process is necessary, and AS-type fiber is tensed at high-temperature conditions above 450 °C [29].

11.4.1 Spinning Dope

A PBO polymer solution in polyphosphoric acid (10–15 % concentration by weight) is assumed to form the polydomain structure (Fig. 11.10). Due to the poor optical transparency of polymer solution, the optical observation is difficult. Therefore, Odell et al. studied the relaxation of molecular orientation after shear deformations by X-ray diffraction method [53]. The PBZT/PPA = 9.8/90.2 solution (polymer IV = 31 dL/g) at 80 °C between gaps of 50–100 μm for enough time, then stopped shear deformation. One-hundred twenty seconds after ceasing flow, the orientation of polyphosphoric acid relaxed and it took 20 min to relax rigid-rod PBZT molecules. As polyphosphoric acid is an oligomeric material, the flow-induced orientation of polyphosphoric acid molecules can be also observed. However, the 116 % polyphosphoric acid contains more than 40 % chain-formed phosphoric acid molecules which are longer than a heptamer (Fig. 11.11), and the relaxation of orientation of PPA easily occurs. The relaxation of oriented PBZT polymer chain is quite slow in this system.

Fig. 11.10 Schematic image of polydomain structure

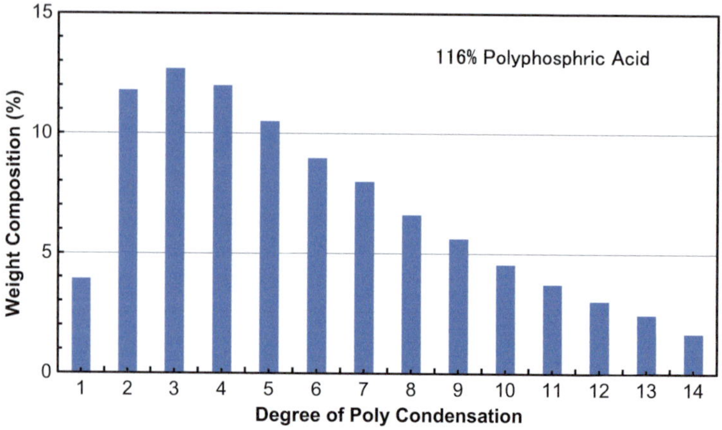

Fig. 11.11 Polycondensation of polyphosphoric acid [54]

The rheological properties of the rig-rod polymer solution complexity fluids those cause easily the flow instability. When the dope is extruded from a tube at slow flow rate, the millimeter-sized texture is observed on the surface of dope. This is considered to be the trace of the shear flow field at the wall surface accompanying secondary flow. This nature of the dope makes the fiber spinning process difficult.

Ernst et al. studied the coefficient of viscosity of the polyphosphoric acid solution of PBO polymer concentration higher than 10 % suitable for fiber spinning. They compared complex viscosity and steady shear viscosity. The complex viscosity was measured by the linear viscoelastic measurement. The steady shear viscosity at low shear rate was measured by the rotary rheometer. Shear viscosity at high shear rate was measured by a capillary rheometer. Three viscosities were in the same line. This means Cox-Merz's empirical rule was applicable [55]. The lyotropic liquid crystal has characteristic of shear thinning. Viscosity greatly decreases at high shear rate. The complex viscosity of 1000 s^{-1} at 150 °C is approximately 100 Pa.s. On the other hand, the viscosity of PPTA spinning solution of sulfuric acid is some 20 Pa.s by the steady shear viscosity of 1000 s^{-1} at 83 °C [56]. The viscosity of PBO spinning dope is several times higher than that of PPTA spinning dope. Figure 11.12 shows the linear viscoelastic data that were collected at Yamagata University. Temperature to time conversion is valid in the temperature range from 140 to 185 °C. In this temperature range, spinning dope composition stays in nematic phase without solid portion. Shear viscosity data from 24 to 6100 s^{-1} were measured by a capillary rheometer eliminating entry pressure drop from Bagley plot. These data were consistent with the complex viscosity data.

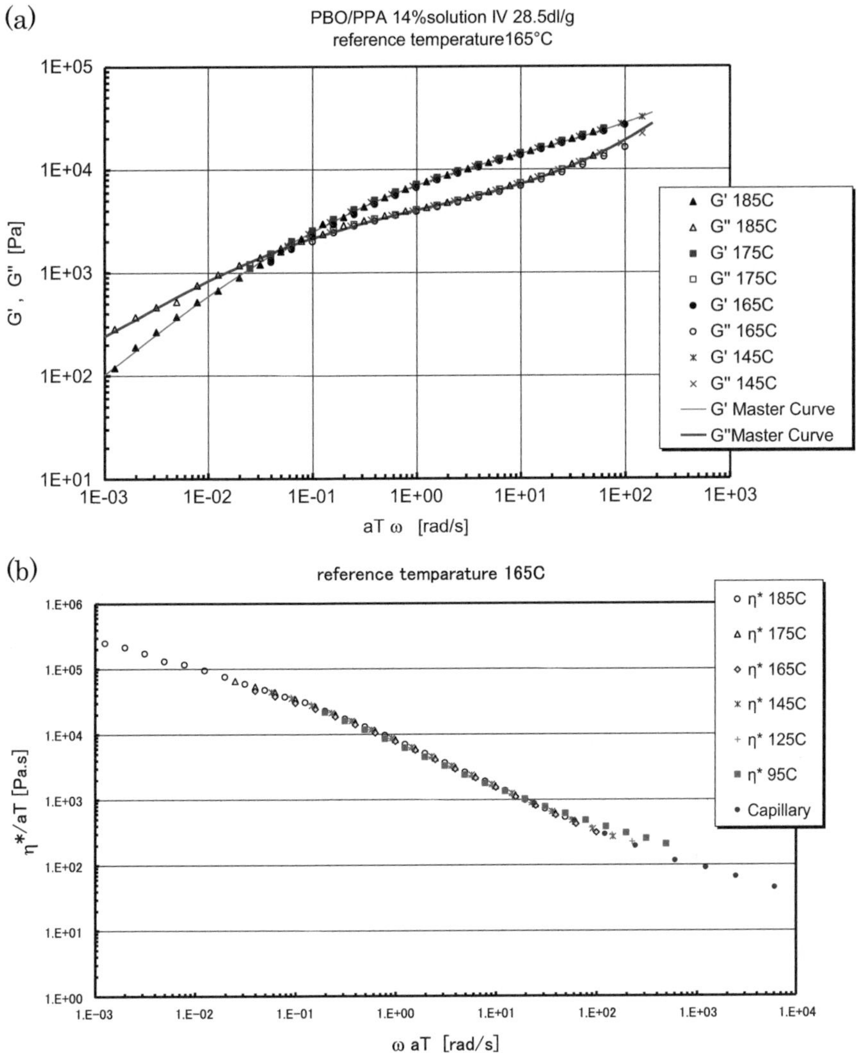

Fig. 11.12 Linear viscoelastic data of PBO/PPA solution. (**a**) Storage modulus (G′) and loss modulus (G″). (**b**) Complex viscosity. Parallel plate, sample thickness 0.8 mm; strain amplitude 10 %

11.4.2 Fiber Processing

11.4.2.1 Spin-Drawing

The spinning dope for PBO fiber is extruded from a capillary and stretched in the air gap. The polymer rods are aligned to axial direction during this process. Researchers of Dow set up a compact spinning line at the synchrotron radiation

X-ray source of the Cornell University. They analyzed the orientation parameter using the equatorial [100] (10.6 Å) reflection peak and the coherence length using equatorial [100] peak and meridian [002] layer line (Fig. 11.13 [57]). Spin-draw ratio (SDR) is defined as winding speed divided by average velocity at the capillary. When the SDR went up to 20, the orientation parameter reached 0.95 and the orientation parameter was almost saturated. The domain size (coherence length) of axial direction calculated from the spread of the [002] peak increased when SDR exceeds 20.

11.4.2.2 Coagulation

Ran et al. reported in situ analyses on the structure formation in the coagulation process. A 240 μm diameter filament of 14 % polyphosphoric acid solution was used for the analysis. At the time of 2 s after dipping in the water bath, there was no big change in the scattering pattern both of WAXD and SAXS. During this coagulation time, the portion that had developed fiber structure was little [58]. The swollen microfibrillar network structure which is proposed by Martin and Thomas [59] would appear later, when most of the phosphoric acid is removed from the fiber. Ran et al. reported a follow-up analysis. At 0.3 s of coagulation, the [010] reflection as the cohesion of the polymer appeared at the fiber surface [60]. Daves et al. proposed a structure development model, which generated crystals at the fiber surface regulating the lattice direction of [010] plain and increasing toward the inner part of the fiber [61]. At present, there is no evidence we can ascertain as to whether or not molecules in solution are oriented radially before coagulation. Preferential orientation of PBO fiber was reported as the *a*-axis of crystal lined up in radial direction [29].

11.4.2.3 Washing and Neutralization

The fiber swelled with water is formed after polyphosphoric acid is extracted from the stretched solution filament. The volume fraction of the water in this soaked fiber is approximately 50 %. The phosphoric acid can be extracted rapidly in a state containing enough amount of water. Approximately 0.6 % of phosphoric acid stays in fibers even after a prolonged extraction time. Phosphoric acid molecule coordinates to an amino phenol end group or to an oxazole ring, and such a "trapped" phosphoric acid is not easy to extract by washing and stays in the fiber as a residual acid [62]. The washed fiber is soaked in an alkaline water to neutralize the residual phosphoric acid.

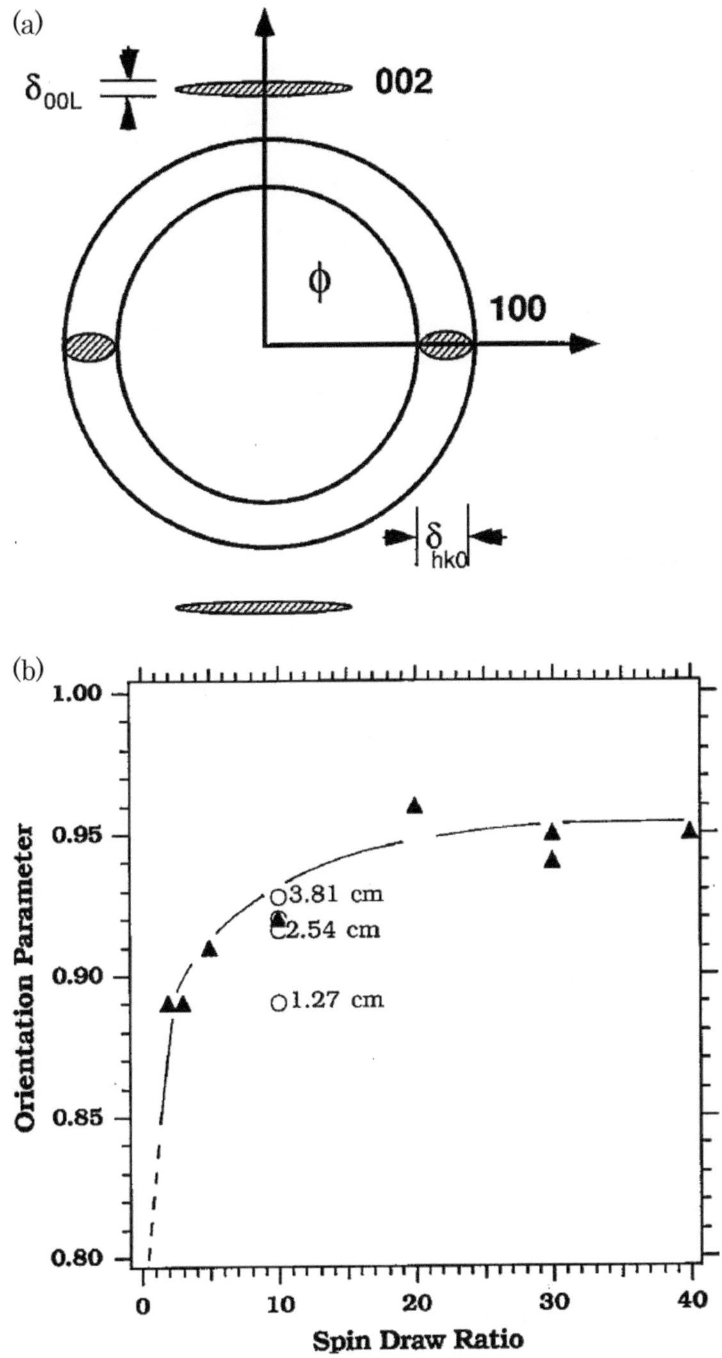

Fig. 11.13 In situ WAXD analysis of PBO/PPA solution spinning [57]. (**a**) Principal diffraction spots. (**b**) Orientation development in spin line

11.4.2.4 Drying

The moisture content of the washed fiber can reach an equilibrium moisture content, even if the fiber is simply left under normal ambient atmosphere. The modulus of dried fiber, reflecting the orientation degree of the molecules, can be changed by controlling tension and temperature during the drying process [63].

11.5 Applications

Having an extremely high performance as a novel super fiber, Zylon® has explored a lot of applications during the last 15 years, since it was launched in 1998. Its applications are categorized into three groups.

11.5.1 Heat-Resistant Materials

Zylon® is used in industrial materials and personal protective equipment (PPE). For these applications, its excellent mechanical strength is highly appreciated, in addition to its superior properties such as heat dissipation, flame resistance, and heat resistance. Examples of application for industrial materials are cushions used in extrusion process of aluminum materials, cushions for processing heated glasses, and protective materials against sputtered metals for welding robots in car production lines. Major applications for PPE are garment and accessories for firefighters and protective materials used in metal manufacturing process and metal welding work.

11.5.2 Fiber-Reinforced Composites

The fiber reinforcements using Zylon® are intended to utilize its tensile properties (strength and modulus) or its abrasion resistance. In the former case, Zylon® is used for carbon fiber-reinforced plastics to improve fracture toughness and retain configurations at crash destruction. Some Formula One cars have introduced a chassis of carbon fiber-reinforced plastics hybridized with Zylon®. The materials reinforced with Zylon® show superior penetration resistance (Fig. 11.14). Zylon® is used for high-end applications requiring high specifications, such as racing helmets and turbine containment.

In the latter case, the very high abrasion resistance of Zylon® is appreciated, as the abrasion wear is minor, when the cross sections of fibers are rubbed by metals. The life of the rubber belts for continuously variable transmission (CVT) improves

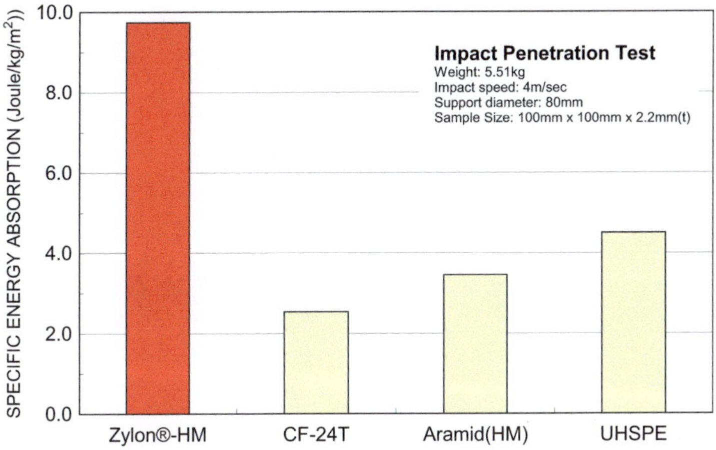

Fig. 11.14 Impact penetration property

a lot, when Zylon® chopped fibers are added to the rubber as fillers. The Zylon® filler is used in CVTs for motorcycles and snowmobiles.

11.5.3 Rope and Cables

There are two unique applications where Zylon® is used as a tether for safety in racing cars and space exploration vehicles. In 1999, Formula One World Championship started to take safety measures to avoid accidents caused by scatter of suspension parts of crashed cars. Since then, Zylon® has been used for tethers tying monocoque bodies and suspensions or uprights.

Another interesting application is for Mars Exploration Rover (MER). In 2004, NASA successfully landed a MER on Mars. For a smooth and safe landing, the rocket and the rover were tied with a tether to control the posture of the three bodies (Rover wrapped by air bags, solid rocket motor, and parachute). Zylon® was used for this tether application, due to the highest properties in strength and heat resistance among the super fibers existing on the earth [64].

Zylon® has been used for a lot of sports goods so far, such as tennis rackets, table tennis rackets, and strings for the Japanese art of archery, "wakyu." Another unique example is a tire or wheel for bicycles. Spinergy company commercialized a lightweight wheel for racing bicycles [65], in which metal spokes have been replaced by Zylon® spokes (Fig. 11.15). Figure 11.16 shows creep properties with safety factor 2 (50 % load of the break). Such excellent creep resistance will further realize even more unique applications.

Fig. 11.15 Zylon spokes for racing bicycle [65]

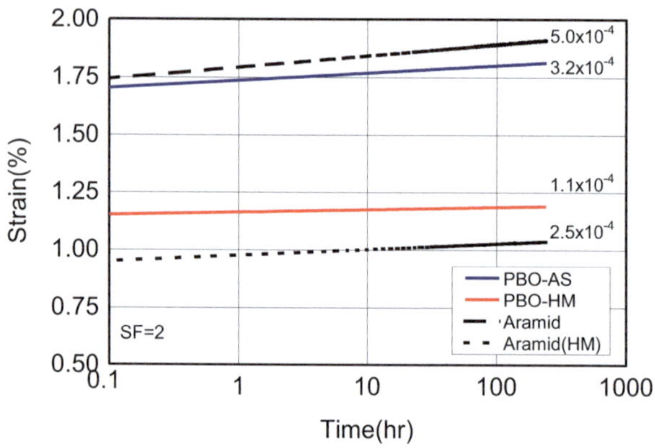

Fig. 11.16 Creep parameters of PBO (Zylon®) and *p*-aramid (PPTA)

11.6 Conclusions

Many researchers and engineers in universities, USAF, SRI, Dow Chemical, and Toyobo contributed to the birth of the world's strongest fiber. It is noteworthy that each party played a different key role at a different stage, and the total development finally bore fruits under the continuous "teamwork" performed over 20 years.

As described and discussed in this chapter, Zylon® has outstanding characteristics in strength, modulus, flame resistance, creep resistance, and thermal conductivity among the existing organic fibers. However, high-strength fibers have the tendency to be affected by the defect formed by various deterioration modes, as they have very few defects before use. Further efforts to control defect generation will make Zylon® more useful and attractive in the existing applications and will also open a door for new applications. Meanwhile, for applications where heat resistance is appreciated, further development to improve the performance targeting inorganic materials is highly anticipated.

References

1. K. Yabuki, K. Kato, Sen'i Gakkaishi J. **52**, 143 (1996)
2. T. Kuroki, K. Yabuki, Sen'i Gakkaishi J. **54**, 16 (1998)
3. K. Kato, Kobunshi J. **48**, 20 (1999)
4. K. Yabuki, in *Progress in Textiles: Science and Technology*, ed. by V.V. Kothari, vol. 2 (IAFL Publications, New Delhi, 2000), pp. 615–651
5. F. Kubota, Kobunshi J. **54**, 246 (2005)
6. Y. Imai, Sen'i Gakkaishi J. **61**, 278 (2005)
7. Y. Iwakura, K. Uno, Y. Imai, J. Polym. Sci. Part A: Polym. Chem. Macromol. Chem. **2**, 2605 (1964)
8. S. Inoue, Y. Imai, K. Uno, Y. Iwakura, Macromol. Chem. **95**, 236 (1966)
9. Y. Iwakura, Kobunshi J. **17**, 130 (1968)
10. Y. Imai, Kobunshi J. **20**, 510 (1971)
11. D.R. Ulrich, Polymer **28**, 533 (1987)
12. F.E. Arnold, Mater. Res. Soc. Symp. Proc. **134**, 75 (1989)
13. J.F. Wolfe, Mater. Res. Soc. Symp. Proc. **134**, 253 (1989)
14. J.F. Wolfe et al., Macromolecules **14**, 1135 (1981)
15. SRI U.S. Patent, 4,225,700, U.S. Patent, 4,533,724, U.S. Patent, 4,533,692, U.S. Patent, 4,533,693, U.S. Patent, 4,545,232
16. Z. Lysenko, U.S. Patent, 4,766,244
17. http://www.toyobo-global.com/seihin/kc/pbo/Technical_Information_2005.pdf
18. H.G. Chae, S. Kumar, J. Appl. Polym. Sci. **100**, 791 (2006)
19. H. Murase, Sen'i Gakkaishi J. **66**, 176 (2010)
20. T. Ito, Polym. Prepr. Jpn. J. **42**, 49 (1993)
21. S.G. Wierschke, Mater. Res. Soc. Symp. Proc. **134**, 313 (1989)
22. N. Horikawa, Y. Nomura, T. Kitagawa, Y. Haruyama, A. Sakaida, T. Imamichi, S. Sasaki, H. Takekawa, Nihon Kikai Gakkai Ronbunshu J. **75**, 373 (2009)
23. Z. Chi, T.-W. Chou, G. Shen, J. Mater. Sci. **19**, 3319 (1984)
24. T. Kitagawa, K. Tashiro, K. Yabuki, J. Polym. Sci. Part B **40**, 1269 (2002)
25. C.Y.-C. Lee, U. Santhosh, Polym. Eng. Sci. **33**, 907 (1993)
26. T. Kitagawa, M. Ishitobi, K. Yabuki, J. Polym. Sci. Part B **38**, 1605 (2000)
27. T. Kiagawa, K. Yabuki, R.J. Young, J. Macromol. Sci. Part B **41**, 1 (2002)
28. T. Kitagawa, K. Tashiro, K. Yabuki, J. Polym. Sci. Part B **40**, 1281 (2002)
29. T. Kitagawa, H. Murase, K. Yabuki, J. Polym. Sci. Part B **36**, 39 (1998)
30. C. Burger, S. Ran, D. Fang, D. Cookson, K. Yabuki, Y. Teramoto, P.M. Cunniff, P.J. Viccaro, B.S. Hsiao, B. Chu, Macromol. Symp. **195**, 297 (2003)
31. Y. Yamashita, S. Kawabata, H. Minami, S. Okada, A. Tanaka, Proceedings of the 30th Textile Research Symposium at Mt. Fuji in the New Millennium, Shizuoka, **219**, (2001)
32. A.A. Leal, J.M. Deitzei, J.W. Gillespie Jr., J. Compos. Mater. **43**, 661 (2009)
33. S.J. Deteresa, R.S. Porter, R.J. Farris, J. Mater. Sci. **23**, 1886 (1998)

34. K. Tashiro, J. Yoshino, T. Kitagawa, H. Murase, K. Yabuki, Macromolecules **31**, 5430 (1998)
35. K. Tashiro, H. Hama, J. Yoshino, Y. Abe, T. Kitagawa, K. Yabuku, J. Polym. Sci. Part B **39**, 1296 (2001)
36. E. Lorenzo-Villafranca, K. Tamargo-Maet¡nez, J.M. Molina-Aldareguia, C. Gonz'alez, A. Mat¡nez-Alonso, J.M.D. Tasc'on, M. Gracia, J. Llorca, J. Appl. Polym. Sci. **123**, 2052 (2012)
37. N. Horikawa, Y. Haruyama, A. Sakaida, M. Ueda, J. Soc. Mater. Sci. Jpn. J. **54**, 801 (2005)
38. S. Bourbigot, X. Flambard, Fire Mater. **26**, 155 (2002)
39. S. Bourbigot, X. Flambard, F. Poutch, S. Duquesne, Polym. Degrad. Stab. **74**, 481 (2001)
40. H. Fujishiro, M. Ikebe, T. Kashima, A. Yamanaka, Jpn. J. Appl. Phys. **36**, 5633 (1997)
41. X. Wang, V. Ho, R.A. Segalman, D.G. Cahill, Macromolecules **46**, 4937 (2013)
42. Japanese Laid-open Patent 2010-144299A
43. Y. Kaburagi, K. Yokoi, A. Yoshida, Y. Hishiyama, TANSO **217**, 111 (2005)
44. Y.-H. So, S.J. Martin, K. Owen, P.B. Smith, C.L. Karas, J. Polym. Sci. Part A **102**, 1428 (1999)
45. G. Wu, C.-H. Hung, S.-J. Liu, ICCM12 Conference, Paris, July 1999, paper 616, ISBN 2-9514526-2-4
46. J. Chin, A. Forster, C. Clerici, L. Sung, M. Oudina, K. Rice, Polym. Degrad. Stab. **92**, 1234 (2007)
47. C.-H. Zhang, Y.-D. Huang, W.-J. Yuan, J.-N. Zhang, J. Appl. Polym.Sci. **120**, 2468 (2011)
48. T. Zhang, J. Jin, S. Yang, G. Li, J. Jiang, J. Polym. Adv. Technol. **22**, 743 (2011)
49. Y.-H. So, J.M. Zaleski, C. Murlick, A. Ellaboudy, Macromolecules **29**, 2783 (1996)
50. Y.-H. So, Polym. Int. **55**, 127 (2006)
51. B. Song, Q. Fu, L. Ying, X. Liu, Q. Zhuang, Z. Han, J. Appl. Polym. Sci. **124**, 1050 (2011)
52. E.C. Chenevey, T.E. Helminiak, U.S. Patent 4,606,875
53. J.A. Odell, G. Ungar, J.L. Efijoo, J. Polym. Sci. Part B **31**, 141 (1993)
54. D.F. Toy, *Comprehensive Inorganic Chemistry* (Pergamon Press, New York, 1973), p. 486
55. B. Ernst, M.M. Denn, P. Pierini, W.E. Rochefort, J. Rheol. **36**, 289 (1992)
56. M. Mortie, P. Moldenaers, J. Mewis, Rheol. Acta **35**, 57 (1996)
57. M.J. Radler, B.G. Landes, S.J. Nolan, C.F. Broomall, T.C. Chritz, P.R. Roudolf, M.E. Mills, R.A. Bubeck, J. Polym. Sci. Part B **32**, 2567 (1994)
58. S. Ran, C. Burger, D. Fang, X. Zong, S. Cruz, B. Chu, B.S. Hsiao, R.A. Bubeck, K. Yabuki, Y. Teramoto, D.C. Martin, M.A. Johnson, P.M. Cunniff, Macromolecules **35**, 433 (2002)
59. D.C. Martin, E.L. Thomas, Mater. Res. Soc. Symp. Proc. **134**, 415 (1989)
60. S. Ran, C. Burger, D. Fang, X. Zong, S. Cruz, B. Chu, B.S. Hsiao, Y. Ohta, K. Yabuki, P.M. Cunniff, Macromolecules **35**, 9851 (2002)
61. R.J. Davies, M. Burghammer, C. Riekel, Macromolecules **40**, 5038 (2007)
62. N.V. Lukasheva, Polymer **52**, 1458 (2011)
63. M.E. Mills et. al., U.S. Patent 5,976,447
64. http://marsrover.nasa.gov/mission/spacecraft_edl_parachute.html
65. http://www.spinergy.com

Part III
Functional and Speciality Man-Made Fibers

Chapter 12
Overview of Functional and Speciality Fibers

Togi Suzuki

Abstract The development of functional and speciality fibers, which have equivalent performances of easy-care properties of man-made fibers, superior touch and appearance of silk and wool, and outstanding moisture and water-absorbing abilities of cotton, was started shortly after the birth of man-made fibers. Afterward, the application of functional and speciality fibers has expanded toward high-touch, moisture and water absorbability, heat insulation and exothermicity, enhanced color developability, antistatic and conductivity, antibacterial and deodorant, flame retardant and proofing properties, and have contributed to our comfortable, healthy, safe, and environment friendly life. In the development of the functional and speciality fibers, Japan has cultivated the world's top level of development capability. The driving force of this Japanese R&D originates with the development of technologies that consist of the trinity of polymer, fiber and yarn, and post-processing modifications, together with biomimetic approaches, to learn the structure and functions of living organisms in the natural world. In this chapter, the key production technologies of functional and speciality fibers are presented with comments on representative examples of the fibers.

Keywords Functional and speciality fiber • High-touch fiber • Moisture-control fiber • Heat-controllable fiber • Antistatic fiber • Electroconductive fiber • Antibacterial fiber • Deodorant fiber • Flame-retardant fiber • Biomimetic fiber

12.1 Introduction

It was over 70 years ago that the history of man-made fibers started with the sale of women's stockings made of nylon, the first man-made organic textile fiber. Nylon fiber was an invention of note not only because it is derived from "coal, water, and air" but also because it is "as strong as steel" and is "as fine as the spider web." Since then, macromolecular design has been conducted by human hands to construct various polymer structures that are suitable for making man-made fibers. As a

T. Suzuki (✉)
Teijin Limited, Tokyo, Japan
e-mail: tou.suzuki@kke.biglobe.ne.jp

Society of F.S. Technology, Japan (ed.), *High-Performance and Specialty Fibers*,
DOI 10.1007/978-4-431-55203-1_12

result, various kinds of man-made fibers were created and widely used to various purposes to bring revolutionary changes in our life [1].

Looking back on the history of man-made fibers, their progress can be divided into four stages. (1) The first stage was in "the dawn of man-made fibers." In this stage, the original easy-care properties of man-made fibers such as wash and wear characteristics, wrinkle resistance, and durability were highlighted, accomplishing remarkable developments with the invention of various fibers. (2) The second stage was in "the time for imitating natural fibers." Here, man-made fibers were improved with an idea, "how to make them closer to natural fibers." By using the elaborate structures and ingenious functions of natural fibers as models, much effort has been made to impart functional properties such as superior touch and appearance exhibited by silk and wool and outstanding hydrophilic nature of cotton to man-made fibers. (3) The third stage was in "the time of new man-made fibers called "Shin-gosen"." Here, the fibers having high sensitivity and textures that are intrinsic to man-made fibers and not available even with natural fibers were investigated. For example, bulky yarns exceeding silk fibers and superior yarns having downy hair touch as peach skin were created. (4) In recent years, we have entered the fourth stage, i.e., "the time of market-oriented functional and speciality man-made fibers." People have been placing so much importance on advanced functionalities that are required for better quality of life and environment, for improved safety, and for saving and recycling of energy and natural resource. Such a surge of social needs to a better life urges man-made fibers to be more comfortable, healthy, safe, and environment conscious, calling for a big change in the social trend from the product-oriented development to the market-oriented development. Table 12.1 summarizes a genealogy of the man-made fibers mentioned above, whereas Fig. 12.1 shows a map of functions and specialities that have been accomplished by various functional man-made fibers [2–4].

In the development of man-made fibers, Japan has cultivated and accumulated a strong potential, which should be of top rank in the world, particularly in developing those of high-touch and elaborate functionalities. The driving force of the Japanese R&D ought to be based on the trinity of technological developments in polymer synthesis, fiber fabrication, and fiber processing/dyeing, as well as on the biomimetic or bioinspired approaches that can be attained by learning the structures and functions of living things in nature.

This chapter offers a basic knowledge on the functional and speciality man-made fibers by reviewing (1) the world production amounts of man-made fibers in 2013, (2) the modification technologies of man-made fibers (chemical modification of polymers, structure modification of fibers, and post-processing technologies), and (3) the functional and speciality man-made fibers developed with advanced technologies as well as those developed by the biomimetic approaches.

Table 12.1 Genealogy of the man-made fibers

Stage	Feature, structure, technology	
1st	The dawn of man-made fibers (1958 ~)	Easy-care properties (wash and wear characteristics, wrinkle resistance, durability)
2nd	The time for imitating natural fibers (1964 ~) [silky man-made fibers]	Triangular cross-sectional fiber, alkali-reduction treatment, composite yarn combining filaments having different shrinkages, high-count yarn, thick and thin yarn, irregular cross-sectional filament yarn
3rd	The time of "Shin-gosen" (1988~) [sensitivity intrinsic to man-made fibers]	New silky fabric (bulky, exceeding silk), rayon-like dry fabric (dry touch and high drape), peach-faced fabric (downy hair touch as peach skin), new worsted yarn fabric (soft-touch and drape heterogeneous with wool)
	Evolutionary "Shin-gosen" (1993 ~) [sensitivity + function]	New "Shin-gosen" (linen-like, new worsted yarn fabric (tailoring brilliancy), complex new sensitivity, fibril), comfortable "Shin-gosen" (quickly absorb and dry sweats, lightweight and heat retaining, stretchable)
4th	The time of market-oriented functional and speciality man-made fibers (1997 ~) [comfortable, healthy, safe, and environment-conscious]	Moisture-absorbing and sweat-absorbing, self-regulating, moisture-retaining, water-repellent, antifouling, moisture-permeable waterproofing, lightweight and heat retaining, heat reserving, infrared ray related, antistatic, electroconductive, electromagnetic wave shielding, antibacterial, deodorant, flame-retardant, ultraviolet protection, gentle to skin, aromatherapeutic

12.2 Production Amount of Man-Made Fibers [5]

The total amount of fibers produced in the world in 2013 was 84,494,000 tons in which the amount of chemical fibers reached 57,615,000 tons (68.2 %) with the rest 26,879,000 tons (31.8 %) occupied by natural fibers. The production of man-made fibers amounted to 52,706,000 tons, dominating 91.5 % of the production of chemical fibers. Among the man-made fibers, polyester fibers amounted to 45,284,000 tons in total including both filament and staple forms, whereas nylon and acrylic fibers amounted to 4,135,000 and 2,001,000 tons, respectively. Namely, the three major man-made fibers, i.e., polyester, nylon, and acrylic, occupied 97.6 % of man-made fibers (Fig. 12.2).

While most of the acrylic and nylon fibers are utilized as staple fibers and filament yarns, respectively, the polyester fibers are used both as filament yarns and staple fibers in large quantity. Polyester fibers have the widest application not

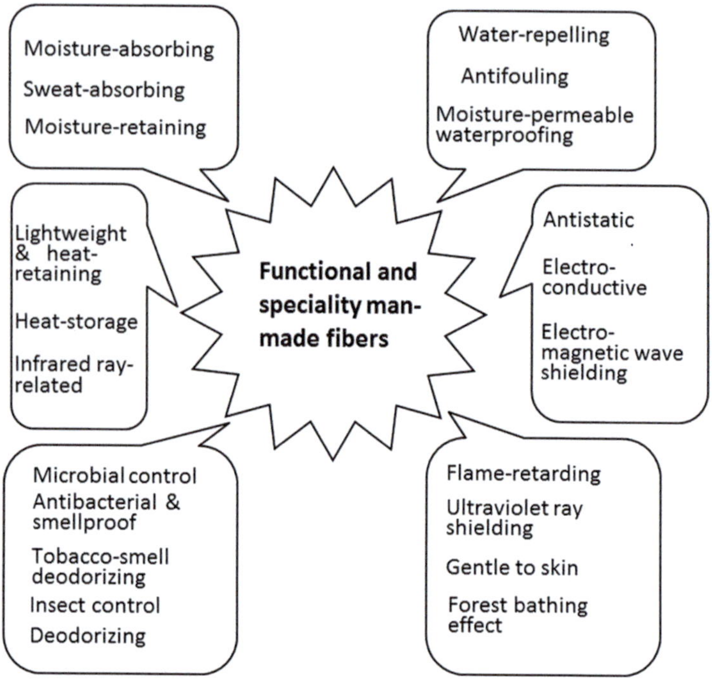

Fig. 12.1 Map of functions and specialties

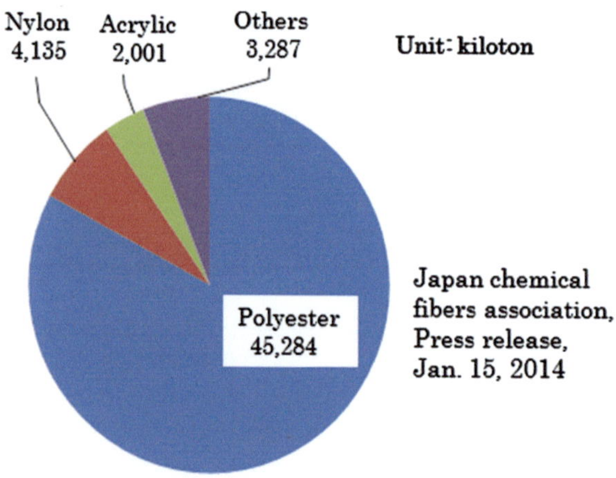

Fig. 12.2 Man-made fiber production in 2013

only to textile use but also to home and industrial uses. This large share of polyester fibers can be attributed to the following reasons:

1. Polyester fibers are superior to other fibers in physical and chemical properties. In particular, they show advantages in strength, heat resistance, chemical resistance, etc.

2. Their raw materials are inexpensive, getting an economical advantage.
3. The productivity is very high because of their excellent melt spinnability. Differing with dry and wet spinning methods, melt-spinning method needs no organic solvent and is preferred in terms of environmental load.
4. Various kinds of technologies are available for modification of polyester fibers. The modification can be done at any stage of fiber production, i.e., polymer production, fiber making, and post-processing such as dyeing and finishing.

One of the most famous examples for the modification of polyester fibers and textiles is "alkali-reduction treatment" by which textiles of polyester fiber are allowed to become softer but stiffer with anti-drape and resilient properties. Accordingly, the textiles of alkali-treated polyester fibers can give a feeling similar to that of silk textiles from which the surface sericin is removed by the similar alkali treatment. The man-made silky fibers thus developed have created a new market. During the alkali-reduction treatment, polyester fibers are hydrolyzed with alkali and narrowed sequentially with dissolution of the fiber surface. The fibers become thinner without losing mechanical properties, which are close to those of native silk fibers. Since this technique is used only for polyester, big advantage has been taken by polyester fibers over the other man-made fibers [6]. In addition to this easy modification, the overwhelming share, reaching 88 % of the three major man-made fibers, has made use of polyesters for developing functional and speciality fibers.

In spite of these advantages of polyester fibers, nylon fibers, consisting of aliphatic polyamides, are used in considerable amount for moisture-absorbing application, because nylon shows highly hygroscopic nature in comparison with polyester (official moisture regain: nylon 4.5 % and polyester 0.4 %).

In addition, cross-linked acrylic fibers containing a large number of hydrophilic groups are made from acrylonitrile fibers, whose primary structure is shown by $^{-}CH_2CH_2CN^{-}$. By converting the pendent cyano groups (CN group) to hydrophilic groups (metal salts of carboxylic acids), the acrylic fibers can attain such a highly hygroscopic nature as to exceed that of natural fibers. Some of the acrylic fibers currently manufactured are known to show a moisture absorption coefficient of about 40 % under the standard environment of 20 °C and 65 %RH.

12.3 Modification Technologies of Man-Made Fibers [7, 8]

Key technologies that have driven the evolution of man-made fibers are divided into three categories: "technology for chemical modification of polymers," "fiber modification technology," and "post-processing technology." Below, these three technologies are explained by taking polyester fibers as examples.

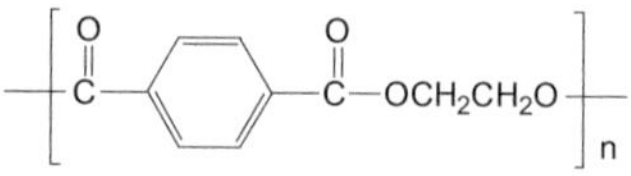

Fig. 12.3 PET structural formula

12.3.1 Technology for Chemical Modification of Polymers

Although "polyester" generally represents a class of polymers connected with ester bonds ($^{-}CO^{-}O^{-}$), it is also used to specify poly(ethylene terephthalate) (PET) consisting of terephthalic acid and ethylene glycol as dicarboxylic acid and diol components, respectively (Fig. 12.3). Thus, when one simply says "polyester fiber," it means PET fiber.

Generally, two methods are utilized in the chemical modification of polymers, copolymerization and polymer blend (with organic and inorganic modifiers). In the copolymerization method, it is a merit that the function of the modifiers lasts semipermanently because an appropriate amount of modifying units are directly connected to the PET main chains as comonomers. However, it involves demerits that the melting temperature and degree of crystallization likely decrease in as much as to make the heat resistance decrease. Accordingly, the variety of the comonomer design is not high.

In the polymer blend method, on the other hand, the choice of the modifiers is wider because of requiring no chemical reaction. However, the durability of the function of modifiers is inferior to that realized by the copolymerization method, because the modifiers are not connected to the polymer backbones. In particular, inorganic modifiers likely cause many problems about processability, for example, filter choking during spinning. Even organic modifiers cause thermal degradation somewhere in the thermal history of polymerization and melt-spinning processes.

Table 12.2 compares the principal technologies for chemical modification of polymers used for polyester fibers. The cationically dyeable polyester fiber is made by copolymerization of sodium 5-sulfoisophthalic acid as a modifier, whereas the polyester fiber dyeable with disperse dye at atmospheric pressure is made by copolymerizing polyalkylene glycol as a modifier. Since the latter copolyester is easily alkali soluble, it can be utilized for manufacturing special modified cross-sectional fibers with sharp edge and super extra-fine fibers. Here, sea-island or splittable conjugate yarns are made by the conjugate melt-spinning process with the copolyester as one component, then fabricated into woven or knitted fabrics, and finally subjected to dyeing and finishing to dissolve the easily alkali-soluble polyester to obtain fibers having such finest structures.

12.3.2 Fiber Modification Technology

Different from the polymer modification technology depending mainly on functionalization with chemical elements, the fiber modification technology utilizes

Table 12.2 Principal technologies for chemical modification of polymers used for polyester fibers

Class	Modified polyester	Function, sensitivity	Concrete examples of modifier
Copolymerization	Cationically dyeable polyester	Brilliant color development with use of cationic dyes, different color effect in blends with disperse-dye dyeable regular polyester	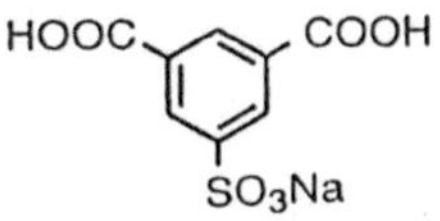
	Cationically dyeable polyester at atmospheric pressure	Dyeable with cationic dyes at 100 °C, blends with fibers weak in heat resistance, such as wool	Increased copolymerization amount of the compound mentioned above
	Disperse-dye dyeable polyester at atmospheric pressure	Dyeable with disperse dyes at 100 °C	$HO\{AO\}_nH$ (A: alkylene group), aliphatic dicarboxylic acid
	High shrinkage polyester	Creating textiles with excellent bulkiness by use of the composite yarns combined with low shrinkage polyester, improved amorphousness	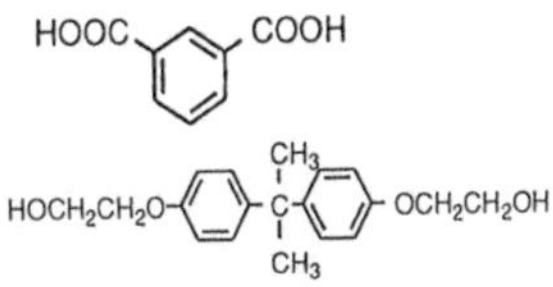
Polymer blend	Antistatic polyester	Avoiding static electricity, preventing clothes from crackling when putting on and taking off them and skirts from clinging to the body	Polyalkylene glycol polymer + ionic compound
	Microporous polyester	Microporous fibers exhibiting improved color depth and brilliance when dyed, water-absorbing microporous hollow fibers, dry-touch fibers having microgrooves	Inorganic microparticles such as colloidal silica R-SO_3Na (R: aromatic or long-chain aliphatic group)
	High functions and specialities	High density (drape)	Titanium dioxide, barium sulfate
		Ultraviolet protection	Titanium dioxide
		Solar energy storage	Zirconium carbide
		Antimicrobial	Silver-bearing zeolite
		Electroconductive	Carbon black

the functionalization by changing the physical properties and morphology of fibers, for example, fiber diameter (from super-extra-fine to ultra-thick), mechanical properties (such as strength elongation, thermal shrinkage, thermal stress, etc.), fiber cross-sectional shape (triangular, multilobed, hollow, flat), and conjugate structure (such as sheath core, side by side, sea island, etc.). One of the key technologies in melt-spinning process consists in the design of spinning machine, in particular, cap design and pack channel design. Furthermore, a wide variety of modification technologies have been developed in the drawing and yarn texturing processes. For example, composite yarns combining filaments having different shrinkages and those interlacing filaments having different physical properties are manufactured. Core-sheath double-layered yarns made by the process of false-twist texturing the yarns consisting of filaments having different physical property are also manufactured. Table 12.3 compares the representative technologies for fiber modification of polyester using specific mechanical processes (spinning, drawing, and yarn texturing).

The filament yarns and textured yarns described above are fabricated into woven or knitted fabrics and subjected to post-processings such as dyeing and finishing to finally attain the functional properties engineered by the polymer chemical modification and/or fiber modification.

12.3.3 Post-processing Modification Technology

In the post-processing of woven or knitted fabrics, dyeing is the main process. The dyed fabrics are then subjected to various chemical and/or physical processings such as refinement, relaxation, preset, alkali-reduction, and other functionality imparting treatments and finally heat set under dry or moist heat conditions. Table 12.4 summarizes the main technologies for the post-processing of polyester woven or knitted fabrics.

As mentioned above, the alkali-reduction treatment is a process for manufacturing silky man-made fiber textiles and is utilized only for silk and polyester fibers. The weight reduction of polyester fibers can easily be performed by contacting with concentrated aqueous solution of alkali, because the main-chain ester bonds are easily hydrolyzed. Since polyester is hydrophobic and does not swell in an alkaline solution, the hydrolysis proceeds from the fiber surface in a stoichiometric manner, and the hydrolysis products (alkali salt of terephthalic acid and ethylene glycol) are allowed to dissolve away and diffuse in the alkali bath. Such a mechanism working on polyester fiber is favorable for the homogeneous thinning of fibers from the surface. The process control of weight reduction is possible by measuring the quantity of alkali consumption even in the industrial production (Fig. 12.4). The fiber thinning can bring a fine space between the woven fibers of fabric and an increase in mobility of the fibers. In consequence, the handling of the fabric becomes soft and flexible, while the physical properties such as strength do not substantially decrease in values per cross-sectional area of a fiber [6].

Although the bath absorption method similar to the dyeing process with disperse dyes is utilized to impart functional properties to fibers, its applicability is rather

Table 12.3 Representative technologies for fiber modification of polyester using specific mechanical processes (spinning, drawing, and yarn texturing)

Process		Technologies	Content	Application cases
Spinning		Modified cross section	Triangular, multilobed, flat, Y, U, M, W, 	Sensitivities, functions
		Hollow section	Hollow section, large hollow section, modified hollow section	Lightweight and heat-retaining fibers
		Conjugate spinning	Multilayer type	Super extra-fine fibers (mechanical splitting)
			Sea-island type	Super extra-fine fibers (dissolution splitting)
			Sheath-core type	Antistatic fibers, electroconductive fibers
			Side-by-side type	Latently crimpable fibers (stretchable fabric)
Drawing		Combined-filament yarns	Low shrinkage filaments + high shrinkage filaments (interlacing)	Composite yarns combining filaments having different shrinkages (different polymers, different fiber deniers, different fiber cross sections, etc.)
		Impartment of unevenness	Low-ratio drawing (necking drawing)	Thick and thin yarns (natural unevenness visual feeling)
Yarn texturing	False-twist texturing	Double-layer structure	Core-sheath double-layered yarns (false-twist textured yarns consisting of filaments having different physical property)	Spun-like appearance and handle
		Three-layer structure	Uneven texturing	Tsumugi-like textured yarn and fabric (tsumugi, hand spun silk woven fabric)
		Special conditions	Patial melt binding	Melt-spot false-twist textured yarns, pri-twist false-twist textured yarns (crispy-touch fabric)
	Fluid texturing	Interlaced-combined filaments	Air jet interlacing	Combined-filament yarns with another filament material (regular polyester, cationically dyeable polyester, cellulose, nylon)
		Bulk-texturing	Air jet texturing (Taslan process, loop formation)	Bulky yarns

Table 12.4 Main technologies for the post-processing of polyester woven or knitted fabrics

Main technologies	Means, equipment	Application, purpose
Relaxation	Rotary washer, wince, jet dyeing machine, continuous relaxer	By relaxing the fabric in hot water, the distortions are relieved and such properties as bulkiness and stretch-ability are imparted. (Potential functions of the combined filament yarns with the different shrinkage, latently crimpable fibers, etc. are revealed)
Alkali-reduction treatment	Jet dyeing machine (with weight reduction control mechanism), continuous weight reduction machine (pad steam method, etc.)	Improvement of handle or touch (a process for manufacturing silky man-made fiber textiles)
Dyeing	High temperature high pressure jet dyeing machine (most commonly applied)	Level dyeing with disperse dyes or cationic dyes (pale color to deep color), accompanied with crumpling, relaxing, and also high productivity effects
Functionality imparting treatments	Pad method (most generally applied)	Pad-dry-cure method or pad-steam method in which a modifier is fixed onto the fiber surface with a binder resin
	Bath absorption method (partially applied)	Dyeing and modifier-treatment simultaneous processing, etc., treating in the same dyeing bath, the modifier is absorbed into the core of fibers like dyes (bromine-based flame retardant, phosphorus-based flame retardant, specific bacteriostatic agents, etc.)

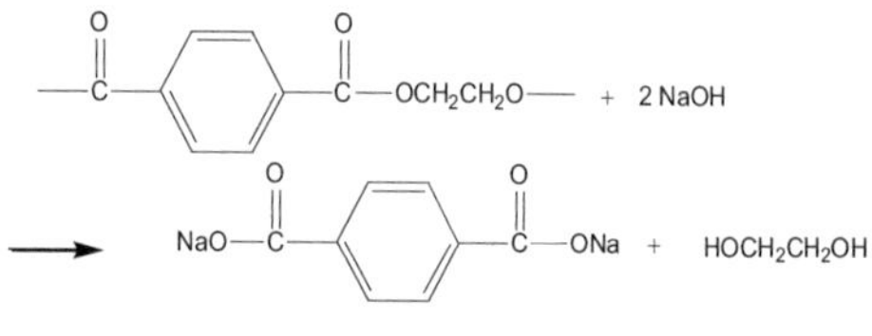

Fig. 12.4 Alkaline hydrolysis of PET

limited because the modifiers having a solubility parameter close to that of polyester are limited in number.

The most generally applied to fiber modification is the pad method in which a modifier is fixed onto the fiber surface with a binder resin. In this pad method, the handling of fabrics likely becomes harder, and the durability of functional properties is not as high as that imparted by the aforementioned fiber modification.

However, the pad method shows advantages over the polymer and fiber modifications in easiness of production in small lots and in variety of modifiers available. The requirements for the modifiers are milder, particularly concerning the heat resistance, particle size, and so on.

12.4 Biomimetic Man-Made Fibers Having Specific Structures and Functions [7, 8]

The potential functionalities and specialties are first contrived to polymers and/or fibers by using the aforementioned polymer and/or fiber modification technologies and intensified in the post-processing stages, including the dyeing step to produce highly functionalized and specialized fibers.

Table 12.5 shows the high-level functional and speciality man-made fibers that can be differentiated by the modification stages, i.e., polymer, fiber, and their elemental technologies. It is evident that many of the speciality fibers were developed by envisaging the trinity of polymer, fiber, and post-processing technologies dealing with high sensitivities and functionalities.

The blend and copolymerization with special modifiers are used in the polymer technology, while the fibers with modified cross section, conjugate fibers, composite yarns combining filaments with different shrinkage, and composite yarns of false-twist texturing filaments are elements of the fiber modification technologies. The alkali-reduction treatment, dyeing, and functionality imparting processings are important in the post-processing technology.

Simultaneously, another important key concept that has led the development of these speciality fibers is the "form of synthetic fibers." It represents that the structure and function of living things in nature can serve as the sources of idea, which is sometimes represented by a word "bioinspired" or "bio-mimic." For example, such inspiration is gotten from the function that the cornea of a moth having a submicron concave-convex surface structure does not reflect light for camouflage. Similarly, the capillary action of trees for absorbing water, squeaky feeling of a tussah silk fabric, sand-washed silk fabric, natural and suede leathers, and the color depth of black-dyed wool fabric and the function of the concave-convex surface structure (waxy substance) of a lotus leaf or a taro leaf that repels drops of water are examples of functions and structures from which new ideas are inspired in designing man-made fibers.

Table 12.5 High-level functional and speciality man-made fibers that can be differentiated by the modification stages

Materials, products	Characteristics	Polymer/fiber	Post-processing	Resources of idea
Micro-crater fiber	Antireflective structure on the fiber surface, enhanced color depth and brilliance	Blending of micropore-forming agents	Alkali-reduction treatment	Cornea of a moth having a submicron concave-convex surface structure
Microporous hollow fiber	Absorbing water into the hollow via the pores by capillary action, quick drying			Water-absorbing action of trees using capillary phenomenon
Microgroove fiber	Squeaky feeling of a tussah silk fabric			Tussah silk (Indian silk, Chinese silk)
Fibrillated fiber on its surface	Fibril touch of lyocell or sand-washed silk fabric	Blending of fibril-forming agent	Application of stress to fabric surfaces, alkali-reduction treatment	Sand-washed silk fabric
Dyeable polyester fiber at 100 °C (disperse dyes, cationic dyes)	Blending use with natural fibers and/or cellulosic fibers, garment dyeing	Copolymerization	Dyeing technology	Natural fibers
"Shin-gosen"	Bulkiness exceeding silk, downy hair touch as peach skin	Composite yarns combining filaments having different shrinkages (self-elongating yarn + high shrinking yarn)	Control of heat shrinkage behavior in dyeing and finishing process	From imitating natural fibers to exceeding natural fibers
Super extra-fine fibers (less than approximately 3 μm in diameter)	Artificial leathers, wiping cloths (for spectacles, etc.), filters	Conjugate spinning (sea-island type, multilayer type)	Dissolution-splitting, mechanical splitting	Super extra-fine fiber bundle of natural leather, Swede leather with a napped finish
Sweat-absorbent, quick-drying fiber materials (sportswear, innerwear)	Sweat-disposing function	Modified cross-sectional filament yarn (W-shaped flat section,)	Perspiration absorptive finish	Water absorbing by capillary action
Water-repellent fiber materials	High-density woven fabric with concave-convex surface structure	Combined-filament yarns comprising microfiber filament yarn	Water-repellent finish (fluorine-based water-	Concave-convex surface structure (waxy substance) of a

(continued)

Table 12.5 (continued)

Materials, products	Characteristics	Polymer/fiber	Post-processing	Resources of idea
(umbrellas, rainwear)	(surface structure of a lotus leaf)		repelling agent)	lotus leaf or a taro leaf
Fiber materials having deep black color (formal wear, student uniform)	Spun-like appearance and handle + deep black color	Composite false-twist textured yarns consisting of filaments having different shrinkages	Dyeing technology, deep coloring technology	Color depth of black-dyed wool fabric

12.5 Conclusion

This chapter includes the following contents:

1. Most of the functional and speciality man-made fibers are composed of polyester, which occupies a predominantly high share in the world production of man-made fibers.
2. The modification technologies are based on the trinity of polymer, fiber and yarn, and post-processing modifications, for which Japan has taken a leadership.
3. The functional and speciality fibers developed in Japan depend on the modification technologies based on biomimetics.

In the following chapters, functional and speciality man-made fibers are explained in detail in the following order related with the functions:

1. High-Touch Fibers and "Shin-gosen" (see Chap. 13)
2. Moisture and Water Control Man-Made Fibers (see Chap. 14)
3. Heat-Controllable Man-Made Fibers (see Chap. 15)
4. Light-Control Man-Made Fibers (next issue)
5. Antistatic and Conductive Man-Made Fibers (next issue)
6. Antibacterial and Deodorant Man-Made Fibers (next issue)
7. Flame-Retardant Man-Made Fibers (next issue)

References

1. T. Saegusa, Kagaku **54**(4), 46–51 (1999)
2. T. Suzuki, 35th *Sen'i Gakkai* Summer Seminar Proceedings, 6–11 (2004)
3. T. Suzuki, 38th *Sen'i Gakkai* Summer Seminar Proceedings, 35–38 (2007)
4. T. Suzuki, Sen'i Kikai Gakkaishi **54**(7), P277–P282 (2001)
5. T. Suzuki, Sen'i Gakkaishi **59**(7), P220–P227 (2003)
6. M. Okamoto, N. Minemura, Sen'i Gakkaishi **63**(4), P84–P88 (2007)
7. T. Suzuki, Sen'i Kikai Gakkaishi **61**(2), 123–128 (2008)
8. T. Suzuki, Kako-gijutu **44**(11), 669–679 (2009)

Chapter 13
High-Touch Fibers and "Shin-gosen" (Newly Innovated Fabrics)

Hiroshi Takahashi

Abstract The history of high value-added synthetic fibers for clothes began with the development of silky polyester fibers in Japan. Many new technologies such as noncircular cross-sectional shape, alkali-reduction treatment, differently shrinking combined-filament yarn, etc., were invented by the imitation of the shape of each individual silk fiber and the characteristics of silk fabrics. These technologies, combined with those for ultrafine fibers, have brought novel, refined-taste, and distinguishing high-grade feel texture for the newly innovated synthetic textiles, which were not attained by natural and conventional synthetic textiles. These new textiles created a boom of "Shin-gosen" during several years from 1988, and this boom played a leading part to restore synthetic fiber industries. These new technical capabilities have been upgraded ever since and are still supporting Japanese textile industries as non-price competitiveness even now.

Keywords High-touch fiber • Shin-gosen • Silky polyester • Ultrafine fiber • Cross-sectional shape • Alkali-reduction treatment • Conjugate fiber

13.1 Technology of High Value-Added Synthetic Fibers

A million and hundreds of thousands years before, human ancestors covered their bodies by the game fur and plant leaves to keep out the cold. Wearing clothes is one of the important cultural habits to distinguish human beings from other animals. At the beginning, our ancestors used natural fibers for their clothes. However in the early modern times, according to the inventions of regenerated and synthetic fibers, and the developments of their mass production technologies, the main materials for clothes have shifted from natural to regenerated and synthetic fibers.

Polyester fiber, one of the typical synthetic fibers, was characterized by the absence of unevenness, that is, each filament had a round cross-sectional shape and a uniform diameter. The polyester staple fiber has established stable applications for shirt fabric, work clothes, etc., by blending with cotton, because the

---

H. Takahashi (✉)
Toray Science Foundation, Tokyo, Japan
e-mail: Hiroshi_Takahashi@tgnts.toray.co.jp

Society of F.S. Technology, Japan (ed.), *High-Performance and Specialty Fibers*,
DOI 10.1007/978-4-431-55203-1_13

polyester staple is not only superior in strength and durability but also in easy-care properties, e.g., wash-and-wear and non-iron features. On the other hand as for the polyester filament, which has a similar mechanical characteristic with that of silk when compared to other synthetic fibers such as nylon or acrylic, the possibility was mainly considered for the application of silk fabric area.

By the way, fibers have characteristics of fine and long shape, flexible and tough properties, and large specific surface area. Fibers are one-dimensional materials; however, they can be two-dimensional fabrics by weaving and knitting and further processable for three-dimensional clothes by sewing. If one can create new features and functions to each one of the fibers, it is possible to realize new clothes with many novel features and functions, by the appropriate combination of weaving, knitting, and sewing.

Polyester fiber is easily processable because it is a thermoplastic polymer, and the polymer properties can be modified by using copolymerization and blending techniques. Further, other techniques, e.g., modification of fiber cross section by deforming the hole shape of spinneret in melt spinning, conjugate spinning with a different kind of polymer, and combined-filament yarn by yarn processing, were developed, and many new characteristic commercial products with various kinds of feelings and functions have been created. Particularly, many commercial high-touch products for clothing have been developed under the concept of "learn from nature." This concept is based on an advanced technological development philosophy that natural feelings come from the inhomogeneous character of the products; however, this inhomogeneous character should be under control.

The first attempt to create high added value for fibers is the development of natural fabric-like, especially silk-like, synthetic fibers. This project aimed to produce synthetic fibers with the high-class feeling like silk, keeping with advantages of synthetic fibers. Almost all Japanese synthetic fiber companies launched the research project to find out the origin of characteristics of natural fibers, especially the feature of silk, in order to add those characteristics to synthetic fibers of polyester and nylon filaments. Then before long, the silk-like materials were materialized.

13.2 Development of Silky Polyester

Silk has always been admired by mankind. The delicate texture and elegant gloss of silk have attracted people from the age of Silk Road, and it has been the queen of fibers even now. Therefore, the production of synthetic silk, without relying on nature, is a human dream like alchemy, and the history of the development of silky polyester started only 6 years later from the beginning of polyester production in Japan.

13.2.1 The First Stage: Imitate the Shape of Silk Fiber

The first step to the synthetic silk started from imitating the characteristics of each silk fiber. The representative techniques for that purpose are the change in cross-sectional shape and alkali-reduction treatment.

Silkworms produce cocoons by raw silk fibers spewed from a pair of silk gland. The raw silk has a structure of packaging two fibroin fibers, consisted of fibrous protein and covered by sericin as another kind of protein which dissolves in hot water. The cross-sectional shape of this fibroin fiber is just about triangle, and this triangle shape is important for the gloss of silk. Polyester fiber is produced by melt-spinning method extracting molten polymer through a spinneret. Here the circular cross-sectional shape is the easiest to produce because it is stabilized due to the symmetrical balance. In case of emulating silk, however, polyester with a triangle cross section was initially produced by changing the hole shape of the spinneret, according to the idea policy that the first step starts from imitating the shape. This emulation attained the silky polyester fibers which have the gloss of silk, and since then triangle cross-sectional shape became the mainstream of the silky polyester fiber [1]. Afterward, filament with many other types of cross sections, e.g., heptalobed or octalobed, hollow or flat type, appeared in the market, and at present this is the base technique in providing various feelings and functions to the polyester fiber (Fig. 13.1).

On the other hand, almost all of sericin component, which surrounds two fibroin fibers of raw cocoon filament, is dissolved and removed by hot water through a scouring process. The combined two fibroin fibers of raw silk are divided into two independent silk fibers by this refinement process. The silk thread thus produced has void, which was filled by sericin component before the scouring process, and this void makes it possible for silk fibers to move freely in woven textile. The unique soft texture and draping properties to form graceful curves of silk fabrics result from this void.

By the way, it had been known among engineers of ICI that polyester undergoes alkalinity-induced hydrolysis, and polyester filaments can be made thin because the decomposed component is soluble in water [2]. The engineers of ICI, however, intended this knowledge to make fabrics consisted of fine thread, because it was difficult to make fabrics with fine thread, not to improve textile texture. Therefore, they weaved fabrics with thicker thread for the first and then choked the thread by the dissolution treatment. Japanese engineers applied this technique for the improvement of draping properties, which corresponded to the removal of sericin for natural silk fibers. Recently, this alkali-reduction treatment is widely applied not only for the production of silky polyester textiles but also for the finish processing of general polyester fabrics. Normally, from 20 to 30 % of the reduction treatment affords good draping properties and soft feeling for the fabrics [1].

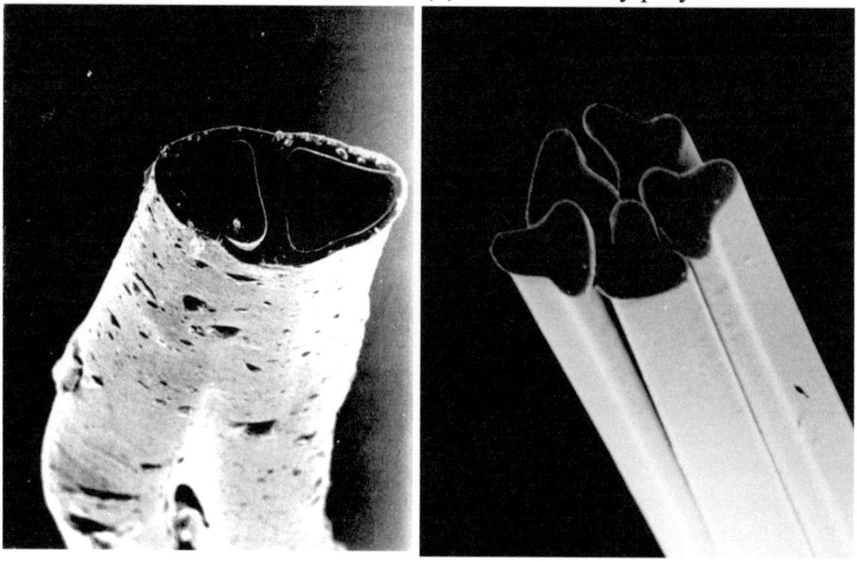

Fig. 13.1 Cross-sectional shape of (**a**) raw silk fiber and (**b**) silky polyester fibers

13.2.2 The Second Stage: Imitate the Features of Silk Fabrics

The second step of making synthetic silk was imitative of the features of the silk textiles. The natural silk fiber has random curvature, twist, and uneven contractility in the direction of the fiber length, because when a silkworm makes a cocoon, it spews raw silk with waggling the head in the shape of "8" character. These properties afford unique soft swollen feeling for silk textiles. In order to attain this soft swollen feeling by synthetic polyester fibers, differently shrinking combined-filament yarn technique and random heat-treatment technique, which produces fibers with randomly different degree of shrinkage, in the fiber longitudinal direction and/or for each single yarn, by randamly heating of firaments after drawing, were developed. The differently shrinking combined-filament yarn technique affords swollen feeling to textiles, which were weaved by two kinds of yarns with different degree of heat shrinkage, due to the difference in the length of the two yarns caused by the heat treatment to the textile. DuPont originated this technique in the late 1960s for the polyamide fiber named "Qiana." In Japan, similar technique was developed using the difference in the degree of heat shrinkage between normal polyester and copolymerized polyester, which has larger degree of shrinkage than the former [3]. This differently shrinking combined-filament yarn technique realized smooth lustrous silky polyester fabrics with soft swollen feeling.

13.2.3 The Third Stage: Imitating the View of Nature and the Inhomogeneousness of Silk Fabrics

The third stage for the approach to manufacturing artificial silk is to afford controlled inhomogeneousness, which is the view of nature of silk fabrics, to the polyester fabrics. Natural silk fiber has unevenness in diameter, cross-sectional shape, and dyeing, among different fibers and/or for the longitudinal direction in one fiber, which are characteristics to natural fibers. People who use the silk accept the unevenness as desirable feelings of warm and comfort. On the other hand, the uniformity is desired for synthetic fibers to facilitate mass production, and this uniformity is one of the features of synthetic fibers. However, this uniformity of synthetic fibers gives people monotonous and non-warm feelings. Therefore, many engineering developments have been made to incorporate this view of nature fibers into synthetic fibers. For example, "thick and thin yarn," which has thickness unevenness in the longitudinal direction of each yarn, was developed using special drawing technique to afford drawn and undrawn part alternatively. When the "thick and thin yarn" is dyed, drawn and undrawn parts become deep and shallow dying, respectively, and the dyed fabrics afford visually shades of colors and dry-touch feelings. Further, other types of techniques have also been developed to afford yarn bulkiness and longitudinal fluffiness, e.g., combined false-twist texturing by yarns with widely different degree of elongation, and combined air texturing by yarns interlacing treatment by two kinds of yarns supplied with different speed.

To approach the features of silk, not only by senses of sight and touch but also by sense of hearing, an engineer challenged to the artificial squeak sound of silk. Fujimoto analyzed the friction sound of silk fabrics acoustically and studied the relationship between the cross-sectional shape of fiber and the squeak sound in detail. He found finally that a fiber of the cross-sectional shape with three petals, whose tip has slit-type concave in submicron scale size, produces the best squeak sound resembling silk [4].

13.2.4 From the Natural-Fiber-Like Materials to the Synthetic Fiber Original Materials

By the pursuit of technology, as abovementioned, many new commercial products have been developed, in order to afford deep feelings of natural fibers to synthetic fibers which are characterized by their uniformity. However oneday, engineers realized the limitation of the imitating technology, in spite of the many efforts to imitate natural fibers, that synthetic fibers are after all synthetic fibers and not natural fibers. So, the momentum was rising to search for the new feeling and high-quality feeling that is original and unique to synthetic fibers, which are not attained by natural fibers and conventional synthetic fibers, not only by the imitation of natural fibers. This motivation created the "Shin-gosen" (literally new fabrics). However, in the process of research and development of the Shin-gosen, other important factors were necessary, which were the invention of ultrafine fibers and the merchandise developments by them.

13.3 Development of Ultrafine Fibers and Their Evolution

There has been no clear definition to the ultrafine fibers, but in the broad sense it is thought that the diameter of ultrafine fibers is less than that of silk (diameter ca. 10 μm) which is the finest natural fiber as a standard. However, the diameter at which many features of ultrafine fibers can be recognized, e.g., special tactual sense (touch) and flexibility which are different from the conventional fibers, is less than 7 μm, and the features are particularly significant at the diameter less than 3 μm. Therefore sometimes, fibers with the diameter less than 3 μm are called as ultrafine fibers. Table 13.1 shows the diameters of some familiar fibers to grasp an image of fineness of ultrafine fibers.

13.3.1 Manufacturing Process of Ultrafine Fibers

The simplest way to make ultrafine fibers is the direct spinning using spinneret with small holes. However, this method is applicable for the fibers up to 3 μm in diameter, and at present there are many problems in technology and manufacturing cost to make thinner fibers less than 3 μm by this method.

On the other hand, there is an alternative method. For the first a conjugate fiber which consists of two different kinds of polymers is produced. Then, woven or knitted fabric made from the conjugate fibers is processed to remove one polymer component or peel off both components of the conjugate fiber. There are two morphologies for the conjugate fibers, e.g., sea-island type and peeling and splitting type. As for the production method of the sea-island-type conjugate fibers, combined polymer flows where many island components, which will be finally the ultrafine fiber and are covered by sea component part, are bundled and pushed out

Table 13.1 Comparison of fiber thickness

Type of fibers	Circular approximated diameter (μm)
Human hair	50~60
Wool	20~26
Polyester fiber for men's shirt	15
Silken thread	12~20
Ultrafine fiber for suede-like artificial leather	3
Ultrafine fiber for spectacles wiping cloth	2
Super ultrafine fiber attained in laboratory	0.1

Fig. 13.2 Various types of ultrafine fibers

as a filament. Then the sea component part was removed by dissolving in order to produce the ultrafine fibers. On the other hand, as for the peeling- and splitting-type conjugate fibers, two components are peeled off by chemical or mechanical treatment to produce the ultrafine fibers. Figure 13.2 shows the various types of ultrafine fibers.

The largest advantage of the conjugate fiber technology is the complete avoidance of fiber twinning and yarn breakage in the processing stage of ultrafine fibers. That is, in the processing stage, the ultrafine fibers are still thick conjugate fibers and it is possible to handle them in the same way as other general fibers. After that at the stage when there is no possibility of problems in treating ultrafine fibers, the conjugate fibers are changed to ultrafine fibers by chemical and/or physical treatment.

13.3.2 Further Evolution of the Ultrafine Fibers

The ultrafine fibers have some happenstance effects such as biocompatibility in addition to the original features which come from the geometrical fineness, e.g., pliability, flexibility, tactual sense (touch), luster, large specific surface area, and fillability (flatness on the fabric surface). Therefore, the commercialized technology to draw forth these features of the ultrafine fibers well is important, and thereby many high-quality products have been developed unique to the ultrafine fibers. The typical examples are the suede-like artificial leathers and the high-performance wiping clothes for optical lens, jewelies, cameras, mobile phone, and so on. Especially, it was the development of suede-like artificial leather that attracted a great deal of attention. Natural leather consists of entangled ultrafine fiber bundles of collagen. The real suede-like artificial leather has a single-layer nonwoven fabric structure, very similar to the entangled structure of natural ultrafine fiber bundles, and has characteristics equal to or above those of natural leathers. Figure 13.3 shows the structure comparison between suede-like artificial leather produced by

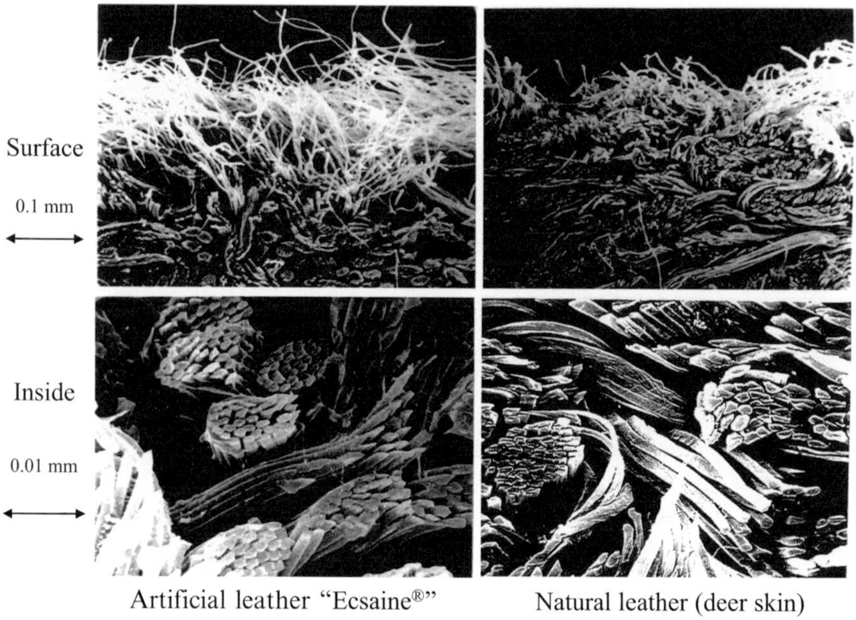

Fig. 13.3 Comparison between suede-like artificial leather and natural leather (deer skin)

the abovementioned method and natural leather. The figure clearly shows the similar structures not only on the leather surface but also at the inner part.

The reason why this Japanese original artificial leather could dominate the global market is due to the success in developing ultrafine fibers with sea-island structure or conjugated sea-island structures, and in duplicating the natural leather structures, which was analyzed thoroughly, by synthetic fiber technology. Here too the policy of Japanese synthetic fiber companies, "learn from nature," is reflected.

13.4 The Birth of "Shin-gosen"

13.4.1 What Is "Shin-gosen"?

Shin-gosen was produced by the general technology where many Japan's boasting higher-order processing skills from upstream to downstream were combined and by various techniques established in the development of silky polyester fibers and ultrafine fibers as the basis. Shin-gosen is not a single commercial product. It is generally defined as "refined-taste synthetic textiles with distinguishing feeling which has not been attained by the conventional synthetic fibers and also by the newly created natural fibers, in other words, the drape property (drapey) that can endure for the exhibition to the Parisian haute couture." It played a leading part and caused a boom for the rehabilitation of synthetic fiber industries during several years from 1988. At that time many new materials and commercial products have been produced by the categorized development of the following four items, e.g., (1) new silky materials with swollen feeling, soft feeling, and resilient feeling based on the technology of silky polyester technique, (2) slightly nap-raised (peach face) materials which have soft tactual sense (touch) made of ultrafine fibers, (3) - dry-touch materials which have rayon-like touch attained by unevenness on the fiber surface, and (4) new worsted materials made by the advanced combined texturing technique.

The following are the typical "Shin-gosen" materials and the technologies applied to them.

13.4.1.1 New Silky Materials

New silky materials are defined as the artificial fabrics afforded with enough swollen and soft feelings that cannot be attained by conventional silk fabrics. To afford new silky fabrics with these characteristics, the use of differently shrinking combined-filament yarn, which is composed of low shrinking (sheath) filament yarn and high shrinking (core) filament yarn, is known. There are some types of the new silky materials, e.g., (1) large shrinkage of the fabrics is attained by thermal treatment of the core filament yarn in a high-order processing step using a high shrinkage filament which was made by copolymerization of a monomer with bulky molecular structure as the third component; (2) larger differential shrinkage is attained even under the binding of fabric structure by the spontaneous elongation yarn as a low shrinking yarn; (3) combination of these some techniques; etc. These techniques are used not only for new silky materials but also for other Shin-gosen as the basic technology (Table 13.2).

13.4.1.2 Slightly Nap-Raised (Peach Face) Materials

Slightly nap-raised materials can express soft tactual sense (touch) like peach fuzz by the alignment of ultrafine fibers on the fabric surface at the final stage of weaving textiles. There are some types of them, e.g., materials that use ultrafine fibers made by peeling- and splitting-type conjugate fibers, or fine denier filament made by direct spinning or elution-type ultrafine fibers as the sheath yarn of the new silky materials (Table 13.3).

13.4.1.3 Dry-Touch Materials

Dry tactual sense (touch) like rayon can be expressed by the alignment of surface uneven fibers on the fabric surface at the final stage of weaving textiles. There are some methods to form unevenness at the fiber surface, e.g., materials that use

Table 13.2 New silky materials [5]

Manufacturer	Product name	Technical content
Toray	Sillook Sildew	Multiple high shrinkage yarn + Differently shrinking combined filament yarn
	Sillook Royal S	Three-petal cross-section yarn + Multiple shrinkage combined yarn
Teijin	Mixel VII	Ultrafine yarn + Differently shrinking combined filament yarn
	Ajenty	Spontaneous elongation yarn + Differently shrinking combined filament yarn
Unitika	Mixy	Multiple filament with multiple cross-section blended yarn
	Efin	Multiple filament with multiple cross-section blended yarn
Kuraray	Nymphus	Spontaneous elongation yarn + Multi-denier and Differently shrinking combined filament yarn
	Silmars	Different cross-section shrinkage combined yarn + Randomly crimped yarn
Mitsubishi Rayon	Criseta	Fine thick and thin yarn
	Audi	Differently shrinking combined filament yarn
Toyobo	Geena	Spontaneous elongation yarn + Differently shrinking combined filament yarn
Kanebo	Treview	Randomly conjugated yarn + Elution processing
Asahi Kasei	Sowaie Duo II	High speed spinning yarn + Differently shrinking triple combined yarn

Table 13.3 Slightly nap-raised (peach face) materials [6]

Manufacturer	Product name	Technical content (P:Polyester, N:Nylon)
Toray	Reebarg P	Ultrafine yarn + High shrinkage yarn
	Piceme	P/N Splittable conjugate yarn
Teijin	Asty	P/N Splittable conjugate yarn
	Easel	Ultrafine yarn + Thick fineness combined false-twisted yarn
Unitika	Rominar	Ultrafine yarn + Thick fineness combined air-mixed yarn
Kuraray	Wramp	P/N Splittable conjugate yarn
Mitsubishi Rayon	Micloop	Ultrafine yarn + High shrinkage yarn
Toyobo	Riviera III	Ultrafine yarn + High shrinkage yarn
Kanebo	Savina Peach-Skin (PS)	P/N Splittable conjugate yarn
	Nazca	P Elution type conjugate yarn

filament with largely modified cross-sectional shape, or filament with cratered surface by alkali-reduction treatment in a high-order processing step after the addition of microparticles such as titanium dioxide or silica (Table 13.4).

Table 13.4 Dry-touch materials [7]

Manufacturer	Product name	Technical content
Toray	Sillook Chateauleine	Particle addition + Modified cross-section shape + Differently shrinking combined filament yarn
	Ceo, Ceo α	Modified cross-section shape + Multi denier + Differently shrinking combined filament yarn
Teijin	Aerocapusule Dry	Particle addition + Large hollow cross-section fiber
	Emoule II	Particle addition + Multi denier + Differently shrinking combined filament yarn
Unitika	Mixy	Easily soluble component blend + Multi denier + Modified cross-section shape
	Earis 5	Particle addition + Differently shrinking combined filament yarn
Kuraray	SN2000	Microparticle addition
	XT-E	Particle addition
Mitsubishi Rayon	Suprema	Particle addition
	Lisny	Modified cross-section shape + Multi denier + Differently shrinking combined filament yarn
Toyobo	Louvro	Particle addition
Kanebo	Vivan α	Particle addition + Modified cross-section shape
Asahi Kasei	Gulk	Particle addition + Multi denier + Thick and thin

13.4.1.4 New Worsted Materials

Woollike materials made of spun-like textured yarn had existed before the emergence of the Shin-gosen. Woollike materials aimed the complete imitation of real wool. New worsted materials, on the other hand, bringing the feeling of natural wool in, still have peculiar taste of synthetic fibers. It was made by the full use of combined texturing technique (see following section), and there are combined false-twist textured yarn type and combined-fluid textured yarn type (Table 13.5).

However, since 1992 the "Shin-gosen" boom declined and the leading part of the clothing materials shifted to natural and regenerated fibers by the casualization of fashion. Under such business environment, many technologies created and established in the development of the "Shin-gosen" have been still sophisticated after that. These technologies have also been applied to the development of new commercial products such as comfortable materials aiming for lightweight, sweat and moisture absorption properties, and stretchability, combined with the advance of evaluation technology such as artificial weather room (Technorama).

Table 13.5 New worsted materials [8]

Manufacturer	Technical content and the product name	
	Combined false-twist texturing	Combined fluid texturing
Toray	New Moranna, Malor	Reebarg F, Conclaire
Teijin	Milpa, Delite, Lukaro	Ajenty μ, Consoff
Unitika	Matolesta	Clement
Kuraray	Microcarrot, Lomouna	Junoir
Mitsubishi Rayon	Jenner	Sanstoi, Senege
Toyobo	Pommier, Ohdus	Rosally
Kanebo	Walk	Cashmeena, Sanstoi
Asahi Kasei	Senetta	

13.4.2 Higher-Order Processing Technology Which Supported Shin-gosen

Polyester filaments are occasionally used as textured yarn in order to afford stretchability and bulkiness (swollen feeling). Textured yarn is the fibers which were afforded two- or three-dimensional bulkiness. These fibers are categorized into two, e.g., false-twist textured yarn, which has stretchability, made by the untwisting after the simultaneous twisting and heat setting, and structurally bulky yarn, which has no stretchability but has bulkiness by the loop structure, made by fluid processing. Normally, both false-twist processing and fluid processing use one kind of source filament. However, many combined textured yarns have been produced by using multiple source filaments, e.g., multiple filaments with different feed speed, or simultaneous feeding of multiple filaments which have different diameter, degree of heat shrinkage, and/or elongation. This technique has been sophisticated in Japan, and various combined textured yarns have been produced and has become one of the important base technologies to create "Shin-gosen."

Fibers are used as one-dimensional string such as sewing threads, fishing lines, and ropes. However, they are more commonly used as two-dimensional fabrics. Normally, woven or knitted fabrics for clothing are finished to commercial textile products after dyed by colorants such as dyes and pigments and treated by processing agents. Further, the dyed textiles are subjected to finishing processes, e.g., physical processing to change tactile impression by the physical change of the fiber structure caused by heating, beating, and loosening treatment, and/or to chemical processing to afford new functions by the attachment of chemical reagents such as resins. In these dyeing and finishing processes, heating (wet and dry heating) and cooling cycles are repeated, and the feeling (fabric feelings: surface feeling and tactile impression) of final textile cloth products are largely affected. Especially the abovementioned "Shin-gosen" fabrics, which consist of multiple fibers with different properties of degree of heat shrinkage and/or elongation and thickness, change their feelings largely by the heating, cooling, and tension (external forces applied to the fabrics) applied at dyeing and finishing processes.

No newly developed fiber with special features can produce textiles with high-quality feeling, without the higher-order processing skills to make the best use of the features. Japanese local companies of textile finishing have extremely excellent higher-order processing techniques, and it is no exaggeration to say that "Shin-gosen" fabrics were created by their higher-order processing techniques. Their technical capabilities are still supporting Japanese textile industries as non-price competitiveness even now.

References

1. M. Fukuhara, H. Takahashi, Funct. Mater. (Kinou Zairyou) **12**, 17–23 (1992)
2. U.S. patent No. 2590402
3. Jpn. Kokai Tokkyo Koho JP 950012
4. M. Fujimoto, Funct. Mater. (Kinou Zairyou) **39**, 371–377 (1986)
5. T. Okada, Sen'i Gakkaishi **50**, 354–357 (1994)
6. M. Morioka, Sen'i Gakkaishi **50**, 361–363 (1994)
7. T. Tanaka, Sen'i Gakkaishi **50**, 358–360 (1994)
8. M. Matsumoto, Sen'i Gakkaishi **50**, 364–367 (1994)

Chapter 14
Moisture and Water Control Man-Made Fibers

Togi Suzuki and Sonoko Ishimaru

Abstract Moisture and water absorption properties of clothes are very important for the excellent wearing comfortability. Moisture-absorbing materials absorb gas-phase water and suppress humidity rise within clothing. Water-absorbing materials absorb and diffuse liquid-phase perspiration, increase the amount of evaporation heat transfer, and expedite the heat dissipation. At one time, many research projects were conducted to provide man-made fibers, which are hydrophobic innately, with moisture and water absorption ability as well as natural fibers. At the present, however, moisture and water absorption abilities of some man-made fibers exceed those of natural fibers. And also, according to the development of assessment technologies of consumption science for wearing comfortability, man-made fibers with excellent moisture and water absorption abilities, which cannot be attained by natural fibers, have been developed and used for sports and inner wears. Specifically, perspiration-absorbent and quick-drying materials, polyacrylate fibers which have 3.5 times larger moisture absorption ability than that of cotton, man-made fibers with improved moisture-absorbing and moisture-releasing abilities, materials with self-controlling functions, and yarn and knitted or woven fabric composite structural materials, are examples. In this chapter, basic scientific knowledge and the evaluation methods of moisture and water absorption, current moisture- and water-absorbing materials, imparting method of these abilities, etc., are presented.

Keywords Moisture absorption • Water absorption • Insensible perspiration • Sensible perspiration • Humidity within clothing • Highly cross-linked acrylate fiber • Efficiency of moisture absorption-release • EVOH • Capillary effects • Self-regulating function

T. Suzuki (✉)
Teijin Limited, Tokyo, Japan
e-mail: tou.suzuki@kke.biglobe.ne.jp

S. Ishimaru
Toyobo Co., Ltd., Osaka, Japan
e-mail: Sonoko_Ishimaru@toyobo.jp

Society of F.S. Technology, Japan (ed.), *High-Performance and Specialty Fibers*,
DOI 10.1007/978-4-431-55203-1_14

14.1 Why Moisture and/or Water Absorption Is Important for Fibers

Water escapes from human body via insensible and sensible perspiration. Through the former, water goes away by the diffusion through the skin tissue. In the ordinary atmosphere, where thermal environment is neutral, around 25 g/h of water evaporates from the total body, and water amount changes depending on the body's metabolic rate, atmospheric temperature, and humidity [1]. Water escaping from the latter way is of larger amount than the former. Another important role of perspiration is, of course, lowering of body temperature using the heat dissipation by evaporation. Fibers must play an important role for quickly removing the high moisture from the body to avoid the humid feeling. When the evaporation of perspiration is slow and water vapor turns to liquid water, we feel wet and sticky. Here, wet and sticky feelings are the typical ones of humid and wet sensation. We feel hot at the same time. We should note here that the cold feeling also takes place, when the liquid perspiration which remains on the skin or in clothes is cooled. In conclusion, good fibers take in moisture quickly from the body and also send out the moisture toward the atmosphere from the fibers themselves. Highly moisture-absorbing fiber materials suppress the humidity rise within clothing over the skin surface (see Fig. 14.1) [2, 3]. Highly water-absorbing fiber materials absorb and diffuse liquid-phase perspiration, increase the amount of evaporation heat transfer, and expedite the heat dissipation [4, 5].

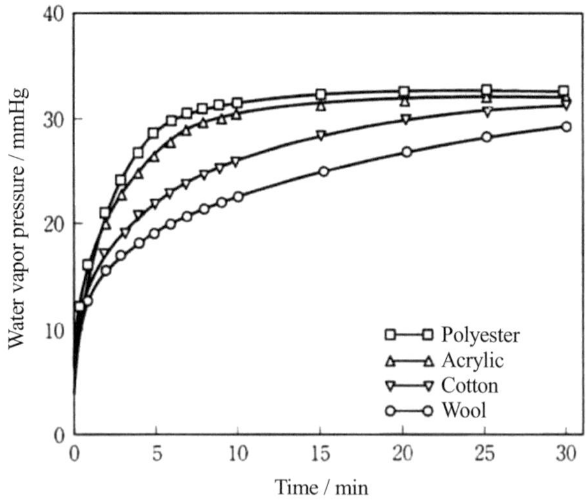

Fig. 14.1 Humidity within clothing for some fabric materials

14.2 Moisture and Water Absorption of Fibers

Moisture and water absorption mean the absorption of water in gas and liquid phases, respectively. Thus, the moisture absorption of clothes will be influenced strongly by the hydrophilic moieties of fibers, such as hydroxyl, amino, sulfonic acid, carboxyl, and ether chemical groups.

When fibers are kept in air atmosphere, the gas-state water, moisture is adsorbed on the surfaces of the fibers and then followed by the diffusion and absorption of the water molecules into the interiors of the fibers. Thus, moisture absorption is composed of the processes of adsorption and diffusive absorption. Figure 14.2 shows one of the examples of the absorption isotherm of various fibers [6]. Ability of the moisture absorption is highest for wool and decreases in the order wool > rayon > silk > cotton > acetate > nylon > acrylic > polyester. Clearly from the figure, magnitude of moisture absorption is sensitive to the relative humidity of atmosphere. Standard moisture regain of various fibers and polymers is given by JIS standard for fair trading as is shown in Table 14.1 [7]. The fraction of water content absorbed in the absolutely dried-state fibers at the standard state (20 ± 2 °C in the atmospheric temperature and 65 ± 2 % in the relative humidity) is called to be the standard moisture regain value. These values of standard moisture regain are multiplied with the absolutely dried-state weight of fibers and used to guarantee the official weight of fibers for commercial trading. Table 14.2 shows water molar absorption contents estimated from each functional chemical group of polymers at various relative humidity levels [8]. Substantial large moisture absorption ability was observed for salts of carboxylic acids and ammonium bases.

Now, water absorption means absorption of liquid-phase water into fibers, where the interfacial tension between the fibers and water plays an important role to the

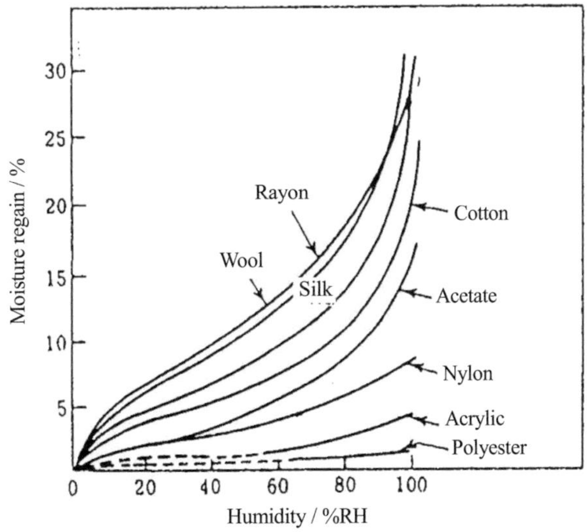

Fig. 14.2 Absorption isotherm of various fibers

Table 14.1 Official moisture regain of fiber materials

Fiber	Official moisture regain (%)
Ramie, linen	12.0
Cotton	8.5
Wool	15.0
Silk	11.0
Viscose rayon	11.0
Cupro	11.0
Acetate	6.5
Triacetate	3.5
Promix	5.0
Nylon	4.5
Polyester	0.4
Acrylic	2.0
Modacrylic	2.0
Vinylon	5.0
Polyurethane	1.0
Polypropylene	0
Polyethylene	0
Polyvinyl chloride	0
Vinylidene	0
Polyclar	3.0
Benzoate	0.4
Glass	0

Table 14.2 Molar moisture content of functional groups

	Relative humidity				
Group	0.3	0.5	0.7	0.9	1.0
$-CH_3$, $-CH_2-$, $-CH<$	(1.5×10^{-5})	(2.5×10^{-5})	(3.3×10^{-5})	(4.5×10^{-5})	(5×10^{-5})
$-C_6H_5$	0.001	0.002	0.003	0.004	0.005
$>C=O$	0.025	0.055	(0.11)	(0.20)	(0.3)
–COO–	0.025	0.05	0.075	0.14	0.2
–O–	0.006	0.01	0.02	0.06	0.1
–OH	0.35	0.5	0.75	1.5	2
$-NH_2$	0.35	0.5	0.75	(1.5)	(2)
$-NH_3^+$			2.8	5.3	
–COOH	0.2	0.3	0.6	1.0	1.3
$-COO^-$	1.1	2.1	4.2		
–CONH–	0.35	0.5	0.75	1.5	2
–Cl	0.003	0.006	0.015	0.06	(0.1)
–CN	0.015	0.02	0.065	0.22	(0.3)

absorption, and liquid-phase water is kept in the space between fibers and voids within fiber structures. When fiber assemblies such as cloths are put into water, water penetrates into interior of the fiber and also into the space between fibers simultaneously, replacing air therein with water. These processes are called water absorption, and water absorption property is influenced greatly with the state of assemblies of fibers. In other words, amount of water absorption is influenced significantly with chemical kind of fibers, twist of yarn, structure of cloths, and the applying finishes. Water absorption is often evaluated by the water absorption rate in addition to the amount or percentage of absorbed water. Here, the water-absorbing rate of clothes is measured by water suction (Byreck), titration (falling-drop), and precipitation methods. The water suction method (or Byreck method) measures the water absorption height of the cloths as a function of time, while the cloth strip samples are hung vertically and their lower ends are dipped in the distilled water by 1-cm depth. In JIS standard, the average height of water rise in the warpwise/wale and weftwise/course directions ten minutes after the dipping is measured. In the Byreck method, water transport in the parallel direction to the cloth surface is measured. On the other hand, Larose method measures water transport in the direction perpendicular to the cloth surface. Now, the falling-drop method (or wicking speed method) measures the time until disappearing of 0.1 ml of water (20 °C) dropped on the cloth from 1-cm up. Precipitation method measures the time before falling of the test piece of cloths 1×1 cm floating on water surface at 20 °C.

The speed of water absorption is sensitive greatly to the contact angle of cloths to water and the porosity of the fiber assemblies. Table 14.3 [9] and Fig. 14.3 [10] show the contact angles and the amount of water absorption, respectively. Table 14.3 demonstrates that the fibers having high contact angles except wool are in the low levels of standard moisture regain values. This means that fibers of high moisture absorption ability get wet easily by the effect of low contact angles. Wool was exceptional from this relationship. This will be ascribed to the existence of cuticles or scales on the fiber surface, and of epicuticles which have water-repellent property, at the outmost layer. Figure 14.3 demonstrates that the amount of water absorption of water-repelling fibers such as polyesters, polypropylene, and wool products is insensitive to their porosity, whereas the amount of water absorption of the hydrophilic fibers is sensitive to their porosities and comes to be maxima at 75–85 % of the porosities.

Table 14.3 Contact angle and moisture regain of fiber materials

	Contact angle/degree				Moisture regain/% (at 20 °C,
Measured by	Tachibana	Nemoto	Hollies	Stewart	60 % RH)
Rayon	38	–	–	39	13.0
Nylon	64	61	83	70	4.5
Polyester	67	64	79	75	0.4
Acrylic	53	53	–	48	1.6
Polypropylene	90	88	–	–	0.1
Cotton	59	–	–	47	7.0
Wool	81	78	85	–	16.0

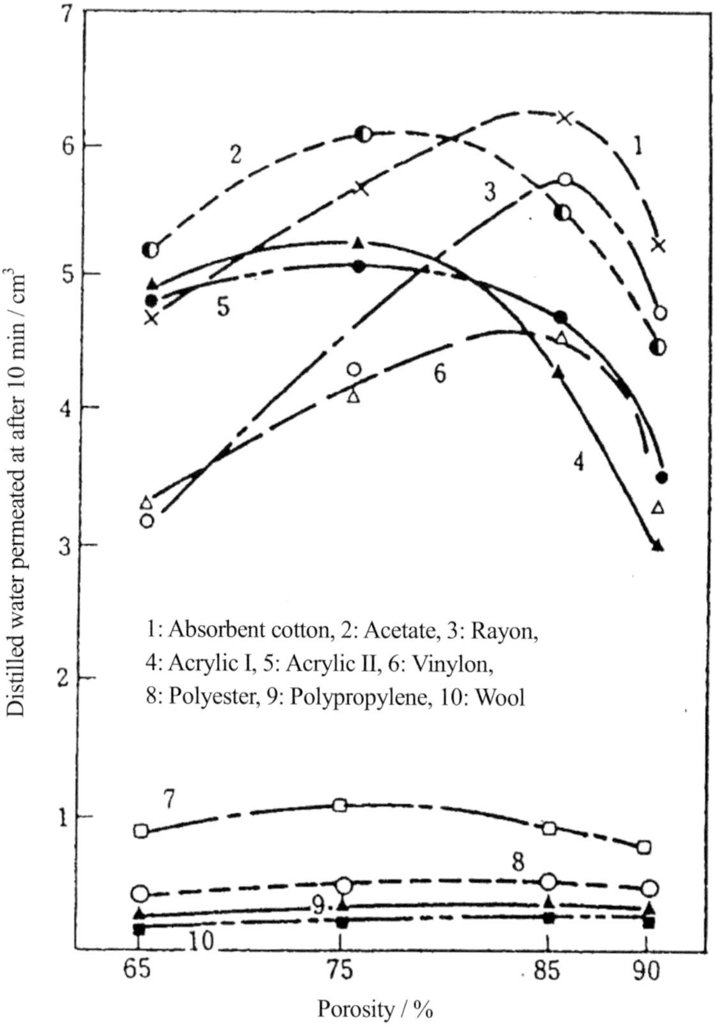

Fig. 14.3 Amount of water absorption for various fiber assemblies

14.3 Moisture Absorption Fibers and Moisture Absorption Modification Methods

Natural fiber having the highest standard moisture regain value is wool, 15.0 %. Nowadays, many synthetic fibers having high and even higher regain values have been produced. Main methods for obtaining high moisture-absorbing synthetic fibers are (1) design of polymer main chain, (2) usage of conjugated fiber manufacturing technique, (3) “blending spinning” method, and (4) usage of graft reaction technique. (1) to (3) are the modification in the fiber manufacturing process, and (4) is that of post-processing stage. Typical examples of synthetic fibers having high moisture-absorbing ability are described below.

14.3.1 *Cross-Linked Acrylate Fiber MOISCARE^R (Toyobo)*

MOISCARE^R is one of the highly cross-linked acrylate fibers having many hydrophilic moieties such as metal salt of carbonic acid. The moisture gain of MOISCARE^R is 40 % (at 20 °C and at 65 % RH), higher than that of wool [11]. It should be mentioned that MOISCARE^R evolves heat by absorbing moisture. When atmospheric humidity changes from 0 to 85 % RH, 1 g of MOISCARE^R staple fibers evolves heat of 230 kcal, corresponding to the raising temperature of a cup of water by 1 °C. The heat of MOISCARE^R is ca. twofold larger compared with that of wools and/or down feathers. Percentage of water absorption of MOISCARE ^R is ca. 3.5-fold higher than that of cottons. Furthermore, heat generation/release and moisture absorption/release rates of MOISCARE^R are regulated as around half of those of natural fibers [12].

14.3.2 *High Moisture-Absorbing Nylon Fiber QUUP^R (Toray)*

QUUP^R is obtained by blending nylon 6 with high moisture-absorbing special polymer, melting the mixtures homogeneously and further spinning the resulted polymer alloy. ΔMR of QUUP^R is 4.0, close to that of cotton (5.0) [13]. Here, ΔMR is given by equilibrium water absorption at 30 °C and at 90 % of RH minus that at 20 °C and at 65 % of RH. For comparison, ΔMR of nylon is 2.0. ΔMR is a convenient parameter showing the humid feeling, and larger value means less humid feeling. It should be noted here that the equilibrium water absorption value at 20 °C and at 65 % of RH is often used as a standard.

Figures 14.4 and 14.5 show the humidity within clothing for panty stocking and underwear, respectively. Humidity of QUUP^R is clearly lower than that of nylon, [14] and rather close to cotton.

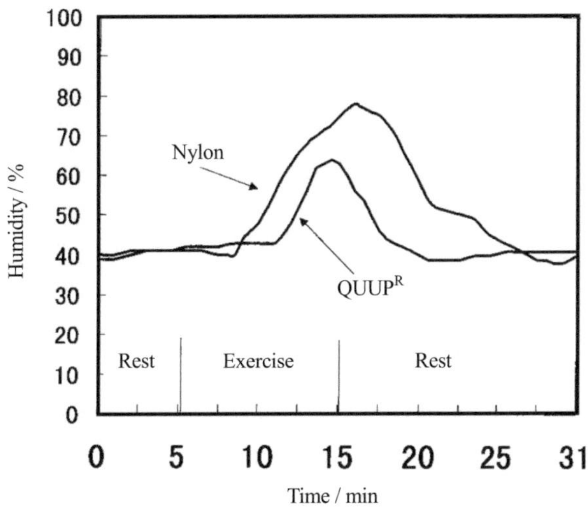

Fig. 14.4 Wearing test of panty stockings

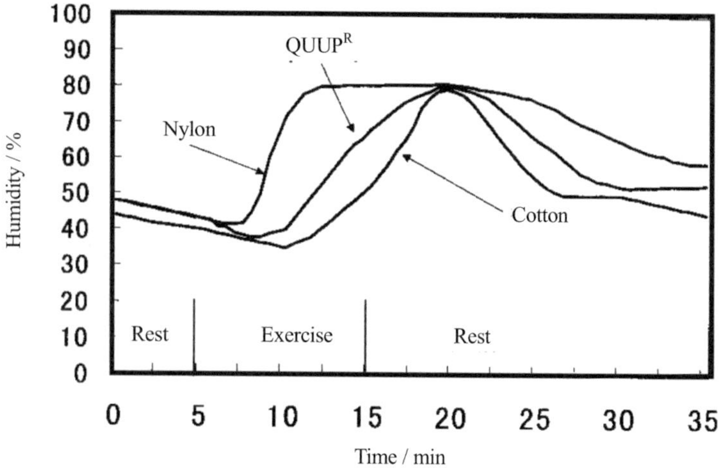

Fig. 14.5 Wearing test of underwears

Table 14.4 Moisture-absorbing and moisture-releasing properties of fiber materials

Fiber	Moisture-absorbing and moisture-releasing ability (%)	Moisture-absorbing and moisture-releasing rate/10^{-2} %/min
HYGRA^R	4.8	7.3
Cotton	4.1	3.8
Nylon	2.4	4.6
Polyester	0.4	0.4

14.3.3 Sheath-Core Structural Nylon Fiber Demonstrating Moisture Absorption and Release HYGRA^R (Unitika)

HYGRA^R is a nylon fiber having network structure composed of core of highly hygroscopic thermoplastic polymer and sheath of conventional nylon. Typical example of core polymer is cross-linked poly(ethylene oxide) [15]. Efficiency of moisture absorption-release of various fibers is compiled in Table 14.4 and Fig. 14.6. HYGRA^R is high in the moisture absorption-release effects compared with cotton and polyester, for example [16].

14.3.4 Sheath-Core Structure Conjugated Fiber SOFISTA^R (Kuraray)

Sheath and core parts of SOFISTA^R are EVOH (ethylene vinyl alcohol) copolymer and polyester, respectively, as is shown in Fig. 14.7. Figure 14.8 demonstrates that

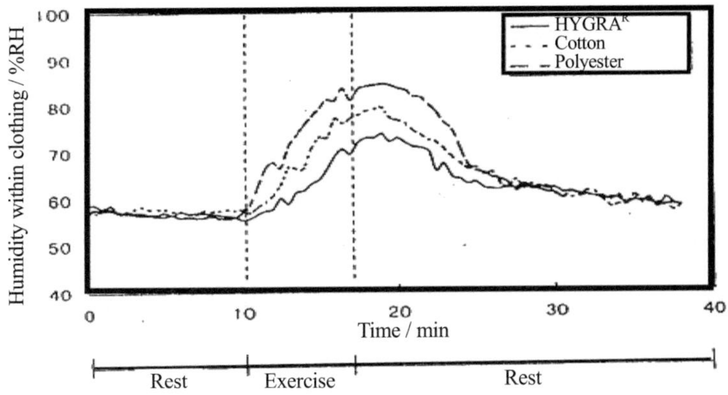

Fig. 14.6 Wearing test of HYGRA[R]

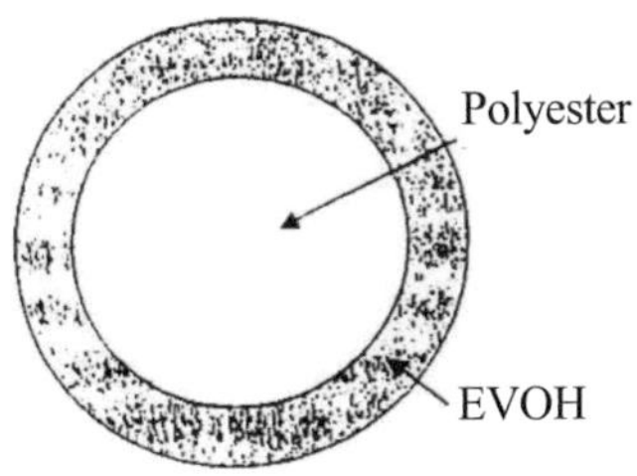

Fig. 14.7 Cross-sectional model of SOFISTA[R]

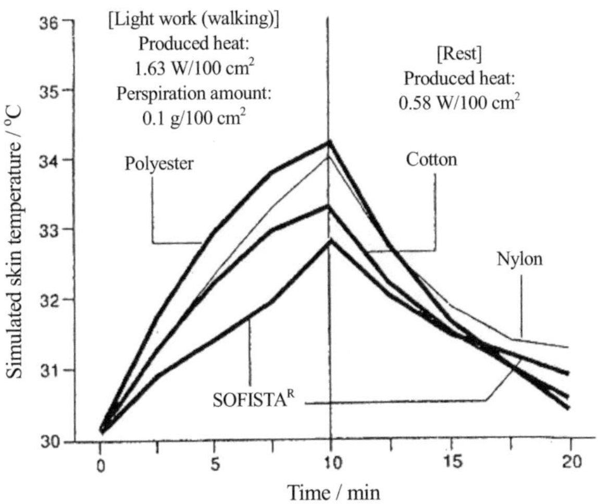

Fig. 14.8 Skin temperature change during wearing. Environmental conditions: temperature, 21–22 °C; humidity, 63 ± 3 % RH; wind speed, 0.1 m/s

Table 14.5 Thermal conductivity of fiber materials (at 35 °C)

Fiber	Thermal conductivity/10^{-4} cal/cm s °C
EVOH	8.7
Polyester	5.3
Nylon 6	6.9
Polypropylene	5.5

magnitude of raise in skin temperature during light work is smaller than those of other synthetic fibers. The thermal conductivity of EVOH is much higher compared with other fibers as is shown in Table 14.5 [17]. As is described above, there exist several methods to rise up the moisture-absorbing ability. However, we should note that most of these methods often induce lowering of the heat-resistant property, change in dyeability, and lowering of the mechanical properties of the fibers and of processability, for example.

14.4 Water Absorption Fibers and Water Absorption Modification Methods

Cotton fiber is one of the typical cellulose fibers composed of glucose units containing many hydroxyl groups and very high in their hydrophilicity. So, cotton fibers have been used conveniently as towel, underwear, and socks, for example. However, hydrophilicity of cotton fiber is too strong to permit penetration of water molecules into the interior of the fiber structure. Then, cotton fibers swell by absorbing water, and drying velocity comes to be low, resulting in discomfort feelings such as being wet, sticky, and cold, especially after hard work. On the other hand, conventional synthetic fibers of polyester, nylon, and acrylic ones are composed of hydrophobic polymers and have a very hard time to absorb water but dry up fast. Recently, the perspiration-absorbent and quick-drying materials, composed of water-absorbing synthetic fibers, are developed and used often as good feeling fibers in the fields of sports, inner wears, blouses, dresses, and uniforms, for example.

Typical methods of enhancing water-absorbing ability of fibers are grouped into two, i.e., *chemical* method and *physical* one supported by chemical modification. The principal examples of the chemical methods are (1) addition of hydrophilic substances on to the fiber surfaces and (2) graft copolymerization with hydrophilic monomers. Addition of hydrophilic substances such as poly(ethylene glycol) during the dyeing processes of yarns and/or fabrics and also addition of hydrophilic substances during polymerization and spinning processes are typical ones for the former. The typical latter method is the chemical activation by plasma and electron beams on fiber surfaces.

Typical examples of the physical methods are (1) hollow or porous fibers, (2) modified cross-section fibers, (3) extra-fine fibers, etc. (1) One of the typical methods for obtaining hollow and porous fibers is dissolution method, where fine particles, which have been added beforehand, are dissolved away and form pores after spinning. A special method is also known to prepare the fibers having ca. 20 hollows within a fiber. The method of making hollow and also porous fibers is developed by applying the pore-forming techniques to the hollow fibers. Furthermore, there is the technique of fixing a huge number of voids which are formed just after wet spinning, i.e., void fixing method. In general, different from melt spinning method for polyester or nylon fibers, wet spinning and dry spinning methods for acryl fibers are rather easy and convenient for void formation during solvent removal process. Besides, blending and removing of additives are performed relatively easily. Polyester microporous hollow fiber WELLKEY^R (Teijin) is a typical example of the hollow and porous fibers. As is shown in Fig. 14.9, many fine pores reach the hollow core and effective removal of perspiration water by the capillary effects is expected. Figure 14.10 left and right shows the properties of wet/dry and hot/cold feelings of WELLKEY^R, respectively [18] (2). The modified cross-section fibers mean noncircular concave cross-section fibers. Substantial increase in the inter-fiber space, surface area of fibers, and the capillary effects are expected. L-, + (cross)-, Y-, X-, C-, W-shaped and their mixed shaped sections are available. (3) The extra-fine fibers give huge surface area within yarn resulting in the enhanced capillary effects.

Several other methods obtaining highly hydrophilic fibers are developed hitherto. Firstly, ultra-hydrophilic acrylic fibers have been prepared using sheath of highly water-swelling polymers. They are able to absorb several tenfold weight of water compared with the fiber's weight. Secondly, highly hydrophilic fiber, MRT^R fiber (Teijin), where elongation and shrinking take place by wetting and drying,

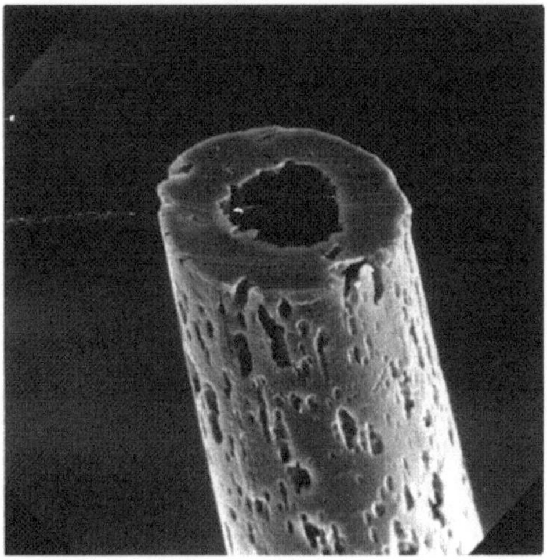

Fig. 14.9 Electron micrograph of WELLKEY^R

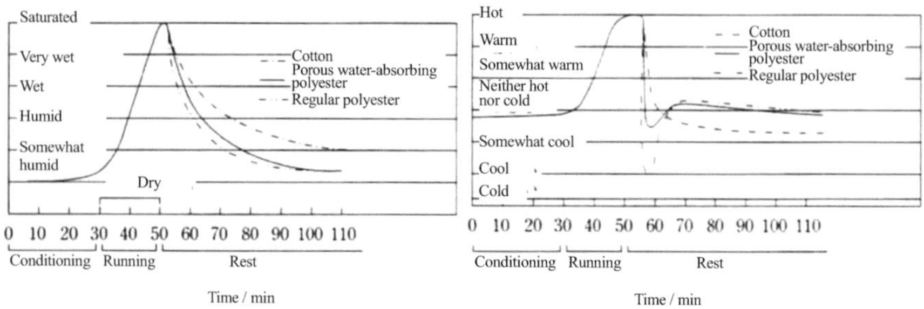

Fig. 14.10 Wearing test of T-shirts. *Left*: test for humid feeling. *Right*: test for hot and cold feelings

Fig. 14.11 Surface irregularity control type (*left*) and ventilation control type (*right*)

respectively, has been developed. A drop of water induces ca. 22 % elongation of the fiber, but shrinks reversibly by drying the fiber (see Fig. 14.11). MRT^{R} fibers enable to prepare several kinds of so-called smart textiles which have self-regulating functions, i.e., (i) bumpy deformation of cloth surface takes place by perspiration and then the clothes do not contact with skin densely. (ii) When a person perspires, openings are formed through cloth with quick expansion of stitch length, and these openings enhance perspiration evaporation and airflow [19].

14.5 Composite Structures of Yarns and Knitted or Woven Fabrics

Not only imparting moisture- and/or water-absorbing properties to single yarn material but also combining yarns having different properties can add special functions to fabrics. For example, knitted and/or woven fabrics, which consist of moisture-absorbing staple fiber spun yarn for the outer layer and non-moisture-absorbing staple fiber spun yarn for the inner layer, are able to keep lower humidity within clothing, compared with those consist of homogeneously blended spun yarn of moisture-absorbing staple fiber and non-moisture-absorbing staple fiber. Furthermore, using non-moisture-absorbing yarn made by extra-fine staple fibers for inner layer enhances further lower humidity and removing the humid feeling substantially [20].

Another example of the functionalization of fabric is the knitted and/or woven fabric composed of water-absorbing and non-moisture-absorbing filament yarn such as water-absorbing polyester filament yarn. When the single fiber fineness of the yarn is smaller for the atmospheric side than that of the skin side, perspiration transfer is enhanced by the capillary effects in the thickness direction of the knitted and/or woven fabrics. Then, the wet and sticky feeling is minimized. This strategy is applicable to the structure of individual yarns. Specifically, an example is the core-sheath double-layered yarn composed of water-absorbing fiber and non-moisture-absorbing fiber, where the single fiber fineness of the core layer is smaller than that of the sheath layer.

Furthermore, moisture-absorbing fibers can bring about the similar effects. The composite fabrics, whose skin side is composed of water-absorbing and non-moisture-absorbing fiber and the atmospheric side is composed of highly water-absorbing and moisture-absorbing fiber, also enhance the perspiration transfer from skin side to atmospheric side and give an example of the fabric of low sticky and low wet feeling [21, 22].

In conclusion, it is effective to combine the fibers so as to enhance the perspiration transfer in the thickness direction of knitted and/or woven fabrics in order to attain the minimized sticky and wet feeling.

Further, in order to increase evaporation efficiency of perspiration to exhibit cooling function, it is effective to combine the methods of quick absorption and diffusion of perspiration with those of increasing the evaporation area. For these purposes, the use of water-absorbing and non-moisture-absorbing fibers, especially modified cross-section fibers, is largely effective because of the enhanced capillary effects [23]. We should note here that moisture-absorbing fiber also affords the similar effects by mixing the water-absorbing and non-moisture-absorbing fiber in part. Figure 14.12 shows the examples of the cooling effect to suppress body temperature rise during exercise by the use of clothes composed of water-absorbing and non-moisture-absorbing fiber (solid curve) and of water- and moisture-absorbing fiber (broken curve), respectively.

Lastly, one of the effective methods to reduce the discomfort feeling of clothes clinging to skin, when a person perspires excessively, will be the use of the core-

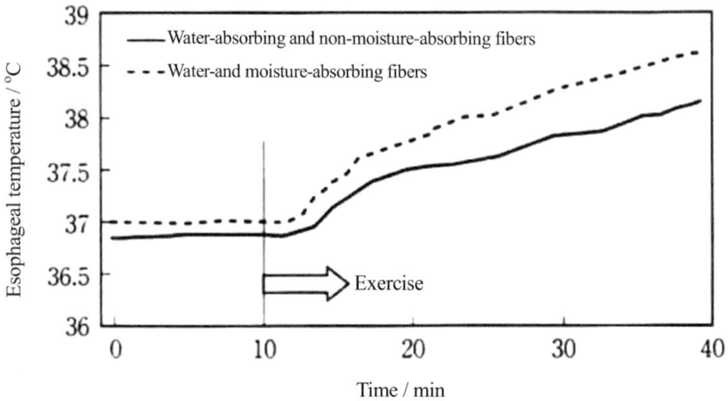

Fig. 14.12 Esophageal temperature during exercise

sheath double-layered yarn, which consists of water-absorbing and non-moisture-absorbing fibers for the sheath layer and moisture-absorbing fibers for the core part.

References

1. A. Nakayama (ed.), *Onnetsu Seirigaku* (Rikogakusha Publishing Co., Ltd, Tokyo, 1985), pp. 13–188
2. T. Yasuda, M. Miyama, H. Yasuda, Text. Res. J. **62**(4), 227–235 (1992)
3. T. Harada, K. Tsuchida, Sen'i Kikai Gakkaishi (J. Text. Mach. Soc. Japan.) **36**(12), 586–595 (1983)
4. H. Morooka, M. Niwa, Jpn. Res. Assoc. Text. End. Uses. **27**(11), 495–502 (1986)
5. I. Nomiyama, T. Tamura, Jpn. Res. Assoc. Text. End. Uses. **36**(10), 633–640 (1995)
6. W.E. Morton, J.W.S. *Hearle, Physical Properties of Textile Fibers* (Butterworths, 1962), p. 164
7. *The Society of Fiber Science and Technology*, Japan Ed., "Sen'i Binran" (Maruzen, Tokyo, 2004), p. 220
8. D.W. van Krevelen, *Properties of Polymers* (Elsevier Science B. V, Amsterdam, 1990), p. 572
9. H. Sanuki, Jpn. Synth. Text. Mon. (Kasen Geppou) **29**(9), 58–63 (1976)
10. H. Sanuki, K. Kuri, K. Ōta, Sen'i Gakkaishi **21**, 91 (1965)
11. Japanese Patent No. 3196855.
12. Y. Ohi, T. Nakagawa, M. Wakitani, Text. Proc. Technol. **33**(7), 428–433 (1998)
13. Japanese Patent No. 3716517.
14. N. Kinoshita, Sen'i Gakkaishi **56**(5), 153–154 (2000)
15. Japanese Patent No. 3705644.
16. H. Yamaguchi, Sen'i Gakkaishi **56**(5), 155–156 (2000)
17. E. Akiba, Sen'i Gakkaishi **56**(6), 177–178 (2000)
18. T. Suzuki, O. Wada, Sen'i Gakkaishi **41**(11), 401–409 (1985)
19. T. Suzuki, Kobunshi **58**(8), 534–535 (2009)
20. T. Harada, K. Tsuchida, Sen'i Gakkaishi **45**(3), 131–133 (1989)
21. S. Ishimaru, *Dyeing and Finishing (Senshoku)*, **17** (1) (65), 5–14 (1999)
22. Y. Suzuki, S. Ishimaru, Text. Proc. Technol. **29**(8), 39–41 (1994)
23. S. Ishimaru, Sen'i Gakkaishi **56**(9), 264–267 (2000)

Chapter 15
Heat-Controllable Man-Made Fibers

Sonoko Ishimaru

Abstract Thermal-retaining properties of clothing materials are indispensable for a comfortable life in the winter season. In general, thermal retention of clothes is enhanced by the amount of stationary air stored in space among the constituent fibers. Specifically, thinning the fibers, crimping the fibers, and increasing the number of fluffs are effective methods to enhance the thermal retention ability.

In this chapter, some methods are introduced to make clothes warm more positively. For examples, following attempts are described, e.g., lowering the thermal conductivity of the fiber itself by the introduction of air in it, increasing radiant heat flux using heat generated by the human body, improving the amount of heat storage using solar light, enhancing hygroscopic exothermicity of cloth using moisture generated from the human body by insensible perspiration, and canceling the environmental temperature change to keep the temperature within clothing constant using the heat energy of phase-change materials. The important policies to enhance the thermal retention ability are explained by specific examples of newly developed materials by some companies.

Keywords Thermal retention • Heat generation • Thermal insulation • Thermal storage • Thermal conduction

15.1 Methods for Imparting Heat-Retaining Ability

A general way to maintain body temperature in a cold environment is to reduce the heat emission through radiation, convection, and conduction by increasing the thermal insulation. For this purpose, several methods have been utilized thus far: (1) a method of retaining a static air layer inside the garment or in the surface of clothes because of the low thermal conductivity of air, (2) a method of making the garments of fibers having low thermal conductivity, (3) a method of laminating metallic cloth or film on the fabric surface to reduce the thermal radiation from the human body, and (4) a method of closing the garment openings for preventing the

S. Ishimaru (✉)
Toyobo Co., Ltd., Osaka, Japan
e-mail: Sonoko_Ishimaru@toyobo.jp

Society of F.S. Technology, Japan (ed.), *High-Performance and Specialty Fibers*,
DOI 10.1007/978-4-431-55203-1_15

heat dissipation by convection. Another active heat-retaining method is to gain heat from the energies existing in the surroundings. For example, sunlight (optical energy) is converted to heat for warming, and the heat of moisturizing fibers is used for heat retention in which the moisture generated by the insensible perspiration of the human body is responsible for the hygroscopic exotherm. In some cases, the heat generation is controlled in response to the temperature change.

15.2 Units of Heat Retention

The thermal retention (or thermal insulation) is quantified by clo value, which is defined by the unit of thermal resistance.

One unit clo corresponds to the thermal resistance of the clothes that are necessary for an adult male who is sitting on a chair and lying quietly in a room regulated at 21.2 °C in temperature, 10 cm/s in air flow, and 50 %RH or less in humidity, to wear in order to maintain a comfortable state by keeping a mean skin temperature at 33 °C. Here, the heat generation of the human body is assumed to be 50 kcal/(m^2·h), 75 % of which is emitted by convection and radiation. Consequently, 1 clo is estimated to be 0.18 °C/(kcal·m^2·h). This value approximately simulates the warmth that one can feel when wearing an all-wool business suit for the autumn season. The clo value is generally measured by using a thermal manikin.

15.3 Heat-Retaining Materials

15.3.1 Thermal Conduction

Table 15.1 shows the thermal conductivities of various fibers as compared with that of air [1]. Since air shows an excellent insulating property that is about ten times higher than those of fibers, it is an effective way for increasing the heat retention (thermal insulation) to put much air inside garments and form a static air layer between the garment surface and skin. Namely, the specific volume of garments must be increased for thermal insulation.

Stuffing with down (feather) is a general method for increasing the interfiber gap for better thermal insulation. This method is also imitated by using synthetic fibers

Table 15.1 Thermal conductivities of various materials

Material	Thermal conductivity/kcal m^{-1} h^{-1} °C^{-1}
Polyester	0.25
Nylon	0.22
Cotton	0.56
Wool	0.19
Air	0.02

such as ultrathin fibers, cross-section-modified fibers, hollow fibers, etc. that can help holding much air in garments, more specifically, in between and inside the component fibers. There is another method to increase the static air layer by specific structural design of woven/knitted fabrics and nonwoven fabrics.

A typical example for the hollow fibers is the Aero Capsule® produced by Teijin (Japan) (Fig. 15.1) [2]. The Aero Capsule® has a hollow ratio of ca. 35–40 vol.%, which refers to the cross-sectional area ratio of the hollow part in the whole cross-section of the fiber. The core technologies for manufacturing such fibers consist of a suitable polymer synthesis, a spinneret design, and spinning/drawing technologies by which a high hollow ratio can be maintained in the stage of fiber spinning. If the weight of the fiber per unit area is identical, the Aero Capsule® hollow fibers can make the heat retention 60–70 % higher and the weight 60–70 % lighter than the ordinary non-hollow fibers (Fig. 15.2).

15.3.2 Radiation

An efficient way for retaining heat in garments is to reflect the radiant heat coming out from the human body back on the side of the human body. For example, the fiber surface is overcoated with a resin containing fine metal powders such as aluminum or covered with a thin metal film by deposition technique. The metallic layers perform the heat reflection. In another way, the far-infrared ray emitted from the human body is absorbed and reemitted to the human body side to improve the heat retention to a higher degree than that by simply reflecting the radiant heat from the human body. It is known that far-infrared rays are mostly absorbed before reaching a depth of about 200 μm from the skin surface and converted into heat. This generated heat is transmitted inside the body (body core) by the blood stream or the like to warm the whole human body. For this purpose, ceramics are generally

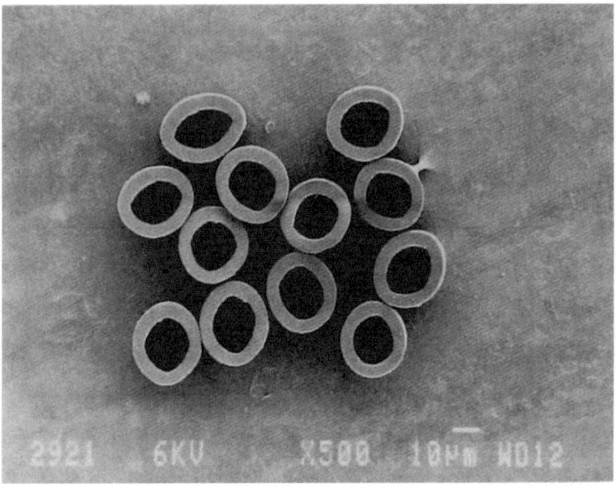

Fig. 15.1 Micrograph of the cross-sections of Aero Capsule®

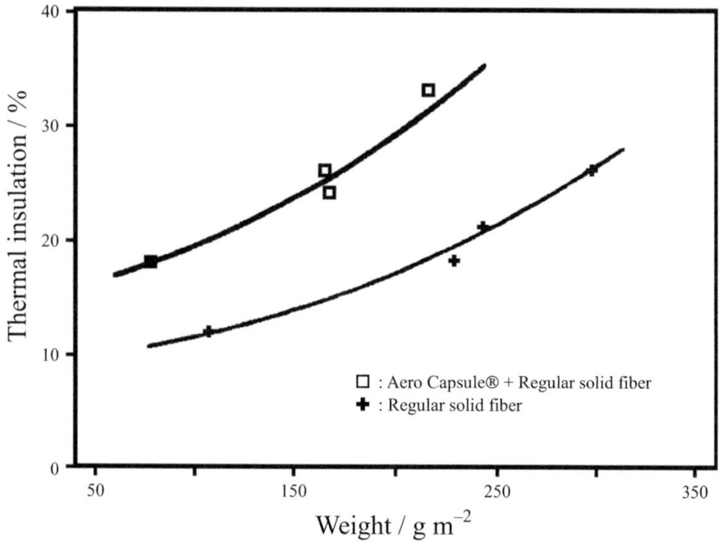

Fig. 15.2 Heat retention of Aero Capsule®. Knitted structure: interlock knit, measurement: JIS L1096-1990 6.28-1A, air flow: 1 ms^{-1}, temperature difference: 10 °C, environmental: 20 °C, 65 % RH, apparatus: Thermo Lab II®

used. The addition of ceramic components to fibers is established by (1) blend spinning of fine ceramic powders having specific properties, (2) dispersing ceramic fine particles in the resins used for surface coating and laminating, and (3) fixing ceramic powders with binder resin. A typical example is the Long Wave® produced by Kuraray (Japan) based on method (1) [3].

15.3.3 *Thermal Storage of Solar Energy*

It has been known that some materials can convert sunlight to thermal energy and store it to increase the heat retention. They are called heat storage materials of sunlight. For example, carbides of transition metals of group IV are capable of absorbing light with energy higher than 0.6 eV (<2 μm in wavelength) and converting it into thermal energy while reflecting light with a lower energy without absorption. These properties of carbides, typically, zirconium carbide (ZrC), are suitable for heat storage of sunlight. Furthermore, sunlight shows a peak around 0.5 μm in wavelength in its emission spectrum, involving more than 95 % of the total energy in the wavelength region from 2 to 0.3 μm. Figure 15.3 compares the reflectance of carbides of transition metals of group IV. Note that the heat wave emanating from the human body is approximately 10 μm in wavelength.

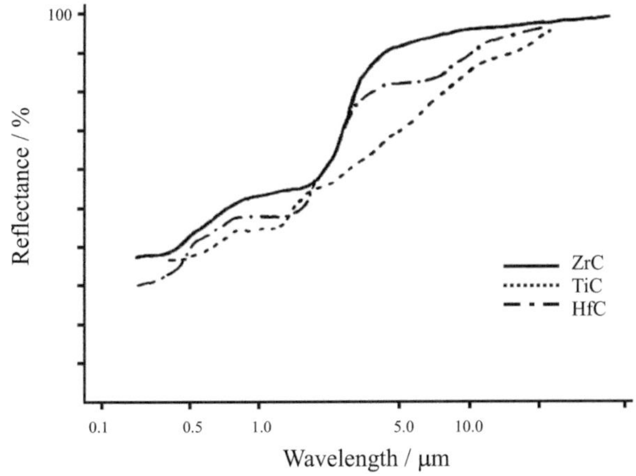

Fig. 15.3 Light reflections of transition metal carbides (the reflectance of the optically polished surface)

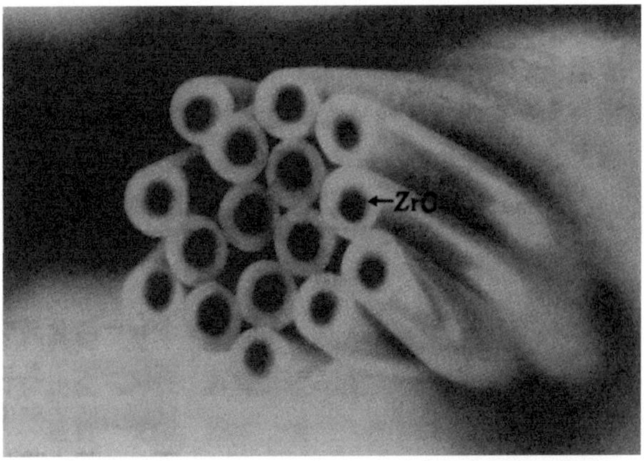

Fig. 15.4 SEM micrograph of Solar α®

For example, carbides are used as a core part of fiber. Solar α® marketed by Descente (Japan) and Thermotron® produced by Unitika (Japan) are fiber materials consisting of a polyester sheath and a core part in which zirconium carbide is blended (Fig. 15.4) [3]. It cannot only absorb and convert visible and near-infrared lights to heat but also reflect the heat generating from the human body, resulting in efficient heat retention. Figure 15.5 shows the changes in temperature inside clothes when one skis by wearing a one-piece skiing suit wadded different downs made of different fibers. The temperature inside clothes increased by about 8 °C with Solar α® than with the conventional fibers.

Some carbon materials as well as organic dyestuffs are also utilized as light-heat conversion materials instead of ceramics as zirconium carbide. ***Figure 15.5 shows the effect of the zirconiumu carbide by the subjects test***. These materials are usually applied by blend spinning technique, occlusion method, coating method, and padding method.

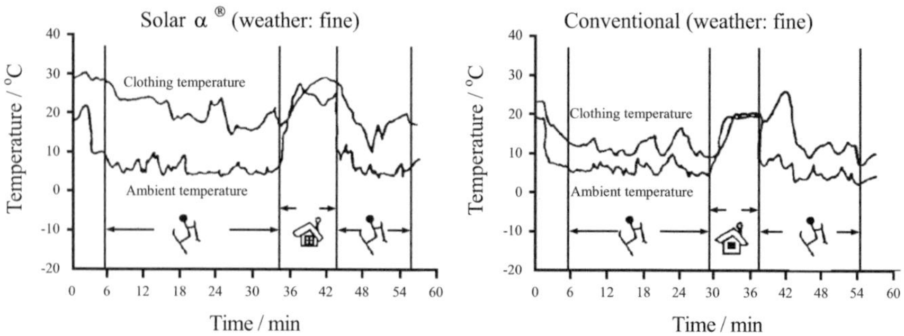

Fig. 15.5 Results of wearing test of a skiing suit made of Solar α® (Data provided by Descente Ltd.)

15.3.4 Response to Temperature Change

Kelvin Thermo® produced by Toray (Japan) is known to show suppressed heat storage under direct sunlight but enhanced heat storage under less solar radiation in cloudy weather. It consists of both thermochromic layer that can control the absorption of solar light by changing the color with the environmental temperature and ceramic layer that can convert sunlight into thermal energy and radiate it to the human body. This two-layer structure makes it possible to absorb heat at outside temperature below 0 °C where the color of the thermochromic layer becomes black and in contrast suppresses the heat absorption above 5 °C where the thermochromic layer changes its color to white.

15.3.5 Hygroscopic Exotherm

A considerable amount of moisture is always evaporated from the human body by insensible perspiration. The insensible perspiration rate from the human skin is said to be 600 ml/day (or 25 g/h). An effective method to increase the heat retention has been established by absorbing this perspiration moisture to generate heat from moisture adsorption.

Any type of hygroscopic fiber generates heat from moisture adsorption, more or less. A particularly high heat of moisture absorption is shown by acrylate fibers. Figure 15.6 shows the heats of moisture adsorption of various fibers in different environmental humidities [4]. Table 15.2 compares the heats of moisture adsorption of an acrylate-based fiber and other fibers in an environment of 25 °C and 80.5 % RH. The acrylate-based fiber selected here is Moiscare® produced by Toyobo (Japan) [5, 6]. This fiber has a highly cross-linked and hygroscopic nature that is obtained by the unique polymer modification technique incorporating many

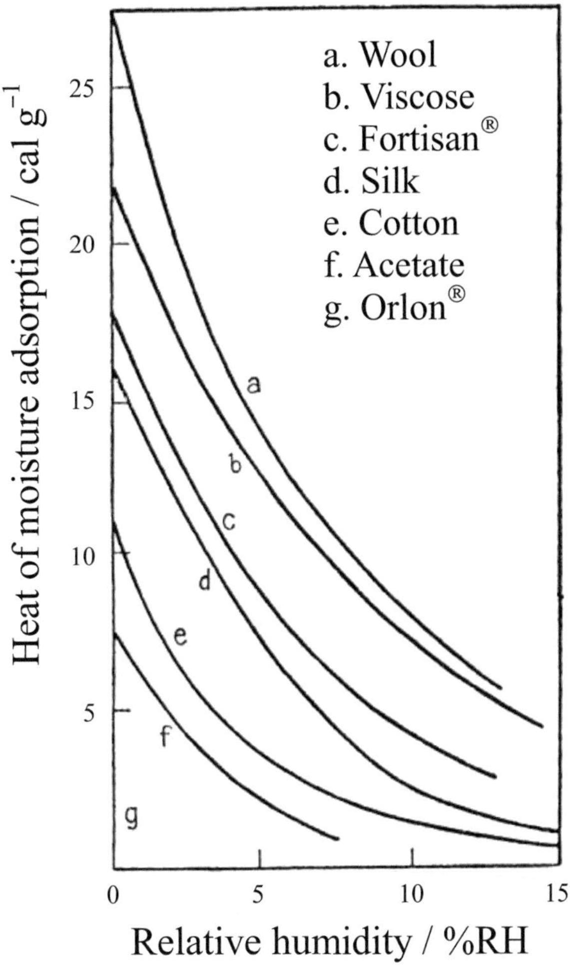

Fig. 15.6 Relationship between humidity and heat of moisture adsorption for various fibers

Table 15.2 Heats of moisture absorption of various fibers

	Polyacrylate	Wool, down	Cotton	Polyester
Heat of moisture absorption[a] /J g^{-1}	800–2000	350–450	40–50	~0

[a]Measured by a C80 calorimeter using absolutely dried samples at 25 °C, 80.5 %RH

hydrophilic groups and cross-linking sites into the polymer molecules. It exhibits a moisture-absorbing property that exceeds that of natural fibers.

As shown in Fig. 15.7, Moiscare® represents a water content of 41 wt.% in an environment of 20 °C and 65 %RH and a high moisture absorption that is about 5 and 2.5 times the absorptions of cotton and wool fibers, respectively [7]. There is a tendency that the more hydrophilic the fibers, the higher the heat of moisture adsorption. The acrylate-based fiber indicates a heat of moisture adsorption two

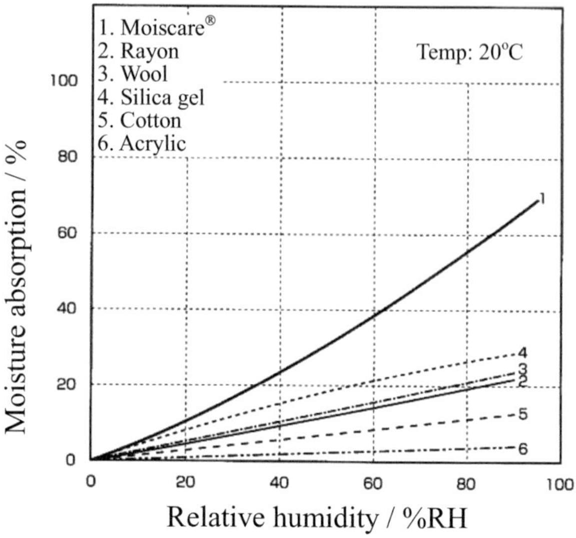

Fig. 15.7 Moisture absorptions of various fibers

to three times larger than that of wool fibers that show the highest heat of moisture adsorption among natural fibers.

15.3.6 Phase Change

It is possible to firmly fix microcapsules containing phase-change materials (PCM) on the fiber. The PCM-containing microcapsules can generate and absorb heat in response to temperature change. Generally, paraffin is encapsulated in the microcapsules, inducing a phase change from solid to liquid and vice versa with absorption and release of latent heat, respectively. The principle that the heats of crystallization and fusion evolve at cooling liquid and heating solid, respectively, is utilized for the heat and cold storage effect of the heat-controlled fibers. Their effective temperature can be adjusted by changing the melting point (phase change) of the paraffin encapsulated as the core material. A typical example is Thermo Capsule® produced by Daiwabo Neu (Japan).

References

1. H. Ohmura (ed.), *New Technologies and Markets of the Fibers for Health and Comfort* (in Japanese). Osaka Chemicals Research Series, vol. 3, No. 205 (Osaka Chemicals Marketing Center, Osaka, 2000),p. 113
2. A. Matsunaga, Sen'i Gakkaishi **51**(6), 245–248 (1995)
3. T. Furuta, Sen'i Gakkaishi **49**(11), 399–404 (1993)

4. Kiyoichi Matsumoto, Wu Ying, Yumiko Izumi, Hiroyuki Uejima, *Sen'i Kikai Gakkaishi (J. Text. Mach. Soc. Japan)*, **52** (7), 105–112 (1999)
5. The Society of Fiber Science and Technology, Japan (ed.), *Sen'i Binran* (Maruzen, Tokyo, 2004), p. 465
6. Yasuhiro Oi, Mitsuru Wakitani, *Sen'i Kikai Gakkaishi (J. Text. Mach. Soc. Japan)*, **49** (6), 324–331 (1996)
7. T. Ogino, T. Sumitani, Takamasa Sakamoto: Sen'i Gakkaishi **57**(12), 320–323 (2001)

Part IV
Ultrafine and Nano Fibers

Chapter 16
Nanofibers

Akihiko Tanioka and Mitsuhiro Takahashi

Abstract We explained the definition of nanofiber and extend the fiber diameter in the nanofiber definition from 1–100 nm to 1–1000 nm because various interesting phenomena are observed between 100 and 1000 nm. We developed the novel electrospinning which is called the blow electrospinning (BES), where we can spin solvent-dissolved polymers. On the basis of BES, we also developed the melt air spinning (MAS), where we can spin molten polymers. Both are categorized into cross air blow spinning (CABS), and their nanofiber productivities are very high so that we can realize the industrialization of nanofibers. The potential applications of the nanofibers by CABS expand to the fields of health and biomedical care, energy and environment, and robot and information technologies.

Keywords Definition of nanofiber • Nano-sized fiber • Nanostructured fiber • Electrospinning • Blow electrospinning • Melt air spinning • Cross air blow spinning • Potential applications

16.1 Introduction [1, 2]

The nanofibers consist of "nano-sized fiber" and "nanostructured fiber" as shown in Fig. 16.1. Fiber thickness of the former one is controlled in nano- or submicron order, and that of the latter one is not necessary to be in nano- or submicron order, but the surface or the internal structure should be controlled in nano-order. The most common recognition to nanotechnology is "the technology which produces the novel function and the more excellent characteristics by operating and controlling the atom and molecule in the scale from 1 to 100 nm to change the structure and arrangement of a substance." According to the recognition, "nanofiber" is defined as the fiber-like substance whose diameter is 1–100 nm or 1000 nm, which is comparable to wavelength of light, and length is more than 100 times of a diameter as shown in Fig. 16.2. Though it is different from the general definition of nanotechnology, we had better extended the fiber diameter in the nanofiber

A. Tanioka (✉) • M. Takahashi
Department of Organic Polymeric Materials, Tokyo Institute of Technology, Tokyo, Japan
e-mail: tanioka.a.aa@m.titech.ac.jp

Society of F.S. Technology, Japan (ed.), *High-Performance and Specialty Fibers*,
DOI 10.1007/978-4-431-55203-1_16

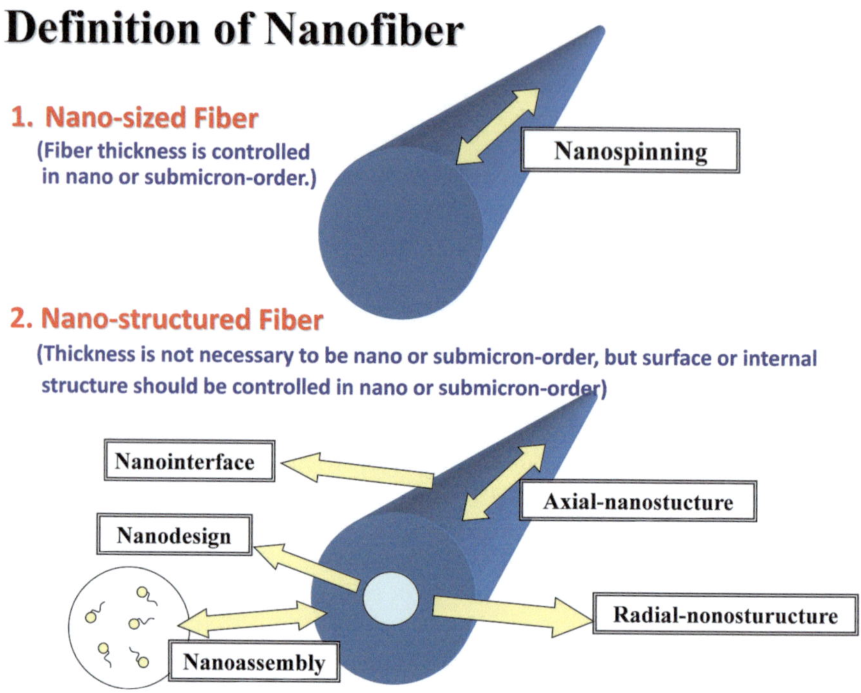

Fig. 16.1 Definition of nanofiber

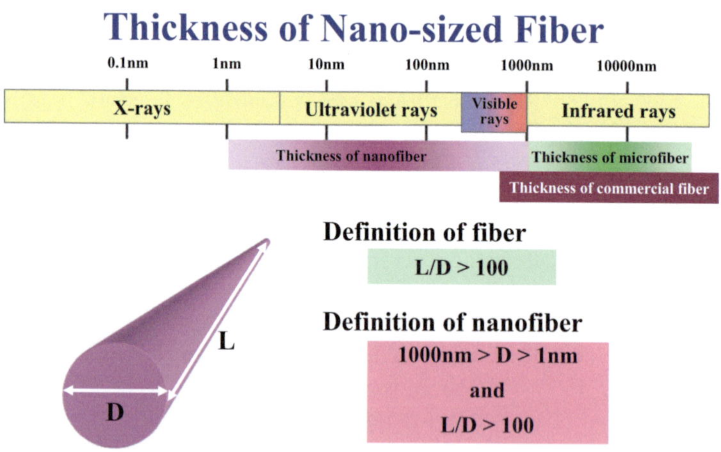

Fig. 16.2 Thickness of nano-sized fiber

definition from 100 to 1000 nm positively, because the structural color phenomena are observed if the fiber diameters are located between 200 and 800 nm, the antimicrobial phenomena are shown in the common polymers if the fiber diameters are less than about 400 nm, and so on.

There are various manufacturing processes for nanofiber production, which are shown as nanospinnings, CVD for carbon nanotubes, self-organization for natural nanofibers such as collagen nanofibers, grinding procedure for nanocellulose, and the processes for nanowires. Among those manufacturing processes cited above, the nanospinnings, such as electrospinning, melt air spinning, conjugate spinning, meltblown, etc., are the most attractive technologies which are realizing the practical utilization. We can spin ca. 5 nm diameter of nanofiber by using electrospinning, ca. 30 nm by melt air spinning, ca. 10 nm by conjugate spinning, and ca. 200 nm by meltblown. Especially, electrospinning and melt air spinning are the fascinating procedures to produce nanofibers, because they can produce them relatively simply in an industrial scale.

The productivity of normal electrospinning in a nozzle is, however, very low because the fluid flow rate from it is about 2 μl/min so it is very difficult to apply the nanofibers in an industrial scale, because if the number of nozzles is increased for high productivity, it causes some troubles such as electric field interference, electric shock, solvent explosion, etc. Recently, we have developed two high-speed nanofiber spinning systems by using cross air blow spinning (CABS: Zs) which are applied to polymer solution and molten polymer, respectively. The former one is called "blow electrospinning (BES: Zs-solvent)" and the flow rate of polymer solution from a nozzle is more than 2 ml/min to produce nonwoven fabrics, and the latter one is called "melt air spinning (MAS: Zs-melt)" and the flow rate of molten polymer is more than 15 ml/min to produce raw cotton-like material, which has been developed during the BES development. The former one is available for all kinds of solvent-soluble polymer and the latter one for molten polymer.

The development of high-speed production systems of nanofibers expands their potential applications, which are oil-water separation, oil fence, flood protection, membrane distillation, water retaining in soil, air and water purification filters, acoustic insulation, protection against cold, snow shoveling, battery separator, flying carpet, superconducting fibers, scaffold for tissue engineering, cosmetics, disposable diaper, and so on. In this section, the nanofiber production systems such as electrospinning, blow electrospinning, and melt air spinning are compared, and application of nanofibers is discussed.

16.2 Nanospinnings [1, 2]

There are several methods to produce nanofibers, such as electrospinning, melt air spinning, conjugate spinning, meltblown, etc. In this section we mainly explain the electrospinning and the melt air spinning.

16.2.1 Electrospinning

Electrospinning is a straightforward and versatile method for forming nonwoven fabrics composed of thin fibers. This spinning procedure has the following advantages: (i) applicability for a broad spectrum of materials such as synthetic polymers, bio-related materials, and inorganic molecules and (ii) ability to spin thin fibers or nonwoven fabrics with nano-microscaled structures, under atmospheric pressure.

The simple electrospinning is carried out using the device equipped with a capillary nozzle made by the metal. Polymer solutions were sprayed through the nozzle whose inner diameter is about 50 μm. Applying the voltage between nozzle and counter electrode substrate, the polymer solution is discharged from it and the rate is about 2 μl/min. Usually, the sprayed polymer forms a fiber-like structure and flies to the opposite side to be collected by the counter electrode substrate made from the metal plate or mesh. The solvent in the fiber evaporates during spraying or after detecting by the counter electrode substrate. The distance between the nozzle and the counter electrode substrate depends on the scale of the system. The voltage applied between the capillary and the counter electrode substrate is ranged from 1 to 100 kV.

If we produce a large amount of nanofiber nonwoven fabric sheets, it is necessary to use a number of nozzles. It, however, causes plenty of problems, which are shown as follows:

1. A large amount of nozzles is needed because of the low productivity of nanofiber.
2. Maintenance becomes very hard by using many nozzles.
3. The use of multi-nozzles causes the electric field interference so the spray becomes unstable.
4. There is a possibility of causing electric shock by the application of high voltage between nozzles and counter electrode substrate.
5. There is a possibility of exposure to organic solvent in a large amount.
6. There is a possibility of causing explosion by organic solvent and electrical discharge.
7. Organic solvent exposure is considered.
8. Effect of ionic wind has to be considered.
9. Others.

16.2.2 Novel Electrospinning [3, 4]

In order to avoid those problems cited above, we have developed two novel systems. One is the ultrahigh-performance nozzle which is called the "rotary

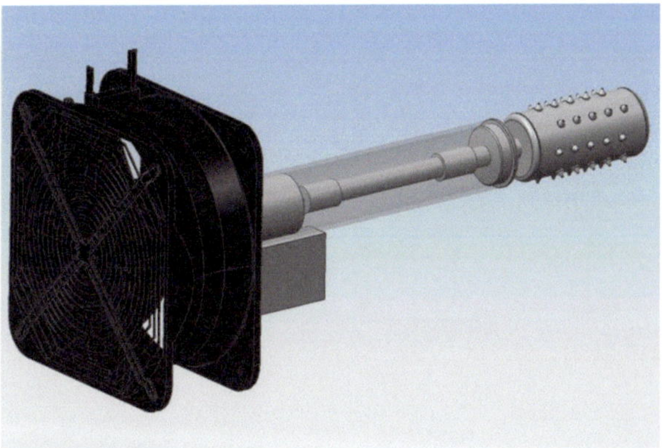

Fig. 16.3 Rotary cylinder nozzle

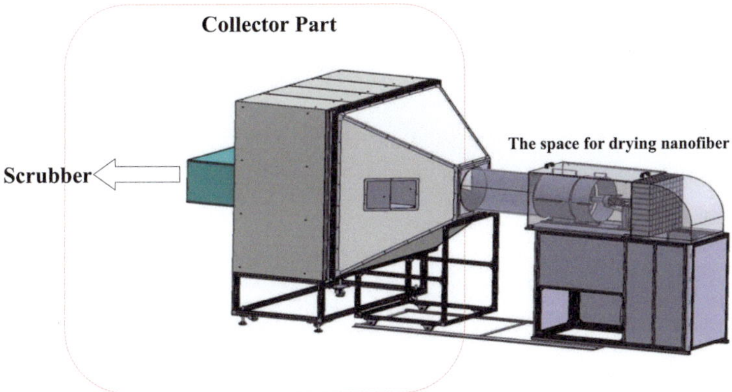

Fig. 16.4 Fiber ring system

cylinder nozzle" as shown in Fig. 16.3, where the nozzle orifices are put in the cylinder surface, polymer solution is set in it, and the cylinder is rotated during spinning, and the other where the spray and collector parts are separated is called the "fiber ring system" as shown in Fig. 16.4, where air is blown to the rotary cylinder nozzle and nanofiber collected by the collector and the evaporated solvent is recovered by the scrubber. The advantage of the "rotary cylinder nozzle" is that the discharge rate of the polymer solution is more than 2 mL/min/orifice, which is 1000 times larger than the traditional nozzle, and that of the "fiber ring system" is that solvent concentration reduces to less than 2000 ppm to avoid explosion, and inside pressure of the system is lower than outside so that organic solvent and nanofiber are not scattered to the outside.

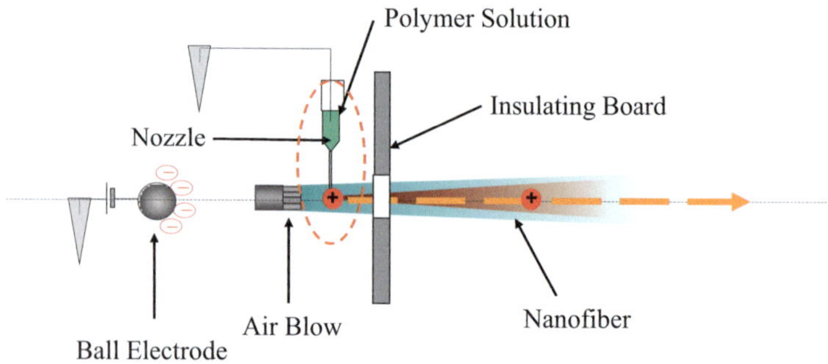

Fig. 16.5 Side nozzle method

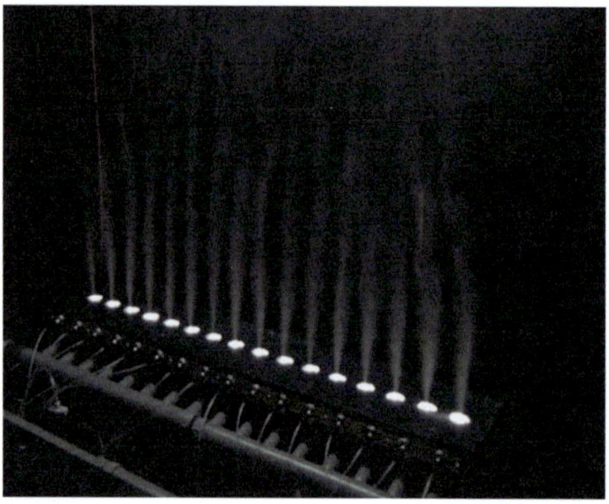

Fig. 16.6 Nanofiber spinning with 16 nozzles by blow electrospinning (BES)

A fiber ring system with a rotary cylinder nozzle is a major advance in electrospinning. Though the productivity is much more increased, thickness control of nonwoven sheets and system simplification are so difficult to be industrialized, especially if the problems are with the rotary cylinder nozzle. Under consideration of air blow to the rotary cylinder nozzle in the fiber ring system, we have developed a new nozzle system, which is called the "side nozzle method," as shown in Fig. 16.5, where a ball electrode, an air blower, a nozzle, and an insulating board are lined from left to right and air is blown to the nozzle in a perpendicular direction. The discharge rate of polymer solution is much more than 2 mL/min and the total system including maintenance becomes very simple. The nanofiber spinning with 16 nozzles is shown in Fig. 16.6, and the system with the side nozzle system is called "blow electrospinning (BES)." In Fig. 16.7 a scanning electron microscope (SEM) image of polyethersulfone (PES) spun by BES is shown, and fiber thickness is less than 100 nm. The most distinctive point is that nanofibers are

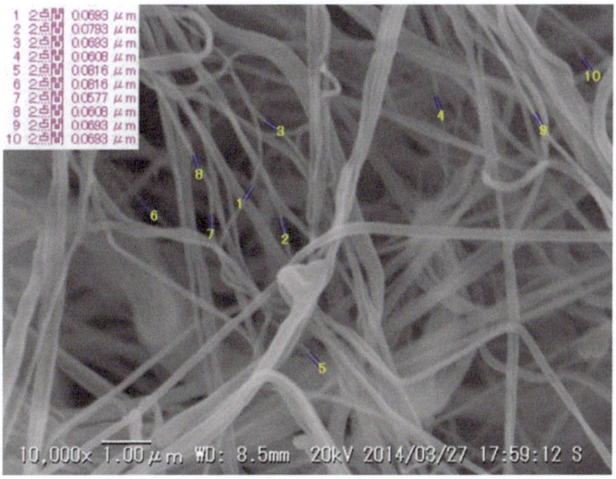

Fig. 16.7 SEM image of polyethersulfone (PES) spun by BES

overlapped three-dimensionally, which is not seen in nanofiber sheets by the traditional electrospinning.

16.2.3 Melt Air Spinning [3, 4]

Though BES is a great advantage in nanospinning because we can spin a large amount of nanofibers of solvent-soluble polymers in a very simple way, we cannot spin molten polymers. In reference to the side nozzle method, we have developed the novel nanofiber spinning system with heating devices for molten polymers shown in Fig. 16.8. Air is blown to the nozzle in a perpendicular direction from a non-frame torch, and the ball electrode is also equipped, but even if it is removed there are no effects to the spinning rate and the produced fiber diameter. This spinning system is called "melt air spinning (MAS)." The discharge rate of polymer is more than 50 g/min and the total system including maintenance is also very simple. The most important advantages are that we can use different kinds of molten polymers with particles and the power consumption is much smaller than that of meltblown. In Fig. 16.9 a SEM image of polylactic acid (PLA) spun by MAS is shown, and fiber thickness is between about 100 and 1000 nm. Also the most distinctive point is that nanofibers are overlapped three-dimensionally.

BES and MAS are categorized into "cross air blow spinning (CABS)." In Table 16.1 cross air blow spinning, electrospinning, conjugate melt spinning, meltblown, CVD, and bionanofiber are compared.

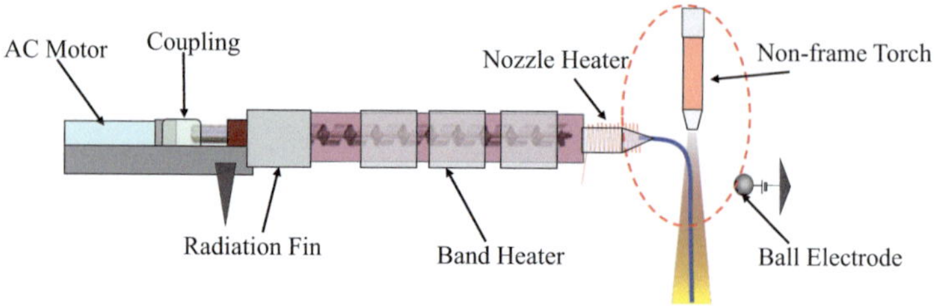

Fig. 16.8 Melt air spinning (MAS)

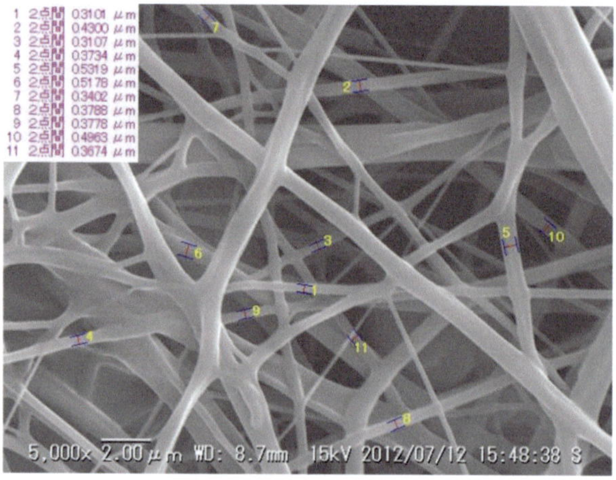

Fig. 16.9 SEM image of polylactic acid (PLA) spun by MAS

16.3 Potential Applications [1]

Characteristics of nanofibers are composed of three scientific effects and seven technological performances. The former three are the ultrahigh specific area, nano-order, and ultrahighly aligned effects, and the latter seven are high molecular recognition, high adsorption, slip flow, optics, high electrical current, high mechanical property, and high heat resistance performances. If the nano-sized fiber and the nanostructured fiber with technological performances are processed in linear, planar, or cubic form, we can obtain various products as shown in Fig. 16.10 after mass production.

Namely, the development of BES and MAS expands the potential applications in the fields of health and biomedical care (stimulation wear for healthcare, skin therapy, drug delivery system, tissue engineering, biosensor, etc.), energy and environment (hydrogen production, secondary battery, nano-solar battery, solar power station, superpower capacitor, filter for water production, high-efficiency catalysis, high-performance nanofilter for air, oil-water separation, membrane

Table 16.1 Comparison of cross air blow spinning, electrospinning, conjugate melt spinning, meltblown, CVD, and bionanofiber

Comparison of the nanofiber production method			
Spinning technologies	Advantages	Problems to be solved	Materials
Cross air blow spinning (CABS)	Very high productivity	Spinning less than 50 nm	[Blow electrospinning (BES)]
	Very simple process	Metal spinning	Solvent-soluble polymer
	Surface structure is easy to be controlled		Protein
	It is easy to be composited		[Melt air spinning (MAS)]
			Thermoplastic polymer
			Inorganic material
			Carbon
			Metal
Electrospinning	Spinnable in room temperature	Very low productivity	Solvent-soluble polymer
	Surface structure is easy to be controlled	High voltage is necessary	Protein
	It is easy to be composited	Solvent recovery is necessary	Inorganic material
Conjugate melt spinning	Spinnable less than 100 nm	Fiber split process is necessary	Limited thermoplastic polymer
		Weak material to heat is inapplicable	(Nylon, PET)
			Two different compatible polymers
Meltblown	Simple process	Spinning less than 200 nm	Limited thermoplastic polymer (PP, PE, PET)
		Weak material to heat is inapplicable	
CVD	CNT	Low productivity	Carbon
	Nanowire	Safety	Metal
Bionanofiber	Medical use	Establishment of extraction method	Biological polymer
	Biodegradability	Quality	(Cellulose, collagen, chitin, chitosan, etc.)
		Biocompatibility	

distillation, water retaining in soil, acoustic insulation, protection against cold, nanofilter in high temperature, etc.), and robot and information technology (high-performance battery separator, high-performance secondary battery, electronic devices, organic EL, wearable electronics, mobile fuel cell, space elevator, space station, robot, smart fabrics, super safety wear, etc.).

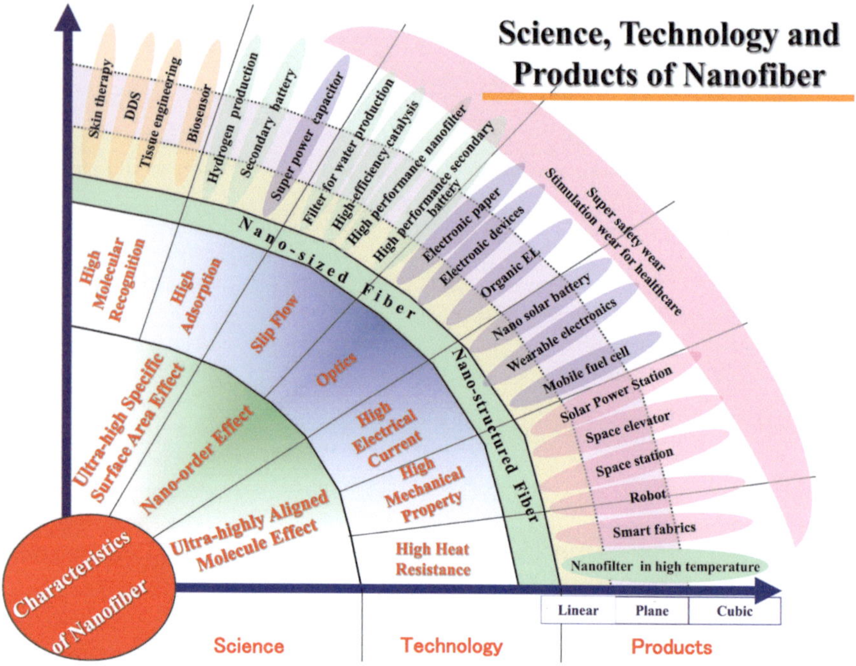

Fig. 16.10 Science, technology, and products of nanofiber

16.4 Conclusions

Firstly, we explained the definition of nanofiber which was consisting of “nano-sized fiber” or “nanostructured fiber.” Secondly, though the most common recognition to nanotechnology is “the technology which produces the novel function and the more excellent characteristics by operating and controlling the atom and molecule in the scale from 1 to 100 nm to change the structure and arrangement of a substance,” we extend the fiber diameter in the nanofiber definition from 100 to 1000 nm because the interesting phenomena such as structural color and antimicrobial effect in common polymer and so on are observed between 100 and 1000 nm in fiber diameter. Thirdly, we developed the novel electrospinning which is called the blow electrospinning (BES). On the basis of BES, we also developed the melt air spinning (MAS), where we can spin molten polymers. Both are categorized into cross air blow spinning (CABS), and their nanofiber productivities are very high so that we can realize the industrialization of nanofibers. The potential applications of the nanofibers by CABS expand to the fields of health and biomedical care, energy and environment, and robot and information technologies.

References

1. Tatsuya Hongu, *Nanofiber*, Nikkann Kogyo, Tokyo (2006)
2. Darrell H. Reneker, Hao Fong, Polymeric nanofibers, ACS Symposium Series 918, (2006)
3. Akihiko Tanioka et al., USP 8,544,410 B2 (2013)
4. Mitsuhiro Takahashi, The future world spun by nanofiber, Web. J., **154**, 1–6 (2014)

Chapter 17
Nanofibers by Conjugated Spinning

Yuhei Maeda and Masato Masuda

Abstract Historically, conjugated spinning technology was proposed more than 50 years ago. Recently, this technology has enabled the production of nanofibers. Using this technology it is possible to produce uniform nanofibers in a filament shape. Thus, produced nanofibers can be processed to woven fabrics and knitted works; this is not possible for other types of nanofibers such as the web-type shape made by electrospinning technology, etc.

In this chapter, two types of conjugated spinning technology, i.e., blend-spinning technology and spinneret technology, are introduced. By the application of the former technology, nanofibers can be thinned to 60 nm in diameter, and by the latter technology, circular and noncircular cross-sectional nanofibers with uniform thickness can be produced.

Keywords Nanofiber • Conjugated spinning • Blend spinning • Spinneret technology • Noncircular cross section

17.1 Introduction

Fibers are fine and long materials, and it is not too much to say that the necessity to utilize the fiber materials originates from these morphological features. To pursue the ultimate limit of "fineness" of fibers, which will greatly improve the unique characteristics of fibers as mentioned below, is important not only from an academic point of view but also for industrial applications.

Y. Maeda (✉)
Toray Fibers and Textiles Research Laboratories (China) Co., Ltd., No.58 Xinkai Road(S), Nantong Economic & Technological Development Zone, Nantong, Jiangsu 226009, People's Republic of China
e-mail: Yuhei_Maeda@toray.cn

M. Masuda
Toray Industries Inc. Fibers and Textiles Research Laboratories, 4845 Mishima, Shizuoka 411-8652, Japan
e-mail: Masato_Massuda@nts.toray.co.jp

Society of F.S. Technology, Japan (ed.), *High-Performance and Specialty Fibers*,
DOI 10.1007/978-4-431-55203-1_17

Remarkable Characteristics of Microfibers

1. Excellent softness, flexibility, smoothness, and ease of twisting
2. Microscopic space formed among the fiber bundles
3. Large specific surface area and enhanced surface properties
4. Microscopic interpenetration with other materials
5. Low kickback property against bending
6. Reduction of concentrated stress

17.2 Spinning Method for Microfiber

The method for manufacturing microfibers is roughly divided into two, i.e., for the production of filament-type and web-type microfibers.

17.2.1 Spinning Method of Filament Type

The filament-type microfibers are manufactured by the shape which can be treated as a string, and they can be processed as a continuous fiber in the manufacturing steps. This type of microfibers is made by the direct spinning method or conjugated spinning method.

As for the direct spinning method, microfibers are spun directly using a spinneret which has small discharge holes. This is the simplest method of making microfibers; however, industrially the thinnest limit of the microfibers is ca. 3 μm in diameter, and there are many problems, such as inferior spinning stability in manufacturing thinner microfibers at present.

On the other hand, as for the production technology of microfibers by the conjugated spinning method, initial conjugated fibers, which consist of two different component polymers, are manufactured for the first. Then textile fabrics are made using the conjugated fibers in the same manner as other fibers, and after that, the conjugated fibers are converted to microfibers by the removal of one component of the conjugated fibers or by the peeling and splitting of both components at their boundaries. There are a few types of morphology for the conjugated fibers that originated from the manufacturing process, e.g., sea-island type and peeling type. In the case of sea-island-type conjugated fibers, the conjugated flow of polymers which consists of many island polymer components (this will remain to be microfibers finally) covered by sea polymer component is prepared for the first. Then the conjugated flows are, while being merged, extruded from one discharge hole and conjugated fibers are formed. Finally microfibers are made by dissolution and removal of the sea component. On the other hand, in the case of peeling- and splitting-type conjugated fibers, after the textile fabrics are made, microfibers are formed by the separation of both components using chemical treatments to swell or shrink one component or mechanical treatments such as beating, frictioning, etc.

As for the manufacturing technology of microfibers by the conjugated spinning method, apart from the abovementioned method having features in the spinneret, the blend-spinning method is also used. In this method, sea-island-type conjugated fibers are spun from the mixture of two incompatible material polymers. After removal of the sea component, microfibers are obtained. This sea-island-type conjugated fiber made by the blend-spinning method has a long extended island component along the fiber axis. However, the length of the island component is finite and the fiber diameter is not uniform, and these points are different from the sea-island-type conjugated fibers made by the spinneret technology. According to the recent technological advancements, every conjugated spinning technology can produce nanofibers as will be detailed later.

The conjugated spinning technology itself was first proposed by DuPont in the 1960s (Tokkou S39-933, Tokkou S39-29636). This technology is characterized a lot by the spinneret. Multiple flows of materials are aligned precisely and, while being merged, extruded from one discharge hole. Toray first applied this conjugated spinning method to make microfibers in the 1970s. Microfiber fabrics with a suede texture named "Ecsaine" (US name "Ultrasuade" and European name "Alcantala"), which are made from microfibers 3 μm in diameter, have been placed on the market, and the basis of the microfibers is founded.

The greatest advantage of this conjugated spinning technology is that in the processing stage, e.g., weaving and knitting, the thick conjugated fibers can be treated in the same manner as conventional fibers. At the processing stage from where the thickness of the fiber does not cause problems, thick conjugated fibers are converted to microfibers by chemical or physical treatment.

17.2.2 Spinning Method of Web Type

Different from the filament-type microfibers, the web-type microfibers are not uniform. In the manufacturing process of the web-type microfibers, polymer material fibers are thinned by miniaturization and drawing using some force after the discharge from a spinneret, and then they are collected in a sheet shape. This method has a merit in producing microfiber nonwoven in one step; however, the demerit is it is difficult to obtain microfibers in a filament shape, and therefore textile fabrics are not possible to make. Further, the diameter of obtained web-type microfibers is not so uniform when it is compared with that of filament-type microfibers.

There are some production methods for the web-type microfibers, e.g., melt-blow method, flash spinning method, electrospinning method, etc. As for the melt-blow method, microfibers are produced by blowing away the discharged polymer fibers using high-temperature fluid. The diameter of obtained microfibers is a few micrometers in general, and it is said that 1 μm diameter microfibers may be possible to make. As for the flash spinning method, a high-temperature and high-pressure polymer solution is discharged from fine nozzles, and the polymer solution is miniaturized and drawn by the explosive power of solvent vaporization. This method has been put to practical use for polyethylene, etc., and the diameter of obtained microfibers is a few micrometers.

As for the electrospinning method, the polymer solution is ejected from a nozzle applying a voltage toward a target electrode. The ejected polymer solution is accelerated toward the electrode and is blown away due to the dispersion force of surface charge of the solution, and hence micronization and solvent evaporation are carried out simultaneously. Though the diameter distribution of obtained fibers is larger than other methods, sheet-shaped product consisting of fibers with a diameter ranging from dozens of nanometers to one micrometer range is obtained.

17.3 Nanofiber Technology by Conjugated Spinning

In the fabrication technology of nanofibers, the conjugated spinning method is superior to the electrospinning method, because as mentioned before the former method makes it possible to process textile structures consisting of thick conjugated fibers before they are converted to nanofibers. Hence, a variety of product forms, e.g., textile fabrics, nonwovens, liquid dispersions, etc., are possible. Nanofibers made by this conjugated spinning method are also advantageous in mechanical and thermal properties because the fiber structure is formed in the spinning process. Further, the conjugated spinning method uses no solvent in spinning, and therefore this method has advantages in environmental and safety points of view. The largest appeal of this method is that it is possible to utilize existing manufacturing facilities, and hence from the industrial point of view, development period and cost are reduced.

From these viewpoints, the detailed explanations of the nanofiber technologies using the recent conjugated fiber method are given below especially for both blend-spinning method and conjugated spinning method by spinneret technology.

17.3.1 Nanofiber Technology Using Conjugated Spinning by Blend Spinning

The blend-spinning method is superior to the conjugated spinning method because in the former method it is possible to produce nanofibers by conventional melt-spinning manufacturing facilities only using different materials, whereas the latter method is inferior because the designing of the spinneret is difficult and restricted. However, the spinnability for the former blend-spinning method tends to decrease greatly by the instability in discharging and thinning the fiber because two incompatible polymer components have each specific deformation behavior. Especially to produce nanofibers by the blend-spinning method, the island component should be finely divided in the initial stage just after discharge. This necessity also decreases spinnability due to the increase in boundary surface area and increase in boundary energy when the island component is drawn in the direction of the fiber axis.

In answering this technological problem, Toray found a solution to attain stable spinnability despite the nano-size dispersion of the island component by the accurate control of the deformation processes of fibers after the discharge and by that of the flux in the spinneret hole, and in 2002 the nanofiber technology using the melt-spinning method was established for the first time in the world. Figure 17.1 shows the TEM photograph of the cross section of nanofibers obtained by this method. The diameter of this nanofiber is extremely fine as 60 nm in average. This nanofiber is a short-fiber type whose fiber length is finite, and as shown in Fig. 17.1 short-fiber-type nanofiber are self-assembled to form a nanofiber aggregate, and then their assemblage results in nanofiber yarn hierarchically. By this hierarchical structure, many forms of products are attained and various new properties which could not be attained by former nanofibers have been revealed, e.g., as shown in Fig. 17.2, short-fiber-type nanofibers can easily be deformed by the water absorption and by the swelling of the nanofiber aggregate. Therefore, for example, if they are applied to beauty products such as skin care clothes, nanofibers can wipe off human skin stain with reduced damage by the "soft hand" (touch feeling) due to the swelling by water absorption. Therefore, if they are compared with conventional nylon fibers, some superior effects become apparent, e.g., three times larger water retentivity and adsorption ability of gases such as ammonia and VOC and, further, good workability and supporting ability of functional agent.

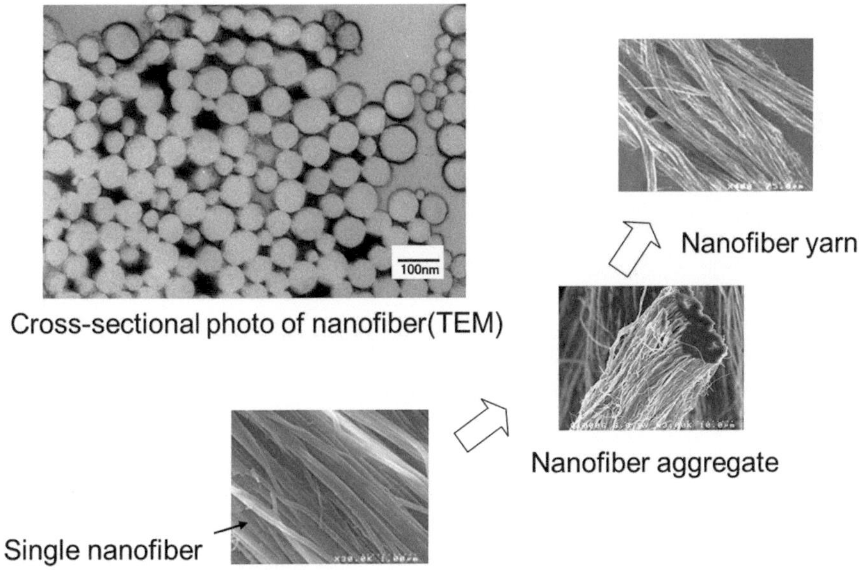

Fig. 17.1 Hierarchical structure of short-fiber-type nanofiber

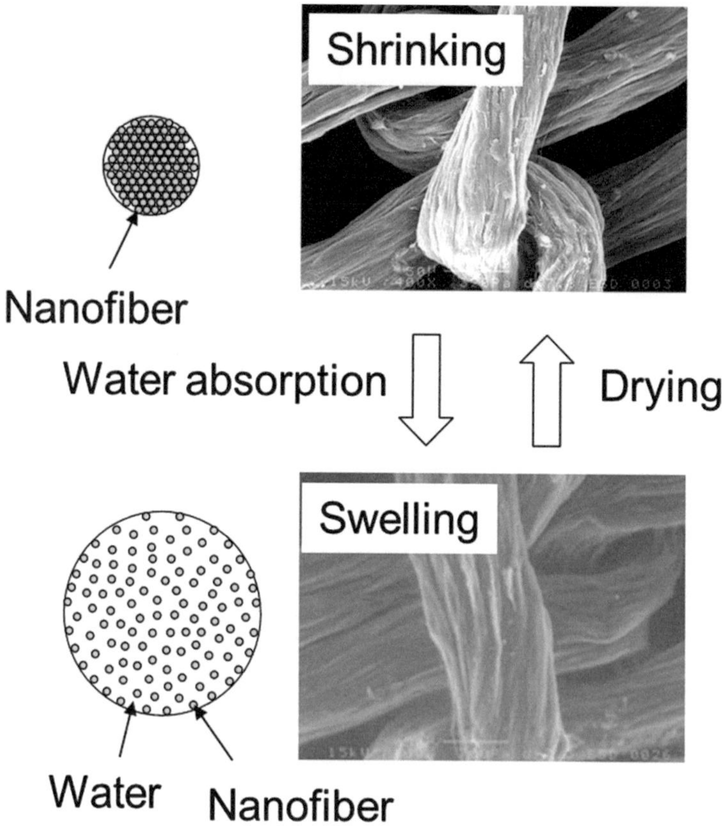

Fig. 17.2 Swelling behavior of short-fiber-type nanofiber aggregate

These short-fiber-type nanofibers can take a variety of product forms depending on the processing method as shown in Fig. 17.3, and therefore by utilizing this feature, the commercial demands in the future are expected for the beauty and textile products as well as industrial materials.

17.3.2 Nanofiber Technology Using Conjugated Spinning by Spinneret Technology

Nanofiber technology using conjugated spinning by spinneret technology has been considered to be disadvantageous in seeking the ultimate thinness of the fibers when it is compared with the blend-spinning method. However, the extreme efforts to pursue the spinneret technology made it possible to obtain fibers whose diameter is less than 1 μm. When the nanofibers obtained by this method are compared with

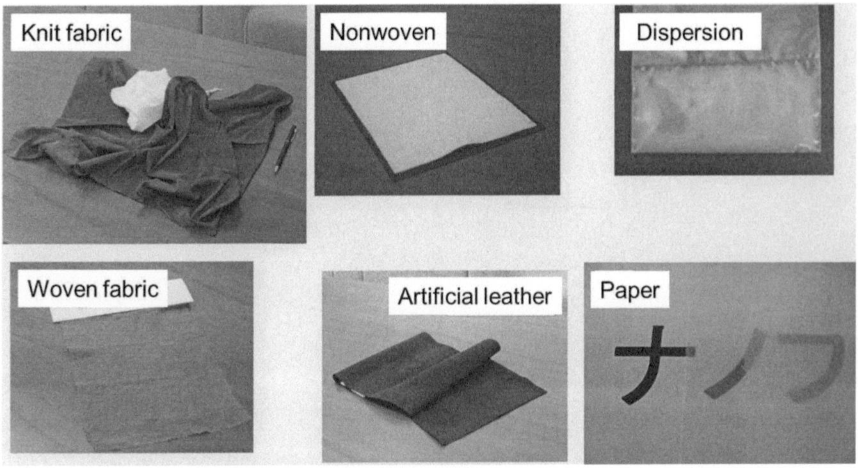

Fig. 17.3 Diversity of nanofiber products

those by other methods, the uniform fiber diameter is the outstanding feature. Further, this nanofiber is obtained as a continuous fiber and is most superior in the endurance of fiber structures including textile fabrics and in the processability of such as texturing or textile-processing, etc.

In order to fabricate nanofibers by the conjugated spinning method, sea-island-type conjugated fibers, for example, the important point is how to increase the number of islands and how to align individual polymers. The basic policy has not changed since a long time ago, and one can increase the number of islands extremely according to this policy. However in this case the island number is limited physically. Further, to begin with, the control of the ultra-miniaturized flow of polymer is very difficult. A slight disturbance of polymer flow causes instability of the whole conjugated fiber cross section. As a result the instability causes the confluence of island component flow, and once the confluence of the island component has occurred, it is very difficult to separate them again. This is one of the biggest problems in obtaining nanofibers by the conjugated spinning method. Against this problem, it is possible to attain good sea-island cross-section profile by the control of combined polymer ratio and viscosity ratio. However in this case, the kinds of available polymers are limited, and due to many other problems such as difficulties to increase island component which influence the productivity, the merit of this method is greatly impaired.

To solve this technical problem, Toray succeeded in developing a new nanofiber technology by pursuing the flux control of these ultrafine polymers together with the further deepening of the conventional conjugated spinning technology. Using this new technology, it becomes possible to produce a wide range of fiber diameters from 1 μm to the ultimate fineness of 150 nm class, whose diameter has been generally said to be difficult to produce.

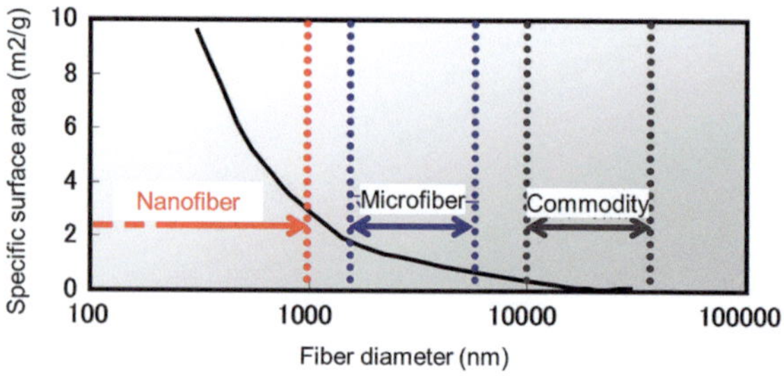

Fig. 17.4 Relationship between diameter and specific surface area

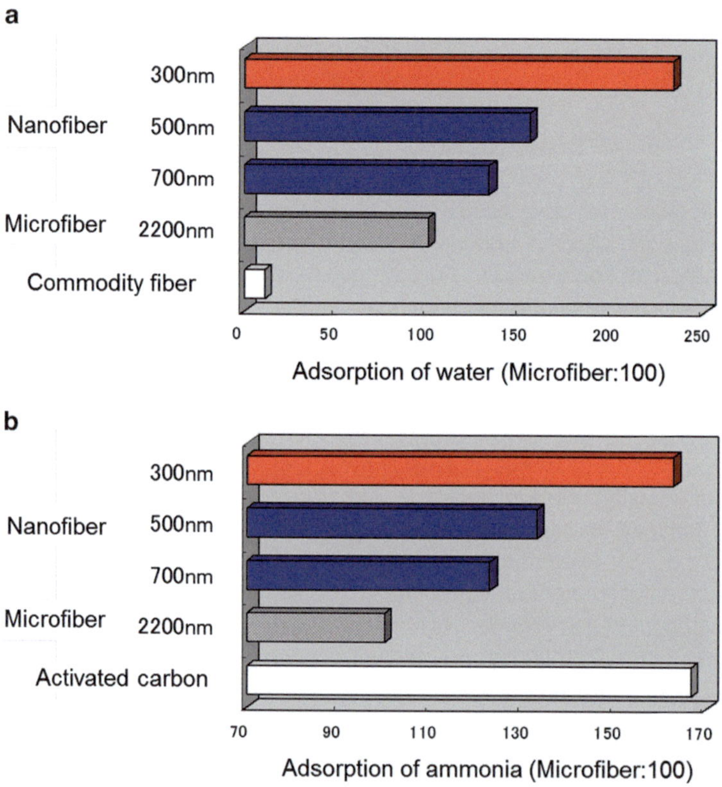

Fig. 17.5 Diameter dependence on functionality. Adsorption of (**a**) water, (**b**) ammonia

This ultrathin nanofiber has a superior flexibilitiyt and attain good "hand." Further, according to the increase in the specific surface area as shown in Fig. 17.4, it has been found possible to add new functions on the fiber surface, e.g., even polyester which is a hydrophobic polymer, good moisture absorption and gas adsorption abilities are expressed without adding functional agent, as shown in Fig. 17.5. These functions are originated from the physical structure, and therefore, there is no fear for the functional deterioration. Therefore the ultrathin nanofiber is expected for the highly functional apparel products with the "good hand."

Thus, in addition to seeking for ultimate thinness of fibers, which has been made for the conventional nanofibers, pursuing flux control of the ultrafine polymers indicated the development and application direction of new nanofiber materials. That is, the flux control of the ultrafine polymers made it possible to control the boundary of conjugated polymers precisely. As a result, the cross-sectional shape of the nanofibers, which has conventionally been circular or distorted circular in principle, can become freely controlled. Using this technology it becomes possible to create noncircular cross-section nanofibers, for the first time in the world as shown in Fig. 17.6.

As for the conventional nanofibers with circular cross section, as shown in Fig. 17.4, the product development is focused on the super-specific surface area effect due to the decrease in fiber diameter and on the improvement of the flexibility due to the decrease in the moment of inertia of area. However, these properties are sometimes overexpressed and turned into a disadvantage, e.g., causing uncomfortable feelings such as caught or entangled feelings to human skin, making textile manufacturing complicated, and lowering processability by the lack of stretch and stiffness. On the other hand, as for the newly developed nanofibers with noncircular cross section, it is possible to enhance stiffness of the individual nanofiber by the control of the morphology, for example. As a result the mechanical properties, such as anti-drape stiffness (HARI) and stiffness (KOSHI), of textile fabrics are improved even in water, as well as in dry condition, as shown in A and B of Fig. 17.7. Therefore, the nanofiber is hardly limited under use environment; even under textile manufacturing which needs frequent aqueous processing such as dyeing, the nanofiber has excellent processability, and hence wide applications are expected also for industrial material use. Further, the nanofibers with noncircular cross section exhibit, by the edge effect, an excellent function of scraping off the contamination, though the surface friction is greatly decreased. As a result, as shown in C and D of Fig. 17.7, it was found that both low friction and greatly increased wiping performance, which have been thought to be mutually contradictory characteristics, can be attained compatibly.

In this way, nanofibers made by the conjugated spinning method by spinneret technology indicate, due to the thinness and the morphology, the completely new

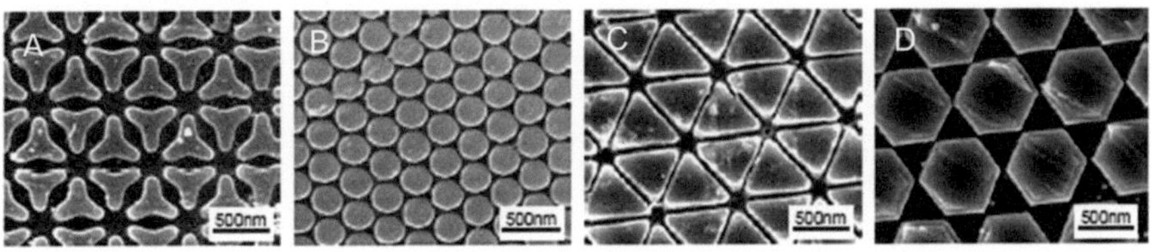

Fig. 17.6 "Shape" control by the innovative nanofiber technology. Expanding cross-sectional photo with island of (**a**) tri-lobe shape. (**b**) circular shape, (**c**) triangle shape, (**d**) hexagonal shape

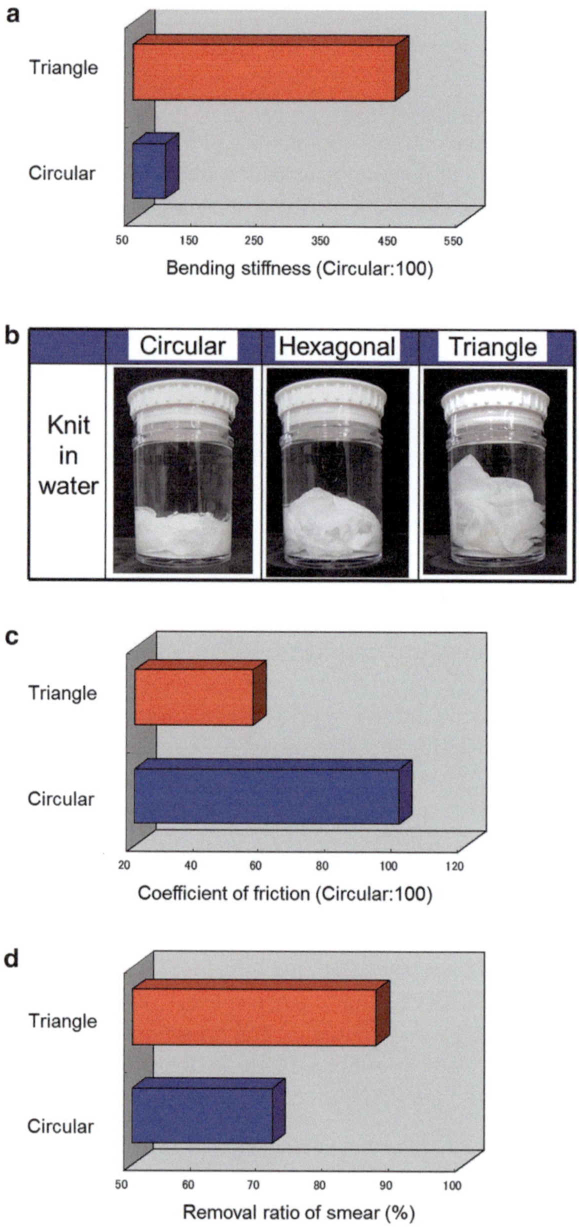

Fig. 17.7 Properties of noncircular cross-section nanofiber. (**a**) Bending stiffness. (**b**) Stiffness of knit in water. (**c**) Coefficient of friction. (**d**) Wiping property

future direction of nanofiber materials, and this technology will be developed further.

17.4 Conclusion

The characteristic features of nanofibers, which are expectable from the morphological characteristics, have been confirmed in the practical use, recognized in the public, and established the position of functional materials. The market of nanofiber materials is considered to be expanding in the future because the features of nanofibers are matched with growing environmental and safety concerns and needs for a more comfortable life, and therefore the engineering developments of them will be enhanced. What is important in this stage is to take a measure of the capability of nanofibers and how to handle them. Specifically, not only pursuing the manufacturing technology of nanofibers but also developing the textile manufacturing technique matched with nanofibers and creating new products in which the advantages of nanofibers are effectively utilized, is necessary. For this purpose, close cooperation and information exchange exceeding a framework of an individual research group are necessary.

In this article, the features are explained for two types of nanofibers made by conjugated spinning technology. This nanofiber technology has a possibility to expand nanofiber applications greatly, by making it possible to control fiber diameter and cross-sectional shape, which has been limited hitherto. A wide range of applications of these nanofibers are expected, e.g., for highly functional apparel products such as comfortable fabrics and functional sports clothes; filters and medical materials; highly functional materials for batteries; environmental, water, and energy purpose; information communication and electronics; automobiles; and life sciences.

Chapter 18
Cellulose Nanofibers as New Bio-Based Nanomaterials

Akira Isogai

Abstract Nanocelluloses have attracted much attention as new bio-nanomaterials, which are prepared from abundant wood biomass resources by downsizing of cellulose fibers with or without pretreatments by mechanical disintegration in water. In this review paper, fundamental and application researches of fibrous TEMPO-oxidized celluloses (TOCs) and TEMPO-oxidized cellulose nanofibers (TOCNs) primarily carried out in our laboratory are reviewed for future prospects of nanocelluloses. The characteristic points of TOCNs different from other nanocelluloses are (A) wood TOCNs are crystalline nanofibers with homogeneous widths of ~3 nm and high aspect rations, (B) TOCNs have abundant sodium carboxylate groups exchangeable to other metal carboxylate groups and alkylammonium carboxylate groups by simple ion exchange, and (C) TOCNs are dispersed at the individual nanofiber level in water, forming nematic-like liquid crystalline or self-aligned structures caused by electrostatic repulsions between TOCN elements. From these aspects, self-standing films, hydrogels, aerogels, and composite materials have been prepared with TOCNs, and some of them showed unique and excellent properties.

Keywords Nanofiber • Cellulose • Nanofibrillation • TEMPO • Nanocomposite • Biodegradation

18.1 Historical Background

Wood cellulose forms a hierarchical structure in wood cell walls to protect living bodies of trees against gravity, outer mechanical stresses, and biological attacks. The smallest elements next to cellulose molecules are "cellulose microfibrils," each of which consists of fully extended 30–40 cellulose chains. Wood cellulose microfibrils have small width of ~3 nm, long lengths > several microns, and high crystallinities of 60–70 % (Fig. 18.1). Thus, cellulose microfibrils are

A. Isogai (✉)
Department of Biomaterials Science, The University of Tokyo, 1-1-1 Yayoi, Bunkyo-ku, Tokyo 113-8657, Japan
e-mail: aisogai@mail.ecc.u-tokyo.ac.jp

Society of F.S. Technology, Japan (ed.), *High-Performance and Specialty Fibers*, DOI 10.1007/978-4-431-55203-1_18

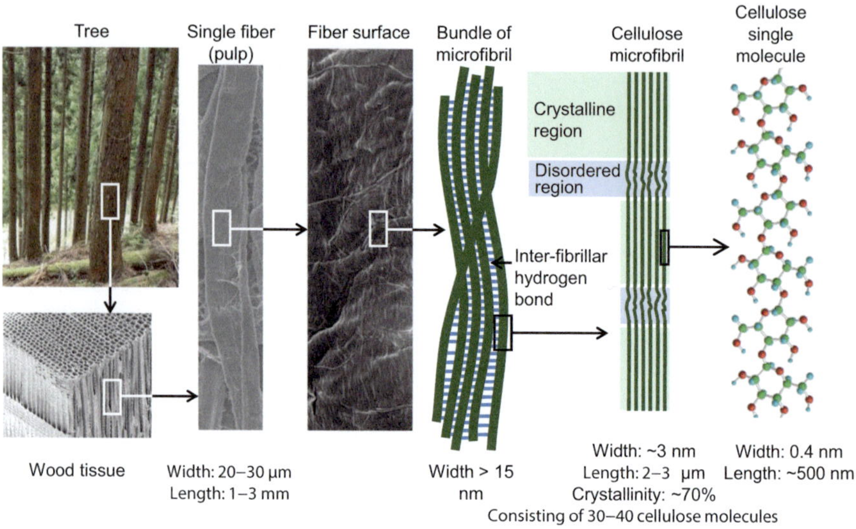

Fig. 18.1 Hierarchical structure of wood cellulose, forming crystalline cellulose microfibrils [2, 24]

bio-nanomaterials most abundantly present and accumulated on earth and are reproducible, biocompatible, biodegradable, and safe to our living bodies. However, wood cellulose microfibrils form complicated natural composite structures at both molecular and nano-sized levels with hemicelluloses and lignin in wood cell walls, and hence wood cellulose microfibrils have not been separated and isolated as individual nanofibers directly from wood cell walls. Although wood cellulose fibers can be isolated as pulps from wood chips by conventional chemical pulping and subsequent bleaching processes, numerous hydrogen bonds are present between cellulose microfibrils. Thus, it has been difficult to separate cellulose microfibrils also from wood pulps without serious yield losses or molecular-weight decreases.

In 1980s, highly fibrillated cellulose/water dispersions, i.e., microfibrillated celluloses (MFCs), were developed by Turbak et al. [1]. MFCs are prepared from wood dissolving pulp by repeated homogenization in water and therefore require high energy consumption in preparation process. In 1990s, some advantageous points to use MFCs in papermaking were reported in Japan, and not only MFCs but also bacterial celluloses produced by *Acetobacter xylinum* showed preferable effects as additives or fillers in papermaking. In 2000s, nanocellulose-related research and development have been remarkably accelerated particularly in developed countries to find out new markets and applications of pulp fibers in academia and pulp and paper industry. The accumulated nanotechnologies and their related science also have promoted nanocellulose researches particularly in academia. As a consequence, four principal breakthrough technologies to prepare nanocelluloses efficiently and to show their excellent potential applications have been reported.

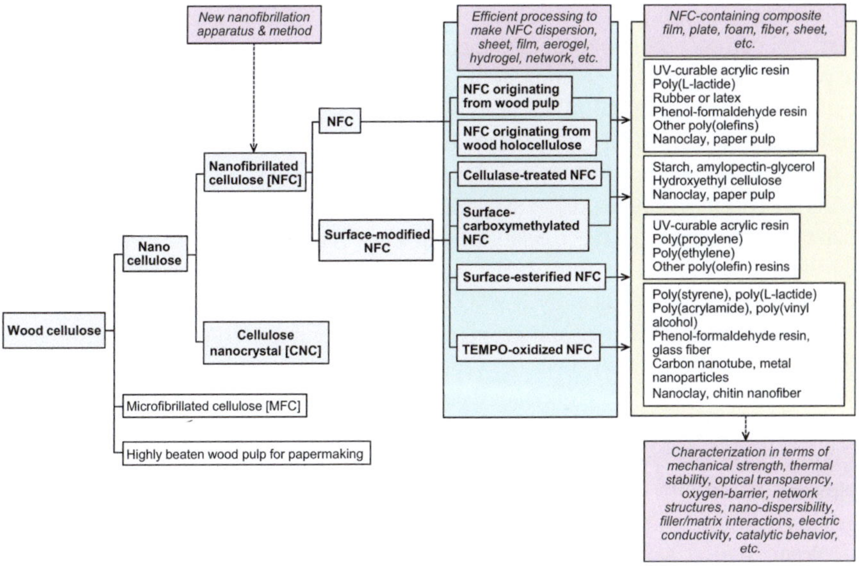

Fig. 18.2 Classification of nanocelluloses and four key breakthroughs developed in nanocellulose researches [2, 24]

First, new apparatuses to efficiently produce nanocellulose/water dispersions without clogging problem have been developed, such as high-pressure homogenizers, water collision-type homogenizers, super fluidizers, and grinder-type supermasscolloiders. Second, some pretreatments of wood celluloses (pulps) to reduce energy consumption in nanofibrillation process have been reported, such as mild endo-glucanase treatment, mild carboxymethylation under aqueous conditions, esterifications, and TEMPO-mediated oxidations. Third, the obtained nanocelluloses have been converted to new films, fibers, plates, aerogels, hydrogel, packages, etc. from nanocellulose/water dispersions or by compositing with conventional polymers or nanomaterials. Lastly, the obtained bulk nanocellulose materials and nanocellulose-containing composite materials were found to have excellent mechanical, optical, thermal, oxygen barrier, conductive, catalytic, and other properties, compared to the conventional nanocellulose-free materials (Fig. 18.2) [2]. These breakthroughs have caused driving forces to extend the nanocellulose-related research and development in both academia and industry.

18.2 TEMPO-Mediated Oxidation of Cellulose

In our laboratory, we have started fundamental researches of TEMPO-mediated oxidation of cellulose and other polysaccharides from 1995, based on a paper reported by de Nooy et al. in 1995 [3]. TEMPO is an abbreviation of 2,2,6,6-

tetramethylpiperidine-1-oxyl radical, which is water soluble, negative to the AMES test, and commercially available. When cellulose/water slurries are stirred in water at pH ~10 and room temperature in the presence of NaClO used as a co-oxidant and catalytic amounts of TEMPO and NaBr, the C6-primary hydroxyl groups of cellulose are selectively converted to sodium C6-carboxylate groups though C6-aldehydes by the TEMPO-mediated oxidation (Fig. 18.3) [4]; the glucose units in cellulose are oxidized to sodium glucuronate units. TEMPO and NaBr behave as catalysts in the oxidation, and only inexpensive NaClO is consumed during the oxidation to form NaCl. Some of the C6-aldehydes are oxidized to C6-carboxylate groups also by NaClO and NaBrO coexisting in the system.

The chemical structures and properties of the obtained TEMPO-oxidized celluloses are significantly different between native and regenerated celluloses used as starting materials. When regenerated celluloses with cellulose II crystal structures, such as viscose rayon fibers, are used, the water-insoluble celluloses turn to water-soluble products by the TEMPO-mediated oxidation. The oxidation converts almost all C6-OH groups of regenerated celluloses to sodium C6-carboxylate groups, forming water-soluble polyglucuronate Na salts [5].

When native celluloses consisting of cellulose microfibrils (Fig. 18.1) with cellulose I crystal structures, such as bleached kraft pulps, cotton, and bacterial and tunicate celluloses, are used as starting materials, the original fibrous forms or

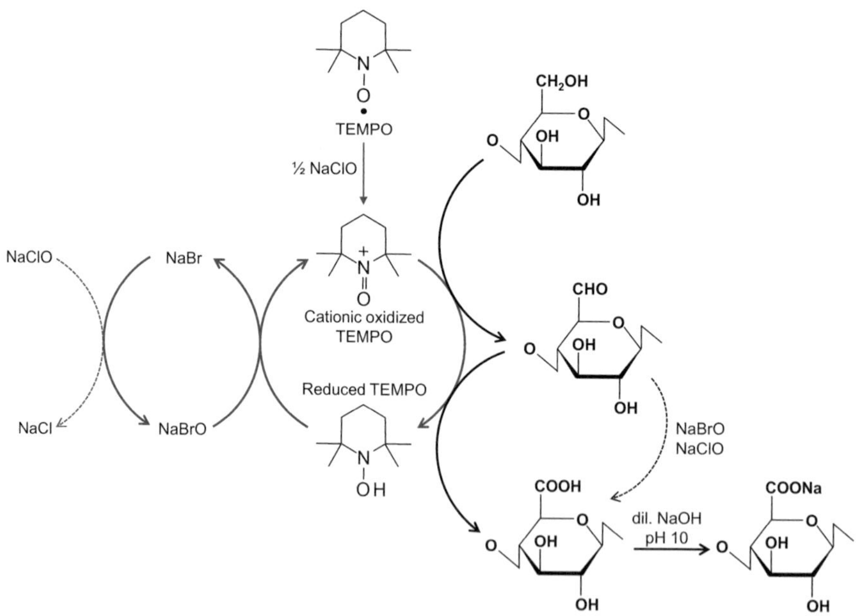

Fig. 18.3 TEMPO-mediated oxidation of cellulose with TEMPO/NaBr/NaClO system in water at pH 10 and room temperature [4]

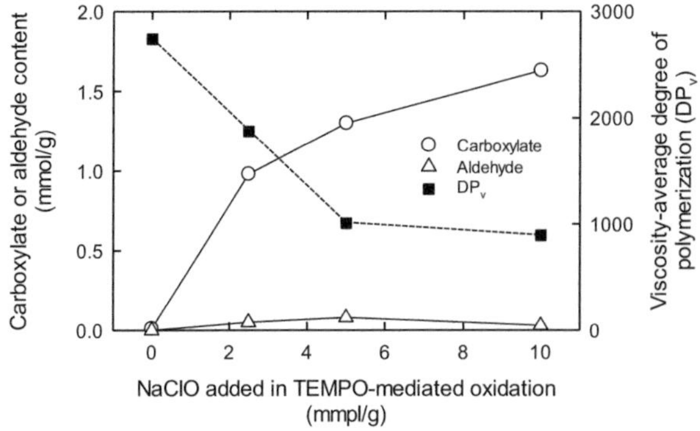

Fig. 18.4 Relationships between the amount of NaClO added in TEMPO/NaBr/NaClO oxidation of softwood bleached kraft pulp suspended in water at pH 10 and either carboxylate or aldehyde content and viscosity-average degree of polymerization of the oxidized celluloses [4]

fibril-like morphologies are maintained during and after TEMPO-mediated oxidation. However, significant amounts of carboxyl groups are formed in the TEMPO-oxidized native celluloses. For instance, carboxylate contents of softwood bleached kraft pulps (SBKP) for papermaking grade increased up to ~1.7 mmol/g as the amount of NaClO added was increased to 10 mmol/g by oxidation with the TEMPO/NaBr/NaClO system in water at pH 10 (Fig. 18.4) [4, 6]. Thus, in this case, the carboxylate content of the TEMPO-oxidized SBKPs increases up to 170 times as much as that of the original SBKP with a carboxylate content of 0.01 mmol/g by oxidation. Yields of the TEMPO-oxidized SBKPs were ~90 %, indicating that most of hemicelluloses present in the original SBKP were degraded during oxidation and were removed into filtrates during washing process with water by filtration. Small amounts of C6-aldehyde groups were present in the oxidized SBKPs. The degree of polymerization of the original SBKP remarkably decreased from 2800 to 900 during oxidation, which is one of the shortcomings of the TEMPO-mediated oxidation of SBKP in water at pH 10.

Although significant amounts of C6-carboxylate groups were formed in the TEMPO-oxidized SBKPs (Fig. 18.4), their X-ray diffraction patters showed that the original cellulose I crystal structure, crystallinity, and crystal width were unchanged before and after the oxidation. Hardwood bleached kraft pulps, cotton linters, cotton lint, and bacterial and tunicate celluloses showed similar results, although the TEMPO-oxidized cellulose prepared from these native celluloses had different carboxylate contents, depending on their crystal sizes [7]. These results indicate that the oxidation occurs only in C6-OH groups present on the crystalline cellulose microfibril surfaces (Fig. 18.5) [4, 7]. Because every one of two glucose units present on the crystalline cellulose microfibril surfaces is exposed to the outer surface, only these C6-OH groups are position-selectively oxidized to sodium C6-carboxylate groups by the TEMPO-mediated oxidation, resulting in the formation of anionic carboxylate groups densely and regularly on the cellulose microfibril surfaces.

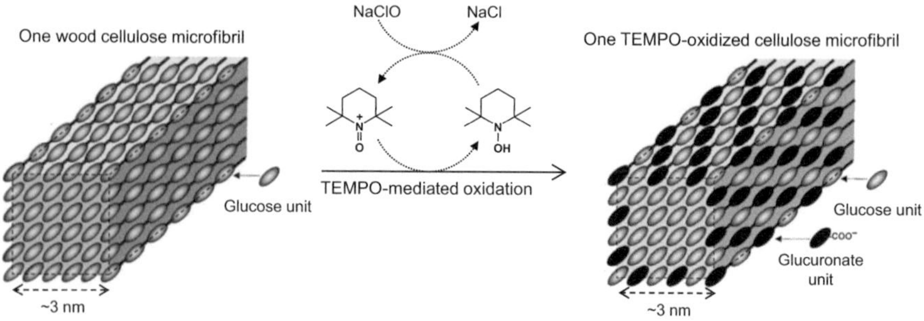

Fig. 18.5 Schematic model of TEMPO-mediated oxidation of wood cellulose microfibrils. Position-selective surface oxidation results in formation of sodium carboxylate groups densely and regularly present on crystalline cellulose microfibril surfaces [2, 4, 24]

18.3 Characteristics of TEMPO-Oxidized Wood Celluloses

Because the TEMPO-oxidized wood celluloses prepared from bleached softwood and hardwood kraft pulps for papermaking grade maintain the original fibrous morphologies, i.e., 20–40 μm in width and 1–3 mm in length, the fibrous TEMPO-oxidized celluloses (TOCs) are able to be obtained as pure TOCs through washing with water by simple filtration. The TOCs with sodium carboxylate structures (TOCs-COONa) sometimes strongly swell in water at the final washing stage, in which the salt concentration remarkably decreases. At the initial stage of water washing, filtration of TOCs-COONa is rather smooth because of some salting-out effects. When the TOCs-COONa are converted to TOCs-COOH by soaking in acidic water, the following washing of fibrous TOCs-COOH with water by filtration can be carried out smoothly as a result of suppression of fiber swelling in water.

Since abundant sodium carboxylate groups are present on crystalline cellulose microfibril surfaces in TOCs, various metal ions and alkylammonium ions can be efficiently introduced into the microfibril surfaces in fibrous TOCs as counterions of carboxylate groups by simple ion-exchange treatments. When the TOCs-COONa were soaked in aqueous metal salt solutions followed by washing with water, various metal ions can be introduced into fibrous TOCs (Fig. 18.6) [8]. In particular, not only monovalent Ag^+ ion but also divalent Ca^{2+} and Pb^{2+} ions were introduced to TOCs, forming metal carboxylate structures with almost 1:1 molar ratios. Because these Ca and Pb ion-containing TOCs had both metal ions and chloride ions with almost the same molar ratios to each other (when $CaCl_2$ and $PbCl_2$ were used in ion exchange), these TOCs probably have structures like TOC-COOCaCl and TOC-COOPbCl [8].

The fibrous TOCs-COONa are convertible to fibrous TOCs-COOH by soaking in acidic water and subsequent washing with water by filtration. Then the TOCs-COOH can be efficiently and stoichiometrically converted to fibrous TOCs with quaternary alkylammonium carboxylate-type structures, TOCs-$COO^{-}{\cdot}N^{+}R_4$, by

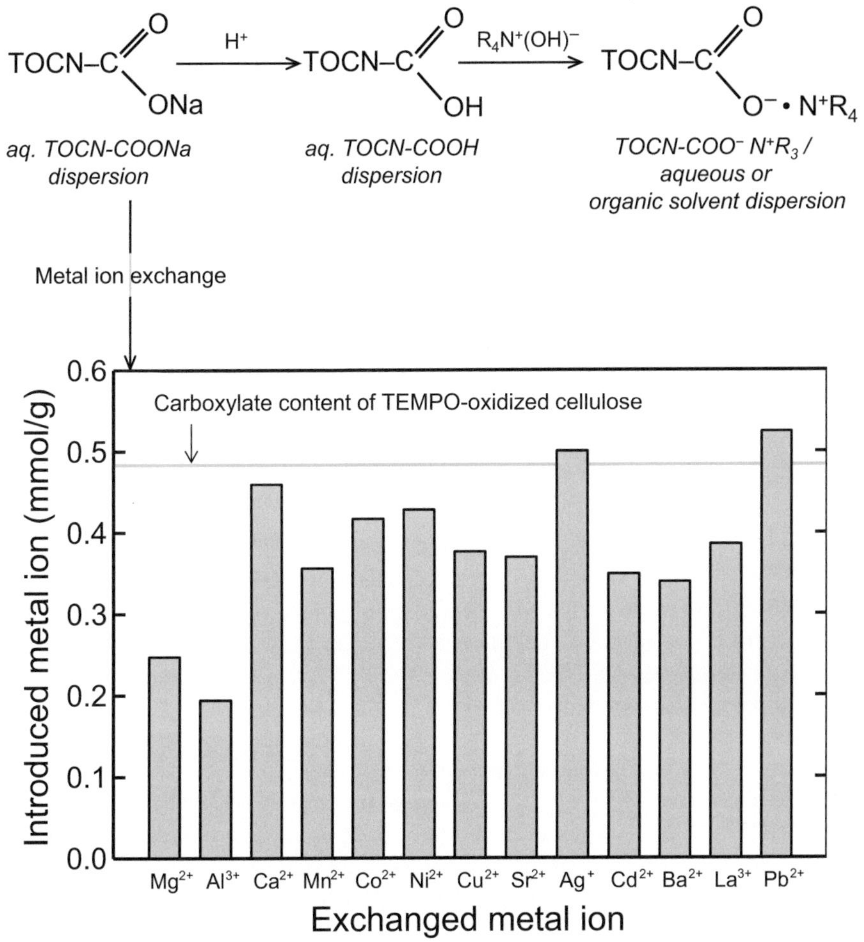

Fig. 18.6 Ion exchanges of sodium carboxylate groups in TEMPO-oxidized cellulose to quaternary alkylammonium carboxylate groups and metal carboxylate groups [8, 9]

neutralization of the protonated carboxyl groups of TOCs with quaternary alkylammonium hydroxides ($QA^+ \cdot OH^-$) [9, 10]. As described later, these fibrous TOCs-$COO^- \cdot QA^+$ are convertible to individual nanofibers by mechanical disintegration in not only water but also some organic solvents. Thus, hydrophilic TOCs-COONa are switchable to hydrophobic TOCs-$COO^- \cdot QA^+$ by simple counterion exchange, which is characteristic for TOCs.

One of the promising applications of nanocelluloses is preparation of nanocellulose-containing composites, which have light weights but superior mechanical, optical, and thermal properties in comparison to nanocellulose-free neat materials [11]. In the case of compositing of nanocelluloses with plastic matrix polymers, thermal molding is the most common and cost-effective process. Native

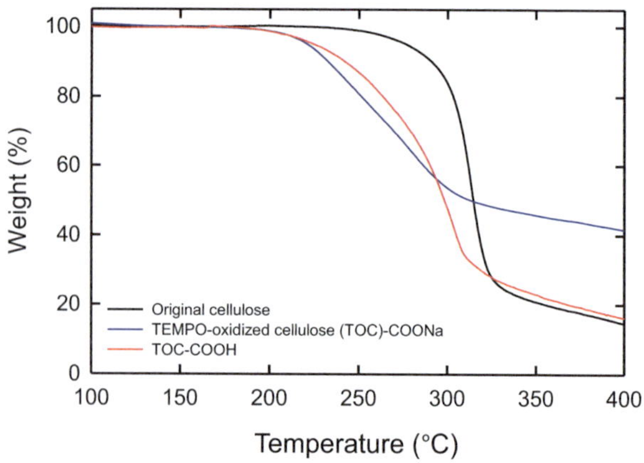

Fig. 18.7 Thermogravimetric curves of the original cellulose and TEMPO-oxidized celluloses with sodium carboxylate and protonated carboxyl groups [12]

celluloses do not have any melting points but thermal degradation points around 300 °C in inert gas. However, TOCs-COONa and TOCs-COOH have thermal degradation points of ~200 °C; degradation points remarkably decrease by ~100 °C from that of native celluloses (Fig. 18.7) [12]. Even under the thermal degradation temperatures, TOCs often discolor by heating. As in the case of thermal degradation of carboxymethyl cellulose [13], carboxyl groups in TOCs likely degrade in association with decarboxylation by heating. These properties of thermal instability of TOCs are disadvantageous in some cases.

Some alternative TEMPO-mediated oxidation systems other than the TEMPO/NaBr/NaClO system in water at pH 10 have been proposed to prevent remarkable depolymerization and formation of small amounts of C6-aldehydes and C2/C3 ketones in TOCs (Fig. 18.4). When TEMPO/NaClO/$NaClO_2$ oxidation system in a buffer at pH 4.8 or 6.8 is applied to native celluloses, TOCs with significantly higher molecular weights and carboxylate contents slightly lower than those prepared by the TEMPO/NaBr/NaClO system in water at pH 10 are obtained [14]. In this system, $NaClO_2$ behaves as a co-oxidant with catalytic amounts of TEMPO and NaClO. Because $NaClO_2$ selectively oxidizes C6-aldehyde groups to C6-carboxyls, TOCs without any C6-aldehydes can be obtained [6, 14, 15]. When TOCs prepared by the TEMPO/NaBr/NaClO system in water at pH 10 are consecutively post-reduced with $NaBH_4$ in the same reaction container at pH 10 using one-pot process, TOCs with neither C6-aldehydes nor C2/C3 ketones can be obtained. These TOCs have no discoloration by heating at 105 °C for 3 h [16].

18.4 Preparation of TEMPO-Oxidized Cellulose Nanofibers (TOCNs)

When the fibrous wood TOCs have sodium carboxylate contents >1 mmol/g by TEMPO-mediated oxidation, they can be converted to transparent and highly viscous gels by gentle mechanical disintegration in water. Transmission electron microscopy (TEM) images of dried materials of the diluted gels after staining with uranyl acetate and lead citrate showed that the gels consisted of individual TOCNs with homogeneous widths of ~3 nm and lengths of several microns (Fig. 18.8). The crystalline cellulose microfibrils present in wood cell walls (Fig. 18.1) are, therefore, completely separated, and individual cellulose microfibrils are obtained in the first time by the TEMPO-mediated oxidation of wood celluloses and subsequent gentle mechanical disintegration in water [4, 6, 14–18]. Because sodium carboxylate groups are present densely and regularly on the crystalline cellulose microfibril surfaces in TOCs (Fig. 18.5), osmotic effects and electrostatic repulsion efficiently work between the oxidized cellulose microfibrils in water owing to dissociation of

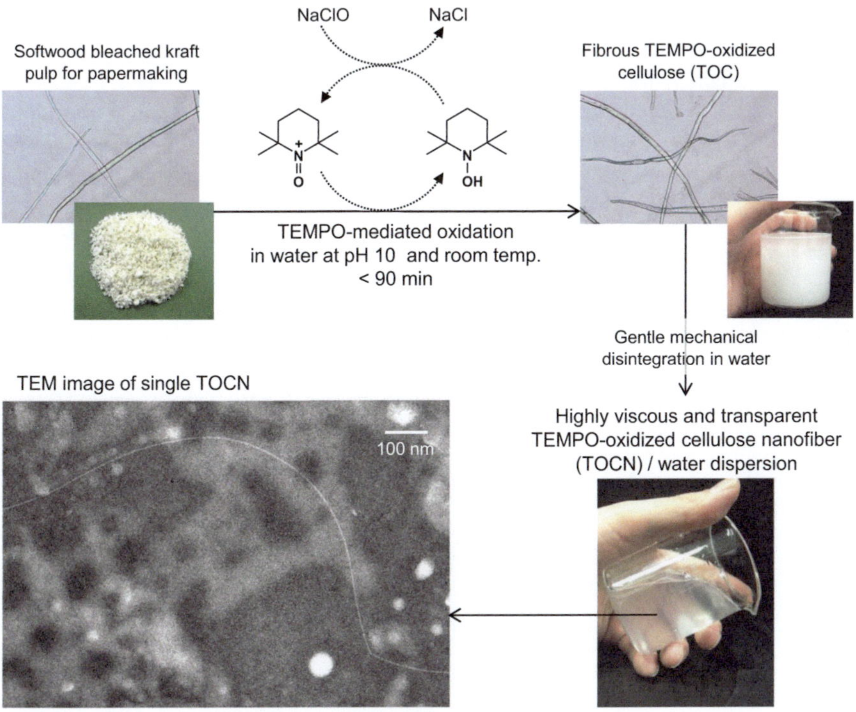

Fig. 18.8 Preparation of TEMPO-oxidized cellulose nanofibers (TOCNs) from softwood bleached kraft pulp (for papermaking grade) by TEMPO-mediated oxidation and subsequent gentle mechanical disintegration in water. TEM image shows complete individualization of wood cellulose microfibrils with ~3 nm width [2, 24]

sodium carboxylate groups, resulting in the complete individualization of cellulose microfibrils to TOCNs. TOCs-COOQA with quaternary alkylammonium counter-ions were converted to individual TOCNs-COOQA by mechanical disintegration in not only water but also methanol, *N,N*-dimethylformamide, and other organic solvents [9, 10].

18.5 Characterization of TEMPO-Oxidized Cellulose Nanofibers

Although the wood TOCNs have always homogenous widths of ~3 nm, the lengths and length distributions vary, depending on the TEMPO-oxidation and mechanical disintegration conditions. Shear viscosity measurements of diluted TOCN/water dispersions and other analytical methods have been carried out to determine the lengths and length distributions of TOCNs, which are fundamental and significant data required for applications to high-tech materials [19, 20].

Once completely individual TOCNs without any agglomeration in water are obtained, tensile strengths and Young's modulus of single TOCN elements are measurable using cavitation forces during ultrasonic treatments and stress-train curves obtained using cantilevers in atomic force microscopy (AFM) analysis, respectively [21, 22]. The results showed that wood TOCNs had tensile strengths of ~3 GPa, which are almost the same as those of multi-wall carbon nanotubes and Kevlar fibers and are approximately five times higher than that of steel, although the densities of TOCNs are approximately one fifth of steal. The tensile Young's modulus of tunicate TOCNs measured by AFM was ~140 GPa. Thus, TOCNs are new bio-nanofibers with remarkably high tensile strength and Young's modulus.

Figure 18.9 shows TEM images of wood TOCNs and cellulose nanocrystals (CNCs) [23] and an optical microphotograph of commercial MFC. Explicit differences in morphology are observed between TOCNs and CNCs. TOCNs have homogeneous widths of ~3 nm and have high aspect ratios. In contrast, CNCs have rodlike spindles in shape with maximum widths of ~15 nm and low aspect

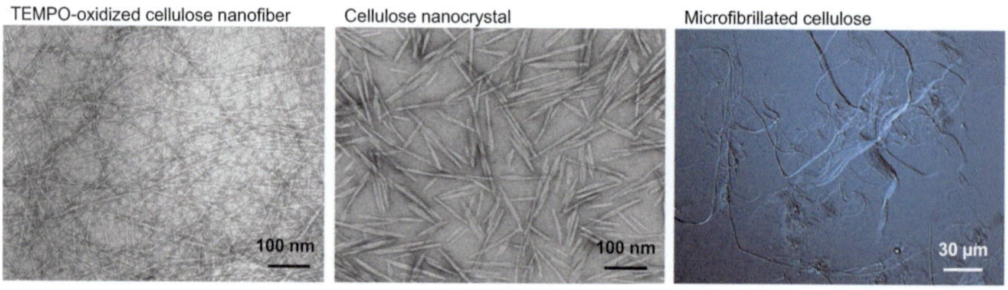

Fig. 18.9 TEM images of TOCNs and cellulose nanocrystals [23] and optical microphotograph of microfibrillated cellulose [24]

ratios <20. Meanwhile, MFCs consist of not only nano-sized fibrils but also large fibers and hence have heterogeneous morphologies.

18.6 Properties of TOCN-Containing Composite Materials

The characteristic points of TOCNs different from other nanocelluloses are (A) wood TOCNs are crystalline nanofibers with homogeneous widths of ~3 nm and high aspect rations, (B) TOCNs have abundant sodium carboxylate groups exchangeable to other metal carboxylate groups and alkylammonium carboxylate groups by simple ion exchange, and (C) TOCNs are dispersed at the individual nanofiber level in water, forming nematic-like liquid crystalline or self-aligned structures caused by electrostatic repulsions between TOCN elements. Owing to these characteristics of TOCNs, self-standing films, hydrogels, aerogels, and composite materials prepared with TOCNs have quite unique and excellent properties in some cases (Fig. 18.10) [24].

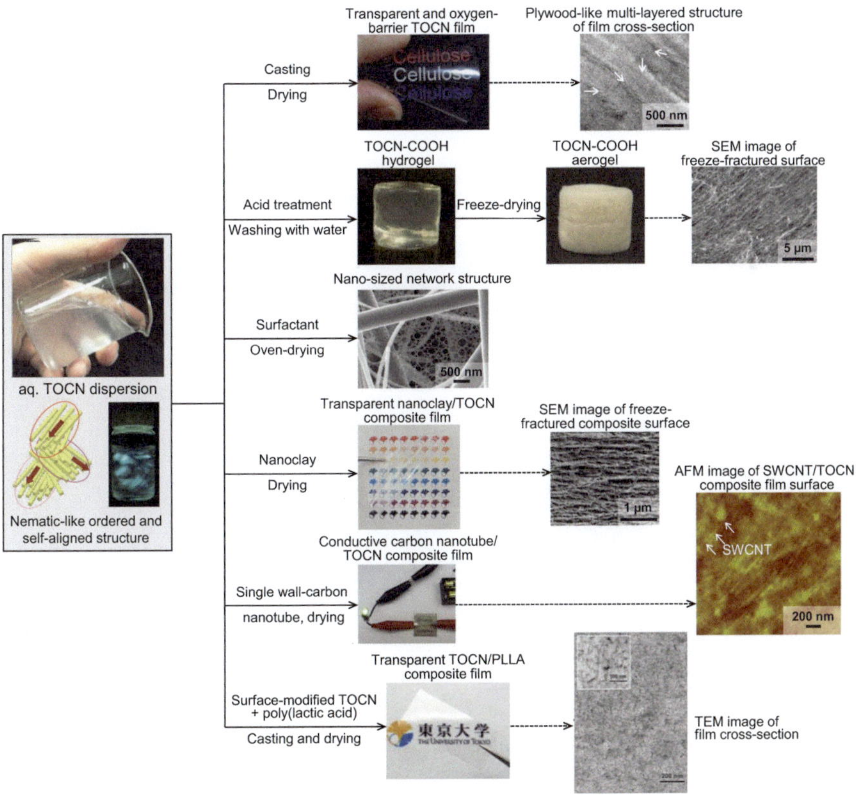

Fig. 18.10 Self-standing films and TOCN-containing composites prepared from aqueous TOCN dispersions, in which TOCNs form nematic-like ordered and self-aligned structures, owing to electrostatic repulsion between TOCN elements in water [24]

The self-standing and transparent TOCN films form plywood-like multilayered and dense structures, resulting in extremely high oxygen-barrier properties under dry conditions [25, 26]. These films had tensile strengths and Young's moduli of 200–300 MPa and 8–15 GPa, respectively. The TOCN hydrogels are stiff, and TOCN aerogels prepared from hydrogels by freeze-drying or supercritical drying had large surface areas and quite low thermal conductivities [26, 27]. When nanoclays were incorporated into TOCN matrix, highly ductile properties appeared on the transparent clay/TOCN composite films at clay contents of 5–10 %, which had works of fracture of 26–30 MJ/m^3 [28, 29].

TOC and TOCN films are switchable from biologically stable properties in use to biodegradable nature after used by controlling counterions of TOCNs. When fibrous TOCs with various counterions were treated with a commercial crude cellulase in water, TOC-$COONH_4$ and TOC-COONa were easily degraded, partly forming water-soluble compounds with low molecular weights, and yield as water-insoluble TOC fractions decreased to ~40 % after treatment for 12 days. In contrast, almost no degradation occurred on TOC-COOH by the cellulase treatment [30]. When TOCN films were embedded in soil, TOCN-COONa films were clearly degraded to small pieces after 30 days, whereas films of TOCN-COOH, TOCN-COOCaCl, and TOCN-COOCuCl were relatively stable and maintained the original film forms after 30 days (Fig. 18.11) [31]. Thus, the quantitative and qualitative designs of counterions of carboxylate groups in TOCs and TOCNs allow them to control their biodegradability.

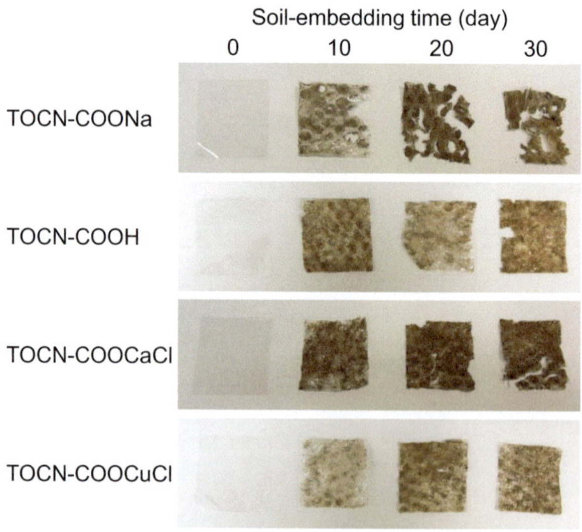

Fig. 18.11 Biodegradability of TOCN films with different carboxyl counterions [31]

18.7 Conclusions

TOCs and TOCNs have unique chemical and nano-sized structures, different from other cellulose nanofibers or nanocrystals. Because TOCs can be converted to cellulose nanofibers with various degrees of nanofibrillation by controlling TEMPO-mediated oxidation conditions and mechanical disintegration conditions in water, versatile applications of TOCs and TOCNs are available. In particular, when TOCs have carboxylate contents >1 mmol/g, they are convertible to crystalline TOCNs with homogenous widths of ~3 nm and high aspect ratios by mechanical disintegration in water under adequate conditions. This key point of the homogeneous widths with very narrow deviations is probably required for applications of bio-based TOCNs to high-tech materials. The overall concept to establish a sustainable society using TOCs and TOCNs produced from abundant forest biomass resources is illustrated in Fig. 18.12.

In 2014, the Nanocellulose Forum has been established in Japan to accelerate practical applications and commercialization of nanocelluloses and nanocellulose-containing value-added materials, supported by the Ministry of Economy, Trade and Industry [32]. This all-Japan-type framework of Nanocellulose Forum is quite characteristic, and so far there are no similar organizations in other countries. New perpendicular and horizontal collaborations between academia and various industries covering from forestry to high-tech materials are going on via the Forum platform. Because the nanocellulose-related research and development of both

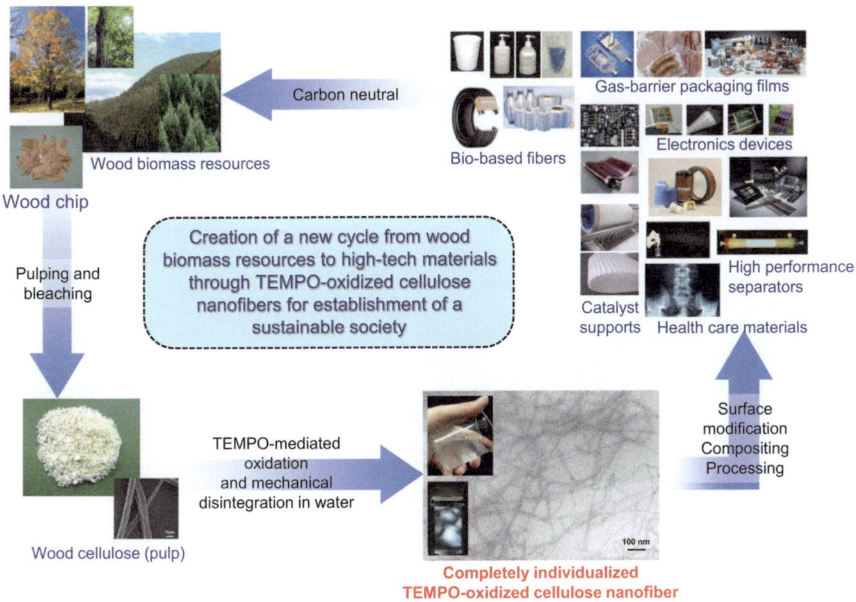

Fig. 18.12 Overall concept of creation of a new wood biomass utilization cycle with TOCNs for establishment of a sustainable society [32]

academia and industry in Japan precede other countries at present, further advances of these research fields would be expected.

Acknowledgment The researches of TOCNs have been supported by Grants-in-Aid for Scientific Research from the Japan Society for the Promotion of Science (JSPS) and by Core Research for Evolutional Science and Technology (CREST) of Japan Science and Technology Agency (JST).

References

1. A.F. Turbak, F.W. Snyder, K.R. Sandberg, Microfibrillated cellulose, a new cellulose product: properties, uses, and commercial potential. J. Appl. Polym. Sci. Appl. Polym. Symp. **37**, 815–823 (1983)
2. A. Isogai, Wood nanocelluloses: fundamentals and applications as new bio-based nanomaterials. J. Wood Sci. **59**, 449–459 (2013)
3. A.E. de Nooy, A.C. Besemer, H. Bekkum, Highly selective nitroxyl radical-mediated oxidation of primary alcohol groups in water-soluble glucans. Carbohydr. Res. **269**, 89–98 (1995)
4. A. Isogai, T. Saito, H. Fukuzumi, TEMPO-oxidized cellulose nanofibers. Nanoscale **3**, 71–85 (2011)
5. A. Isogai, Y. Kato, Preparation of polyuronic acid from cellulose by TEMPO-mediated oxidation. Cellulose **4**, 153–164 (1998)
6. R. Shinoda, T. Saito, Y. Okita, A. Isogai, Relationship between length and degree of polymerization of TEMPO-oxidized cellulose nanofibrils. Biomacromolecules **13**, 842–849 (2012)
7. Y. Okita, T. Saito, A. Isogai, Entire surface oxidation of various cellulose microfibrils by TEMPO-mediated oxidation. Biomacromolecules **11**, 1696–1700 (2010)
8. T. Saito, A. Isogai, Ion-exchange behavior of carboxylate groups in fibrous cellulose oxidized by the TEMPO-mediated system. Carbohydr. Polym. **61**, 183–190 (2005)
9. M. Shimizu, T. Saito, A. Isogai, Bulky quaternary alkylammonium counterions enhance the nanodispersibility of 2,2,6,6-tetramethylpiperidine-1-oxyl-oxidized cellulose in diverse solvents. Biomacromolecules **15**, 1904–1909 (2014)
10. M. Shimizu, T. Saito, H. Fukuzumi, A. Isogai, Hydrophobic, ductile, and transparent nanocellulose films with quaternary alkylammonium carboxylates on nanofibril surfaces. Biomacromolecules **15**, 4320–4325 (2014)
11. D.R. Paul, L.M. Robeson, Polymer nanotechnology: nanocomposites. Polymer **49**, 3187–3204 (2008)
12. H. Fukuzumi, T. Saito, Y. Okita, A. Isogai, Thermal stabilization of TEMPO-oxidized cellulose. Polym. Degrad. Stab. **95**, 1502–1508 (2010)
13. D. Britto, O.B.G. Assis, Thermal degradation of carboxymethylcellulose in different salty forms. Thermochim. Acta. **494**, 115–122 (2009)
14. R. Tanaka, T. Saito, A. Isogai, Cellulose nanofibrils prepared from softwood cellulose by TEMPO/NaClO/$NaClO_2$ systems in water at pH 4.8 or 6.8. Int. J. Biol. Macromol. **51**, 228–234 (2012)
15. T. Saito, M. Hirota, N. Tamura, S. Kimura, H. Fukuzumi, L. Heux, A. Isogai, Individualization of nano-sized plant cellulose fibrils by direct surface carboxylation using TEMPO catalyst under neutral conditions. Biomacromolecules **10**, 1992–1996 (2009)
16. S. Takaichi, T. Saito, R. Tanaka, A. Isogai, Improvement of nanodispersibility of oven-dried TEMPO-oxidized celluloses in water. Cellulose **21**, 4093–4103 (2014)
17. T. Saito, Y. Nishiyama, J.L. Putaux, M. Vignon, A. Isogai, Homogeneous suspensions of individualized microfibrils from TEMPO-catalyzed oxidation of native cellulose. Biomacromolecules **7**, 1687–1691 (2006)

18. T. Saito, S. Kimura, Y. Nishiyama, A. Isogai, Cellulose nanofibers prepared by TEMPO-mediated oxidation of native cellulose. Biomacromolecules **8**, 2485–2491 (2007)
19. R. Tanaka, T. Saito, D. Ishii, A. Isogai, Determination of nanocellulose fibril length by shear viscosity measurement. Cellulose **21**, 1581–1589 (2014)
20. R. Hiraoki, Y. Ono, T. Saito, A. Isogai, Molecular mass and molecular-mass distribution of TEMPO-oxidized celluloses and TEMPO-oxidized cellulose nanofibrils. Biomacromolecules **16**, 675–681 (2015)
21. S. Iwamoto, W. Kai, A. Isogai, T. Iwata, Elastic modulus of single cellulose microfibrils from tunicate measured by atomic force microscopy. Biomacromolecules **10**, 2571–2576 (2009)
22. T. Saito, R. Kuramae, J. Wohlert, L.A. Berglund, A. Isogai, An ultrastrong nanofibrillar biomaterial: the strength of single cellulose nanofibrils revealed via sonication-induced fragmentation. Biomacromolecules **14**, 248–253 (2013)
23. Courteously provided by Ruide A. W. Pilot scale production of CNC and CNF. Presented in ISWFPC pre-symposium on cellulose nanocrystals. Victoria, 9–10 June (2013)
24. Isogai A. Cellulose, in *Encyclopedia of Polymeric Nanomaterials* (Springer, Berlin, 2014). doi:10.1007/978-3-642-36199-9_320-1
25. H. Fukuzumi, T. Saito, T. Iwata, Y. Kumamoto, A. Isogai, Transparent and high gas barrier films of cellulose nanofibers prepared by TEMPO-mediated oxidation. Biomacromolecules **10**, 162–165 (2009)
26. T. Saito, T. Uematsu, S. Kimura, T. Enomae, A. Isogai, Self-aligned integration of native cellulose nanofibrils towards producing diverse bulk materials. Soft. Matt. **7**, 8804–8809 (2011)
27. Y. Kobayashi, T. Saito, A. Isogai, Aerogels with 3D ordered nanofiber skeletons of liquid-crystalline nanocellulose derivatives as tough and transparent insulators. Angew. Chem. Int. Ed. **53**, 10394–10397 (2014)
28. C.N. Wu, T. Saito, S. Fujisawa, H. Fukuzumi, A. Isogai, Ultrastrong and high gas-barrier nanocellulose/clay-layered composites. Biomacromolecules **13**, 1927–1932 (2012)
29. C.N. Wu, Q. Yang, M. Takeuchi, T. Saito, A. Isogai, Highly tough and transparent layered composites of nanocellulose and synthetic silicate. Nanoscale **6**, 392–399 (2014)
30. I. Homma, T. Isogai, T. Saito, A. Isogai, Degradation of TEMPO-oxidized cellulose fibers and nanofibrils by crude cellulase. Cellulose **20**, 795–805 (2013)
31. I. Homma, H. Fukuzumi, T. Saito, A. Isogai, Effects of carboxyl-group counter-ions on biodegradation behaviors of TEMPO-oxidized cellulose fibers and nanofibril films. Cellulose **20**, 2505–2515 (2013)
32. https://unit.aist.go.jp/brrc/ncf/

Chapter 19
Forefront of Nanofibers: High Strength Fibers and Optoelectronic Applications

Hidetoshi Matsumoto

Abstract Nanofibrous materials, which are some of the one-dimensional nanomaterials, have unique properties compared to their bulk solids (e.g., high specific surface area, high electrical conductivity, and good electrochemical activity). They are easy to form a 2-D or 3-D network structure. This network structure enables an efficient charge transport through the network backbone and an efficient chemical reaction at the surface. In addition, nanofiber assemblies (nanofiber yarns or nanofiber webs) provide good mechanical properties and good handling characteristics. This chapter highlights two topics in nanofiber applications, i.e., high strength nanofibers and optoelectronic applications of nanofiber networks.

Keywords Nanofiber • Carbon • High strength fiber • Electrode • Solar cell

19.1 Introduction

Rapid advances in nanotechnology have provided a large variety of nanomaterials with various shapes and superior properties during the past two decades. Among these nanomaterials, nanofibrous materials, which are some of the one-dimensional (1-D) nanomaterials, have unique properties compared to their bulk solids (e.g., high specific surface area, high electrical conductivity, and good electrochemical activity). Nanofibers are nanomaterials known for their cross-sectional direction and a macroscopic material known for its fiber axis. Therefore, nanofibers have both the advantages of functionality due to their nanoscaled diameter and the ease of manipulation due to their macroscopic length. They are also easy to form a 2-D or 3-D network structure. This network structure enables an efficient charge transport through the network backbone and an efficient chemical reaction at the surface (Fig. 19.1). In addition, nanofiber assemblies (nanofiber yarns or nanofiber webs) provide good mechanical properties and good handling characteristics [1]. Electrospinning is a straightforward and versatile method of forming

H. Matsumoto (✉)
Department of Materials Science and Engineering, Tokyo Institute of Technology, Meguro-ku, Tokyo, 152-8552 Japan
e-mail: matsumoto.h.ac@m.titech.ac.jp

Society of F.S. Technology, Japan (ed.), *High-Performance and Specialty Fibers*,
DOI 10.1007/978-4-431-55203-1_19

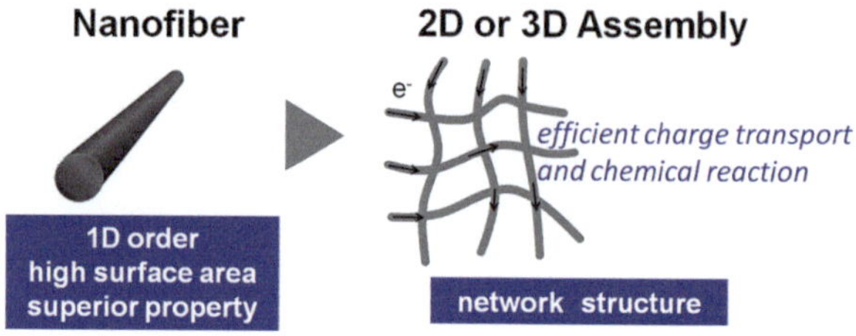

Fig. 19.1 Characteristics of nanofibers

continuous thin fibers from several nanometers to several tens of micrometers in diameter and can be used for the one-step formation of a 2-D or 3-D nanofiber network structure (nanofiber coatings and nanofiber webs) [2]. Electrospun nanofiber networks with high surface areas have drawn significant attention for their practical applications, such as high-performance filter media, battery separators, electrode materials, composites, drug delivery systems, and biomaterial scaffolds for tissue engineering [3]. This chapter highlights two topics in recent nanofiber applications, i.e., high strength nanofibers and optoelectronic applications of nanofiber networks.

19.2 High Strength Fibers

Continuous strong, tough, lightweight, thin fibers are of significant interest for various applications ranging from composites to bulletproof vests [4]. In particular, carbon nanofibers (CNFs) with lightweight, high electrical and thermal conductivities, mechanical strength, and chemical stability have attracted a considerable amount of attention for many applications such as structural materials for transportation equipment and energy devices. CNFs have been produced by a variety of methods that include catalytic chemical vapor deposition (CCVD), 1-D nanotemplate-assisted carbonization, polymer blends, and electrospinning [5].

The carbon nanotube (CNT), one of the CNFs, is a promising material for strong and lightweight fibers with high electrical and thermal conductivities because of its unique structure and superior physical properties, i.e., high aspect ratio, high tensile strength (~150 GPa), high Young's modulus (~1 TPa), high electrical and thermal conductivities, low density, and good chemical stability [6]. However, to transfer these superior properties into practical applications, such as reinforcements and electronic devices, it is necessary to find ways to process nanotubes into bulk forms or device structures. There are two main approaches, the "top-down" and "bottom-up" approaches, to fabricate the CNT assemblies. The top-down fabrication techniques from CNT dispersions have been reported. Poulin et al. drew CNT fibers by direct spinning from an aqueous CNT suspension into a poly(vinyl alcohol) solution [6]. Pasquali et al. processed fibers and sheets from a lyotropic liquid crystalline CNT suspension in strong acids (the best tensile strength was 1.3 GPa for scalable CNT fibers [7]). On the other hand, some researchers combined the spinning process with a bottom-up synthesis approach. Zhang et al. reported a "dry spinning process" to produce CNT yarns from "CNT forests" in which the CNTs were grown and vertically aligned on a substrate by CVD [6]. Windle's group developed the "direct spinning from CVD" [6]. In this process, the CNT yarn is directly spun from the aerogel of the CNTs formed in a CVD reactor (the centimeter-long CNT yarns had a 2 GPa strength, corresponding to the gravimetric strength of a commercially available Kevlar aramid fiber (DuPont)) [8].

As one of the top-down approaches, electrospinning, which is based on an electrohydrodynamic phenomenon under high voltage (typically 5–30 kV), can exert a directional shear force coupling with the external electric field to the flow of the spinning solution (the estimated draw ratio reaches the value of 25,000) (Fig. 19.2) [9]. The fiber strength and toughness are remarkably influenced by the internal structure of the fibers. Dzenis et al. demonstrated that the reduction of the diameter of the electrospun poly(acrylonitrile) (PAN) nanofibers from 2.8 μm to ~100 nm resulted in the simultaneous improvement of the elastic modulus (from 0.26 to 48 GPa), true strength (from 0.015 to 1.75 GPa), and toughness (from 0.25 to 605 MPa) for the 5–10 mm long individual nanofibers. This result can be explained by the hypothesis that the reduced crystallinity in the ultrathin electrospun nanofibers due to fast solvent evaporation and rapid solidification during electrospinning is responsible for preserving the high nanofiber ductility, and the increased chain molecular orientation due to directional shear force during electrospinning is responsible for the high strength and modulus [10].

The major advantage of the top-down approach is its straightforwardness because of utilizing commercially available nanocarbons without any special synthesis techniques. One-dimensional nanocarbons, such as CNTs and graphene nanoribbons (GNRs), were highly oriented along the fiber axis of the electrospun nanofibers by the synergistic effect of the external electric field and shear force. For example, multiwalled carbon nanotube (MWNT)/poly(vinyl butyral) (PVB) composite nanofibers were prepared by electrospinning, successive twisting, and heat treatment. The MWNTs were highly oriented in an electrified thin jet during

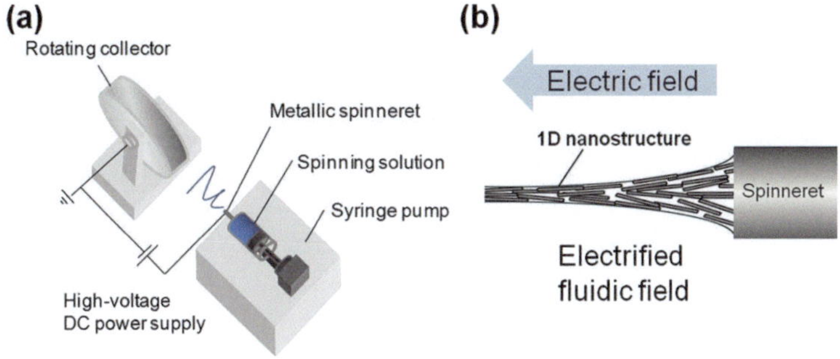

Fig. 19.2 Schematic diagrams of (**a**) the apparatus used for electrospinning with a rotating disk collector and (**b**) the orientation of 1-D nanomaterials in the electrified thin liquid jet during electrospinning

electrospinning (Fig. 19.3a). The heat treatment of the twisted electrospun nanofiber yarns produced the characteristics of the CNT in the composite nanofiber yarns and enhanced their electrical properties (the best electrical conductivity: 15,400 S m^{-1}), mechanical properties (the best tensile strength: 0.14 GPa), and thermal properties (the best thermal conductivity: 16.8 W m^{-1} K^{-1}) [6]. The GNR/carbon-composite nanofiber yarns were also prepared by electrospinning from poly(acrylonitrile) (PAN)-containing graphene oxide nanoribbons (GONRs) and successive twisting and carbonization. During electrospinning, the well-dispersed GONRs were highly oriented along the fiber axis in an electrified thin liquid jet (Fig. 19.3b). The addition of GONRs at a low weight fraction significantly improved the mechanical properties of the composite nanofiber yarns. In addition, the carbonization of the matrix polymer simultaneously enhanced both the mechanical (the best tensile strength: 0.38 GPa; the best modulus: 53.6 GPa) and electrical properties of the composites. The electrical conductivity of the carbonized composite yarns containing 0.5 wt % GONR showed the maximum value of 16,500 S m^{-1}. Interestingly, it is higher than the conductivities of both the PAN-based pristine CNF yarns (7,700 S m^{-1}) and the monolayer GNRs (5,400 S m^{-1}) [11]. Our X-ray diffraction analysis supported that the oriented GONRs contained in the PAN nanofibers effectively functioned as not only a 1-D nanofiller but also as the template agent for the formation of the ordered graphitic structure during carbonization. These results indicated that the 1-D nanocarbons are a promising filler for electrospun thin fibers. At present, the optimum functional design of ultrathin nanofibers and 1-D nanocarbon-composite nanofibers has not been accomplished. It is expected that much better physical properties can be attained by optimization of the internal structure of the nanofibers. High strength nanofibers and their assemblies could be applied as reinforcements for lightweight composites. In addition, 1-D nanocarbon-composite nanofibers could be utilized as high-performance electrodes for fuel cells, secondary batteries, and supercapacitors and fiber- and textile-shaped solar cells, including flexible and wearable electronic devices and implantable medical devices.

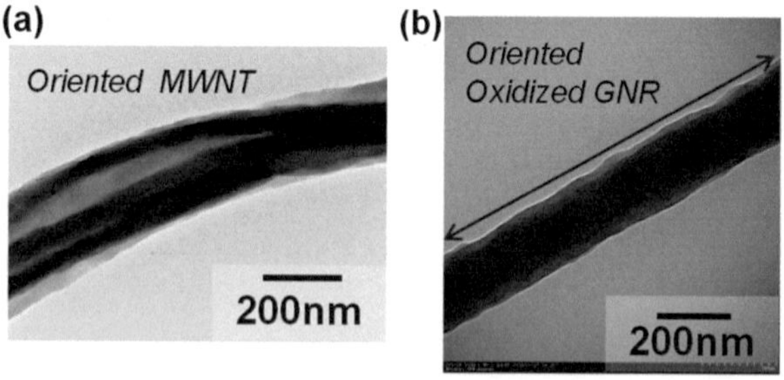

Fig. 19.3 Transmission electron micrographs of (**a**) the MWNT/PVB composite nanofiber and (**b**) the GONR/PAN composite nanofiber

19.3 Carbon Nanofiber Networks for Electronic Applications

Electrospun CNFs with diameters from several tens of nanometers to several microns are a new class of 1-D carbon nanomaterials between conventional carbon nanotubes and carbon fibers [12]. Electrospun CNF nonwoven webs with a thinner diameter have a higher specific surface area, a higher electrical conductivity and flexibility (Fig. 19.4), and higher battery performances [13]. The surface functionalization of the CNF webs also broadens their applications; the introduction of 1-D nanostructures, such as CNTs and semiconductor nanowires (NWs), on the surface of CNFs realizes highly functional electrodes. CNF webs have been used as the substrate of CNT field emitter arrays because of their high thermal stability, electrical conductivity, and flexibility. CNTs were densely grown on the surface of the CNF webs by CVD. Figure 19.5a shows that a large number of vertically aligned CNTs were grown on the surface of the CNF. The diameters of the CNTs ranged from 50 to 100 nm with most of the CNTs being about 80 nm in diameter. In addition, the CNT field emitter arrays on the CNF webs produced a higher current density at a lower turn-on voltage compared to the ones on Si substrates [14]. Zinc

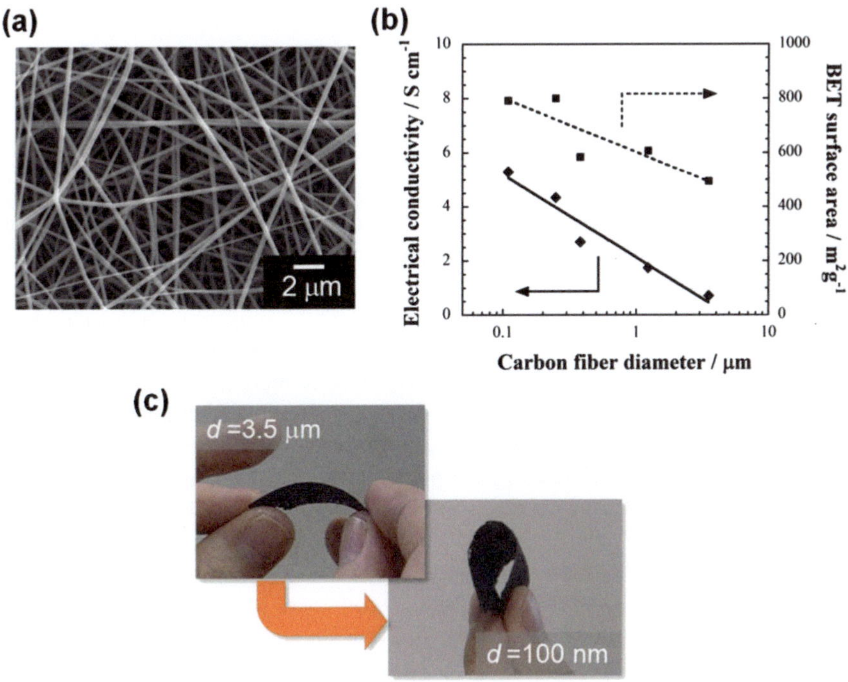

Fig. 19.4 (**a**) Surface electron micrograph of CNF web; (**b**) effect of fiber diameter on electrical conductivity and specific surface area of CNF webs; (**c**) effect of fiber diameter on flexibility of the CNF web

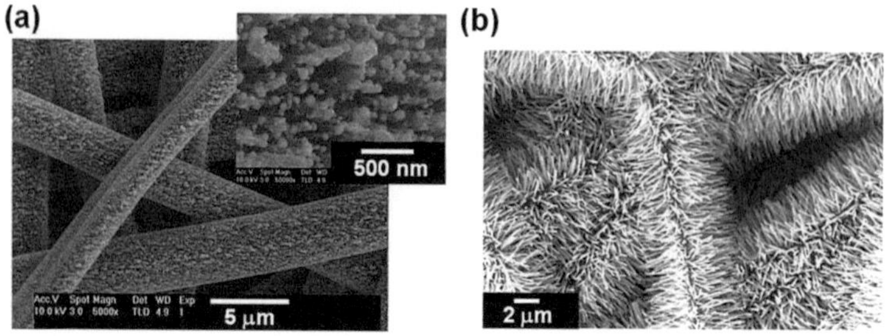

Fig. 19.5 Surface electron micrographs of (**a**) the CNTs grown on CNF webs and (**b**) the ZnO NWs grown on CNF webs

oxide nanowires (ZnO NWs) were also densely grown on the CNFs by CVD. An SEM image of the hierarchy-structured nanofiber web, ZnO NWs on a CNF web, is shown in Fig. 19.5b. The implementation of this composite in a dye-sensitized solar cell (DSSC) in which the ZnO NWs and CNF web were used for photoinduced charge separation/charge transport and current collection, respectively. This dense and flexible 3-D nanostructures would enable a high dye loading which leads to efficient exciton generation under illumination, making it a promising flexible anode material for DSSCs [15].

19.4 Non-carbon Nanofiber Networks for Optoelectronic Applications

Not only carbon nanofibers but also nanofibers or nanofiber networks composed of other materials improve the performance of the optoelectronic devices. The use of a polymer single nanofiber for subwavelength-scale optical waveguiding provides the sensor with a small footprint, fast response, and high sensitivity. Since polymers are good hosts for a wide range of dopants and are easily functionalized, polymer single-nanofiber optical sensors for a variety of specimens can be realized. More recently, photonic applications of polymer single nanofibers, such as electroluminescent devices, lasers, and detectors, have been reported [16]. Electrospun nanofibers are also very promising as optical materials. For example, by precisely controlling the diameter of an electrospun nanofiber in the range of visible light wavelengths (380–780 nm), it is possible to change the coloring of the nanofiber-based coatings due to the thin-film interference [17]. It is also known that a single electrospun nanofiber shows a strong light scattering. In addition, for nanofiber networks, the scattering phenomenon also occurs due to spaces between the nanofibers, which can be controlled by the alignment of the nanofibers (Fig. 19.6a, b). Uniaxially aligned nanofibers showed a diffraction pattern, and this property can be applied to optical polarizers. More recently, light-scattering

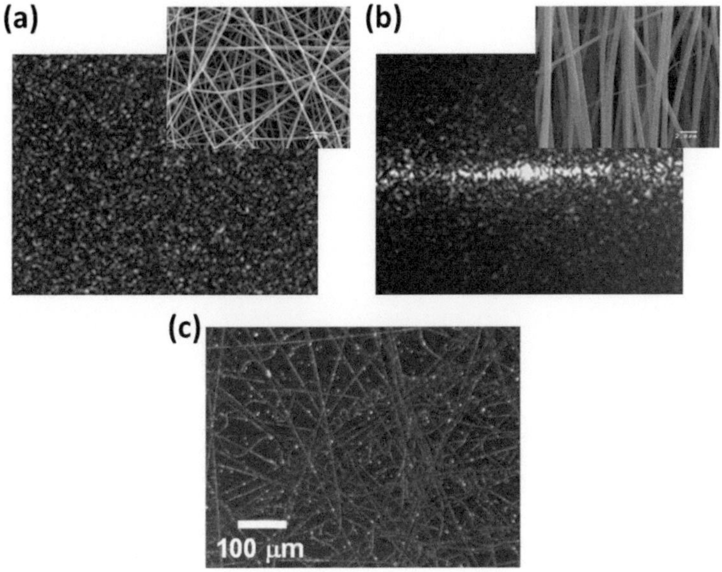

Fig. 19.6 Light-scattering patterns and electron micrographs (*Inset*) of (**a**) randomly deposited nanofibers and (**b**) aligned nanofibers; (**c**) optical micrograph of poly(vinylpyrrolidone) nanofiber-coated gold surface illuminated by Grazing incidence

assisted surface plasmon resonance at an electrospun nanofiber-coated gold surface has been reported (Fig. 19.6c) [18].

Energy harvesting is the process extracting power from ambient sources (e.g., light, thermal gradients, and motion). Recently, this technology has attracted a great deal of interest as a potential inexhaustible source for low-power electronic devices (e.g., wireless sensor networks and wearable electronics). 1-D nanostructures, such as nanofibers, are a promising material for energy-harvesting applications. To improve the performance of bulk heterojunction (BHJ) organic solar cells, a 3-D nanofiber network composed of a semiconductor is a promising option for nanoscaled morphology control and molecular ordering in the active layer [19]. For example, poly(3-hexylthiophene) (P3HT) nanofibers were fabricated by self-assembly (Fig. 19.7a, the average width is about 20 nm; the average length >500 nm). The OSCs containing the self-assembled P3HT nanofiber network (Fig. 19.7a) showed a 20 % better cell performance (power conversion efficiency, PCE) compared to the cells without the nanofiber network; the open-circuit voltage (V_{oc}) slightly decreased by 6 %, but the short-circuit current (J_{sc}) and fill factor (FF) were increased by 18 % and 5 %, respectively (Fig. 19.7b, c, and Table 19.1). In addition, the introduction of a nanofiber network into the active layer improved the hole transport mobility from 2.1×10^{-6} cm^2 V^{-1} s^{-1} (for the conventional BHJ OSCs) to 8.6×10^{-6} cm^2 V^{-1} s^{-1} [20]. These results indicated that semiconductor nanofiber networks in the active layer functioned as a continuous carrier path and allowed the efficient extraction of charges, resulting in a better *FF* and increasing

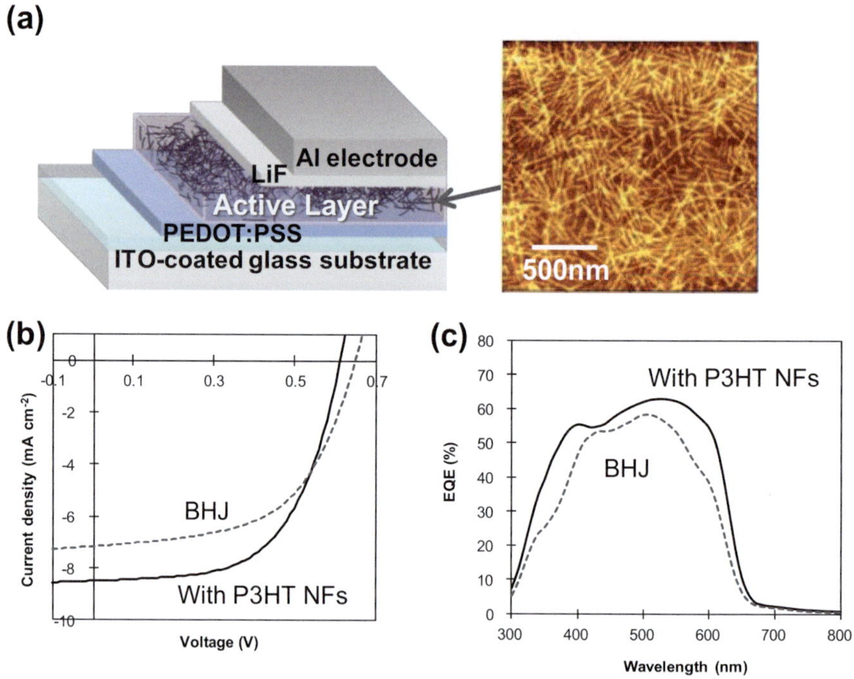

Fig. 19.7 (**a**) AFM image of P3HT nanofibers prepared by self-assembly and structure of organic thin-film solar cell containing nanofiber network in the active layer, (**b**) the current density–voltage (J–V) characteristics, and (**c**) external quantum efficiency (EQE) spectra of organic thin-film solar cells containing nanofiber network

Table 19.1 Performance of organic thin-film solar cells[a]

	J_{sc} (mA cm^{-2})	V_{oc} (V)	FF (–)	PCE (%)
With P3HT NF	8.5	0.61	0.59	3.1
BHJ	7.2	0.65	0.56	2.6

[a]Device structure: ITO/poly(3,4-ethylenedioxythiophene):poly(styrenesulfonic acid) (PEDOT:PSS)/P3HT:[6, 6]-phenyl-C_{61}-butyric acid methyl ester ($PC_{61}BM$)/LiF/Al. The current density–voltage (J–V) measurement was conducted under the illumination of AM 1.5G, 100 mW/cm^2

J_{sc}. In addition, organic solar cells with a high-refractive-index nanowire coating showed improved power conversion efficiencies due to their antireflection and light trapping effects in the active layer [21]. More recently, transparent and conductive Al nanowire networks were fabricated by the wet chemical etching of a 50 nm-thick Al thin layer deposited on a poly(ethylene terephthalate) film with an electrospun polymer nanofiber mask (Fig. 19.8) [22]. For energy storage devices, the effect of adding metal oxide nanofiber networks to a polymer electrolytes for a Li-ion secondary battery on improvement of both the ionic conductivity and mechanical strength has been reported [23].

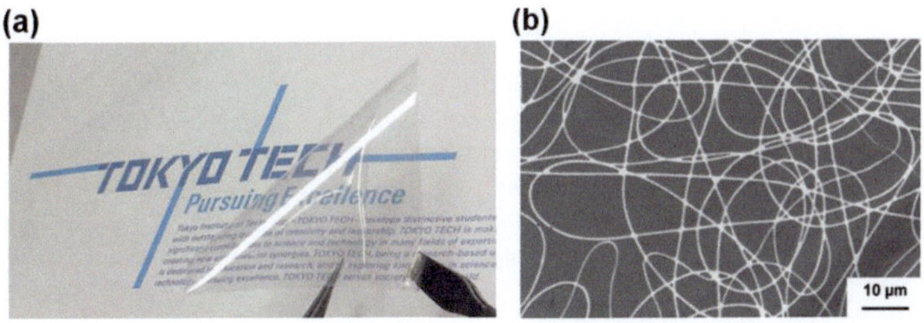

Fig. 19.8 (**a**) Flexible, unbreakable, and transparent electrodes with conductive nanofiber networks; (**b**) micrograph of aluminum nanofiber networks fabricated on a poly(ethylene terephthalate) film by wet chemical etching with an electrospun polymer nanofiber mask

19.5 Summary and Perspective

Nanofibers and nanofiber networks show unique functionalities based on their nanoscaled size, high specific surface area, and highly molecular orientation. The functionalities of nanofibers and the performances of nanofiber networks can be improved by controlling their fiber diameter, surface morphology, and internal structure of the nanofibers. These nanofibers or nanofiber networks are promising materials for the realization of a safe, sustainable, and healthy society [3], in particular, for efficient charge collection, charge transport, and electrochemical reaction in energy conversion and storage devices [24].

Nanofiber technologies can lead to quantum leap innovations for the creation of materials with unique functions such as a nanosize effect, superhigh surface area, and molecular orientation effect. Since the research field of nanofibers contains both scientific and technological frontiers, it requires interdisciplinary backgrounds, such as physics, chemistry, polymer science, energy and environmental technologies, biological technology, and optoelectronics to extensively exploit the nanofibrous materials to their full potential.

References

1. H. Matsumoto, A. Tanioka, Surface electrochemistry of electrospun nanofibers, in *Electrical Phenomena at Interfaces and Biointerfaces: Fundamentals and Applications in Nano-, Bio-, and Environmental Sciences*, ed. by H. Ohshima, 1st edn. (Wiley, Hoboken, 2012)
2. A.L. Yarin, B. Pourdeyhimi, S. Ramakrishna, *Fundamentals and Applications of Micro-and Nanofibers* (Cambridge University Press, Cambridge, 2014)
3. H. Matsumoto, A. Tanioka, Functionality in electrospun nanofibrous membranes based on fiber's size, surface area, and molecular orientation. Membranes **1**, 249–264 (2011)
4. M.K. Shin, B. Lee, S.H. Kim, J.A. Lee, G.M. Spinks, S. Gambhir, G.G. Wallace, M.E. Kozlov, R.H. Baughman, S.J. Kim, Synergistic toughening of composite fibres by self-alignment of reduced graphene oxide and carbon nanotubes. Nat. Commun. **3**, 650 (2012). doi:10.1038/ncomms1661

5. E. Yasuda, A. Oya, S. Komura, S. Tomonoh, T. Nishizawa, S. Nagata, T. Akatsu, Single domain oval carbon nanofiber prepared from a polymer blend process. Carbon **50**, 1432–1434 (2012). doi:10.1016/j.carbon.2011.10.019
6. S. Imaizumi, H. Matsumoto, Y. Konosu, K. Tsuboi, M. Minagawa, A. Tanioka, K. Koziol, A. Windle, Top-down process based on electrospinning, twisting, and heating for producing one-dimensional carbon nanotube assembly. ACS Appl. Mater. Interfaces **3**, 469–475 (2011). doi:10.1021/am101046v
7. N. Behabtu, C.C. Young, D.E. Tsentalovich, O. Kleinerman, X. Wang, A.W.K. Ma, E.M. Bengio, R.F. ter Waarbeek, J.J. de Jong, R.E. Hoogerwerf, S.B. Fairchild, J.B. Ferguson, B. Maruyama, J. Kono, Y. Talmon, Y. Cohen, M.J. Otto, M. Pasquali, Strong, light, multifunctional fibers of carbon nanotubes with ultrahigh conductivity. Science **339**, 182–186 (2013). doi:10.1126/science.1228061
8. M.F.L. De Volder, S.H. Tawfick, R.H. Baughman, A.J. Hart, Carbon nanotubes: present and future commercial applications. Science **339**, 535–539 (2013). doi:10.1126/science.1222453
9. M. Richard-Lacroix, C. Pellerin, Molecular orientation in electrospun fibers: from mats to single fibers. Macromolecules **46**, 9473–9493 (2013). doi:10.1021/ma401681m
10. D. Papkov, Y. Zou, M.N. Andalib, A. Goponenko, S.Z.D. Cheng, Y.A. Dzenis, Simultaneously strong and tough ultrafine continuous nanofibers. ACS Nano **7**, 3324–3331 (2013). doi:10.1021/nn400028p
11. H. Matsumoto, S. Imaizumi, Y. Konosu, M. Ashizawa, M. Minagawa, A. Tanioka, W. Lu, J.M. Tour, Electrospun composite nanofiber yarns containing oriented graphene nanoribbons. ACS Appl. Mater. Interfaces **5**, 6225–6231 (2013). doi:10.1021/am401161b
12. M. Inagaki, Y. Yang, F. Kang, Carbon nanofibers prepared via electrospinning. Adv. Mater. **24**, 2547–2566 (2012). doi:10.1002/adma.201104940
13. P. Hiralal, S. Imaizumi, H.E. Unalan, H. Matsumoto, M. Minagawa, M. Rouvala, A. Tanioka, G.A.J. Amaratunga, Nanomaterial-enhanced all-solid flexible zinc-carbon batteries. ACS Nano **4**, 2730–2734 (2010). doi:10.1021/nn901391q
14. K. Suzuki, H. Matsumoto, M. Minagawa, A. Tanioka, Hayashi et al., Carbon nanotubes on carbon fabrics for flexible field emitter arrays. Appl. Phys. Lett. **93**, 053107 (2008). doi:10.1063/1.2967868
15. H.E. Unalan, D. Wei, K. Suzuki, S. Dalal, P. Hiralal, H. Matsumoto et al., Photoelectrochemical cell using dye sensitized zinc oxide nanowires grown on carbon fibers. Appl. Phys. Lett. **93**, 133116 (2008). doi:10.1063/1.2978957
16. H. Cho, S.Y. Min, T.W. Lee, Electrospun organic nanofiber electronics and photonics. Macromol. Mater. Eng. **298**, 475–486 (2013). doi:10.1002/mame.201200364
17. H. Kuwayama, H. Matsumoto, K. Morota, M. Minagawa, A. Tanioka, Control over color of nanotextured coatings by electrospray deposition. Sen'i Gakkaishi **64**, 1–4 (2008). doi:10.2115/fiber.64.1
18. K. Tsuboi, H. Matsumoto, M. Minagawa, A. Tanioka, Light scattering assisted surface plasmon resonance at electrospun nanofiber-coated gold surfaces. Appl. Phys. Lett. **98**, 241109/1–3 (2011). doi:10.1063/1.3601465
19. Z. Yin, Q. Zheng, Controlled synthesis and energy applications of one-dimensional conducting polymer nanostructures: an overview. Adv. Energy Mater. **2**, 179–218 (2012). doi:10.1002/aenm.201100560
20. H. Matsumoto, Y. Konosu, S. Inagaki, Y. Saito, Functional design of nanofibrous materials and their applications in energy devices. IEICE Technical Report **114**, OME2014-86 (2015)
21. K. Tsuboi, T. Fukawa, Y. Konosu, H. Matsumoto, A. Tanioka, Solution-processed nanowire coating for light management in organic solar cells. J. Nanotechnol. **2012**, 387586 (2012). doi:10.1155/2012/387586
22. K. Azuma, K. Sakajiri, H. Matsumoto, S. Kang, J. Watanabe, M. Tokita, Facile fabrication of transparent and conductive nanowire networks by wet chemical etching with electrospun nanofiber mask template. Mater. Lett. **115**, 187–189 (2014). doi:10.1016/j.matlet.2013.10.054

23. K. Kimura, H. Matsumoto, J. Hassoun, S. Panero, C. Scrosati, Y. Tominaga, A quaternary poly (ethylene carbonate)-lithium bis(trifluoromethane-sulfonyl)imide-ionic liquid-silica fiber composite polymer electrolyte for lithium batteries. Electrochim. Acta **175**, 134–140 (2015). doi:10.1016/j.electacta.2015.03.117
24. S. Cavaliere, S. Subianto, I. Savych, D.J. Jones, J. Rozière, Electrospinning: designed architectures for energy conversion and storage devices. Energy Environ. Sci. **4**, 4761–4785 (2011). doi:10.1039/C1EE02201F

Part V
Carbon Fibers

Chapter 20
Carbon Fiber

Makoto Endo

Abstract Since invented in 1960s, carbon fibers have been widely used in sporting goods, aerospace, automotive, and civil engineering. A mechanical property of carbon fiber is determined by the raw material and production conditions. This paper briefly explains kinds of carbon fiber and these production methods, historical developments of its mechanical properties, commercialization, and CFRP applications.

Keywords Carbon fiber • PAN[*1] • Pitch • CFRP[*2] • [*1]PAN (polyacrylonitrile) • [*2]Carbon fiber-reinforced plastic

20.1 Introduction

Carbon fiber (CF) is composed of carbon atoms, and it is defined by the International Organization for Standardization (ISO) as "the fiber based on a pyrolytic carbon synthesized from organic precursor with more than 90 % of carbon content." The origin of carbon fiber is known to be first considered by Thomas Edison [1]; however Rayon was the first *carbon fiber* precursor material for composite materials. Rayon-based carbon fibers were first available to the US market in 1957 [2]. Polyacrylonitrile (PAN) and pitch-based carbon fibers are currently the major industrial *carbon fiber* precursors. *Carbon fiber* in general also comprises of vapor-grown carbon nano-fibers (CNF) and carbon nanotubes (CNT); however in this paper mass production technology and its historical background of PAN and pitch-based carbon fiber are mainly presented.

M. Endo (✉)
Composite Material Research Laboratories, Toray Industry, Inc., 1515, Tsutsui Masakicho, Iyogun 791-3193, Ehime, Japan
e-mail: Makoto_Endo@nts.toray.co.jp

Society of F.S. Technology, Japan (ed.), *High-Performance and Specialty Fibers*,
DOI 10.1007/978-4-431-55203-1_20

20.2 Properties and Production Methods of Carbon Fiber

20.2.1 General Properties of Carbon Fiber

Carbon fiber offers exceptional advantages in specific strength, specific elastic modulus, as well as fatigue characteristics compared to other fibrous materials as shown in Fig. 20.1. In addition, superior thermal resistance, chemical resistance, and dimensional stability of carbon fibers exhibit remarkable environmental tolerance. Figure 20.2 displays mechanical properties over a diverse range of carbon fibers. PAN-based high-strength carbon fibers are mainly used as structural material for aircrafts, sporting applications, and general industrial applications, while PAN and pitch-based high-modulus carbon fibers are utilized in aerospace and sporting applications. General-purpose pitch-based carbon fiber is utilized in brakes, clutches, and heat insulators.

20.2.2 Production Methods for Carbon Fiber

20.2.2.1 PAN-Based Carbon Fiber

In the manufacturing process of PAN-based carbon fibers, PAN which dissolved into solvent is polymerized from acrylonitrile monomer firstly. The precursor fibers are extracted from a liquid polymer solution through a process known as

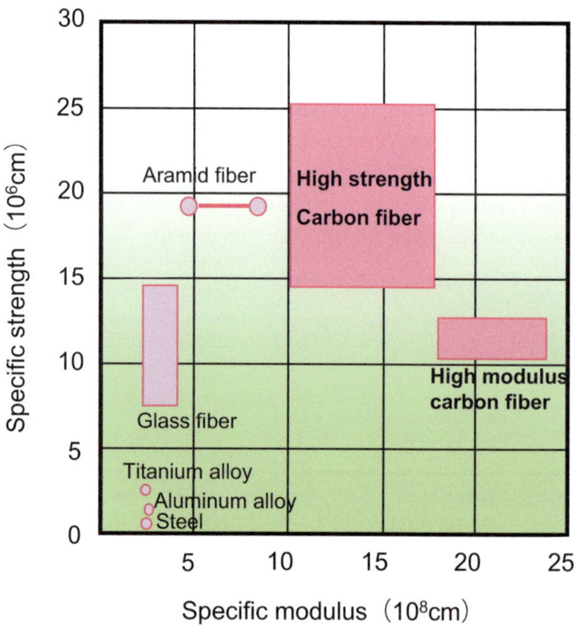

Fig. 20.1 Specific strength and specific modulus of carbon fiber

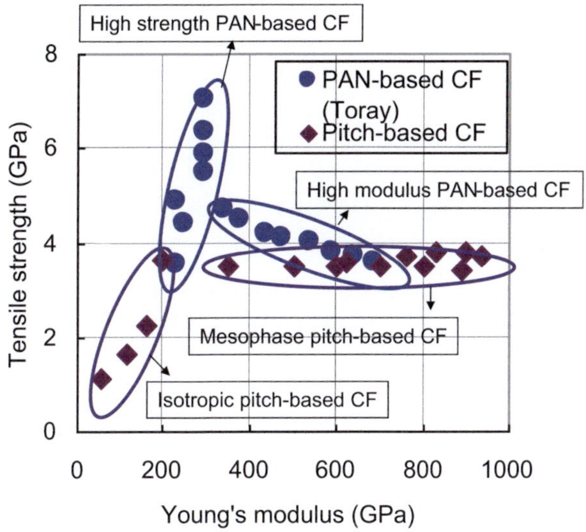

Fig. 20.2 Mechanical properties of carbon fibers

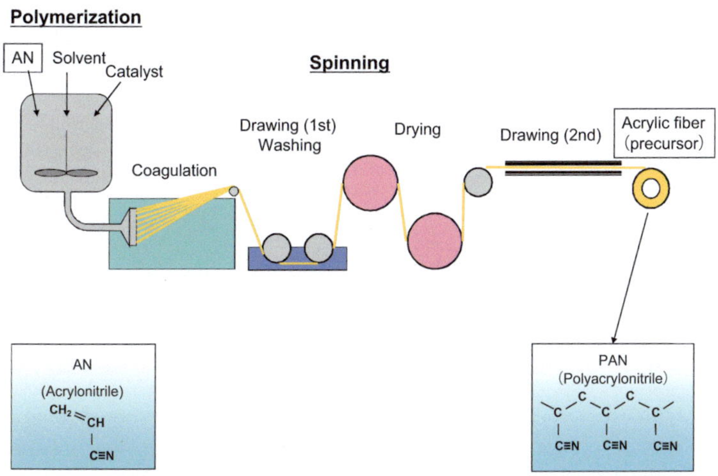

Fig. 20.3 Spinning process for precursor fiber production

wet spinning, illustrated in Fig. 20.3. The precursor fiber is heat-treated at 200–300 °C in air (process known as "stabilization"), followed by a carbonization process at 1000–3000 °C in an inert atmosphere. To acquire desirable bonding properties for the fibers, sizing agents are coated on the surface of the carbon fibers, and these fibers are wound onto bobbins as shown in Fig. 20.4.

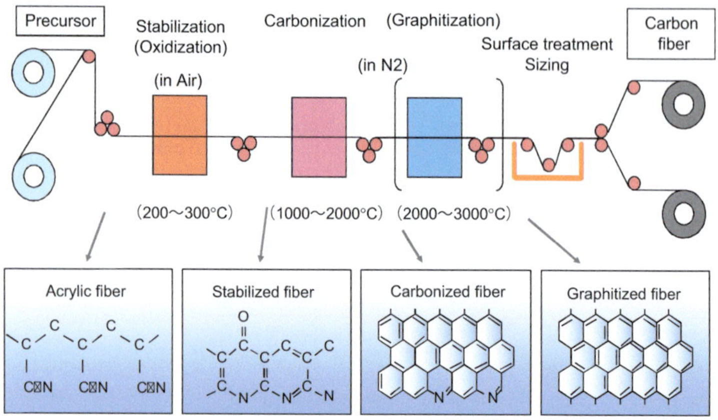

Fig. 20.4 Heat-treatment process for carbon fiber production

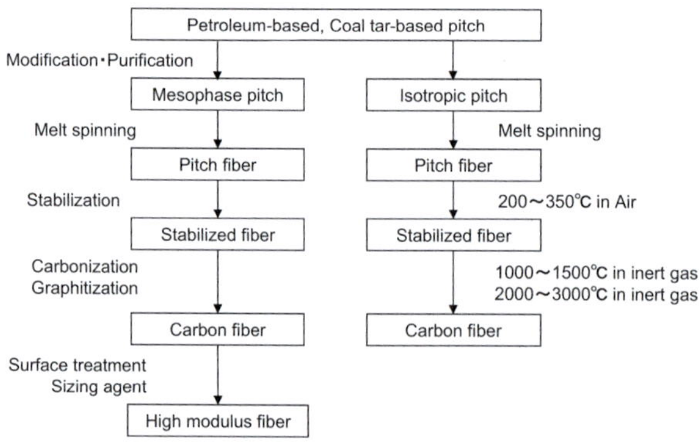

Fig. 20.5 Production scheme of pitch-based carbon fiber

20.2.2.2 Pitch-Based Carbon Fiber

There are typically two kinds of pitch-based carbon fiber, isotropic (general grade) and anisotropic (high grade with high elastic modulus), which depends on the raw materials of the aromatic organic compound essentially petroleum or coal-tar pitch. Pitch for carbon fiber is modified and purified by filtration, hydrogenation, and thermal treatment to improve its spinnability. These fibers are spun from a process known as melt-spinning method. Pitch-based fibers are stabilized in air and carbonized at 1000–1400 °C, followed by a graphitization process at 2500–3000 °C. The sizing process is similar to the PAN-based fiber to further improve the bonding property. Figure 20.5 shows the production scheme of pitch-based carbon fiber.

20.2.3 Commercialization of Carbon Fiber

20.2.3.1 Commercialization of PAN-Based Carbon Fiber

The process of PAN-based carbon fiber was developed by Dr. Akio Shindo at Government Indusrial Research Institute, Osaka. He explored potential carbon fiber precursors by heating various remnant textiles bought at department stores, and he examined the change in mass after heat-treating the textiles to 1000 °C in nitrogen using quartz thermal balance. Through these examinations, only a PAN-based fiber "Orlon" made by DuPont remained in a black fuzz ball after the heating. This was the first PAN-based carbon fiber to be identified. From this discovery Dr. Shindo filed a Japanese patent in 1959 and presented the results at the US International Carbon Conference in 1963. This recognition initiated the research of carbon fiber [3, 4].

Industrialization of PAN-based carbon fiber was first attempted by carbon manufacturers, RAE (Royal Aeronautical Establishment, UK), Nippon Carbon Co., Ltd. (Japan), and Tokai Electrode Mfg. Co., Ltd. (Japan). Nippon Carbon Co., Ltd. launched a preliminary sale in 1969 at a scale of 500 kg per month by operating a pilot plant. Tokai Electrode Mfg. Co., Ltd. started sales in 1968 by a product called "Thermolon S" [5].

Toray started a development of carbon fiber precursor in 1961 with a newly discovered compound, hydroxyl acrylonitrile, which greatly shortened the oxidation process of PAN-based fibers and in addition improved the mechanical properties by copolymerizing with acrylonitrile. These findings accelerated the industrialization of carbon fiber since the optimal polymer composition and production process to the carbon fiber performance had not been recognized. The development of the industrialization of carbon fiber in Toray progressed as a companywide project from fundamental polymer research to carbon fiber plant manufacture engineering [6, 7]. In June 1970, Toray obtained a license from Osaka National Research Institute and cross-licensed with Union Carbide Corporation in the USA for the carbonization technology. In February 1971, a pilot plant was started to operate with 1 t per month, and the product "Torayca®" was on the market in July [7]. The carbon fiber production was followed by Toho Rayon Co., Ltd. in 1975 and Mitsubishi Rayon Co., Ltd. in 1983.

20.2.3.2 Commercialization of Pitch-Based Carbon Fiber

The invention for the manufacturing process of pitch-based carbon fiber began in 1965 by Prof. Ohtani at the University of Gunma, Kureha Chemical Industry, who industrialized commercial production of carbon fiber utilizing an isotropic pitch-based precursor as the raw material [8]. In 1976, UCC (present Cytec Industries) commercially produced carbon fiber from an anisotropic (mesophase) pitch-based precursor, and thus research and development of high-performance pitch-based carbon fiber commenced in the 1980s. The current dominant producers of pitch-based carbon fiber are the USA (Cytec Industries) and Japan (Nippon Graphite Fiber Co., Ltd.; Mitsubishi Plastics, Inc.; Kureha Corp.; and Osaka Gas Chemicals Co., Ltd.).

20.3 Improvement of the Performance of Carbon Fiber

20.3.1 Basic Structure of Carbon Fiber

The structure of carbon fiber generally consists of imperfect graphitic structure oriented in the axial direction of fibers as shown in Fig. 20.6, while various structural models of carbon fiber are proposed [1, 2, 9]. The realization of the tensile strength of PAN-based fiber is relatively high because the molecular orientation of PAN as a precursor reflects the structure of carbon fiber.

A remarkable structural feature of carbon fiber is in the surface layer, where the graphitic basal plane is oriented in the radial direction of fibers. The surface structure of carbonized precursors has considerable amount of graphite edge and amorphous regions, which are most likely modified by surface treatments. Meanwhile for the surface structure of graphitized precursors, the graphite edge and amorphous regions are drastically decreased and transformed to the graphitic basal plane which is highly resistant to the surface treatments as a sufficient growth of the crystallite of graphite in the graphitization process. Consequently the bonding property of carbon fiber is degraded due to low surface energy of the graphitic basal plane formed on the carbon fiber surface.

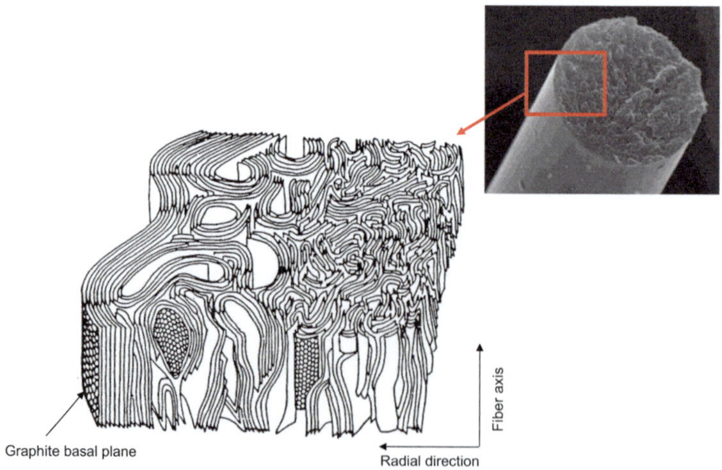

Fig. 20.6 Typical structural model of PAN-based carbon fiber

20.3.2 *Improvement of the Performance of PAN-Based Carbon Fiber*

In the 1980s, investigations to enhance the tensile strength of the PAN-based carbon fiber were engaged more strenuously. Carbon fiber is characterized as a brittle material like ceramics; thus these tensile strengths are strongly influenced by defects. From polymerization to burning process, various techniques to minimize the size and the total abundance of the surface defects which is the dominant factor of the degradation of tensile strength were introduced. In terms of the spinning technique, the continuous processes of synthetic fibers were diverted on the basis of the wet-spinning process of acrylic fibers for fabric purpose. Toray developed a high tensile strength glade product "T800H" by upgrading the tensile strength and elastic modulus at 50 % and 30 % compared to the basic grade product in 1984 "T300"; its historical improving the strength of carbon fiber is shown in Fig. 20.7. Fracture mechanisms of the fibers were analyzed by extensive observations of tensile-fractured monofilaments, submicron in diameter, using electron microscope. It leads to the miniaturization of the surface defects as small as not visible by electron microscope, consequently improved tensile strength of carbon fiber. In 1986 the evolutions of the technique to enhance the tensile strength enable us to produce carbon fiber with 7.1 GPa tensile strength "T1000" which improves the strength of "T800H" more than 30 % [10]. In laboratory scale, carbon fiber with more than 10 GPa in strength was obtained.

High elastic modulus was achieved by the orientational control of the graphitic crystallite. Torayca® "M65J" has 640 GPa in elastic modulus which is 60 % of the theoretical value of graphite at 1,020 GPa.

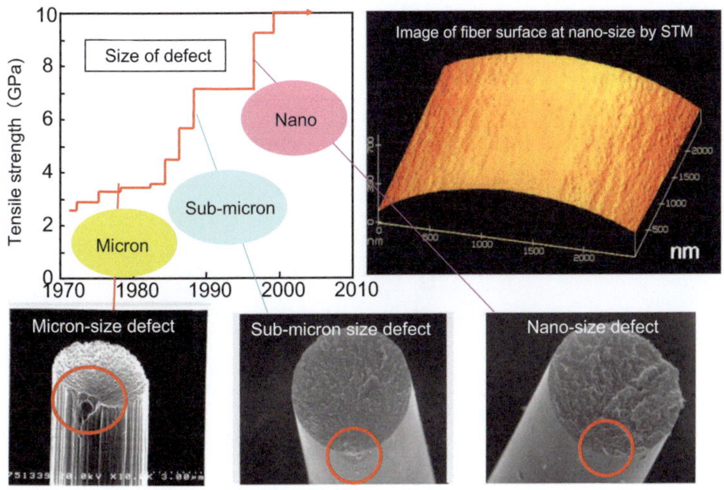

Fig. 20.7 Improving the strength of carbon fiber

20.4 Production of CFRP from Carbon Fiber

20.4.1 Importance of Matrix Resin in CFRP

Carbon fiber is a flexible thread itself, without any compressive load-bearing capacity, and fibers which are loose or broken have no tensile load-bearing capacity. Carbon fiber reinforced with a matrix resin (carbon fiber-reinforced plastic, CFRP) is a composite material which allows a load-bearing capacity and is considered as a structural material. The role of the matrix resin is illustrated in Fig. 20.8.

20.4.2 Production of Carbon Fiber Temporary Materials

The intermediate processing products of carbon fiber and its production methods are illustrated in Fig. 20.9. The primary processing products are "woven fabric," "prepreg," and "chopped fiber." "Woven fabric" is weaved carbon fibers in different directions. "Prepreg" is a sheet consisting of fibers that are arranged in one direction and then impregnated with uncured resin. "Chopped fiber" is bundled carbon fibers that are absorbed into well-converged sizing agent and then cut into 3–25 mm length.

Figure 20.10 summarizes the method of a molding process of carbon fibers and the properties of these processed materials. CFRPs produced from continuous fibers display good properties due to the high stiffening effect between fibers. However, they have difficulty in mass production and are expensive due to the complicated molding process. In contrast, CFRPs produced from short fibers are more suitable for mass production but however demonstrate insufficient properties compared to continuous fibers. It is critical to select appropriate primary products and molding process for the exact applications.

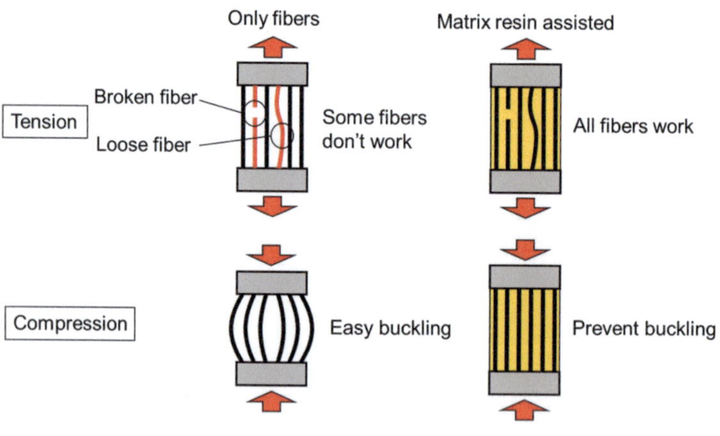

Fig. 20.8 Role of matrix resin in CFRP

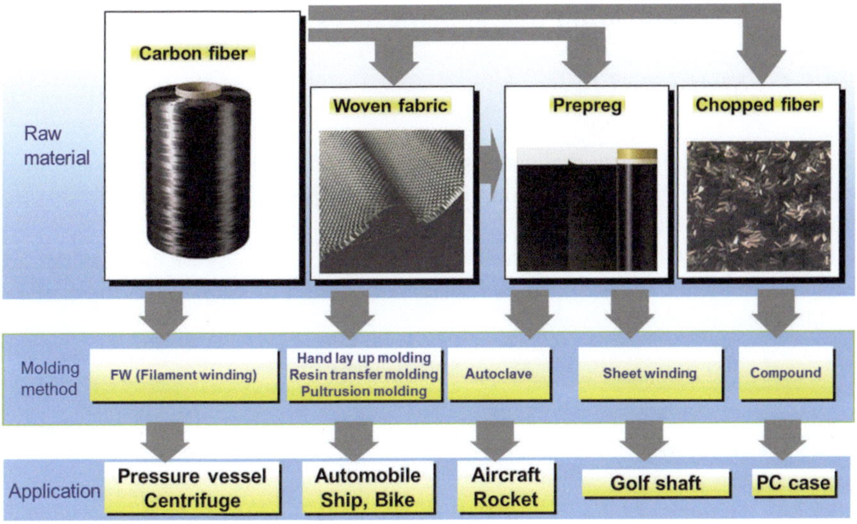

Fig. 20.9 Product lineup of advanced composites

Molding method		Material		Part	Feature	Performance	Productivity
Injection		Pelet	Non-continuous fiber	Interior/ Exterior	Complex shape Mass production Small part	×	◎
Press		sheet SMC Prepreg	Non-continuous fiber	Interior/ Exterior/ Structural material	Flat shape Mass production Large part	×−△	◎
Splay-UP		Cut-fiber	Non-continuous fiber	Interior/ Exterior Structure	Complex shape Medium-Mass production Medium-Large part	×−△	○
RTM Resin Transfer Molding	resin, injection	Fabric	Continuous fiber	Exterior/ Structural material	Complex shape Medium volume production High performance	○−◎	△
FW Filament Winding		Filament	Continuous fiber	Propeller shaft/ Tank	Cylinder shape Medium volume production High performance	○−◎	△−○
Oven		Semi-preg Prepreg	Continuous fiber	Exterior/ Structural material	Flat shape Small quantity production High performance	○−◎	×−△
Auto-clave		Prepreg	Continuous fiber	Exterior/ Structural material	Flat shape Small quantity production High performance	◎	×

Fig. 20.10 Production method and related characteristics of CFRP

20.5 The Application of Carbon Fibers

20.5.1 Sports and Leisure

For sporting and leisure applications, the three most major products made from carbon fibers are golf shafts, fishing rods, and tennis rackets. Nearly 100 % of the golf shafts for 1-woods and 60 % of the driving irons produced in Japan are fabricated from CFRP. Recent trends show that applications for bicycle use are rapidly expanding and demand has increased enough to surpass the three major applications mentioned above.

20.5.2 Aircraft

20.5.2.1 History of CFRP Application

Various kinds of plastic have been utilized in aircraft materials. Among them, fiber-reinforced plastic (FRP) is one of the most important materials. It is said that FRP has been applied in aircraft since World War II. At that time, the kind of FRP used was glass fibers, and it was applied in auxiliary fuel tank and radome (weather proof that protects radar system or antenna). Due to its poor elasticity, glass fibers have only been utilized in secondary structural elements such as fairings and interior materials like ceiling board, overhead bins, sidewall, and divider plates of the bathroom [1]. From the end of 1960s toward the beginning of 1970s, carbon fibers known to have superior elasticity compared to glass fibers were industrialized. From then carbon fiber was firstly utilized in military aircrafts and applications for civil aircrafts started in 1980s. Since carbon fibers were first utilized by Airbus in the vertical tail of A310-300 and the horizontal tail of A320, carbon fibers have been utilized in principal structural elements such as the main wing, tail surface, and body to reduce the weight of the aircraft [1]. As shown in Fig. 20.11, recent aircraft trends show increase in carbon fiber structural components [2].

Due to its high dimensional stability (heat expansion is about one tenth of metals) as well as high specific intensity and specific modulus, carbon fibers are also applied in aerospace applications such as rockets (satellite-mounted part, interstage, solid rocket booster case) and artificial satellite (body frame, solar cell paddle, antenna support, arm).

20.5.2.2 Production Method of Aircraft Elements

CFRPs for aircrafts are mainly produced from a intermediate material called “prepreg.” As shown in Fig. 20.12, “prepreg” is a thin layered sheet consisting of carbon fibers that are arranged in one direction and impregnated with uncured resin.

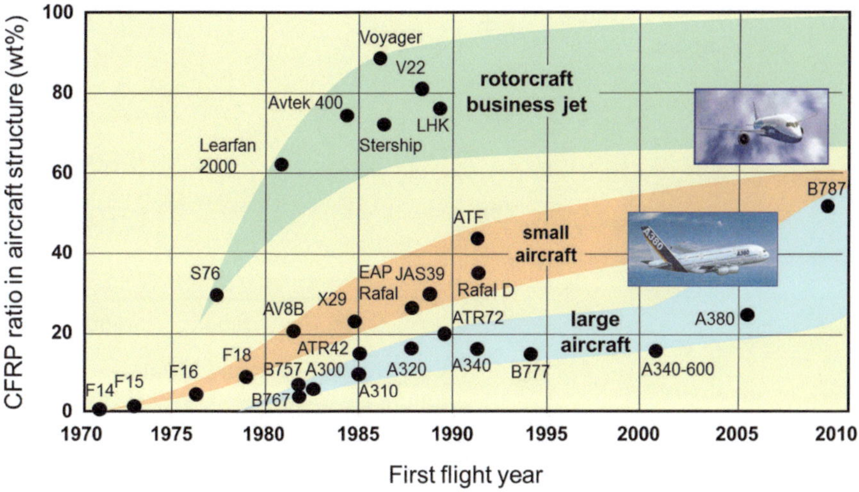

Fig. 20.11 History of CFRP application in aircraft structure

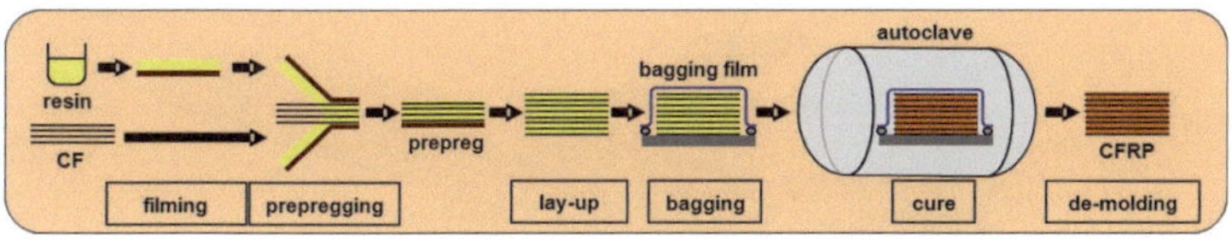

Fig. 20.12 Prepreg/autoclave process (conventional)

CFRPs for aircrafts that are fabricated when multilayered prepregs are heated in high pressure with an autoclave to cure the matrix.

20.5.2.3 Development of High-Impact Resistance Composite Materials

CFRP composite is a layered structure and there are not any carbon fibers to pierce the layers. CFRP is sometimes delaminated when, for example, engineers who fall off their tools on CFRP at the processing or hailstones or pebbles hardly hit CFRP during flight, without dent or damage on the surface. This delamination decreases the strength of CFRP, even if no unusual point is found at visual inspection. The improvement of damage tolerance was one of the main issues when CFRP was applied in Boeing 777, which is the first model throughout Boeing aircrafts whose primary structural elements contain CFRP. Toray focused on the "compression after impact" specification. The key point is satisfying toughness, heat resistance, and modulus at the same time. Toray overcame this issue with the novel CFRP that selectively toughened interlayer with thermoplastic particles. The properties of T800H/#3900-2, in which this technology is used on high-strength carbon fiber, satisfied Boeing specification as shown in Fig. 20.13 [11], resulting in the certification for the Boeing 777. This "interlayer-toughened" CFRP is also applied in

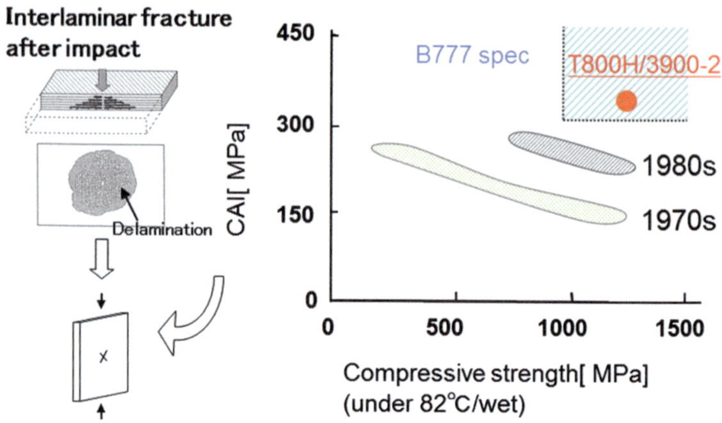

Fig. 20.13 Improvement of CAI

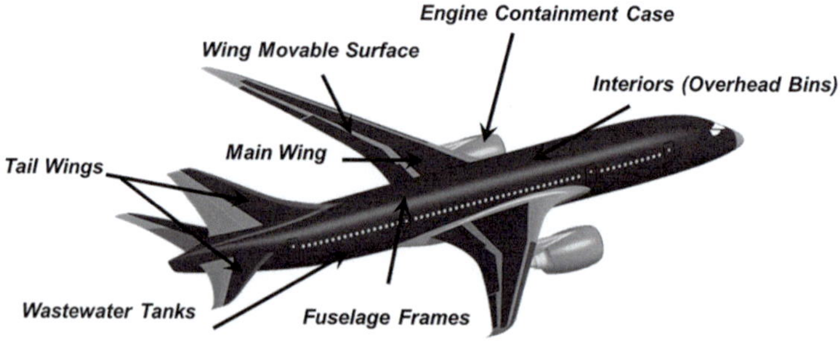

Fig. 20.14 Toray's carbon fiber used in Boeing 787

Boeing 787, which is a middle-sized aircraft and has entered commercial service since 2011. In Boeing 787, CFRPs are also used in its main wing and body. Boeing 787 contains no less than 35 t of CFRP, which is 50 % of its whole weight (see Fig. 20.14). It is recognized as a "black aircraft" since 80 % of the surface of Boeing 787 is covered with carbon fiber. Boeing 787 offers comfortable room for passengers, for example, large windows and relatively high pressure and humid, owing to high fatigue durability and environment resistance of CFRP.

20.5.2.4 Novel CFRP Molding Process

Low-cost CFRP molding process is being developed as alternative method of the prepreg/autoclave processing. Figure 20.15 shows vacuum-assisted resin transfer molding (VaRTM) method, in which layered carbon fibers without resin impregnation are vacuumed after being enveloped into a bag film and then liquid matrix resin is injected. The matrix is cured with heat after the matrix is fully impregnated.

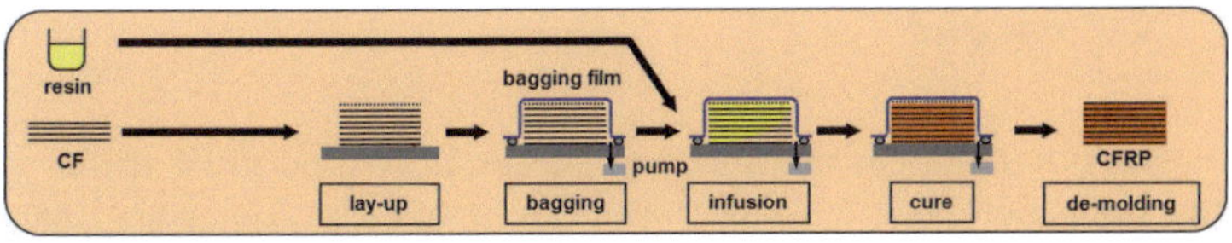

Fig. 20.15 VaRTM process

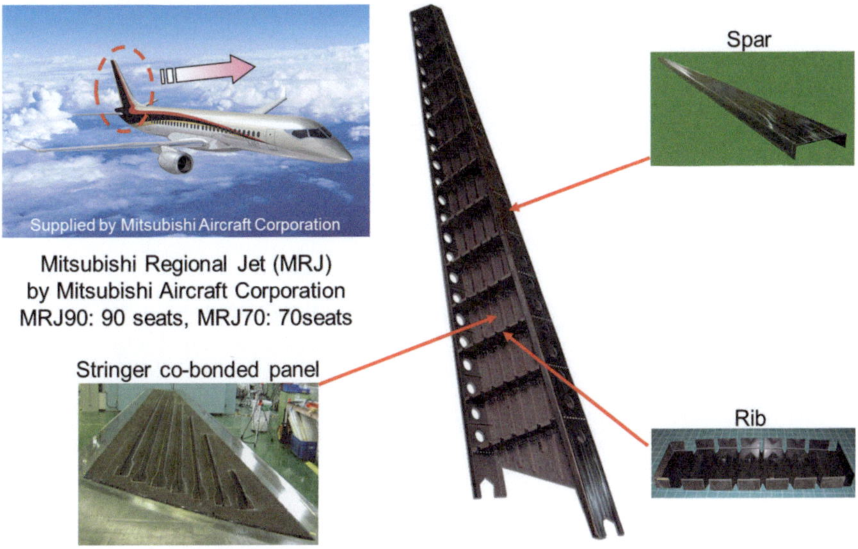

Fig. 20.16 Structural parts of MRJ by A-VaRTM

In this method, application of pressure with autoclave is not required because deaerated low-viscosity resin is utilized.

In VaRTM method, handling-available materials like fabric are required because carbon fibers are layered without resin. Therefore, the mechanical properties are relatively low compared to those of prepreg because the mechanical properties of CFRP depend strongly on direction of the fiber. Moreover, selection of low-viscosity resin limits its composition design that resulted in low elasticity and ductility of CFRP. Hence it had been thought that CFRP produced by means of VaRTM method could hardly be applied in primary structure of aircraft. However Toray and Mitsubishi Heavy Industries overcame this problem with a novel method named "A-VaRTM." CFRP fabricated by means of A-VaRTM method is intended to be utilized in a regional jet aircraft, "Mitsubishi Regional Jet" in Fig. 20.16 [12].

20.5.3 Automobile

In the automobile field, carbon fiber was successfully used in F1 racing cars and high-end cars in Europe at first. Recently, automobile companies are interested in carbon fiber in terms of reducing vehicle weight. Carbon fibers are now used in propeller shaft, engine hood, rear spoiler, and so on. In the future, CFRP whose matrix resin is thermoplastic will be in high demand among others in terms of recycling. Carbon fiber is also considered to be used in the components of fuel-cell vehicle such as electrode and high-pressure hydrogen storage tank.

20.5.4 Electronic Devices

In electronic device field, carbon fiber is used as electromagnetic shielding materials. For example, some chassis of PC is produced by CFRP. The CFRP is produced by carbon fibers that are compounded with thermoplastic resin followed by molding process. These chassis possesses electromagnetic shielding, lightness, and high elasticity without plating.

20.5.5 Others

Carbon fiber is also applied in civil engineering and construction, centrifugal rotors for uranium enrichment, windmills, and IC trays (for compound). Novel applications of carbon fiber are for core material for supply cable, liquid crystal substrate carrying fork, and the top plate of X-ray CT scan.

20.6 Carbon Fiber Composite Material and Global Environment

20.6.1 Life Cycle Assessment

Weight reduction of the aircraft with application of CFRP not only improves the kinematic performance and lengthens the flight distance but however also conserves energy resulting in reduction of environmental burdens (the reduction of CO_2 emissions). Production of CFRP requires energy resulting in CO_2 emission. In contrast, flight of aircraft constantly requires a lot of energy. The amount of CO_2 emission or energy until the end of the product life, "life cycle assessment (LCA)," should be considered in order to evaluate lightweight material. In the case of a Boeing 767-size aircraft which is normally used in domestic, emission of CO_2 in flight occupies 99 % of that from production to disposal as a whole. In the same case

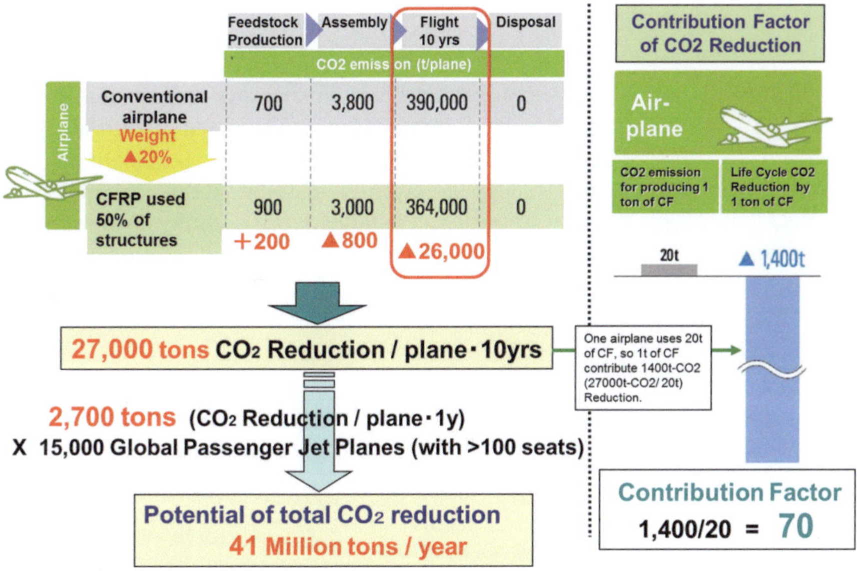

Fig. 20.17 CO_2 reduction by lightweight aircraft

of a Boeing 787, 50 % of whose weight consists of CFRP and 20 % lighter than that of existing aircrafts, 2,700 t per year of CO_2 reduction is expected owing to its lightness as shown in Fig. 20.17, according to "the Japan Carbon Fiber Manufacturers Association." If all aircrafts in Japan became Boeing 787, 1.2 million tons of CO_2 could be reduced.

20.6.2 Recycling

Recycling of carbon fiber is developed into an interesting and essential topic as the amount of carbon fiber consumption increases. In Japan, technological development of carbon fiber recycling has been continuously made by "the Japan Carbon Fiber Manufacturers Association," in the subsidiary project from the Ministry of Economy, Trade, and Industry known as "Verification research and development of carbon fiber recycling technology." A pilot plant was built in Omuta City, Fukuoka Prefecture, and basic technology to produce recycled carbon fiber was studied with the support of Fukuoka Prefecture and Omuta City. Currently, further advancement of carbon fiber recycling is performed by "Carbon Fiber Recycling Technology Development Association," which was established by Toho Tenax, Mitsubishi Rayon, and Toray.

20.7 Summary

The lightweight and high-strength carbon fiber is increasingly applied in various fields such as aircraft, automobile, sporting applications, and industrial applications. To expand the carbon fiber industry, it is particularly important to improve carbon fiber properties and reduce its cost. In addition, further technological developments of new primary products and molding process, carbon fiber recycling techniques, and manufacturing process to minimize energy consumption are all essential from an environmental perspective.

References

1. J. Matsui, The story of carbon fiber #1. Reinf. Plast. **43**(5), 187–191 (1997)
2. J. Matsui, The story of carbon fiber #5. Reinf. Plast. **43**(12), 485–492 (1997)
3. Shindo, A., Chemistry, **65**(22), 22–25 (2010)
4. J. Matsui, The story of carbon fiber #3. Reinf. Plast. **43**(8), 296–300 (1997)
5. S. Takamatsu, Development of PAN precursor carbon fiber. Technol. Civiliz. **12**(1), 1–24 (2000)
6. S. Watanabe, Chem. Eng. **58**(1), 24–27 (1994)
7. J. Matsui, J. Matsui, The story of carbon fiber #6. Reinf. Plast. **44**(1), 29–34 (1998)
8. Otani, S., Pitch-based carbon fiber & COPNA resin. *Carbon*, **2011**(246), 23–37 (2011)
9. D.J. Johnson, C.N. Tyson, J. Appl. Phys. **2**(2), 787–796 (1969)
10. J. Matsui, The story of carbon fiber #8. Reinf. Plast. **44**(6), 237–245 (1998)
11. Odagiri, N., Muraki T. and Tobukuro, K., Toughness improved high performance TORAYCA Prepreg T800H/3900 Series, 33rd International SAMPE Symposium (1988), 272–283.
12. F. Takeda, S. Nishiyama, K. Hayashi, Y. Komori, Y. Suga, N. Asahara, Application of VaRTM for primary materials of aircrafts. Technol. Rep. Mitsubishi Heavy Ind. **42-5**, 220–225 (2005)

Chapter 21
Pitch-Based Carbon Fibers

Yutaka Arai

Abstract Pitch-based carbon fibers, using pitches as raw materials, can have various characteristics derived from the properties of the original pitch and production method of the fibers. This chapter explains the types, manufacturing method, and also characteristics and structural features of the pitch-based carbon fibers, emphasizing the very high elastic modulus and thermal conductivity of them.

Keywords Pitch based • Carbon fiber • Mesophase • Thermal conductivity • Low modulus • High modulus • Graphite

21.1 Introduction

The pitch-based carbon fibers using petroleum residual oil or coal tar as raw materials have a long history. In 1970, Kureha Chemical Industry Co. Ltd. (present Kureha Corporation) was the first in the world to start industrialization of the pitch-based carbon fibers using isotropic pitch as a material. This industrialization was based on the invention of Otani [1], Gunma University, in the 1960s. Otani [2] also invented high-performance pitch-based carbon fibers, whose tensile strength and elastic modulus are enhanced than those of the isotropic pitch-based fibers, for the first time. The industrialization of these high-performance pitch-based carbon fibers was attained in 1975 by Union Carbide Corporation (present Cytec Engineered Materials Inc., via Amoco Corporation) of the United States. This industrialization was based on the anisotropic pitch, which is called as mesophase pitch, as a raw material [3, 4]. Then, after entering the 1980s, development boom of the high-performance pitch-based carbon fibers took place for the purpose of high-level usage of side products of petroleum refinery and coal carbonization, and more than 20 companies entered the market for a time. In the United States, Du Pont (E. I. du Pont de Nemours and Company) placed high-modulus and high-strength carbon fibers, based on the technology of Exxon Corporation, on the market. However, after a while, they withdrew from the business. Afterward, Conoco (Continental Oil

Y. Arai (✉)
Nippon Graphite Fiber Co. Ltd., 1, Fuji-Cho, Hirohata-Ku, Himeji, Hyougo 671-1123, Japan
e-mail: y-arai@ngfworld.com

Society of F.S. Technology, Japan (ed.), *High-Performance and Specialty Fibers*,
DOI 10.1007/978-4-431-55203-1_21

Company) made an announcement of the industrialization of short carbon fibers by the technology purchases from Du Pont; however, they also withdrew from the business in the early 2000s. As of 2014, one in the United States and four Japanese manufactures of pitch-based carbon fibers are known.

21.2 Classification of the Pitch-Based Carbon Fibers [5]

As shown in Fig. 21.1, starting materials of the pitch-based carbon fibers are roughly divided into two categories. The carbon fibers, which derive the isotropic pitch, do not develop graphite structure even by high-temperature calcination and are commonly called as general-purpose type. On the other hand, carbon fibers, which derive the mesophase pitch, develop the graphite structure by high-temperature calcination and are called as high-performance type due to the high-strength and high-modulus properties. The pitch-based carbon fibers are provided by staple fiber form, which is not popular for polyacrylonitrile (PAN)-based carbon fibers, because it is difficult to produce continuous fibers industrially for the former fibers. Nevertheless, the industrialization of continuous pitch-based carbon fibers, though it is more difficult than that for staple fibers, is important for the wide application possibility, e.g., good secondary workability like PAN-based carbon fibers.

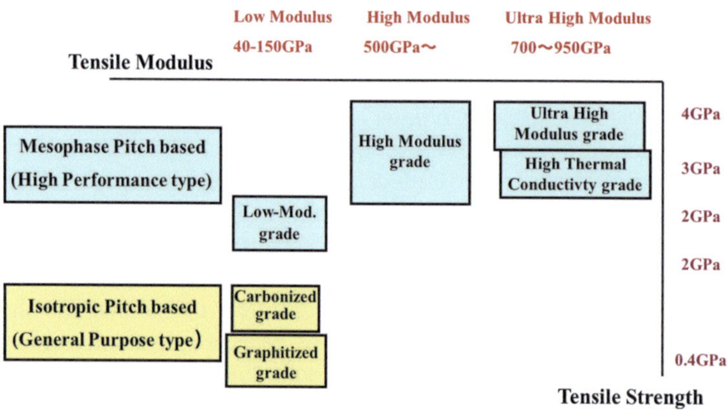

Fig. 21.1 Types of pitch-based carbon fibers

21.3 Production Method of the Pitch-Based Carbon Fibers

The production method of the pitch-based carbon fibers consists of each process as shown in Fig. 21.2, and the details are explained as follows.

21.3.1 Pitch Treatment Process

Raw materials of the pitch-based carbon fibers are obtained from aromatic organic matters such as oil and coal. Generally, they are bottom oil obtained from a fluid catalytic cracking (FCC) apparatus of the petroleum refining process, ethylene bottom oil obtained from naphtha cracker, and coal tar produced in manufacturing coke. Some pitch is also synthesized by the polymerization of naphthalene, methyl-naphtalen, anthracene and et al. using super acid catalysts. Generally, these crude pitches have to be purified by distillation, solvent extraction, and mechanical separation to remove impurities and modified the properties by hydrogenation. If necessary, and finally pitch with high spinnability is obtained by thermal polymerization. In this stage, the softening point of the raw material pitch for spinning has to be adjusted appropriately.

The raw material pitch for spinning is roughly classified into isotropic pitch, without liquid crystallinity, and mesophase pitch, with liquid crystallinity. Figure 21.3 shows the polarized-light microscope photographs of isotropic (Fig. 21.3a) and mesophase (Fig. 21.3b) pitches. The isotropic pitch consists of non-graphitization carbon, and therefore graphite crystals hardly grow by high-temperature heat treatment, whereas the mesophase pitch consists of easy graphitization carbon, and therefore graphite crystals easily grow by high-temperature heat treatment. The properties of carbon fibers thus produced from these two

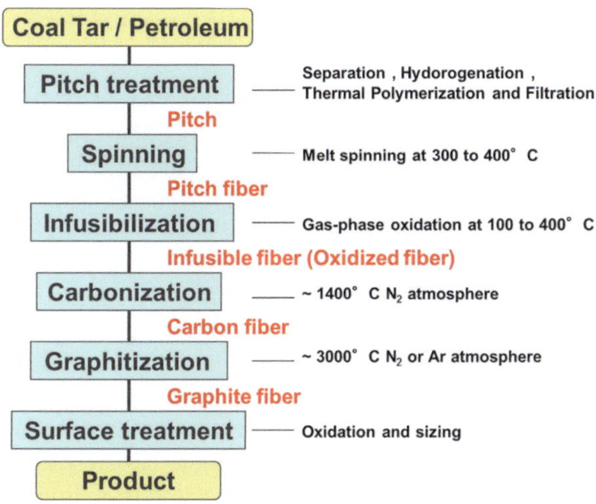

Fig. 21.2 Production scheme of pitch-based carbon fibers

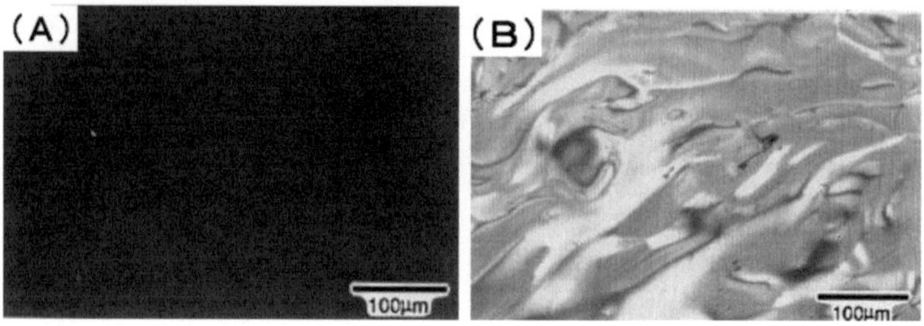

Fig. 21.3 Polarized-light microscope photographs of pitches, (**a**) isotropic pitch, (**b**) mesophase (anisotropic) pitch

pitches are largely different due to the difference of the graphitization properties of the original pitches.

For the case of the isotropic pitch-based carbon fibers, the spinning is easier due to the tendency of lower softening temperature and hence lower spinning temperature. However, in the infusibilization process (corresponding to the stabilization process or oxidization process of PAN-based carbon fibers) of the obtained pitch-based carbon fibers, the infusibilization of the fibers with lower softening point becomes harder. As a compromise, to attain higher softening temperature as possible is important with keeping the isotropy. On the other hand, as for the case of the mesophase pitch, the attempt to improve anisotropic ratio causes the higher softening temperature. The higher the softening temperature is, the higher the operation temperature of the spinning becomes. This results the inferior spinnability significantly due to the decomposition and polymerization during the spinning process. Therefore, it is necessary to prepare mesophase pitch with appropriate softening temperature to keep relatively large anisotropic region.

21.3.2 Spinning Process

Then the pitch is formed fibrously by the spinning process. As for the isotropic pitch as a material, properties and structures of the product fibers rarely affected by the method of spinning and the conditions, because the orientation and the inner structure formation of the fibers are very weak in general. As for the mesophase pitch as a material, however, the structures of the product fibers are largely affected by the method of spinning and the conditions. The polymer liquid crystals of the mesophase pitch are oriented during the flow through a capillary nozzle [6], and planar aromatic groups of the liquid crystals are oriented in the direction of the fiber axis. On the other hand, as same as the case of PAN-based carbon fibers, the mesophase pitch, consisting of nonlinear polymer molecules, can form oriented and ordered structure in the cross-sectional direction orthogonal to the fiber axis. Figure 21.4 shows the characteristic cross-sectional structures, called as radial, random, and onion, observed at the transverse section of mesophase pitch-based

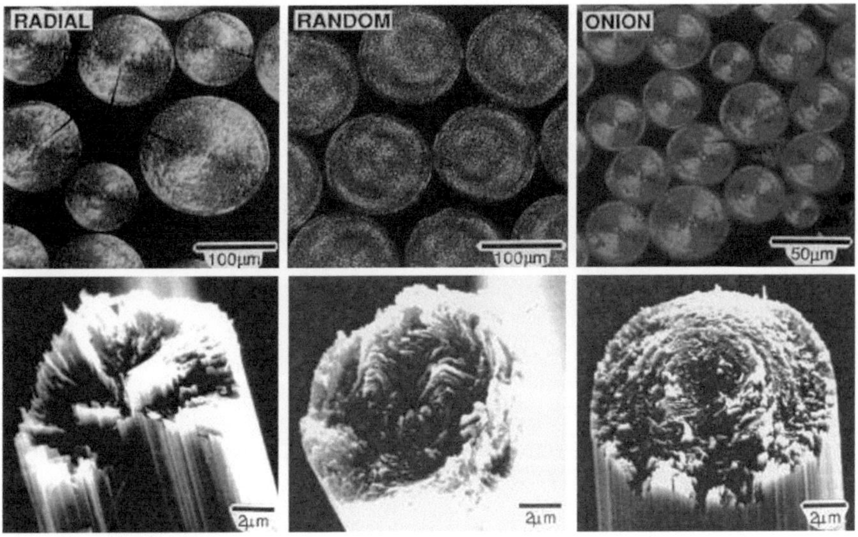

Fig. 21.4 Cross-sectional structures of the pitch-based carbon fibers. Polarized-light microscope photographs of pitch fibers (*upper*) and scanning electron microscope (SEM) photographs of the graphitized fibers (*lower*)

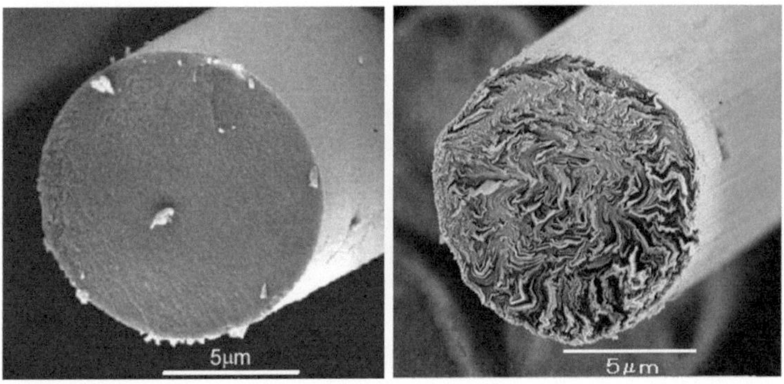

Fig. 21.5 Scanning electron microscope (SEM) photographs of XN-05® amorphous carbon fiber (*left*) and high-modulus type (*right*)

carbon fibers [7]. The type of these characteristic cross-sectional structures for the product pitch-based carbon fibers is determined by the conditions of the spinning process. Upper figures of Fig. 21.4 show the polarized-light microscope photographs of the cross section of the undrawn pitch-based carbon fibers which were flowed out from a capillary nozzle. These photographs show that the oriented structures of radial, onion, etc. are formed by the orientation of the mesophase pitch. On the other hand, as shown in Fig. 21.5 (left), XN-05 carbon fibers [8] (Nippon Graphite Fiber Co. Ltd., hereafter abbreviated as NGF) made from the isotropic pitch do not show the characteristic cross-sectional structures, which are

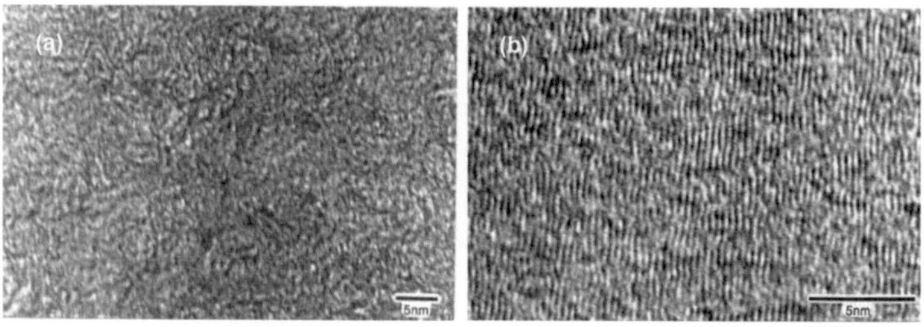

Fig. 21.6 Lattice-fringe images of longitudinal section of pitch-based carbon fibers, (**a**) XN-05® amorphous carbon fiber, (**b**) high-modulus carbon fiber

observed for the mesophase pitch-based carbon fibers, as shown in Fig. 21.5 (right). Figure 21.6 shows lattice-fringe images of longitudinal section of carbon fibers taken by transmission electron microscope (TEM). Though the development of graphite structure is not observed for the TEM photograph of isotropic pitch-based XN-05 carbon fibers (NGF) (Fig. 21.6a), the development of graphite structure along the fiber axis is clearly recognized from the ordered lattice-fringe image (Fig. 21.6b) of high-modulus mesophase pitch-based carbon fibers.

21.3.3 Infusibilization Process

Infusibilization is the process to protect fibers from softening and melting by the high temperature applied at following carbonization process. This infusibilization process causes cross-linking and dehydrogenation to the pitch-based carbon fibers. This process is attained by the oxidation at the lower temperature than the softening temperature of the pitch-based carbon fibers obtained by melt spinning. In other words, infusibilization is an oxygen addition reaction to the pitch-based fibers and is a solid-gas reaction where oxygen reacts with active species of the pitch molecules after the diffusion and infiltration through the pitch solid [9–11]. Therefore, industrially this reaction is a kind of solid-phase diffusion-limited reaction, and hence the control of reaction heat emitted by the oxidation is important.

The oxidation reaction by oxygen substantially starts at the temperature about 150 °C. Hence, the reaction is generally started from the lower temperature at where the pitch-based carbon fibers do not fuse together by the softening and melting. Then the temperature is raised gradually processing the infusibilization of the fibers. Therefore, the infusibilization by oxygen for the pitch whose softening temperature is lower than 150 °C is substantially not possible.

21.3.4 Carbonization, Graphitization, and Surface Treatment Processes

Carbonization is a heat treatment process for the infusibilized fibers in inert atmosphere using nitrogen or argon gas. During this process, the oxygen introduced in the previous infusibilization process is removed as CO_2, CO, etc., and simultaneously, H_2 and hydrocarbons, e.g., CH_4, C_2H_6, etc., are generated. Unlike that of PAN- and rayon-based carbon fibers, the carbonization yield of pitch-based carbon fibers is high as ca. 80 %, depending on the properties of the original pitch and the conditions of infusibilization process. After this carbonization process, the carbon fibers are subjected to a graphitization process if necessary. During several to several tens of seconds of this process, the crystal growth of graphite is attained.

Next, a surface treatment process is carried out for the carbonized or graphitized carbon fibers, aiming mainly for the improvement of adhesiveness of the fibers with resins. This process is indispensable for the continuous-form fibers and generally made by an electrolytic oxidation method as same as the case for PAN-based carbon fibers. Finally, the fibers are subjected to a sizing treatment to improve handle ability, and the pitch-based carbon fibers are thus completed.

21.4 The Structure and Properties of Pitch-Based Carbon Fibers

Figure 21.7 shows the relationship between the tensile strength and the tensile modulus for the PAN- and pitch-based (e.g., NGF and the general-purpose pitch based) carbon fibers. As for the commercially available PAN-based carbon fibers, the highest tensile strength reached to 6500 MPa; however, the tensile modulus is ranged from 230 to 300 GPa for the standard type, and the highest limit is 600 GPa. On the other hand, as for the pitch-based carbon fibers, the variable range of the tensile modulus is wide from 50 GPa for the low-modulus type and more than 900 GPa at the highest-modulus grade. The carbon fibers obtained from mesophase pitch as a starting material have the oriented graphite layers in the direction of the fiber axis. As shown in Fig. 21.8 [12], graphite crystals have extremely high strength and rigidity by the carbon-carbon double bonds located at the expanding direction of the graphite layers (called as "a" direction), and this crystal structure is reflected on the strength and rigidity of final product carbon fibers. Further, a negative thermal expansion coefficient at around room temperature and extremely high thermal conductivity of the pitch-based carbon fibers come from the characteristics of this graphite "a" direction. All these characteristics are the important properties of high-modulus type pitch-based carbon fibers whose graphite crystals are developed.

On the other hand, when an isotropic pitch, which does not exhibit a liquid crystalline property, is used as a starting material, the abovementioned orientation

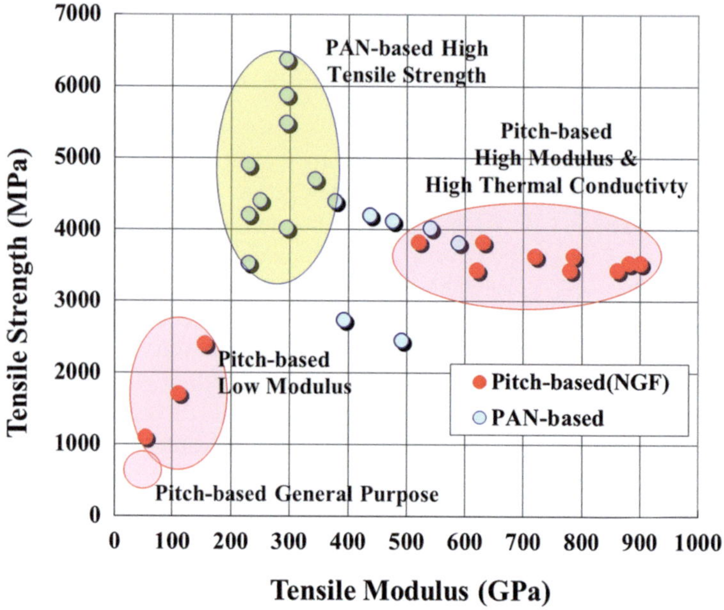

Fig. 21.7 Mechanical properties of carbon fibers

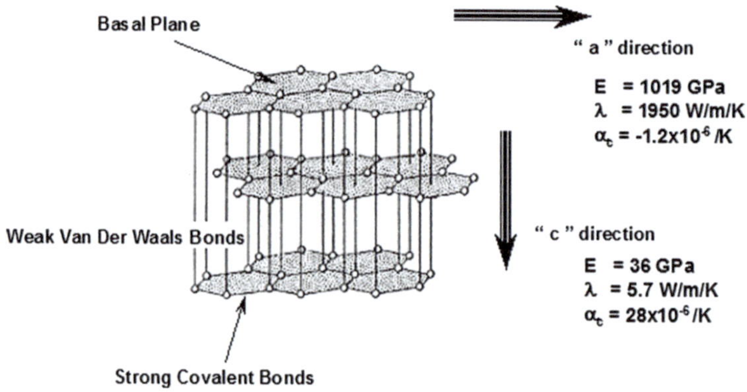

Fig. 21.8 Graphite crystal structure

of graphite layers in the direction of the fiber axis and the development of graphite crystals are not sufficient. Therefore, the product carbon fibers have low tensile modulus and strength, low thermal conductivity, and positive thermal expansion coefficient. These properties are very different from those of the common carbon fibers including the PAN based. As shown by the electron microscope pictures of Figs. 21.5 and 21.6, the isotropic pitch-based carbon fibers have obviously different cross-sectional and fiber structures. The features of pitch-based carbon fibers can be thus variously changed by the properties of original pitches or by the control of crystal growth of graphite crystals during the fiber-forming processes.

21.4.1 Characteristics and New Application Developments of Low-Modulus Carbon Fibers

The high impact-resistant strength and the large fracture absorption energy of materials can be attained by the combination use of the continuous low-modulus pitch-based carbon fibers with the PAN-based carbon fibers, because the former have large compression strain, different from the other carbon fibers and also other organic high-strength fibers. Figure 21.9 shows the results of static and impact bending fracture test for the laminated sheets of XN-05 (NGF, elastic modulus, 55 GPa) pitch-based and other PAN-based (elastic modulus, 230 GPa) carbon fibers. This figure shows that the impact-absorbing ability of laminated sheet which has XN-05 layers at the compression side is improved three times as much when it is compared with the sheet made only by PAN-based carbon fibers [13]. Figure 21.10 shows the cross-sectional pictures of carbon fiber-reinforced plastic (CFRP) test pieces after the bending fracture test. As shown in the figure, the

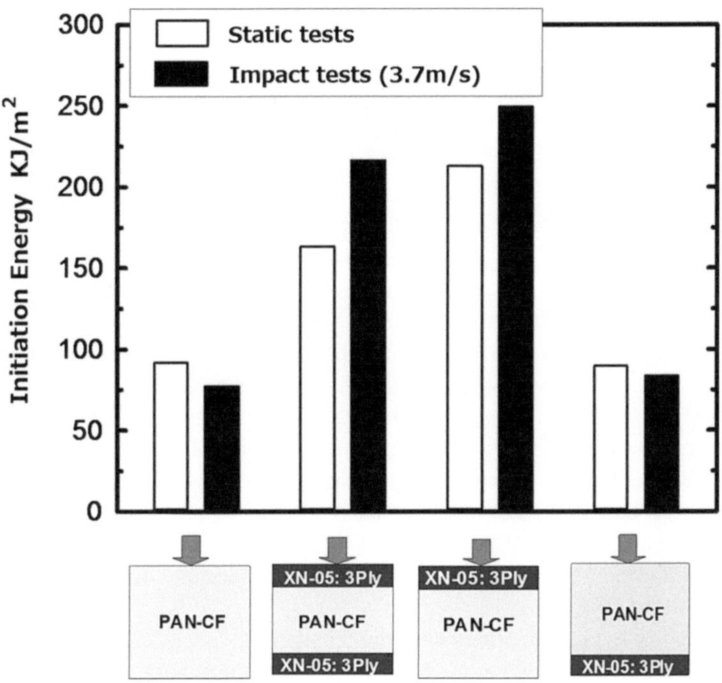

Fig. 21.9 Improvement in initiation energy using low-modulus carbon fibers (XN-05®)

Fig. 21.10 Bending fracture mode after bending test

initial bending fracture for the samples of low-modulus pitch-based carbon fibers (XN-05) is caused at the stretch side, because the compression breaking strain is larger than the stretching breaking strain. On the other hand, as for the general-purpose CRFP (PAN-230 GPa), the compressing breaking strain is lower than that of the stretching, and therefore the initial fracture is caused at the compression side [8]. It is observed from the figure that the total strength and the fracture absorbing energy are improved for the sandwich sample which has the layer consists of low-modulus pitch-based carbon fibers at the surface. This is because the high tensile strength of PAN-based carbon fibers is effectively worked due to the acceptance of maximum compression strain by the pitch-based carbon fibers at the surface. The combination of these complementary characteristic features is applied for the tip reinforcing of golf shafts and the performance improvements of tennis rackets.

21.4.2 High Tensile Modulus and High Thermal Conductivity Carbon Fibers

The greatest features of the pitch-based carbon fibers, mesophase pitch as a raw material, are that the high thermal conductivity and high tensile modulus in the direction of fiber axis are easily attained. As shown in Fig. 21.11, high thermal conductivity of ca. 1000 W/(mK) for some of the pitch-based carbon fibers is far superior to that of metals. Further, the negative thermal expansion coefficient of the pitch-based carbon fibers easily realizes the materials with zero thermal expansion coefficient by the combination of the pitch-based carbon fibers with other matrixes. As new application fields utilizing these characteristics of high thermal conductivity and negative thermal expansion coefficient, various members of satellites such as antenna reflectors and solar arrays are made by the pitch-based carbon fibers. High thermal conductivity which exceeds that of metals is also usefully applied in the field of electronic equipments, e.g., thermal interface and high thermal conductivity circuit board.

The pitch-based carbon fibers with the tensile modulus of ca. 600 GPa are used for various composite rolls, e.g., for printing and films. The weight of the rolls decreases from a half to one third keeping with the same rigidity as iron, and this

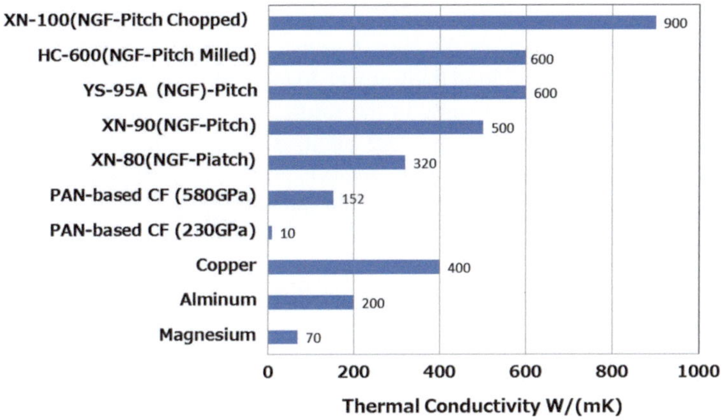

Fig. 21.11 Thermal conductivity of various materials

feature attained the high precision, high speed, and increase in size of the rolls. In addition, the rigidity of the pitch-based carbon fibers with the tensile modulus of more than 800 GPa in the composite form exceeds that of iron also. Therefore, this material is attracting attention as an alternative for ceramic materials, and the applications to various semiconductor manufacturing equipments have started. Furthermore, pitch-based high-modulus carbon fibers are superior for the vibration-damping property. CFRP has good vibration-damping property when compared with that of metals or ceramics., the use of the pitch-based carbon fibers results better damping ability than the use of the PAN-based carbon fibers, and this tendency is more apparent for higher tensile modulus fibers. These features are suitable for robot arms for the conveyance of large-size liquid crystal panels and manufacturing members for machine tools, to shorten the tact time and increase the processing accuracy. This vibration-damping property also makes contribution to sports and leisure area to enhance in the sense, e.g., for the good shot feeling of golf shafts and rackets and usability improvement of bicycles.

21.5 Closing Remarks

The industrialization of the pitch-based carbon fibers was started almost at the same age as that for the PAN-based carbon fibers. However, there is a large difference for the amount of production recently. It is considered that this is because the processes, e.g., material preparation, spinning, and infusibilization, for pitch-based carbon fibers are more difficult when compared with those of PAN-based carbon fibers. However, the market is rapidly increasing by the cost reduction of the pitch-based carbon fiber products attained by the technological improvements. Further, the high rigidity, high thermal conductivity, and high functional properties of the pitch-based carbon fibers match with recent rising demands of labor and energy saving, and the applications of the pitch-based carbon fibers are expanding. Pitch-based carbon fibers take pride in their unprecedented ultimate performance and are very attractive and deep materials. In the future, due to the many possibilities of combination, e.g., selection of material pitch, structure control in spinning, and conditions of infusibilization and carbonization processes, further improvements in the performance and environment load reduction will be expected.

References

1. S. Ohtani, *Japan Patent*, S41–15728 (1966)
2. S. Ohtani, K. Watanabe, T. Araki, *Japan Patent*, S49–8634 (1974)
3. L.S. Singer, Carbon **16**, 409–415 (1978)
4. L.S. Singer, B. Ohio, *U.S. Patent*, 4005183 (1977)
5. Y. Arai, *Tanso*, 2020 (No. 241) 15–20 (2010)
6. L.S. Singer, Fuel **60**, 839–847 (1981)
7. Y. Arai, Nippon. Steel. Tech. Rep. **59**, 65–70 (1993)
8. H. Ohno, M. Shima, S. Takemura, Y. Sohda, *Proceedings of the 44th Japan Int. SAMPE Symposium*, 782–793 (1999)
9. Y. Arai, T. Iwashita, N. Tomioka, *Proceedings of the Society of Chemical Engineers, Japan (SCEJ) 22th Autumn Meeting*, 303 (1989)
10. Y. Arai, T. Iwashita, N. Tomioka, *Proceedings of the Carbon Society of Japan 16th Annual Meeting*, **3A05**, 174–176 (1989)
11. Y. Arai, T. Iwashita, N. Tomioka, *International Symposium on Carbon*, **6PA10**, 782–785 (1990)
12. A. Bertram, K. Beasley, W. Torre, Naval. Eng. J. **104**, 276–285 (1992)
13. N. Kiuchi, Y. Sohda, S. Takemura, Y. Arai, H. Ohno, M. Shima, *Proceedings of the 6th Japan International SAMPE Symposium*, 133–136 (1999)

Chapter 22
Life Cycle Assessment of Carbon Fiber-Reinforced Plastic

Tetsuyuki Kyono

Abstract Life cycle assessment (LCA) is one of the techniques to assess environmental aspects associated with a product or process by identifying energy, materials, and emissions over its life cycle. The energy analysis includes all stages of a life cycle of an end product including the stages of resource extraction, manufacturing of materials, use, and its final disposal. Generally, carbon fiber needs high energy during the manufacturing process. However, carbon fiber composites show outstanding mechanical properties compared with other engineering materials such as metals and others. The use of carbon fiber enables weight reduction and leads directly to improved fuel saving and contributes to reducing CO_2 (carbon dioxide) emission especially in the transportation sector. In this chapter, the examples of LCA on aircraft and automobile are introduced.

Keywords Life cycle assessment (LCA) • Carbon fiber-reinforced plastics (CFRP) • Aircraft • Automotive

22.1 Introduction

Environmental protection is a growing concern for many industries today, with emphasis on the reduction of CO_2 emissions to mitigate climate change. Energy saving is an issue strongly required for the transportation sector.

In Japan, the amount of CO_2 emission is about 1240 million tons per year in total, and that from transportation sector is around 230 million tons whose ratio is 20 % of the total (Fig. 22.1).

T. Kyono (✉)
LCA Division, Technology Committee, The Japan Carbon Fiber Manufacturing Association (JCMA), Tokyo, Japan
e-mail: Tetsuyuki_Kyono@nts.toray.co.jp

Society of F.S. Technology, Japan (ed.), *High-Performance and Specialty Fibers*,
DOI 10.1007/978-4-431-55203-1_22

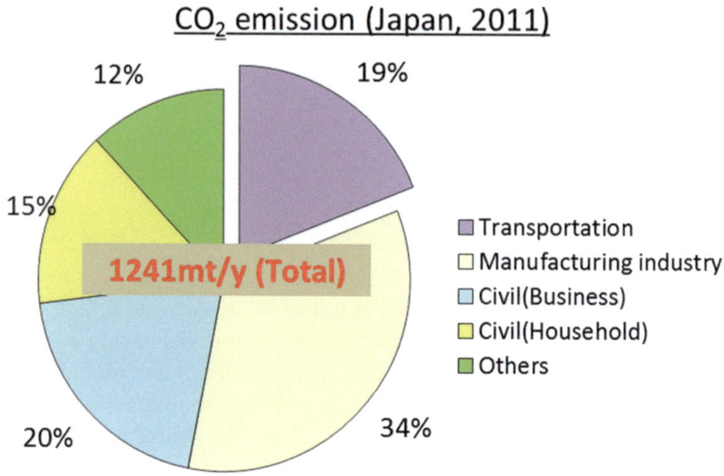

Fig. 22.1 CO_2 emission by sector [1]

22.2 Life Cycle[1] Inventory and Mechanical Properties of Carbon Fiber

More energy is consumed for carbon fiber than conventional material during manufacturing process especially carbonization process (Fig. 22.2, Table 22.1) because raw materials of carbon fiber rely upon fossil resources. Technologies for recycling carbon fiber have not been matured yet. These disadvantages of life cycle inventory for carbon fiber should be improved.

On the other hand, carbon fiber is a promising material due to the excellent properties such as specific strength and modulus (Fig. 22.3), which lead to reduce weight of transport body compared with conventional materials.

It is advantageous that fuel or energy efficiency will be dramatically improved especially during transportation stage by applying carbon fiber to a transport body for its weight reduction.

It is important for LCA to estimate the life cycle load or impact by considering the contribution for each stage, and the considerable benefit of using carbon fiber will be expected mainly during the transportation stage.

[1] Life cycle: Consecutive and interlinked stages of a product or service system, from the resource extraction to the final disposal.

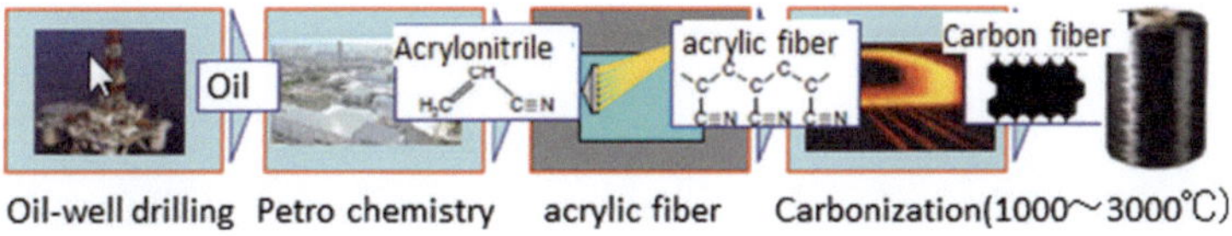

Fig. 22.2 Carbon fiber processing [2]

Table 22.1 Energy content of various materials [4]

Materials	Energy intensity (MJ/kg)
Polymers	
Polyester	63–78
Epoxy	76–80
LDPE	65–92
PP	72–112
PVC	53–80
PS	71–118
PC	80–115
Fibers	
Glass fiber	13–32
Carbon fiber	183–286
China red fiber	3.6
Flax fiber	6.5
Metals	
Aluminum	196–257
Steel	30–60
Stainless steel	110–210
Copper	95–115
Zinc	67–115
Cast ion	60–260

22.3 Fuel Saving Through Weight Reduction

As described above, weight reduction directly leads to fuel saving and reducing CO_2 emission during transportation stage. JCMA implemented two case studies of LCA.[2] One case is for aircraft, and another one is for automobile. The amount of CO_2 emission has been estimated for each case.

The concept of life cycle assessment is shown in Fig. 22.4. In this assessment, the amount of CO_2 emission of each stage is estimated and compared between a conventional design and a carbon fiber-used design. As shown in Fig. 22.4, the amounts of CO_2 emission from the stages of material manufacturing and assembly

[2] LCA: A systematic set of procedures for compiling and examining the inputs and outputs of materials and energy and associated environmental impacts directly attributable to the functioning of a product or service system throughout its life cycle.

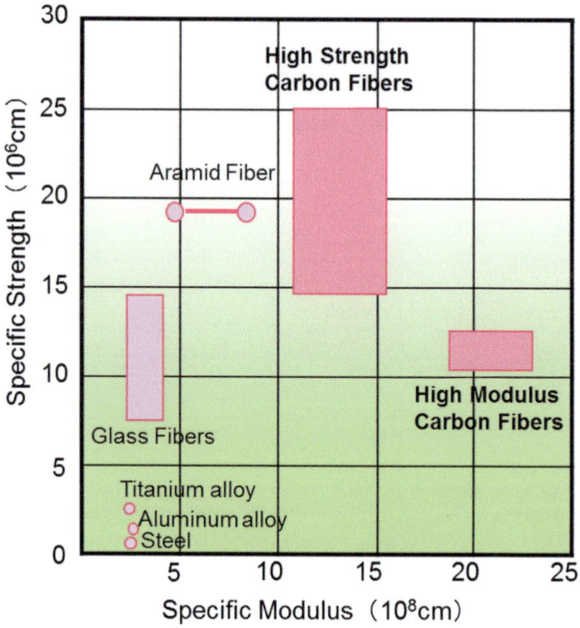

Fig. 22.3 Specific mechanical properties [3]

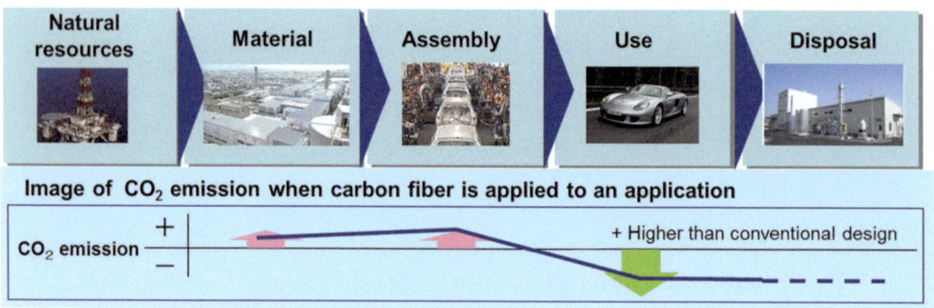

Fig. 22.4 Concept of LCA

are higher for the carbon fiber-used design than a conventional one. On the other hand, the amount from transportation is much lower for the carbon fiber-used design than a conventional one. Finally, the total amount of CO_2 emission over the life cycle is lower for the carbon fiber-used design. As a conclusion, the carbon fiber-used design shows an advantage to reduce the amount of CO_2 emission.

LCAs for aircraft and automobile have been implemented by the Japan Carbon Fiber Manufacturer Association. Its detail is described later.

22.3.1 LCA for Carbon Fiber-Reinforced Plastic (CFRP) Plane

LCAs for conventional aircraft and CFRP aircraft have been implemented according to the following procedure based on the assumption shown in Fig. 22.5.

1. 50 % of aircraft body (airframe) weight is composed by CFRP for CFRP model, and 3 % of aircraft body weight is composed by CFRP for conventional model.
2. Assessments were made for the CFRP model and the conventional model, respectively, and both results were compared by considering energies required at the stages from resource extraction, raw materials manufacture, parts manufacture, and aircraft assembly through flight. Disposal was excluded from the assessment because it is too difficult to have a disposal record for the moment.
3. The following assumptions are applied to estimate emission reduction:

   - Body weight:

Conventional model (767 class)	60 tons/unit	Proportion of CFRP, 3 %
CFRP model	48 tons/unit	Proportion of CFRP, 50 %
20 % reduction compared with conventional aircraft		

   - Fuel economy:

Conventional aircraft	One kiloliter of jet fuel per 103 km
CFRP aircraft	One kiloliter of jet fuel per 110 km

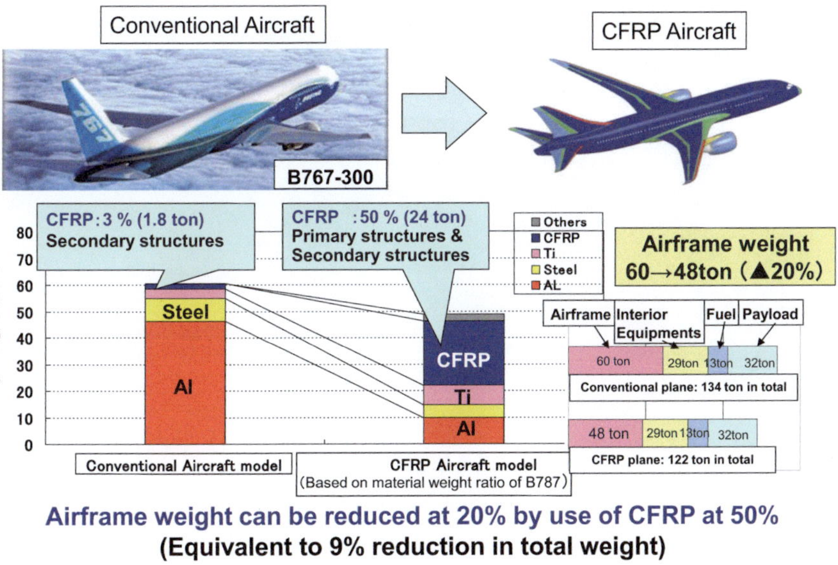

Fig. 22.5 Aircraft life cycle assessment (LCA) of "JCMA model" [3]

- Lifetime mileage is estimated by assuming 2000 flights annually between Haneda Airport and Chitose Airport (500 miles) for 10 years.

4. The CO_2 emission from extraction to assembly is 3.9 ktons/unit. The use of carbon fiber for aircraft structure results in weight reduction of aircraft body, fuel economy improvement, and reduction of CO_2 emission. The difference in CO_2 emission between CFRP and conventional models is actually due to the reduction of body weight and jet fuel consumption by more applying CFRP to aircraft body.

   The CO_2 emission for the CFRP model from resource extraction used in aircraft through the manufacture of materials increases in 0.2 ktons/unit compared with those for conventional model. However, the CO_2 emissions decrease in approximately 0.8 ktons for aircraft assembly and in approximately 26.3 ktons during the flight, respectively.

   Therefore, the reduction of CO_2 emission for the CFRP model over the entire life cycle is approximately 27 ktons/(unit·10 years) compared with the conventional model.

22.3.2 LCA for Carbon Fiber-Reinforced Plastic (CFRP) Automotive

LCAs for conventional automotive model and CFRP automotive model have been implemented according to the following procedure based on the assumption shown in Fig. 22.6:

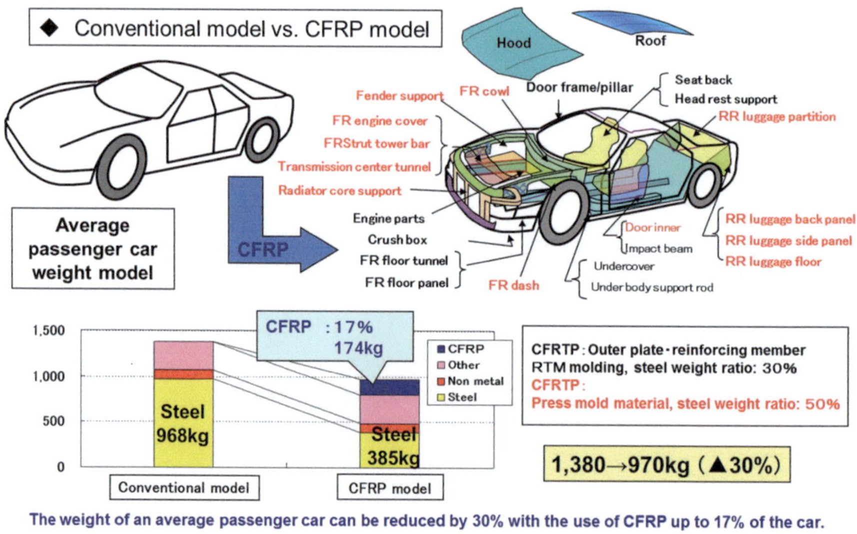

Fig. 22.6 Automobile life cycle assessment (LCA) of "JCMA model" [3]

1. 17 % of automotive body weight is composed by CFRP for CFRP model and no CFRP for conventional model.
2. Assessments were made for the CFRP model and the conventional model, respectively, and both results were compared by considering energies required at the stages from resource extraction, raw material manufacture, part manufacture, and assembly of automobiles, driving through disposal over the entire life cycle of automobile.
3. The following assumptions are applied to estimate the reduction of CO_2 emission.

   - Fuel: Automobile runs only by gasoline.
   - Vehicle weight assumed:

CFRP model	970 kg/unit
Conventional model	1380 kg/unit (average vehicle weight as of 2006)
→30 % reduction in the CFRP model compared with the conventional model. Totally, 174 kg of CFRP is used per unit for CFRP model	

   - Fuel economy assumed:

CFRP model	One liter of gasoline per 12.40 km
Conventional model	One liter of gasoline per 9.83 km

   - The CO_2 emission per automobile unit is calculated for each case of the CFRP model and the conventional model over the entire life cycle including 10 years driving of 94,000 km and disposal.

4. CO_2 emission from extraction to assembly is 6.2 ktons/unit. The use of carbon fiber for automobiles results in the weight reduction, fuel economy improvement, and reduction of CO_2 emission. The difference of CO_2 emission is actually due to the reductions of body weight and gasoline consumption by applying CFRP to automotive body.

   The CO_2 emission for CFRP model from extraction through assembly and disposal increases in 0.8 tons/unit compared with those for conventional model. However, the CO_2 emission reduces in approximately 5.4 tons for CFRP model during the driving compared with conventional model.

   Therefore, the reduction of CO_2 emission for the CFRP model over the entire life cycle is approximately 5 tons/(unit•10 years).

References

1. National Institute for Environmental Studies, Japan, National Greenhouse Gas Inventory Report of JAPAN. http://www-gio.nies.go.jp/aboutghg/nir/nir-j.html
2. METI website. http://www.meti.go.jp/committee/materials/downloadfiles/g80619b05j.pdf
3. Science & Technology in Japan: 2015, No.11
4. Young S. Song, Jae R. Youn, Timothy G. Gutowski, ELSEVIER, Composites: part A, life cycle energy analysis of fiber-reinforced composites

Chapter 23
Recycling Technologies of Carbon Fiber Composite Materials

Yoshitaka Kageyama

Abstract With the expansion of the global carbon fiber market, interest in carbon fiber recycling increases. Especially, concern for the environment, both in terms of limiting the use of finite resources and the need to manage the waste disposal, has led to increasing pressure to recycle materials at the end of their useful life. Therefore, the shift from landfill (current main waste disposal method) to material recycling or thermal recycling is strongly expected. The most important process in recycling technologies is to remove matrix resin and to take out carbon fibers out of carbon fiber composite. In this chapter, the present status of recycling technology development in Japan is outlined. In addition, the environmental and economic loads are introduced by comparing the quantity of LCI and CO_2 emission between virgin carbon fiber and recycled carbon fibers.

Keywords Recycled carbon fiber • CFRP recycling • LCI (life cycle inventory: one of LCA analyses, making the output detailed statement about the environmental impacts) • Recycling technology • Pilot plant

23.1 Introduction

The global demand of carbon fiber was approximately 40,000 metric ton per year in 2013. It is predicted to rise by 15 % over the next 10 years by being used for aerospace, automobile, wind turbine markets, and so on [1]. Accordingly, the amount of waste carbon fiber in the form of offcut in the manufacturing process, out-of-life rolls of prepreg, and end-of-life components is estimated to be significant levels. Therefore, recycling technology becomes very important year by year. Not only indispensable the recycling is for the disposal of CFRP waste but also strongly required by customers for the disposal of CFRP waste.

Y. Kageyama (✉)
Otake Research Laboratories, Carbon Fiber Development Center, Mitsubishi Rayon Co., Ltd., 20-1 Miyuki-Cho, Otake-shi, Hiroshima 739-0603, Japan
e-mail: kageyama_yo@mrc.co.jp

Society of F.S. Technology, Japan (ed.), *High-Performance and Specialty Fibers*,
DOI 10.1007/978-4-431-55203-1_23

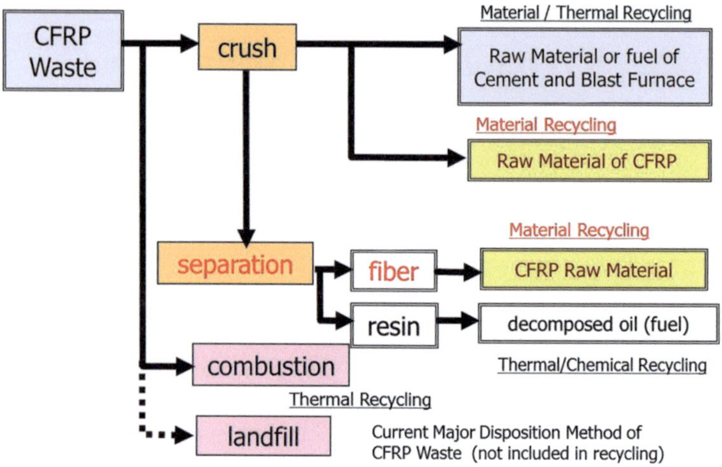

Fig. 23.1 Recycling processes for CFRP

23.2 Classification of Carbon Fiber Recycling Methods

There are some routes for dealing with carbon fiber-reinforced composite (CFRP) waste, material recycling (e.g., powdered fillers, chemical products, and fibers), thermal recycling (combustion with energy recovery), and landfill. These are summarized in Fig. 23.1. As the disposal of CFRP waste has a lot of technical problems, composite waste has been disposed of in landfill in Japan to date. However, the concern for the environment, both in terms of limiting the use of finite resources and the need to manage the waste disposal, has led to increasing pressure on recycling materials for the end of their useful lives. From such a background, many research groups in Japan have been actively researching and developing CFRP waste recycling methods regardless of industry, government, or academia-driven programs.

23.3 Comparison of CFRP Recycling Technologies

There were 197 technical papers (including conference proceedings) regarding CFRP recycling technologies over the past 5 years (2009/1–2014/4) by search engine of JICST (Japan Information Center of Science and Technology). Among them, 135 papers were released by the research institutes in Japan. Moreover, more than half of those were about carbon fiber recovery from CFRP. Figure 23.2 shows the ranking according to the research institute. The following are brief contents about R&D of some research groups.

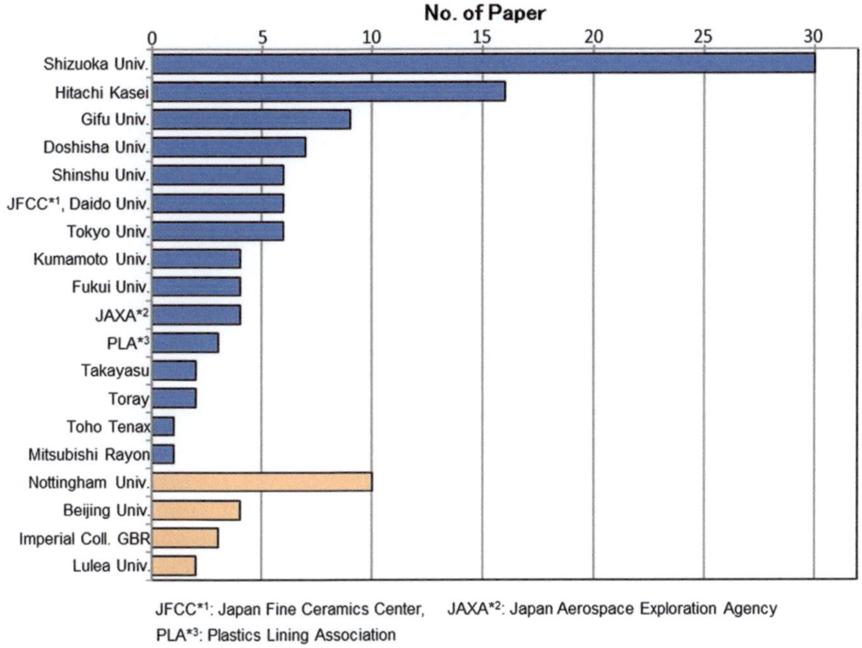

Fig. 23.2 The number of paper on CFRP recycling according to author group

1. Shizuoka University

   Okajima et al. at Shizuoka University have been researching for the decomposition and dissolubility of epoxy resin and the recovery of carbon fiber in CFRP by using supercritical and subcritical fluid. According to their report, the strength of CFRP using recycled carbon fiber by this method was closed to that of virgin CFRP [2].
2. Hitachi Chemical

   Shibata et al. of Hitachi Chemical have developed a CFRP recycling technologies using decomposition of cured epoxy resin under ordinary pressure. Carbon fiber was recovered from CFRP waste by resolving epoxy resin with tripotassium phosphate as a catalyst and benzyl alcohol as a solvent at 200 °C for 10 h. The recycling energy consumption could be reduced by approximately one-fourth [3].
3. Carbon Fiber Recycle Industry and Gifu University

   Itazu et al. have developed a thermal recycling technology. The objective of their research is to optimize the two-staged thermal treatment system, which consists of a carbonizing furnace to thermally decompose the plastic components in the CFRP thermally and an oxidizing furnace to burn the residual charred matrix resin on carbonized carbon fibers [4].

4. Shinshu University

   Mizuguchi et al. discovered the appearance of significant catalytic effects of oxide semiconductor in the 350–500 °C. They applied this phenomenon to the recycling CFRP. The point of this technology is to use thermal activation of oxide semiconductors at high temperature to decompose epoxy resin. This is a 100 %-dry process without any damage of carbon fiber [5].

5. Japan Fine Ceramic Center (JFCC) and Daidou University

   Wada et al. have developed a thermal recycling technology using superheated steam. Especially, superheated steam was effective for decomposing matrix resin when the matrix resin was polyamide 66. They also optimized process conditions to keep carbon fiber strength and to get good adhesive property between carbon fiber and matrix resin [6].

On the basis of these search results, the main recycling technologies are summarized in Table 23.1. Current process conditions, advantages, disadvantages, etc. are presented.

As follows in the next section, Japan Carbon Fiber Manufacturer Association (JCMA) tried small-scale experiments for screening the recycling methods in the New Energy and Industrial Technology Development Organization (NEDO) project and finally adopted thermal decomposition method. The reason was that thermal decomposition was relatively easier method than other methods such as catalyst methods. Although carbon fibers could not be degraded by catalyst method, chemical decomposition rate of resin might not be sufficient and may need pressurizing in the reaction vessel. Furthermore, supercritical fluid was not so easy method that it may need complex process and operation. However, the problem may have been overcome by the progress of the later research and development.

23.4 JCMA Recycling Activities

JCMA serves as a role of leading the CFRP recycling technology development in Japan. Basic recycling technology was established in the NEDO project from 2000 through 2002. This project was sponsored by NEDO and Ministry of Economy, Trade, and Industry (METI). The title was "standardization of crushed recycled CFRP." JCMA members took part in the project and developed the basic technologies. Details of related technologies were as follows:

- Developed crush and classification technology of CFRP
- Decomposition/separation technologies of epoxy matrix CFRP
- Technology converting from intermediate processed material to milled carbon fiber

After the basic research as mentioned above, recycling technology verification project was carried out sponsored by METI (2006–2009). The title was "R&D of production energy reduction in carbon fiber manufacturing," and the object was to

Table 23.1 Comparison of recycling technologies

Recycling technologies	Thermal decomposition	Thermal decomposition using superheated steam	Solvent depolymerization under ordinary pressure	Supercritical/subcritical fluids depolymerization	Thermal activation of oxide semiconductor
Company/Univ.	JCMA CF Recycle Industry Gifu Univ. Takayasu	JFCC Daido Univ. Doshisha Univ.	Hitachi Chemical	Shizuoka Univ. Kumamoto Univ.	Shinshu Univ.
Resin species	All (nontoxic)	All (nontoxic)	Ester-based compound	Ester-based compound	All resin
Recycled CF product shape	Carbon fiber (short-length fiber, milled fiber)	Carbon fiber (long-length fiber)	Carbon fiber recycled resin	Carbon fiber recycled resin	Carbon fiber
Process temperature	500–700 °C	500–700 °C	100–200 °C	Super., 250–300 °C, sub., 300–400 °C	350–500 °C
Pressure	0.1 MPa	0.1 MPa	0.1 MPa	Super., 5–10 Mpa, sub., 1–4 Mpa	0.1 MPa
Atmosphere gas/solvent/catalyzer	Nitrogen, air	Superheated steam	Benzyl alcohol/K_3PO_4	Methyl alcohol, water	Air
Process scale	JCMA, 1,000 t/year (feasibility study), CF recycle industry: 300 t/year (pilot scale)	JFCC: pilot scale	Pilot scale	Laboratory scale	Laboratory scale
Advantages	Relatively low-cost high-throughput rate no use of chemical solvents readily scalable	Relatively low-cost high-throughput rate no use of chemical solvents readily scalable	Good fiber property retention recovery of resin	Excellent fiber property retention no gaseous emissions recovery of resin	Rapid reaction rates: 10 min. no use of chemical solvents
Disadvantages	Degradation by thermal condition chosen CO_2 emission	Degradation by thermal condition chosen CO_2 emission	Wet process (need recycling solvent) long decomposition time, 10 h. Need catalyst	Need complex devices low processability (batch process) difficult scalable	Need to remove catalyst

Note: Edited by author with reference to other materials

verify and enhance the basic recycling technologies in the pilot plant. The project activities were as follows:

- Construct pilot plant and verify basic recycling technologies
- Establish recycling system
- Research market of recycled carbon fiber

The project expectation effect was as follows:

- Promote environmental protection
- Create new market using recycled carbon fiber
- Make sure save energy effect by establishing recycling system

Major carbon fiber manufacturers (Toray, Toho Tenax, Mitsubishi Rayon) in Japan succeeded the Omuta pilot plant which JCMA operated and continued the development jointly. The companies have developed further reduction of recycling energy. Moreover, they have developed the utilization technology of not only recycled milled carbon fiber but also recycled chopped (discontinuous long) carbon fiber, at the moment.

23.5 JCMA Recycled Carbon Fiber Pilot Plant

Figure 23.3 presents JCMA recycled carbon fiber pilot plant process flow. The plant is located in Omuta City, Fukuoka Prefecture. The pilot plant has the facilities necessary to recycle carbon fiber from CFRP waste. The most important process is

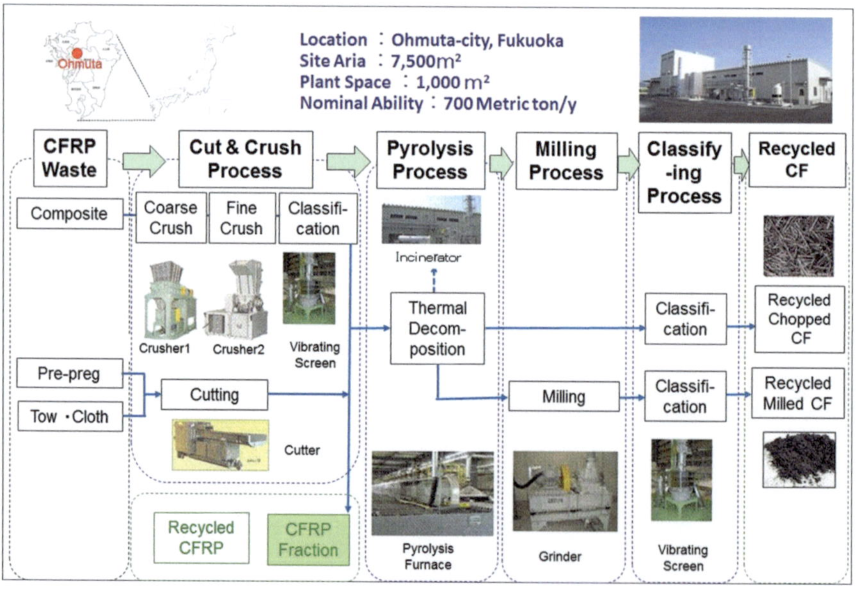

Fig. 23.3 JCMA recycled carbon fiber pilot plant process flow (Note: Edited the METI document [7] by author)

pyrolysis that removes matrix resin from CFRP. In the basic recycling technologies established by JCMA, pyrolysis of matrix resin was implemented under 500–700 °C in nitrogen for around 1 h. Nominal production capacity is about 700 metric tons per year [7].

23.6 Effect of Carbon Fiber Recycling on Environmental Impact

Virgin carbon fiber production energy was reviewed by JCMA in 2008, and the energy level was 290 MJ/kg [2]. Recycling energy, which is consumption energy for carbon fiber recycling, was also evaluated. JCMA estimated LCI of recycled milled carbon fiber and virgin milled carbon fiber as shown in Table 23.2. That was 48 MJ/kg. This value is considered as one of environmental impacts and the sum of consumption energies through the recycling unit processes (e.g., transportation of waste, cut and crush, pyrolysis, milling, screening). This is a first evaluation using an actual recycling process in the pilot plant. Regarding CFRP production energy, it is totally dependent on the carbon production energy. When recycled fiber is used, the CFRP production energy is decreased not considering the retroactive accumulation of production energy for the upstream processes of recycled carbon fiber. The production energy of recycled carbon fiber is approximately 15 % of that of virgin carbon fiber. CO_2 emission is another environmental impact. The major advantage is reduction of waste landfill. Additionally, new carbon fiber market is expected using recycled carbon fiber. On the other hand, the disadvantage is emission of CO_2 by matrix resin decomposition [7].

23.7 Properties of Recycled Milled Carbon Fiber

The specification of recycled milled carbon fiber at Omuta pilot plant is shown in Table 23.3. Additionally, a result of compound test (carbon fiber content, 30 wt.%; matrix resin, polycarbonate) using recycled milled carbon fiber is presented in Fig. 23.4. The mechanical and electric performance of recycled milled carbon fiber compound was equal to or higher than that of virgin milled carbon fiber compound or had further performance. It is suggested that recycled milled carbon fiber shows sufficient performance [7].

Table 23.2 LCI and CO_2 emission of recycled milled carbon fiber and virgin milled carbon fiber

Item	Recycled milled CF	Virgin milled CF
Consumption energy (MJ/Kg)	48	290
CO_2 emission (CO_2·kg/kg)	3.1	22.5

Table 23.3 Specification of recycled milled carbon fiber

Specification		CFRP waste based	CF waste based
Average fiber length (mm)	Nominal	0.06–0.13	0.08–0.15
	Weight	0.10–0.33	0.12–0.40
Bulk density (g/cm^3)	0.34–0.52	0.23–0.50	

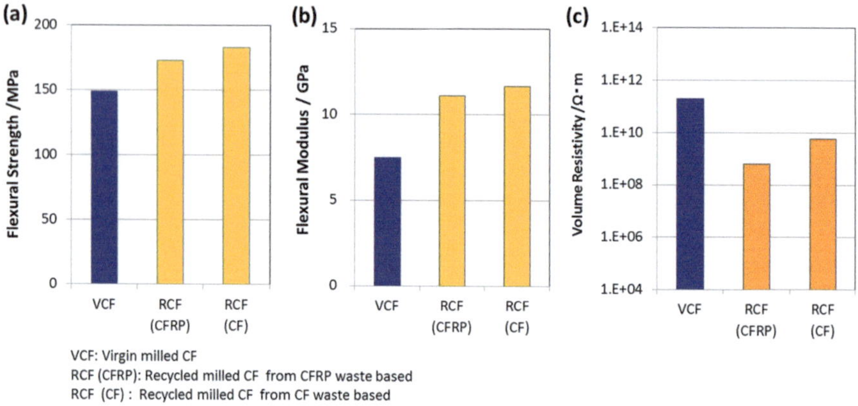

Fig. 23.4 Mechanical and electronically properties of recycled carbon compound. **a** Flexural strength, **b** flexural modulus, **c** volume resistivity; CF content, 30 wt.%; matrix resin, polycarbonate (Source: METI document [7])

23.8 Future Tasks

Researches (by many groups driven by industry, government, or academia) have been conducted and are ongoing. Moreover, some pilot-scale processes (e.g., JCMA pilot plant) have been evaluated. However, many problems will be piled up before CFRP recycling spreads among the society and gets on track. Future tasks considered to be necessary are as follows:

- Securing definite applications
- Reduction of CFRP recycling cost
- Recycling process optimization and scale-up to production
- Quality control and quality assurance of recycled carbon fiber
- Post processing of recycled carbon fiber to new product forms
- Management of collecting CFRP wastes

The CFRP wastes contain a number of different sources of materials that are typically classified to two categories: in-plant and in-service including marketed and utilized products having been used. In-plant materials generally have a clear identification of their sources and history. On the other hand, in-service materials' histories are not well documented. Especially, environmental history (e.g., temperature, humidity, ultra violet ray, mechanical damage, rain, erosion, contamination, etc.) is not clear. The application and quality assurance of recycled carbon fiber seriously depends on this point.

If hundred percent of CFRP waste are converted to recycled carbon fibers as material recycling, it must be the best thing but is deemed impossible. The other recycling method such as thermal recycling should be combined with material recycling method.

23.9 Conclusions

An overview on recycling technologies of carbon fiber composites was presented. Large quantities of carbon fiber will be disposed of as CFRP composite wastes in the near future. Environmental and economic factors have created the interest in carbon fiber recycling. Many R&D programs have been conducted and are ongoing; some pilot-scale processes (e.g., JCMA pilot plant) have been verified, and the effect of carbon fiber recycling on environmental impacts such as LCI and CO_2 emission was evaluated.

Currently, there are several significant recycling programs in Japan, for example, the recycling method of high-volume CFRP waste recycling programs for automotive applications and optimizing the recycling process options and quality control. Key point for CFRP recycling business success relies absolutely on early fulfillment of future tasks as mentioned above.

References

1. O. Tsutsumi, PAN based carbon fiber, J. Jpn. Plast., (1), 98–102 (2014)
2. S. Okajima, T. Sako, CFRP recycling using subcritical and super critical fluids. J. Jpn. Soc. Mater. Cycles Waste Manag. **24**(5), 364–370 (2013)
3. K. Shibata, CFRP recycling technology using depolymerization of epoxy resin under ordinary pressure. J. Jpn. Mater. Cycles Waste Manag. **24**(5), 358–363 (2013)
4. H. Itazu, H. Kanki, H. Moritomi, Recycling of carbon fiber using two-stage thermal treatment system. J. Jpn. Mater. Cycles Waste Manag. **24**(5), 371–378 (2013)
5. J. Mizuguchi, Y. Tsukada, H. Takahashi, Complete recovery of reinforcing fibers from FRP and their recycling with thermally activated oxide semiconductors. J. Jpn. Mater. Cycles Waste Manag. **24**(5), 379–388 (2013)
6. M. Wada, M. Tanaka, K. Kawai, S. Kitaoka, H. Taira, T. Suzuki, Recovery of carbon fibers from CFRP and surface modification using superheated steam. J. Jpn. Mater. Cycles Waste Manag. **24**(5), 389–394 (2013)
7. METI website http://www.meti.go.jp/policy/tech_evaluation/e00/03/h21/356.pdf

Part VI
Nonwovens

Chapter 24
Current Status and Future Outlook for Nonwovens in Japan

Osamu Yaida

Abstract More than half a century has already passed since the production of nonwovens began in Japan. In January 1996, the Ministry of International Trade and Industry (presently the Ministry of Economy, Trade and Industry) modified the contents of production dynamics, and statistical surveys of nonwovens have come to be independently reported.

Up to 2000, the nonwoven industry in Japan had been developing steadily. After the industry experienced a slowdown, it began to grow again and marked historical records consecutively in 2007 and 2008. However, the industry stopped growing and decreased in 2009 due to simultaneous declines of the world's economies. At present, there are certain signs of a recovery, but people engaged in the industry are uncertain whether the business will continue to grow in the future.

Keywords Nonwovens • Technical textiles • Web forming • Web bonding • Spunbonding • Melt blowing • Spunlace • Needle punching • Wet laying • Dry laying • Thermal bonding • Chemical bonding

24.1 Definition of Nonwovens

According to ASTM D-1117-80 of the USA, it is defined as a fibrous structure manufactured by having fibers mutually adhered or entangled or by both, mechanically, chemically, or by using a solvent or by a combination thereof.

ISO 9092:1988 defines it as follows: A manufactured sheet, web, or batt of directionally or randomly oriented fibers, bonded by friction and/or cohesion and/or adhesion, excluding paper and products which are woven, knitted, tufted, stitch bonded incorporating binding yarns of filaments, or felted by milling. The fibers may be of natural or man-made origin. They may be staple or continuous filaments or be formed in situ. Notes: (1) To distinguish between paper and wet nonwovens, nonwovens are defined as fibers having a length-diameter ratio (aspect ratio) exceeding 300 by 50 % or more or as fibers exceeding an aspect ratio of 300 by

O. Yaida (✉)
Adviser, All Nippon Nonwovens Association (ANNA), Osaka, Japan
e-mail: o_yaida@sutv.zaq.ne.jp

Society of F.S. Technology, Japan (ed.), *High-Performance and Specialty Fibers*,
DOI 10.1007/978-4-431-55203-1_24

30 % or more, with a density of 0.40 g/cm^3 or less. In all cases, chemically decomposed vegetable fibers are excluded.

After reviewing ISO 9092-1988 mentioned above, because of its complicated and ambiguous definition, the second edition of ISO 9092-2011 defines it as follows.

Nonwovens are structures of textile materials, such as fibers, continuous filaments, or chopped yarns of any nature or origin, that have been formed into webs by any means and bonded together by any means, excluding the interlacing of yarns as in woven fabrics, knitted fabrics, laces, braided fabrics, and tufted fabrics.

Note: Film and paper structures are not considered as nonwovens.

Simply speaking, however, nonwovens may be regarded as a fibrous sheet manufactured by entangling or adhering fibers by some means. It is a mutually interdisciplinary field existing in and among the fields of film, leather, paper, weaving, knitting, and wool felt.

24.2 Manufacturing Method of Nonwovens

Manufacturing methods of nonwovens are roughly divided into two processes: web forming and interfiber adhering/bonding. In particular, in web forming, it is divided into dry-laying method (producing in air) and wet-laying method (producing in water).

24.2.1 Web-Forming Method

Web forming is classified by fiber length as follows:

Fiber length: 6 mm or less

Wet-laying process and dry-laying process (synthetic fiber, pulp):

Fiber length: 15–100 mm

Dry-laying process:

Fiber length: Continuous

Spunbonding, melt blowing, and others

Due to the web fiber array, consequently, there is a difference in the characteristic values (tensile strength, heat shrinkage, etc.). The random laid web is most preferable, but the parallel laid web is generally used depending on the application where the difference between machine direction and cross machine direction is not important.

In the wet-laying process, the fiber array is random in papermaking process, but, in spunbonding process, the characteristic values are not always equal between the machine direction and the cross direction.

In the dry-laying process using staple fibers, the random laid web is preferred; however, web uniformity is impaired due to the crimp action of fibers when the fibers are departed from the metallic wire of the carding machine and scattered into air in the screen at the suction, and in the case of low-weight web, the product width is broad, and the characteristic value gets stronger in the cross machine direction (CMD), but since the fiber is expanded by the drafter or in the next processing step in the machine direction (MD) and in the final product, the aspect ratio is sufficient, and the dry-laying process is widely used.

24.2.1.1 Wet-Laying Process

The manufacturing equipment in wet-laying process is based on the papermaking technology and is modified and revised depending on the raw materials and applications of products. In the process, raw materials, that is, short fibers and pulp, are uniformly dispersed in water, and the prepared fiber suspension is filtrated on the various types of wire screen belts; then, a thin sheetlike web is formed. Since fibers are dispersed in water, the usable fiber length is limited, and generally it is 10 mm or less, but 20–30 mm may be applicable. Formers include the cylinder mold wire mesh type and the inclined wire conveyor type capable of web forming at low-suspension concentration in order to form a web with comparative long fibers. The formed web is processed such as dewatering and so on and dried by a dryer, and normally the webs are bonded through these processes. The sheet characteristics vary significantly depending on the web-forming method, the web-bonding and web-adhering method, and the drying method.

24.2.1.2 Air-Laying Process

It is one of web-forming processes of dry-laid nonwovens. Pulp sheet, short cuts of fiber with a length of 3–12 mm, synthetic fibers called chops, and other short cut fibers are mechanically opened into single fibers in dry-laying process, and webs are continuously formed on wire mesh running in air as medium and dried to make nonwoven fabrics by employing various adhering methods.

This process was first started by the development of "dry-laid pulp nonwoven" manufacturing process using pulp as the raw material (around 1965), and the demand is high, and dry-laid pulp type is now a representative name of air-laid nonwovens.

Dry laid pulp nonwovens is most excellent in water absorption and oil absorption capacity, and owing to the features of the pulp fiber that is lowest priced among fibers, the bulky and soft touch (apparent density 0.04 to 0.09g/cm^3) is characteristic.

It is used around the world, as absorbing materials for sanitary napkin or other hygienic products and wipes. In Japan, it is mainly used as a cigarette filter, a cooking paper, a paper towel, a food tray mat, sanitary goods, an inkjet printer, adeodorant and aromatic volatile products, and other household goods. Still more, the air-laid nonwovens are composed of short cut or chopped fiber of 100 % synthetic fiber which is called "dry short-fiber synthetic fiber air-laid nonwoven," and its commercial productions began around 1980. These nonwovens are used as automobile filter, insole, wipes, molding, and others.

Webs formed by the air-laying method are high in air contents and are excellent in liquid-holding property, filtering performance, air permeability, touch, and softness, and their pores can be filled with functional powder such as SAP (superabsorbent polymer), activated carbon, and other powders, not to spoil their own functions. The demand field is expanding in the disposable products used for absorption, holding, filtering, evaporation functions of liquid, gas, drip, waste liquid, excrement, and others.

Japan is the first to establish the air-laying method and to start commercial productions of air-laid nonwovens. The manufacturing method of dry-pulp nonwovens was inaugurated by Honshu Paper of Japan, after studies on dry papermaking and cigarette filters using wood pulp, for the first time in the world (Honshu Method 1965), and the company started commercial productions of dry-pulp nonwoven products (1967) (the technology is presently inherited to Oji Kinocloth Co., Ltd. a group company of Oji Paper Group). Other representative methods include Kroyer method of Denmark (presently called M&J method) and Dan-Web method, and together with Honshu method, these are known as the world's three major air-laying methods. Commonly in these methods, pulps are opened by using hammer mill or refiner, sent to a web-forming machine by airstream, and passed through a screen, and a web is formed. Other methods are also patented, for example, a pulp sheet is hooked by a picker rotor and opened, and opened fibers are formed on a net as a web.

Recently, it is a characteristic attempt to introduce only the head of the air-laying process combined with other processes (spunlacing, melt blowing, spunbonding, etc.) to design the equipment and realize multiple functions by combining manufacturing processes such as "high strength + hydrophilic + original hand feeling." Ever since, Honshu has newly developed the TDS process capable of combining high-weight (about 3500 g/m^2) and thick nonwovens (about 40 mm), high blending of powder, and other materials. Furthermore, they developed the air-laid composite type laminated with other low-weight nonwovens composed of long fibers such as rayon and others on one side or both sides. An original field using air-laid nonwovens is built up.

24.2.1.3 Dry-Laying Process

This process is the oldest method; this is still the main process together with spunbonding at the present. Using staple fibers for spinning (fiber length usually

25–76 mm), a conventional spinning machine is used to form nonwoven webs. By the introduction of synthetic fibers, transfer from wool-spinning machine has been promoted from the viewpoint of productivity.

In nonwovens, basically, the synthetic fibers are mainly used, and the machine width is at least 1.5 m or more, and presently carding machines of 2.5–3.5 m in width are used.

A spinning machine mixes raw materials, opens them, bundles webs, twists thinly and gradually, and finally forms spun yarns. In nonwoven processes, raw materials are mixed, opened, and formed into webs, and the subsequent process is directly linked to the bonding process between fibers. At the present, the mainstream is the broad-width machine with high-performance and high-speed production.

24.2.1.4 Spunbonding Method

Spunbonded nonwovens are manufactured by direct coupling of the spinning, drawing, and opening processes with the accumulating process and are significantly different from the staple fiber nonwovens.

Initially, spunbonded nonwovens were patented by DuPont in 1956, further developed, and enterprised. The spunbonded nonwoven "Reemay" was launched by them in 1965, and later in the 1970s, American and European synthetic fiber makers and chemical makers and Japanese synthetic fiber, chemical, and petroleum makers started commercial productions one after another.

The manufacturing method of spunbonded nonwovens (web-forming method) is based on the usual synthetic fiber spinning method and combined with high-speed fiber drawing method such as high-speed airstream. By the way, the process of fiber spinning directly combined with web forming is recently so-called spun-melting process which includes melt-blowing method, flash spinning, and in addition spunbonding. In addition, in the bonding process of the spun web, new methods have been developed, including the thermal-bonding method, mechanical interlacing by needle punching, adhesive agent bonding with resin, and water entangling.

In the aspect of the spunbonding process, many kinds of fibers are used, such as polyester, nylon, polypropylene, and their composite materials.

In the history of production amount of synthetic filament nonwovens, a drastic expansion has been achieved.

24.2.1.5 Melt-Blowing Method

In the 1950s, the US Navy Research Laboratories studied filters of high performance and started basic researches of ultrafine fibers, and in the middle of the 1960s, Exxon began to study melt-blown nonwovens by using polypropylene resin, but Exxon was specialized as a resin supply company and was divided into a company promoting the business by introducing the technology of Exxon and a company promoting development by the own technology.

In the manufacture of melt-blown nonwovens, the polymer is spun and blown from the outlet of a spinning nozzle with high-pressured airstream of high temperature, and the blown fibers are opened and collected on a capturing conveyor.

Its features include characteristics, not obtained in ordinary nonwovens, such as softness, thermal insulation, and high filtering performance because of the microfibers. At present, in the melt-spinning technology of synthetic fibers, a filament of 0.1 dtex is known as the thinnest filament, but the fiber thickness of the melt-blown ranges from submicron to several microns (ordinary fiber diameter 10–30 μm).

The problem was the low productivity in the initial stage of the development, the melting flow rate (FMR) of raw polypropylene resin was low, about 50–100, but recently a high rate of about 1000 is reported. (FMR is an index of start of flow of melted resin in heated state; a higher value shows a better resin fluidity and the productivity.)

Raw materials are mainly polyolefin derivatives (including polypropylene), nylon 6, polyester, low melting point PBT (polybutylene terephthalate), etc. Polyurethane materials are also used partly. The demerit of melt-blown nonwoven is an insufficient strength of sheet, because of its undrawn ultrafine microfibers, and some reinforcement is needed. Since the fiber diameter is very thin, in the case of low fabric weight, for example, at 5 g/m^2, the number density of fiber accumulation is very high; therefore, it is used for such high-performance filter at low fabric weight.

24.2.1.6 Flash Spinning Method

In the 1950s, DuPont developed this technology, and the products using polyethylene resin (TYVEX®) based on this technology were commercialized in the 1960s. Thereafter, Asahi Chemical Industry originally developed a technology and commercialized products (LUXER®). In 1995, in Japan, a joint venture of DuPont and Asahi Chemical Industry was organized as DuPont-Asahi Flash Spun Products Co., Ltd. and started manufacture and sales.

Flash means a sudden gasification phenomenon of liquid polymer solution, and flash spinning is a kind of dry spinning, and a fiber-forming polymer and a special solvent are used basically. The solvent is a liquid and solvent at high temperature and pressure but is a gas and nonsolvent at ordinary temperature and pressure.

In flash spinning, the polymer is dissolved with the solvent to form a uniform solution at high temperature and pressure, and this solution is separated in phase before flashing from the nozzle. This phase-separated solution is flashed from the nozzle into the atmosphere of room temperature and pressure; then by suddenly gasifying and drawing, because of the low boiling point of the solvent, the polymer is drowned and solidified, and a unique structure composed of mesh-like continuous ultrafine filaments is formed.

The jet flow containing the mesh-like continuous filaments is injected against a collision plate and spread. The spread mesh-like ultrafine filaments are charged by corona discharge and electrostatically charged in order to collect stably and uniformly the web on the moving conveyor net. The gas including the solvent used in this process is collected and recovered. The web is sent to a next bonding process and formed into a product. Depending on applications, improvements are needed in print, adhesion, and others, and the specific surface treatment is carried out by additional converting process. Where permeability is needed for such as house wrap, the condition is changed in the bonding process, and the permeability is assured.

24.2.1.7 Tow Opening Method

When producing nonwovens of synthetic filaments, the direct spinning method is employed usually, but this method is to form a web from synthetic filament bundle (tow). In Japan, the process was developed by Unisel Co., Ltd.

24.2.1.8 Film-Drawing Method

In burst fiber method, the raw material polymer mixed with the foaming agent in an extruder is melted and kneaded, and the polymer containing gas is extruded from the slit die and is simultaneously drawn quickly. Then, the compressed gas in the polymer is swelled and ruptured (foaming: bursting) at the die outlet, and extreme micromesh-like continuous fibrous sheet (burst fiber) is formed. This burst fiber sheet is laid up, spread, and finished into a product.

Film split method is said to be developed and commercialized as a result of years of research and development by Polymer Processing Research Institute, Ltd., Japan. The film is drawn, split, and spread at high speed, same as in the burst fiber method above, and the split films are plied up orthogonally to each other in machine direction and cross machine direction. This technology is inherited by JX Nippon ANCI Corporation and commercially produced and is widely sold as Claf®; also the technology is licensed to Amoco of the USA.

24.2.1.9 Electro-spinning Method

Basically, this is a solution spinning method by using electrostatic force. Electric charges that are induced and accumulated on the solution surface by a high voltage repel each other to resist the surface tension; when the electrostatic force exceeds a limit value and the repellent force of electric charges exceeds the surface tension, a jet of charged solution is injected and spun.

The injected jet is larger in surface area as compared with the volume; therefore, the solvent is evaporated efficiently, and the charge density elevates as a result of decrease of volume, and it is divided into a further thinner jet. In this process, a nonwoven of uniform filaments in the diameter of tens to hundreds of nanometers is formed on the collector. In principle, this is a traditionally known technology, but recently application in filters and other examples are realized, and the production technology for practical use has been intensively studied in many research laboratories and companies.

24.2.2 *Web-Bonding Method*

Various methods of web bonding have been developed, and recently the use of thermal bonding and spunlacing is increasing from the viewpoint of environment and energy saving, but along with the advancement of automobile industry and environmental improvement, the needle-punching process is also dominating.

24.2.2.1 Chemical (Binder) Bonding

Earliest in history, it is still widely used at the present. Tending to be decreased in use gradually from the viewpoint of product safety, hand feeling, environmental pollution, and energy-saving policy, it is being replaced by thermal bonding which is easier in process. Chemical-bonding method includes dipping method, foam-bonding method, and spray-bonding method.

Latex adhesive is mainly used for the bonding. Evaporation of moisture is needed except for powder-bonding system. In the dipping method, it is important not to disturb the web uniformity when dipping the web in the adhesive, and various methods are invented. In the method of holding the web with wire mesh and dipping in latex bath, clogging of wire mesh with the binder is a problem. In the foam-bonding method, the web is dipped with binder latex so as to be coated from one side or both sides, but in the case of heavyweight web, the inner layer is hardly immersed, but it is effective because it is less in contamination of binder or residual liquid. In another method, the strength of the web is increased by coating only one side of the web with the foam binder. To increase product durability, a dryer or heat-treating machine is needed to dry and cure the binder. In print-bonding method, the binder resin is printed and bonded partly the web, and the products are widely used in disposable application. The spray-bonding method is effective for production of products of bulkiness and surface fluffiness-preventive property, which is produced by spraying and bonding both sides of the web by spray nozzle. The prevention of the scattering of binder was difficult, but the countermeasures are recently improved, and efficient scattering is realized. The products are used in production of wadding for clothing, air filter, etc.

24.2.2.2 Thermal Bonding

Thermal bonding is used widely instead of chemical bonding method. Recently, along with the development of various thermal-bonding fibers, its use is expected to be increased more and more in the future, from the viewpoint of excellent hand feeling of the product, energy-saving, and environmental advantages. This is the method of bonding between fibers by heating and fusing lower melting portions, and two types are known as follows.

Using heat roll: Divided into the smooth type (contacting with the entire roll surface) and the protruding type (contacting partly with the embossed surface). The embossed type is more used.

Using hot air: Also known as air-through type. Sheath lower melting parts of sheath/core-type bicomponent fibers are fused and adhered by hot air, and bulky products can be produced. When a certain precision in thickness is required, the web is held, and the fibers are adhered between wire mesh nets having a clearance.

In the recent advancement in productivity, wider webs (more than 2 m) are produced, and in the case of low-weight webs, the uniformity of weight, density, and thickness in cross direction is necessary. Accordingly, various methods are developed to apply the pressure uniformly in the entire width. In the case of product width of 4–5 m, and the fabric weight is 50 g/m^2 or less, the gauge uniformity of rolls is very important. Usually, after pressing at room temperature, the uniformity is checked by passing a pressure-sensitive paper. Whether the roll-heating method is steam, oil, heat medium, or electric heat is selected depending on the fiber or final product.

24.2.2.3 Needle Punching

Traditionally used for needle-punched felt and recently owing to the improvement in the machines and the needles, this method is widely used in relatively heavyweight products. Recent machines are high in speed, low in vibration, low in noise, and possible in safety operation. In particular, in heavyweight products, in other bonding methods, bond strength between fibers is not sufficient, and interlayer abrasion is likely to occur. But the problems can be solved easily by needle-punching method. On the other hand, in lightweight products, it is not suitable because the uniform web array is disturbed.

By the development of new fork-type needles, other than plain-tone surface, cord-tone or pile-tone surface products can be also manufactured.

Further, in reducing the needle density (penetration needles/cm^2), the pre-punching is also used. A special type includes a cylindrical bed plate.

Generally, a needle-punching machine moves the needle board (the needle anchored board) by vertical vibration motion, and the vibration during motion may cause a problem, but recent machines have been improved to be stable and be balanced sufficiently if only put in place. For realizing a high-speed rotation

(2000–3000 rpm/min.), the stroke amount of vertical motion (cam eccentricity × two times) must be decreased.

Usually, in the production line directly coupled with the card in the dry-laying process, there are many cross laid webs, and the laid-up web is bulky for needle-punching type, and the needle stroke depth is about 75 mm, but when the number of times the punching increases, the thickness decreases consequently, and the finishing punch is about 40 mm in most stroke depth.

An ordinary needle has a triangular section, and each side has barbs, and when the barb interval is regular (2.1 mm), it is called a regular barb (RB). When the barb interval is half (1.1 mm), it is called a close barb (CB). A special interval (1.6 mm) is called a multi-barb (MB). Recently, special needles are developed to be applicable to new fibers such as high modulus of organic fiber, inorganic fiber, and metal fiber.

The important point in needle-punching process is the needling density per unit area. Various terms are used and a uniform term must be defined. Herein, it is called the needling density which means number of needle penetrations per square centimeter.

24.2.2.4 Hydroentangling (Spunlace Bonding)

This process was developed by DuPont, United States. Fibers are entangled by a water steam jet by injecting high-pressure water to the webs from nozzles, and the fibers are not damaged, and a fabric-like hand feeling is obtained without any adhesive agent. Since the expiration of DuPont's original patent, it is rapidly used. Although the equipment is costly, the productivity is high, and the market is expanding in the field of medical and wiper and other disposable products with unique hand feeling-like woven fabrics. As combination with other materials such as SM or SMS, it is used for such as filter medium, and the products composed of split fibers which are split and entangled by this process are used for precise wipers, abrasive bases, and filter materials. In addition, lace-like-patterned products can be produced. This process is used by combining with wet-laying or dry-laying manufacturing process, and it is also used with spunbonding processes.

Depending on the water pressure, the method is divided into low-pressure type (within 10,000 kPa) and high-pressure type (15,000–20,000 kPa).

By using one manifold stage, the fiber entanglement is insufficient, and several manifold stages are provided in the same number of boards of needle punching. Various characteristics are obtained depending on the web-holding mesh screen or wire net or the type of net conveyor. Recently, various patterned nonwovens are developed, that is, on the patterned suction drum, the fiber webs are entangled and simultaneously patterned to form the nonwoven.

At present, Honeycomb and Perfojet are the leading spunlace machine manufacturers in the world, and this kind of machine is highly demanded recently, and many manufacturers are releasing new machines, such as Courtaulds Engineering. More recently, instead of water steam jet, high-pressure steam jet was developed, and the webs are entangled by the steam jet and produced into bulky nonwovens, which is known as steam jet (SJ) type.

24.2.2.5 Stitch Bonding

Using spun yarn or filament, webs are interlaced by warp knitting. This method is developed in Eastern Europe, such as former East Germany and Czech Republic, and this is not classified into the category of nonwovens in the European Disposables and Nonwovens Association (EDANA), but it is classified in the category of nonwovens in Japan. Basically, this is knitting, and it seems to be different from other nonwoven manufacturing methods because the manufacturing line is stopped even if one interlacing yarn is broken.

Besides, in other methods, without using yarn, fibers are interlaced by needles only, just like needle-punching method. However, the needles of stitch-bonding method are needles used in knitting, and the interlacing strength is weak because many barbs are not provided in one needle, different from the needles used in needle-punching method.

24.3 Applications of Nonwovens

More than a half century has already passed since the nonwoven industry commercially started in Japan. Initially, nonwovens were mainly applied as durability substitutes for woven fabric, felt, leather, and others mainly in the clothing field and are later used in other original fields such as disposable hygiene and medical, and these original products are grown to occupy a major demand. As a result of original product development based on the porous and other features of nonwoven, various fibers, and manufacturing methods, the applications are expanding in all fields.

This advancement has been achieved by developing new materials based on efforts of material makers (fibers and resins) and new designs of facility or equipment. While making use of the features of nonwovens, in the competition of other materials, new developments will be further invented to meet with original needs.

The following explains the feature and history of development process in main applications with the latest information and the market trends.

24.3.1 *Protective Wear*

Protective wear clothes are made of polypropylene spunbonded, SMS, flash spun, polyethylene-coated spunbonded nonwovens, microporous film composite, aramid water-entangled or needle-punched nonwovens, and others.

The protective wear market includes nuclear related, clean room, food preparation, paint manufacture, protection from extreme cold or hot climate, heat or fire, protection bullets or mechanical danger, and protection from harmful microorganisms and chemicals.

In particular, when used in manufacture of semiconductor or pharmaceuticals, it is required to protect the environments from workers. In the infection-preventive measure arising as a problem by the incident of SARS, the role of medical-protective wears has been widely recognized and used. This trend has caused to increase the demand for protective masks.

In facility of nuclear power plants, working clothes exposed possibly to low-level radioactive contamination are provided as disposable wears and stored and sealed in special drums after use. In such applications, nonwoven products are used widely. The permeability and barrier properties are important.

24.3.2 *Medical Care*

In medical care products, products are explained as being classified into medical site and nonmedical site.

24.3.2.1 Medical Site

The use of the nonwoven in the medical and healthcare in Japan is increasing year by year, because of the following reasons:

- Improvement of the usability by increasing variations of items and sizes
- The price down by the increase of production volume with the market expansion and the increase of low-price imported goods
- Increasing the use of disposable or short-life products in medical site, because of the point of view of preventing nosocomial infection

The general characteristics required for medical nonwovens are as follows: bacteria barrier, moderate moisture permeability, water repellency, water absorption, waterproofing, lint-free, drapability, and durability in sterilization.

24.3.2.2 Nonmedical Sites

1. Cataplasm-base material

   Transdermal system is known as a modern cure therapeutic method, and the medicine permeates into the inner parts of the body by way of the skin. It is commonly known as the wet compress or poultice.

   The fiber product used in the cataplasm base was cotton flannel in the past, but the nonwovens are mainly used at present. The usable nonwovens include nonelastic type, lateral elastic type, and longitudinal-lateral elastic type.

   The medicine was formerly a kaolin cataplasm composed of kaolin as the base, mixed with medicine powders and with water. Recently a cross-linking reaction-type ointment is developed, and further the active ingredients are developed in a second generation (including flurbiprofen, ketoprofen, and indometacine as percutaneous absorbing agents), and the coating amount is reduced to less than half. As these result, the nonwoven base materials comes to be used widely.
2. Plaster medicine support-base material

   A therapeutic method of giving medication administration of medicine by way of the skin is highly expected as a new market. By direct administration from the skin of the affected area, the side effects of the internal organ absorbing medicine can be reduced. As the function of nonwoven base material, the medicine adsorption and compatibility are important. In the sealed type, composite with the film is necessary, and as the entire structure, the smaller size and thinner types are demanded.

24.3.3 Architecture

In the field of building and architecture, the nonwoven is used in waterproof roofing, house wrap for prevention of dew condensation, and indoor floor and wallpaper.

The process of waterproof roofing in the building is used in a roll profile of 1 m in product width, for the purpose of roof waterproofing of reinforced concrete building. Waterproofing of a general house (inclined roof) is a method using under-roofing material, and it is not called roofing.

In Japan, from the earlier days, binder-bonded nonwovens, composed of vinylon fibers (PVA fibers) made of its tow, have been used as the base materials. And this technology was exported to Europe and America. Generally, it is divided into new construction and repair, and in repair building, asphalt-melting method has not been used gradually to prevent from environmental deterioration effects due to the malodor and scatter of asphalt particles, and it has been attempted to avoid the harms by the construction method at normal temperature.

24.3.4 Civil Engineering

Nonwovens are very widely used in civil engineering, and the consumption is expected to be increased more henceforth. Mostly, civil engineering works are public construction works, and the quality of works is inspected by the government's board of audit. For the use of new materials, laboratories under direct governmental control are managing, and only the materials approved by the academy can be used in the design. Therefore, this point is radically different from other applications.

Textile products used in civil engineering are called geo-textiles, and including plastic products and others used in civil engineering works, they are called geo-synthetics. Functions of nonwovens include separation, draining, filtering, reinforcing, protection, and water sealing.

24.3.5 Vehicle

Nonwovens are widely used in vehicles. About eight million vehicles are produced a year in Japan, and a half of them is exported. As the overseas productions increase, the domestic consumption tends to be lower recently, but the nonwoven consumption per vehicle is increasing along with the increase in compartments using nonwovens and the increase in the area.

The demand in the automobile industry is in the direction to use nonwovens of easy recycling and lightweight. In the industry, a higher recycling rate of car materials and a higher fuel efficiency are demanded.

As compared with the textiles of the conventional specification, nonwovens were inexpensive and were handled as materials of low-quality level. By overcoming this traditional concept, success and growth has been brought in this field.

In the headlines, visual attraction is enhanced by the print design, Mari-fleece pattern, stitch bond, and other decorations, and the trunk surface material is enhanced in touch by Dilour/shirring process.

By the invention of artificial leather of high quality equal to that of natural leather, choices of materials are expanded in the seat, ceiling, and door panel. It is highly noteworthy that the advanced progress of the nonwoven technology is greatly contributing to the automobile upholstery field.

In the aspect of functions, recent topics are reduction of VOC. New surface materials are developed, such as those not generating VOC or others having functions to take up, adsorb, and decompose VOC.

24.3.6 Hygiene

The largest market of nonwovens in the hygiene industry is the diaper market for both infants and the elderly. The nonwoven and the SAP (superabsorbent polymer) perform synergetic effects and are producing various types of hygiene products.

Disposable diaper is composed of top sheet, acquisition distribution layer, absorbent layer, and water-shielding back sheet. Nonwovens are widely used in the top sheet, acquisition distribution layer, and back sheet laminated with film. The absorbent layer is made of fluff pulp and SAP (superabsorbent polymer), and the nonwoven is used for covering them. The water-shielding back sheet is required to have a texture feeling, and the film is laminated with the nonwoven. The most important growth is spunbonded or SMS for disposable diaper in hygiene field.

In diapers for infants, the pant type and the side gather type are highly reputed, and the consumption of nonwovens is increasing. In diapers for the elderly (for incontinence), the production is rapidly expanding.

Recently, by decreasing the pulp amount in the absorbing material and increasing the SAP, the moisture absorbing quantity is maintained, while the weight and thickness are decreased, in order to save the storage and transportation expense. The trend is the same for the adult market.

Lately, along with the increase of incontinence among elderly people, the use of diapers is estimated to be increase more and more.

24.3.7 Wipes

The wipe market is expanding owing to the function of the nonwovens. In particular, characteristics of microfibers such as enhanced dust-holding performance and liquid-absorbing performance are applied to various fields.

Both dry type and wet type are developed widely. The container ranges from cylindrical type to lunch box type, and the takeout port is developed in various shapes. Display storage and refill types are available. Many portable products are sold.

24.3.8 Filter

The filtration is the market conforming to the inherent function of nonwovens. In other materials except nonwovens, it seems difficult to maintain the low apparent density and design to stack up while giving a density gradient in the thickness direction.

On the other hand, in the recent development of nanofibers, ultrafine particles can be removed at a highest level, and a latent application development is expected in the filter medium based on nanofibers.

For the application of nanofibers, higher filtration efficiency than ULPA or HEPA filter is highly expected in the field of air filters. Thinness, low-pressure drop, large dust-holding capacity, and long filter life will be realized. In the liquid filter, invasion of bacteria and virus will be blocked effectively.

Utilization to filter by using/making use of photocatalyst is one of the technologies highly expected as the indoor VOC-preventive countermeasures.

When filtering, generally, dust particles of various sizes must be captured. In the case of a high-efficiency filter, in particular, first-coarse, dust particles are filtered, and the rest is a high-efficiency filter. Accordingly, a prefilter media are used in front of main high-efficiency ones, and the useful life of the high-efficiency filter is extended.

The manufacturing method is mainly chemical (spray) or thermal bonding of staple fibers. In fiber structure, thick fibers are used at the flow-inlet side, and thinner fibers are used at the flow-outlet side. For example, it is used for the cooking range hood in general household.

For industrial use, it is used in the fresh air intake part of a painting booth for automobiles and electric products, so that foreign substance may not get into the air for the painting.

Filter media of extremely high efficiency are known as HEPA and ULPA, composed of ultrafine glass fibers in general, which are zigzag pleated and formed the assemble structure. Recently, melt-blown products are developing. In the market of semiconductor chip manufacture, the ultrafine glass fiber paper was in the mainstream; also PTFE membranes are used recently. PTFE membranes are made of nanofibers, and it has an advantage that it is free from boron volatilization of the glass fibers.

When static electricity is applied to polyolefin fiber, the filter media have higher filter efficiency by dust-capturing force of static electricity *i*. When combined with this effect, a filter capable of high-capturing dust particles without raising the pressure drop is realized.

As the dust-holding feature of this filter, electrostatic-capturing force begins to decline in a certain time after start of use, but since the media pores become gradually smaller because of the dust clogging, and after a certain limit, the filter efficiency is enhanced by the action of mechanical capturing.

In particular, the electret filter using polypropylene melt-blown nonwoven is not only excellent in mechanical filtration efficiency by ultrafine fibers but also low in pressure drop and high filter efficiency owing to its electrostatic effect, and it is used in a household air purifier filter and so on.

A bag filter is used for prevention of dust scattering and collection of dust particles in the powder-conveying site, powder manufacturing plant, power plant, and others.

Depending on the method of separating air mixed with powder, it is divided into the filter air-flow systems, inside to outside type and outside to inside type. Fibers are mainly synthetic fibers, but glass or metallic fibers may be used depending on the purpose. Recently, since heat resistance is required, functional heat- or chemical-resistant fibers are used. For example, in the atmosphere of high temperature and harmful gas, suitable fiber materials are selected.

Application fields of liquid filters are roughly classified into general industrial, water purifier, and membrane use and classified by shape into cylindrical cartridge type, bag type, or box type. Usable nonwovens are dry-laid, wet-laid, and melt-blown type, which require safety, insolubility oil-resistant, and chemical-resistant properties. In particular, in the semiconductor field, the filter composition is required to have a more pure water characteristic, non-eluting a trace of additive used in the polymerization process of polymer.

24.3.9 Agriculture and Horticulture

The development of nonwovens in the field of agriculture is long in history. In Japan, in the early 1970s, Unitika Ltd. developed polyester spunbonded nonwoven, and it was first used in the weed barrier sheet, and the application in agriculture expanded. A second application was found in the inner curtain for greenhouse. It is because it is evaluated in the effects of thermal insulation, dehumidifying, and water drop prevention.

The materials have been mainly synthesized polymer compounds made from fossil resources such as polypropylene and polyethylene terephthalate, but recently it is demanded to change the direction from the global environmental problems such as "shortage fossil resources" and "refuse disposal problem."

In particular, in the field, an active use of bio-based biodegradable resins (polylactic acid, cellulose, etc.) is expected for solving the problems. Yet, at the present, the mainstream is polypropylene and polyethylene terephthalate materials, but biodegradable fibers and spunbonded nonwovens have come to be used widely in direct covering, mulch sheet, slope protection, and greening sheet.

Along with the development of biodegradable resins, the nonwoven application is promoted in the household, hygiene, and medical or automobile materials; the resins are intensively modified and reformed for practical use.

24.3.10 Artificial Leather

Artificial leather started in shoe materials and spread further in wide fields including bags, furnitures, gloves, apparel material-related precision machines, and abrasives. The nonwoven composing the base material is an important constituent element. At present, there are two major surface-making methods such as polyurethane film laying forming on the base material by dry process and a thin film layer forming on the surface of a foamed layer made by wet process.

Artificial leather achieves excellent properties such as touch and strength of surface state emulating natural leather in structure. Accordingly, in a layered structure of nonwovens using ultrafine fibers, a gradation structure is built so as to be similar to that of natural leather, and ultrafine bundle fibers are used as the outermost layer of the grain side. Also, by the raising process of ultrafine fibers, the artificial suede is realizing appearance and structure similar to those of natural leather.

On the other hand, along with the conventional demand increase, there is a mounting need for lightweight, breathability, humidity permeation, flame resistance, softness, antistatic, and water-repellent properties. By adding these functions not realized in natural leather, an advanced artificial leather is developed and promoted.

By using polyurethane together with water-based emulsion, leather in consideration of environment and ecology is accelerated.

Recently, technical levels are advanced in Korea, Taiwan, and China, and the technology is getting into the age of borderless and major competition.

24.3.11 Others

One of the recent best sellers is Kao's Quickle Wiper®. Static electricity generated in nonwoven is utilized advantageously for capturing hair and various dust, and it has become a phenomenal seller. The wiper itself is a staple fiber nonwoven reinforced with mesh. Initially, a long stick was attached to be used for cleaning the floor, wall, and ceiling, instead of a vacuum cleaner, and a one-hand operation product was developed later. Furthermore, combined with a detergent, products of strong cleaning performance are developed for removing stubborn stains, stickiness, or fine dust particles.

"Kairo" disposable pocket warmer is a Japanese unique product. This is a very simple warming tool as the result of the oxidation of iron powder and it can be used in clothes, shoes, or further stuck to clothes. Nonwovens are also used widely as shopping bags.

For electric appliances of high voltage and large capacity, electric-insulating materials excellent in heat resistance, corona resistance, and arc resistance are needed. These materials are generally known as mica tapes which need the reinforcing materials, such as polyester nonwoven, glass cloth, film, or aramid paper. While polyester nonwoven is lower in tensile strength than other materials, it has advantages such as softness and excellent elongation making adherence or less creases in the winding process and the superior impregnation of insulation resin.

Rechargeable batteries are used widely in mobile phones, laptop PC, or portable communication devices, expanding nowadays to hybrid vehicles, and the development competition is keen, aiming at higher capacity, higher output, or higher reliability. The nonwoven separator is manufactured in dry-laying, wet-laying melt-blowing, or combined processes thereof, while microporous film is also used. Nonwovens are used mainly in the nickel-cadmium or nickel-metal hydride batteries, and fine porous film is preferred in lithium batteries.

In the future, for improvement of electric characteristics of separators, in addition to the absolute requirement of prevention of short-circuiting, dense and homogeneous properties are demanded, and development of thinner separators having higher performance is expected.

Battery-related elements are specialties of makers in Japan, and a further growth is expected by global standardization and global located productions.

Glossary It is expected that the Japanese fiber industry develops mainly on technical textiles. For technical textiles, it is important to push forward the development of new high functional fiber and nonwovens.

Based on a technique and results of research provided so far in the fiber industry, it is thought that it is necessary to push forward a study and the development of the nonwovens.

Chapter 25
Bicomponent Polyester Fibers for Nonwovens

Masatsugu Mochizuki and Nobuhiro Matsunaga

Abstract Bicomponent fibers can be defined as "extruding two polymers from the same spinneret with both polymers contained within the same filament." The main objective of producing bicomponent fibers is to exploit capabilities not existing in either polymer alone. This technique is essential for manufacturing many highly functional nonwovens. Unitika, a pioneer in bicomponent polyester fibers for nonwoven fabrics, is the world's first commercial producer of bicomponent polyester fibers including both side-by-side type and sheath-core type. In this article, we focused on the Unitika's novel technologies and innovative products of polyester bicomponent fibers covering from polyethylene terephthalate (PET) to polylactic acid (PLA) in the history of nonwovens.

Keywords Bicomponent fibers • Conjugate fibers • Side-by-side • Sheath-core • Nonwovens

25.1 Introduction

Today's nonwovens can be called engineered fabrics. They provide specific functions such as absorbency, liquid repellency, resilience, stretch, softness, strength, flame retardance, washability, cushioning, filtering, bacterial barrier, and sterility. Fabrics are created for specific end uses, while achieving a good balance between product use-life and cost [1].

Nonwovens are made by many different processes. All have three general steps: web forming, web bonding, and fabric finishing. Each in turn can be done by a variety of methods. By combining the methods and choosing from an array of synthetic and natural fibers and their blends, nonwoven fabrics can be designed with properties tailored to many applications.

M. Mochizuki (✉)
Center for Fiber and Textile Science, Kyoto Institute of Technology, Kyoto, Japan
e-mail: mmochizuki@vega.ocn.ne.jp

N. Matsunaga
Unitika Ltd., Osaka, Japan

Society of F.S. Technology, Japan (ed.), *High-Performance and Specialty Fibers*,
DOI 10.1007/978-4-431-55203-1_25

The creation of a loosely held together sheet structure, usually by laying down of fibers, is called web forming. The fibers can be in the form of short lengths called staple or continuous length called filament. The web forming step can be done by one of three general methods: dry laid, wet laid, and spunbonded/melt blown.

In most cases, however, the web that is formed is too weak to be used. It requires further processing, web bonding. The individual fibers or filaments must, therefore, be tied together in some way, by gluing, thermally bonding, or mechanically entangling.

Nonwoven staple fiber processes, such as carded, air laid, needlepunched, spunlaced, and wet laid, use an assortment of synthetic and natural fiber types with diverse properties: hydrophilic and hydrophobic, rigid and flexible, thermoplastic and refractory, crimped and uncrimped. In the selection of fibers and their blends, the primary consideration is contribution to end product properties, processability, and cost. Polyester and polypropylene share leadership in the nonwoven staple fibers mix. Smaller quantities of rayon, nylon, and other specialty fibers, such as aramids and cotton, are used in nonwovens. The air laid, wet laid, and spunlaced nonwoven process also use wood pulp.

Increasing quantities of spinning grade polymer resins are used in melt-spun nonwoven processes, such as spunbonded and melt blown nonwovens. Polypropylene is the leading resin type used in melt-spun nonwoven processes, followed by polyester, high density polyethylene, and nylon.

25.2 History of Bicomponent Fibers

Bicomponent fibers can be defined as "extruding two polymers from the same spinneret with both polymers contained within the same filament." The term "conjugate fibers" is often used, particularly in Asia, as synonymous with bicomponent fibers.

The main objective of producing bicomponent fibers is to exploit capabilities not existing in either polymer alone. By this technique, it is possible to produce fibers of any cross-sectional shape or geometry that can be imagined. Bicomponent fibers are commonly classified by their fiber cross-section structures as side-by-side, sheath-core, islands-in-the-sea, and citrus fibers or segmented-pie cross-section types [2, 3].

DuPont introduced the first commercial bicomponent applications in the mid-1960s. This was a side-by-side hosiery yarn called "Cantrese" and was made from two nylon polymers, which, on retraction, formed a highly coiled elastic fiber. However, the crimp characteristic of the above-mentioned DuPont's side-by-side bicomponent fibers was not good and permanent, and thus, it was not commercially mass produced.

The first commercial application of sheath-core binding fiber "Heterofil" (ICI) consisting of a lower melting temperature sheath (Nylon 6, 220 °C) and a higher melting temperature core (Nylon 66, 260 °C) has been in carpets and upholstery fabrics. When 10–20 % "Heterofil" binder fibers are added to wool carpet followed by hot air-through, the wool carpet will improve to be lint-free and bulky.

In the 1970s, various bicomponent fibers began to be made by Japanese producers such as Unitika Ltd., or Chisso Polypro-Fiber Co. Ltd., in Japan. Unitika had targeted on bicomponent polyester fibers for special applications including nonwovens, because Unitika had not been a front-runner of polyester fiber production in Japan. Today worldwide, Japan and Korea led in bicomponent output with a total of 200 million pounds annually.

Unitika, a pioneer in bicomponent polyester fibers for nonwoven fabrics, is the world's first commercial producer of bicomponent polyester fibers including both side-by-side type such as "H38F," "38F," or "C-81" and sheath-core type such as MELTY® or CASVEN®. In this article, we focused on the Unitika's novel technologies and innovative products of polyester bicomponent fibers covering from polyethylene terephthalate (PET) to polylactic acid (PLA) in the history of nonwovens [4].

25.3 Sheath-Core Bicomponent Polyester Staple Fibers, MELTY®, and CASVEN®

The first step of nonwoven making is web forming that is a creation of a loosely held together sheet structure, usually by laying down of fibers. However, most webs have insufficient strength in the unbonded form. The individual fibers or filaments must, therefore, be tied together in some way, most commonly by latex bonding. However, this process requires large amounts of heat to remove water and thereby dry and set the binder to the fabric.

Thereafter, the use of thermal bonding techniques has grown significantly, because the process is cleaner and environmentally friendly without use of latex. Thermal bonding is a technique for bonding fibers of a web, with fusible fibers, when heat from hot air or hot calendars is applied.

Two common thermal bonding methods are through-air heating and calendaring. The through-air method uses hot air to fuse fibers within the web and on the surface of the web to make high loft, low density fabrics. Hot air is either blown through the web in a conveyorized oven or sucked through the web as it passes over a porous drum within which a vacuum is developed.

In calendar point bonding, the web is drawn between heated cylinders that have an embossed pattern so that only part of the web is exposed to extreme heat and pressure. This type of calendaring produces strong, low loft fabrics. Ultrasound in the form of ultrahigh frequency energy, applied to small areas, can also be used to cause localized fusion and bonding of fibers, thereby creating a pattern for web stabilization, lamination, or quilting.

Thermal bonding fibers are made from fusible thermoplastics such as polyethylene, polypropylene, polyamide, and polyester. When calendar bonding is

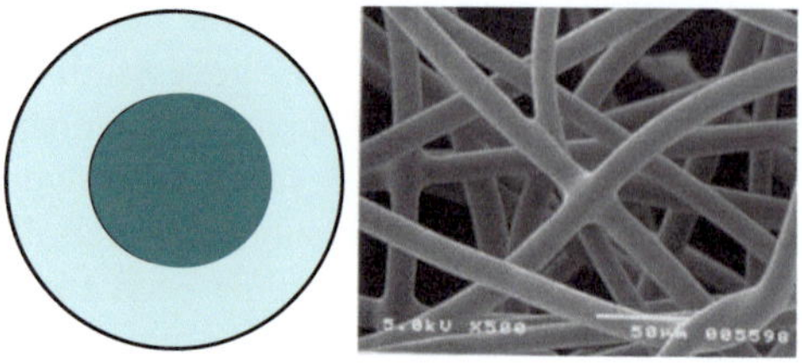

Fig. 25.1 Cross-sectional view of a sheath-core bicomponent fiber for thermal bonding

employed, the binder fibers used are often monocomponent, i.e., they are made of one polymer. The fusing process destroys their shape at the bonding point but retains it in the unbounded regions, as the polymer softens and flows to form the bond between fibers.

In bicomponent fibers made with two polymers, one polymer, the low melting component, either covers all the surface of the higher melting component (sheath-core structure) or is extruded alongside the higher melting component (side-by-side structure). During fusion and subsequent bonding, the low melting component softens and flows to form the bond, while the higher melting component maintains its fiber shape and thereby its structural integrity.

Above all, the sheath-core bicomponent fibers are widely used as bonding/binder fibers in nonwoven industry. The sheath of the fibers with a lower melting point than the core melts at an elevated temperature, creating bonding points with adjacent fibers – either bicomponent or monocomponent (Fig. 25.1) [5].

25.3.1 MELTY®

A first type of sheath-core is self-bonding fibers consisting of a lower melting temperature sheath and a higher melting temperature core. Common sheath-core combinations in such applications include Co-PET/PET, PE/PET, PP/PET, and PE/PP. These products are used in 100 % form as well as in blends with homopolymer fibers. Staple fiber applications of these products in nonwovens are by far today's largest commercial use of bicomponent fibers.

In 1981, the world's first sheath-core type of bicomponent polyester fibers MELTY® for nonwovens and fiber/fill has been launched by Unitika Ltd., Japan. A binder fiber of the sheath-core type MELTY® <4080> has its sheath component comprised of polyester copolymer including ethylene terephthalate and ethylene isophthalate units. This polyester has high rigidity and is a noncrystalline polymer which does not exhibit any definite crystalline melting point Tm but begins to soften at temperatures above the glass transition temperature Tg (about 65–70 °C). After that, a new type of MELTY® <6080> whose sheath component is comprised with high density polyethylene has been developed for the application of baby diapers or sanitary napkins, being essentially high soft touch. MELTY® has many types with different melting point, 110 °C, 130 °C, 200 °C, and 220 °C.

In 1983, Unitika gave an introductory lecture on MELTY® at the Institute of Science and Technology of Manchester University in England, and then MELTY® has been well known in the world. MELTY® (production capacity: 20,000 MT/Y) had been a pronoun as a binder fiber and almost monopoly in the world market for about 10 years since 1981. Hitherto, the sheath-core type of bicomponent polyester fibers like MELTY® had been manufacturing more than 100,000 MT/Y in applications such as filters, interlinings, shoulder paddings, and furniture stuffings like sofa cushion, chair back cushion, and cushion material for beds and automotive seats by many producers more than ten in the world.

Known nonwoven fabrics which are manufactured by subjecting a combination of the principal fiber and the binder fiber to the process of thermal bonding have a disadvantage that they lack handle flexibility and feel rather hard. Another disadvantage is that when subjected to repeated compression and/or bending, the nonwoven fabric is liable to joined-spot fracture, resulting in becoming flattened, or that when used in a high temperature atmosphere, the nonwoven fabric is subjected to bond deterioration, resulting in deformation of the fabric.

25.3.2 CASVEN®

In order to overcome the above-mentioned situations and technical issues, Unitika has originally developed the world's first novel heat-resistant type of binder fiber CASVEN®, the sheath component of which is composed of crystalline polyester with melting point Tm of about 160 °C, and it is less expensive [6].

It is an object of the CASVEN® to provide a novel binder fiber which eliminates the drawbacks and poor heat resistance of nonwoven fabrics using such a known binder fiber. In other words, it is to provide a nonwoven which has soft feel and is highly resistant to flattening during prolonged use or while in use under high temperature atmosphere.

25.3.2.1 Molecular Designing

In an attempt of molecular designing for the targeted characteristics of the sheath component, the relationship between Tm of polyester comprising terephthalic acid (TPA) for a dicarboxyl acid and number of methylene groups for an alkylene diol was first investigated. However, polyester having Tm more than 160 °C (targeted Tm) has not been found, except polyethylene terephthalate (PET), polypropylene terephthalate (PPT), and polybutylene terephthalate (PBT) whose Tm are more than 220 °C, as shown in Fig. 25.2.

Next, copolyesters of PET with incorporating the third components such as dicarboxylic acid or alkylene diol were systematically investigated in respect of their crystalline melting points. In the first case of an incorporation of dicarboxylic acid including isophthalic acid (IP), adipic acid (AD), suberic acid (SU), and

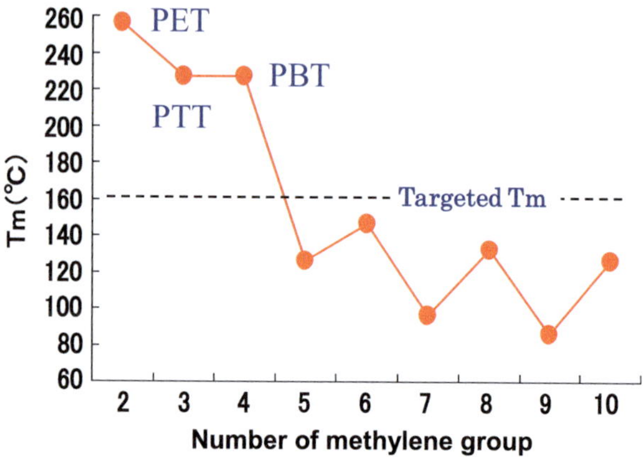

Fig. 25.2 Relationship between Tm and number of methylene group for alkylene diol of polyester comprised of terephthalic acid as a dicarboxylic acid

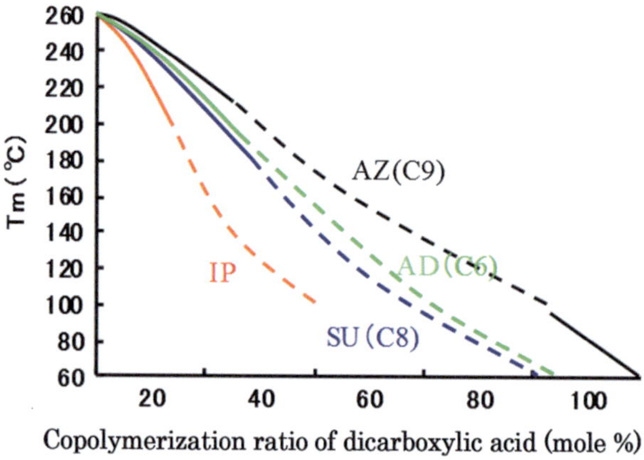

Fig. 25.3 Relationship between Tm and ratio of dicarboxylic acid copolymerized with PET

azelaic acid (AZ), all of the copolymers obtained have decreased rapidly the crystallinity with increasing in the ratio of copolymerization, resulting in amorphous polymer having no melting point above the copolymerization ratio of around 30 %, as shown in Fig. 25.3. Thus, the targeted crystalline copolymer with Tm of 160 °C was not obtained.

In the second case of an incorporation of alkylene diol, such as 1,4-butanediol, 1,6-hexanediol, and 1,9-nonanediol, to PET, the copolymer incorporated with 1,4-butanediol was crystalline all over the ratio of copolymerization ranging from 0 to 100 %, as illustrated in Fig. 25.4. However, the minimum melting temperature 180 °C of the copolymer, at which the ratio of copolymerization of 1,4-butanediol was 50 %, is a little bit higher than the targeted Tm 160 °C. Consequently, the small amounts of the fourth component had to be incorporated to the copolymer to reduce the melting point from 180 to 160 °C.

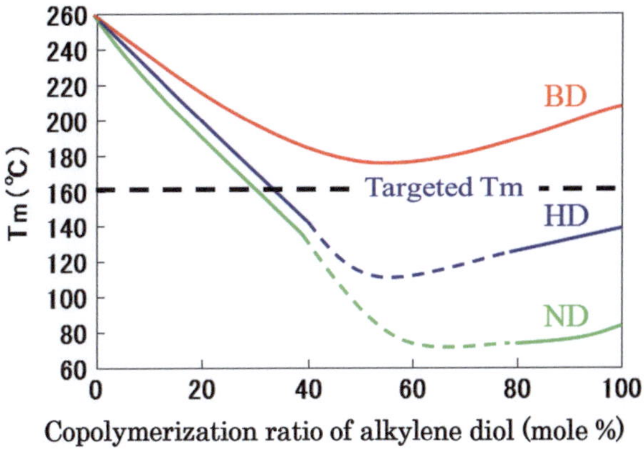

Fig. 25.4 Relationship between Tm and ratio of alkylene diol copolymerized with PET

25.3.2.2 Properties and Potential Applications

The differential scanning calorimetry (DSC) thermograph of the final composition of CASVEN® that is a crystalline PET copolymer with grass transition temperature Tg of 34 °C, crystallization temperature Tc of 86 °C, and melting point Tm of 160 °C is shown in Fig. 25.5. The advantages of CASVEN® <7080> when compared with MELTY® <4080> are shown in Table 25.1, indicating good thermal/mechanical properties, good chemical resistance, good dyeability and spinning property, and excellent durability and heat resistance.

Figure 25.6 illustrates the distortion temperature of nonwoven fabric (800 gsm) under loading (200 g) for CASVEN® and MELTY®, showing the excellent heat resistance of CASVEN® at room temperature of 150 °C. Furthermore, the bonding strength of CASVEN® to various types of fibers including PET, rayon, and cotton is superior to that of MELTY® because of the higher melt flow rate of CASVEN® than that of MELTY®. For instance, the bonding strength retention of PET/CASVEN® (70/30) nonwoven fabric (50 gsm) is 90 % after three times washing, being better than 77 % of MELTY®.

Potential applications of CASVEN® include auto interiors such as headliners, seat components, trunk floor cover or liners, hood silencer pads, or engine air and oil filters, bedding construction sheeting, etc. In addition, a new application for spun yarns has been developed, in which CASVEN® has been specifically engineered to allow spun yarn fabrics or knits to be finished by through-air heating, with long time dimensional stability and good dyeability.

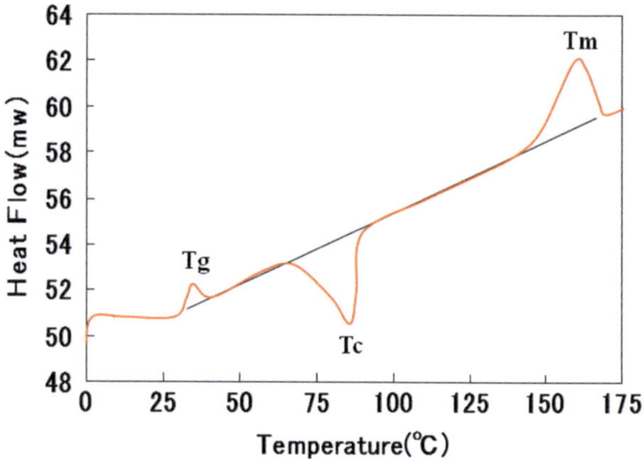

Fig. 25.5 DSC thermograph of CASVEN®

Table 25.1 Comparison of characteristics of sheath-core bicomponent polyester staple fibers MELTY® <4080> and CASVEN® <7080>

Characteristics	MELTY®<4080>	CASVEN® <7080>
Crystalline or amorphous	Amorphous	Crystalline
Melting point (°C)	Not observed	Ca. 160 °C
Softening point (°C)	65–75 °C<	130 °C<
Melt flow rate	Medium	High
Stiffness/Tg (°C)	Rigid/65 °C	Relatively soft/34 °C
Heat shrinkage (%), 100 °C × 15 min	15	0.5
Resistance for alkaline or solvent	Poor less than PET	Better than PET
Dyeing temperature (°C)	< 70 °C	130 °C
Spinnability	Not good	Comparable to PET
Durability of product molded	Not so good	Good
Heat resistance/distortion Temp	Poor/< 70 °C	Good/130 °C

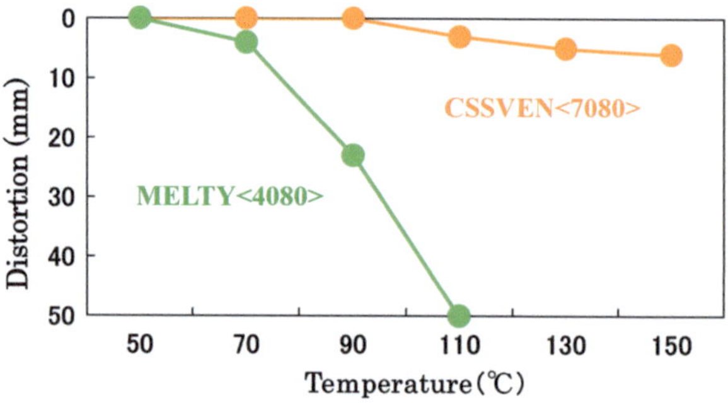

Fig. 25.6 Distortion temperature for CASVEN® and MELTY®

25.4 Side-by-Side Bicomponent Polyester Staple Fibers, "38F," "H38F,"and "C-81"

Side-by-side bicomponent fibers are fibers that have two different polymers or polymer with two different molecular weights running side-by-side along the filaments. The invention of this fiber structure was enlightened by the structure of wool which had been found first by M. Horio in 1950 [7]. The inner structure of a wool fiber is heterogeneous. The cortex of wool is composed of two different types of cortical cells, ortho- or para-cortical cells. The two components attach tightly to each other and rotate helically along the longitudinal axis. Because the two proteins have different mechanical responses, each single filament has a pronounced tendency to coil.

Synthetic bicomponent fibers have been commercially available since the 1960s, when DuPont produced a side-by-side hosiery yarn called "Cantrese." This yarn is made up of two nylon polymers which have different recoil rates. When the yarn retracts under certain conditions, one component shrinks more than the other, which pulls the filament to a permanent crimp. Thus, a highly coiled elastic fiber is formed (Fig. 25.7). Side-by-side bicomponent fibers have more crimps than as-spun fibers, and thus they are highly porous and stretchable.

However, the crimp characteristic of the "Cantrese" was not good and permanent, and thus, it was not commercially mass produced. Thereafter, Unitika, Japan, has launched the structural crimp fibers "H38F" with excellent durability of crimp and high bulkiness, which was composed of side-by-side bicomponent PET hollow fibers with two different molecular weights, as after-mentioned.

25.4.1 *Structural Crimp Fiber: "H38F"*

Side-by-side bicomponent fibers were first used as self-bulking fibers. Self-bulking is created by two polymers within a filament having a different strain level or shrinkage propensity. The different shrinkage rate of the two polymers causes the fiber to curl into a helix when heated under relaxation. Side-by-side self-bulking fibers are now in fairly common use in staple fiber/fill applications.

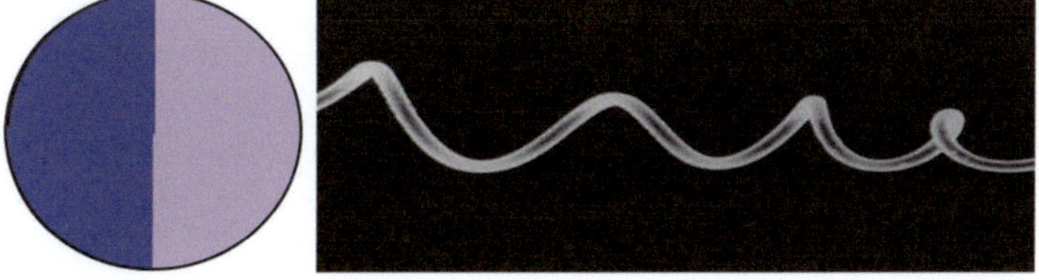

Fig. 25.7 Cross-sectional view of a side-by-side bicomponent fiber with helix coil crimps

This type of side-by-side bicomponent polyester hollow fiber, "H38F," is based on a very simple idea, that is, it is melt-spun by employing a high molecular weight PET to one side and a low molecular weight PET to the other side, then followed by drawing and heat-setting under tension-free. The two PET polymers differing in molecular weight are different from one another in crystallization rate under the spinning conditions; thus, the slower-to-crystallize component, a high molecular weight PET, shrinks more than the other during heat-setting, which causes the fiber to curl into a helix. The structural crimp fiber obtained is bulky with crimps like helix coil spring and has an excellent durability of the crimp during prolonged use. Actually it had been exported to the United States and Europe for long years.

In 1974, the first commercial mass production of "H38F" (4.4T–14T) has been started under the brand name of Airroll DX® mainly for fiber/fill of bedding applications including a mattress and a quilt. Afterwards a new type of feather touch hollow fiber with slippery silicone oil, Youmilan®, has been developed for the application of pillow or cushion in 1978, and it received a Fiber Award from Norway Government in 1979, thus leading to worldwide acknowledgment of the excellent quality and performance.

In 1980, another type of the structural crimp fiber "38F" with non-hollow fine denier (2.2T-3.3T) for hygienic products was started to manufacture commercially.

25.4.2 *Latent Crimp Fiber: "C-81"*

The term "latent crimp" is intended to encompass that situation in which the as-spun side-by-side bicomponent fiber exhibits no crimp, but crimp can be developed upon exposure to heat in a substantially relaxed state.

Around 1982, it was found accidentally by Unitika that the side-by-side bicomponent polyester fiber prepared from PET and cation dyeable polyester having small amounts of 5-sodium sulfo dimethyl isophthalate unit has high latent shrinkage and a high degree of latent crimp in the as-spun bicomponent fiber, in an attempt to get different color dyeable fibers. After dyeing, it unexpectedly happened that the latent shrinkage and crimp were developed to provide fine spiral microcrimps to the bicomponent fibers and textile/nonwoven goods of the spun yarns with excellent stretch and recovery which would be resulted from the crystallization rate difference between regular PET and cation dyeable polyester.

It is a principle of the latent fiber application to provide a bicomponent fiber not manifesting latent crimp in a raw staple-producing process first, and then manifesting the latent crimping capability having excellent elasticity and bulkiness in a secondary finishing at a comparatively low temperature. For example, the fabric is provided by using the spun yarn of the latent crimp fiber, and then the textile or nonwoven product is applied a heat treatment such as a dyeing processing to develop a latent crimp of the side-by-side bicomponent staple fiber on the fabric.

In 1985, Unitika has started commercial mass production of the world's first latent crimp fiber C-81 (production capacity: 6,000 MT/Y) having high stretchability and elongation recovery. It is recognized that high recovery crimped fibers have potential for preparation of high-value nonwovens such as stretchable base fabric of medical adhesive plaster like anti-inflammatory analgesic plaster through needle-punching process without heating, which exhibits sufficient elastic recovery potency even in the case of applying to bending knees or elbows without using an expensive elastic yarn such as polyurethane yarn.

25.5 Sheath-Core Bicomponent Polyester Spunbonded Fabrics: ELEVES®

Unitika has been a Japan's largest producer of polyester-based spunbonded nonwovens (production capacity: 22,000 MT/Y). In terms of new product development, Unitika has developed exclusive products for its core area such as ELEVES®, a sheath-core bicomponent spunbond nonwovens containing a PET core, and a polyethylene sheath with excellent heat sealing due to the low melting point (130 °C or 105 °C) of PE. It is suitable for bag packaging and fit for lamination with PE film for ice pack or house wrapping, or PTFE membrane for HEPA filter.

In addition, ELEVES® is not only highly soft resulted from a PE sheath, but also high strength thanks to a PET core, thus enables to hygienic product applications such as diapers or sanitary napkins.

25.6 Polylactic Acid Fibers for Nonwovens

25.6.1 Introduction

For the past 50 years, textile materials made from synthetic polymers such as nylon and polyester have been used for industrial applications as well as for clothing. In some instances, the excellent durability of synthetic polymers has proved to be a major advantage, but in others the great durability of these materials has led to adverse effects.

The current crisis in solid waste management has, in particular, focused attention on the development of biodegradable/compostable materials in soil or in compost. Polylactic acid (PLA) is a biodegradable aliphatic polyester with thermoplastic processability and has favorable mechanical properties and biodegradation characteristics [8]. Also the flexibility of end-of-life options of PLA bioplastics enables incineration while generating renewable energy or anaerobically digested to yield renewable energy, in addition to the ordinary recycling back into new applications.

Furthermore, in recent years bio-based materials from plants have been highlighted as carbon-neutral materials without increase in greenhouse gas emission in solid waste management, for example, in incineration. PLA is a bio-based, compostable plastic with a low carbon footprint which is used in packaging, disposables, fibers/nonwovens, electronics, and automotive markets. Cargill Dow LLC (now NatureWorks™ LLC) started full commercial manufacture of NatureWorks™ PLA with reducing production costs in 2002.

25.6.2 Bicomponent PLA Fibers

PLA fibers are not only biodegradable/compostable but also highly functional ones, coupled with their intrinsic properties such as bacteriostatic, flame-retardant, and weather resistance properties, when compared with conventional poly(ethylene terephthalate) (PET) fibers [8].

Unitika, a pioneer in PLA fibers/nonwovens in the world, has been manufacturing PLA fibers Terramac® including bicomponent PLA fibers for nonwoven fabrics since 1998. PLA, one of the aliphatic polyesters, is a crystalline thermoplastic having a melting point of 130–180 °C. PLA can be melt-spun into various types of fibers/nonwovens including monofilaments, multifilaments, bulked continuous filaments, staple fibers, short-cut fibers, and spunbond fabrics by conventional melt-spinning machines. The fibers are then processed by drawing and annealing to produce good mechanical properties such as high tenacity, good toughness, and good dimensional stability.

By controlling the ratio of the D- and L-isomers in the PLA polymer chain, it is possible to induce different crystalline melting points. By incorporating polymers with high levels of D-isomer content (10 %) into a sheath component, sheath-core bicomponent fibers with a sheath melting point of 130 °C can be produced. This type of bicomponent fiber is used as binder fibers in PLA nonwovens for thermal bonding.

Furthermore, it can manufacture side-by-side bicomponent PLA fibers in which one component is composed of a lower molecular weight PLA and the other is composed of a higher molecular weight PLA. The fibers conduct fine spiral micro-crimps due to the different thermal shrinkage between each component when exposed to heat treatment. This type of bicomponent fiber is utilized as fiber-fill with high bulkiness and good resilience and as fibers for stretchable nonwovens.

25.6.3 Biodegradable/Compostable Characteristic Features of PLA Fibers

Polylactic acid (PLA) has been highlighted as a carbon-neutral bioplastic because of its availability from agricultural renewable resources like corn and also as a biodegradable/compostable aliphatic polyester in soil or in compost [8].

Fig. 25.8 Aerobic composting of PLA fabric in compost at 58 ± 2 °C in comparison with those of cotton, wool, and PET

Figure 25.8 shows the result of actual aerobic fresh composting of PLA fabric in comparison with wool, cotton, and PET ones at 58 ± 2 °C. The PLA fabric has been degraded more rapidly than other natural fibers such as wool or cotton. It means that PLA fibers have an excellent compostability characteristic, which enables disposable uses of PLA products like wipers or diapers to be composted in industrial composting facility along with natural organic wastes.

In natural environments such as in soil or in water, the degradation of PLA proceeds slowly. It will take around 5 years to break down with fragmentation, although it depends upon soil conditions such as temperature, humidity, pH, and species of microorganisms. The percentage decrease in (A) the tensile strength and (C) the relative viscosity RV, which is closely related to molecular weight of the PLA fibers, are illustrated as a function of time in soil burial in Fig. 25.9, indicating that the degradation of PLA proceeds slowly, but steadily. After 3 years, the RV values of the fibers were reduced to 70–75 % of the initial value, and the tensile strengths have retained only 10–20 % of their initial strength.

However, (B) the weight loss of the fibers was not observed substantially after 3 years (36 months) of soil burial. An increase in weight after 2–3 years might be attributed to experimental errors due to incapability of complete removal of fine sands from the fiber sample. Also a SEM photograph of the PLA fibers after 3 years of soil burial has given smooth surfaces which seemed to be apparently not entirely degraded at all, due to the characteristic features of nonenzymatic bulk degradation different from those of surface erosion of enzymatic degradation.

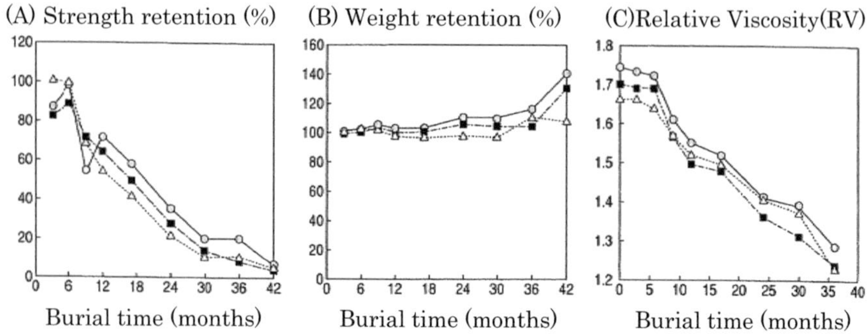

Fig. 25.9 Changes in the tenacity (**a**), the weight (**b**), and the relative viscosity (**c**) of PLA spunbond fabrics with different weight in soil burial test (-○-: 20 gsm, -■-: 70 gsm, -△-: 100 gsm)

It is concluded from the above-mentioned experimental results that the biodegradation of PLA fibers could not be evaluated by weight loss or apparent view, but RV reduction or decrease in tensile strength.

It is convenient for PLA fibers/nonwovens to degrade slowly but steadily in soil for the agricultural, horticultural, and geotextile applications such as anti-erosion sheets/bags for ground stabilization, weed prevention, vertical draining, or plants root covering. They are completely degraded to recover the original ecosystem in nature after their service time of several years has been over.

References

1. *The Nonwoven Fabrics Handbook*, Association of the Nonwoven Fabrics Industry Ed
2. R. Jeffries, *Bicomponent Fibers* (Marrow Publishing Co.Ltd., London 1971)
3. D. Morgan, Indian J. Nonwovens Res. **4**(4), 32 (1992)
4. N. Matsunaga, Sen'i. Gakkaishi. **63**(3), P-72 (2007)
5. N. Matsunaga, Nonwovens Rev. **1**(2), 48 (1990)
6. N. Matsunaga, R. Kinoshita, M. Tokutake, Sen'i. Gakkaishi. **60**(10), P-485 (2004)
7. M. Horio, T. Kondo, Text. Res. J. **23**, 137 (1953)
8. Mochizuki, M. 2009. Poly(lactic acid) fibres. In *Handbook of fibre structure*, ed. J. Hearle, S. Eichhorn, M. Jaffe, and T. Kikutani. Cambridge: Woodhead Publishing

Chapter 26
The World's Only Cellulosic Continuous Filament Nonwoven "Bemliese®"

Eiji Shiota

Abstract In the last 30 years, the market and utilization of cellulosic nonwovens have been remarkably increasing because of progress in productivity and eco-friendship (as a renewable raw material). Most cellulosic nonwovens consist of short fibers from cotton, pulp, viscose, and lyocell that are commonly blended with each other or commodity polymers. In this decade, cellulosic nonwovens are required for some special utilizations such as cosmetics, sanitary products, and medical devices. The world's only cellulosic spunbond "Bemliese®" originated from "Bemberg®", a continuous filament made by cuprammonium solution, invented in the middle of the nineteenth century. "Bemliese®" has some inherent properties which come from the characteristic spinning solution (cuprammonium solution) and sheet-forming technology (wet spunbond process). Recently we have achieved the production of the world's only cellulosic nonwoven made of the finest denier filament.

Keywords Cuprammonium • Spunbond • Microfilament

26.1 Introduction

Regenerated cellulosic materials are widely applied especially for textile and nonwovens because of their advantages on hydrophilicity, softness, dyeability, and so on.

However, their popularity, commercial solvent, and spinning method for cellulose are very limited because enormous energy is required to break their stiff hydrogen bonding, and there are some difficulties for recovering the solving agent. There are also some economical reasons. Nowadays there are only three commercial methods for the production of regenerated cellulose as a viscose method (derivative), NMMO method (solvate), and cuprammonium method (complex) except for some methods using cellulose derivatives.

E. Shiota (✉)
Bemberg Plant, ASAHI KASEI Corporation, Miyazaki, Japan
e-mail: shiota.eb@om.asahi-kasei.co.jp

Society of F.S. Technology, Japan (ed.), *High-Performance and Specialty Fibers*,
DOI 10.1007/978-4-431-55203-1_26

The first report about cuprammonium was in 1857, wherein a German chemist, M.E. Schweitzer, discovered that cotton dissolves in cuprammonium solution at room temperature, though J. Mercer, a caplico printer in the UK, worked on it before Schweitzer's discovery.

In 1890 L.H. Despessis, a French chemist, was the first to make artificial fibers using the cuprammonium method in a laboratory scale. After his invention, his method was industrialized by Vereinigte Glanzstoff-Fabriken A.G. in 1897. His process has two steps, the coagulation part and regeneration part, that are similar to the viscose rayon process that doesn't have the stretching part. Spinning solution came out from a nozzle and was introduced directly into the coagulation and regeneration bath, so it was impossible to produce fine filament and productivity was not so good.

In 1901, a German chemist, E. Thiele, in J.P. Bemberg A.G. invented the "stretch-spinning" process of cuprammonium. It had three parts: stretching, coagulation, and regeneration. The spinning solution that came out from the nozzle was introduced to a funnel filled by water and accompanied the flow, so that the spinning solution could be stretched. His invention enabled the production of fine filament and improved productivity. Thus, the cuprammonium industry was prosperous in the late 1910s–1920s [1].

In 1929, Asahi Kasei introduced the cuprammonium technology, and since 1931, cuprammonium filament was industrialized in Japan (Nobeoka, Miyazaki-Pref.).

Formerly (until 1950s), the cuprammonium method was very popular, and there were six companies producing cuprammonium filament, but they quit production because they had difficulties for high-speed spinning and recovering the solving agent. Now only a Japanese company (Asahi Kasei) can produce cuprammonium filament (since 1931) and nonwoven (since 1974).

Asahi Kasei commercialized cuprammonium nonwoven "Bemliese®" in 1974 after over 10 years' of research and development. The first target was hygiene gauze; at that time, almost all hygiene gauzes were cotton fabric that had some quality problems. They produced a large amount of fiber separation from cut edges and had poor absorbency. In addition, they were not a suitable material for cosmetic puff that had enough softness and smallness of fiber separation. Therefore, technologists tried to improve the quality of cuprammonium filament's coagulation and self-bonding technique (Fig. 26.1).

The cuprammonium method for stretching, coagulation, and regeneration is separated so it enables "self-bonding" under selected sheet-forming conditions (degree of coagulation). On the other hand, the coagulation and regeneration using other regeneration methods occur simultaneously, so it is difficult to produce "self-bonding" between filaments (though lyocell melt blown has been reported and developed recently). That is why almost all cellulosic nonwovens are made from short fibers by the spunlace method.

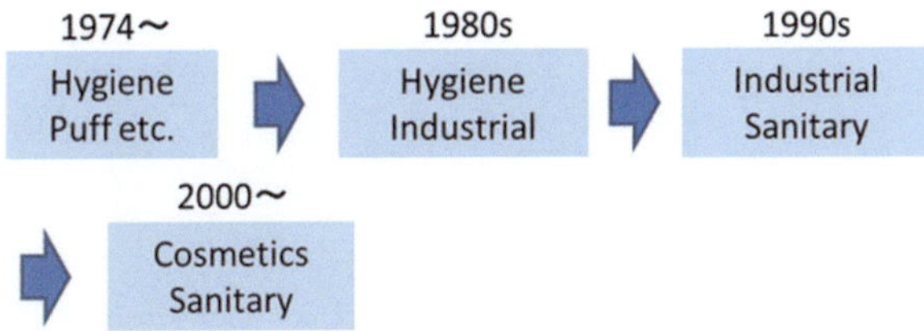

Fig. 26.1 Transition of Bemliese®'s market

26.2 Cuprammonium Solution

Cuprammonium solution has distinctive viscosity (degree of polymerization), spinnability, and coagulation behavior features.

The starting point is cotton linter that has been obtained from the peel of cotton seeds. Cotton linter is refined by steaming under alkaline condition to remove natural ingredients such as hemicellulose, fatty acid ester, lignin, etc. After refining, cotton linter is dissolved into cuprammonium solution and prepared to appropriate DP. Cuprammonium solution has so high viscosity that we can stretch further in spinning section and can produce very fine filaments from a wider spinneret. Cuprammonium solution is discharged from rectangular spinneret apparatus and stretched by water as accompanied flow (same as air flow in the spunbond process using polymers) over 100 times.

By calculating the flow rate of water condition, the size of spinneret, and the distance from spinneret to conveyer net, we can produce various diameters of nonwoven filaments. Furthermore, we can produce nonwoven that has various basic weights by selecting the amount of discharge and speed of conveyer net (Fig. 26.2).

After stretching, the ammonia in the spinning solution is removed into the accompanied water and coagulated gradually showing behavior as a gel. By this time, the nonwoven filaments form "self-bonding," and the mechanical properties of nonwoven are settled. The formed web has some residual copper and ammonia so they are completely removed by acid. After washing by water, regeneration is completed (Fig. 26.3).

High viscosity of spinning solution and accompanied water as stretching agent and coagulant enables high draft and the gradual removal of ammonia. Thus, we can obtain round, porous, and highly oriented (crystal area) filaments. Especially, the surface area of filaments shows a higher orientation and larger pore (Fig. 26.4) [2].

By rheological and spectrometric analysis, cuprammonium filament shows distinctive behavior compared with viscose rayon, such as strength of hydrogen bond in the amorphous region. Figure 26.5 shows that cuprammonium's crystal region is highly oriented by stretching, and the hydrogen bonds in the amorphous region are cleaved and relaxed. Cleavage of hydrogen bonds might be promoted by regeneration on the conveyer net under relaxing conditions.

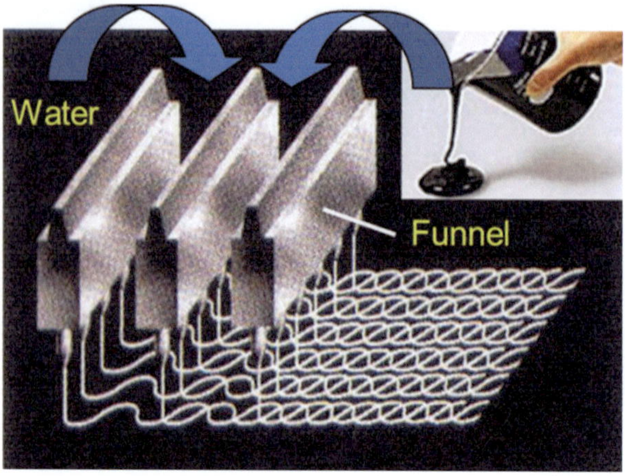

Fig. 26.2 Spinneret apparatus of Bemliese

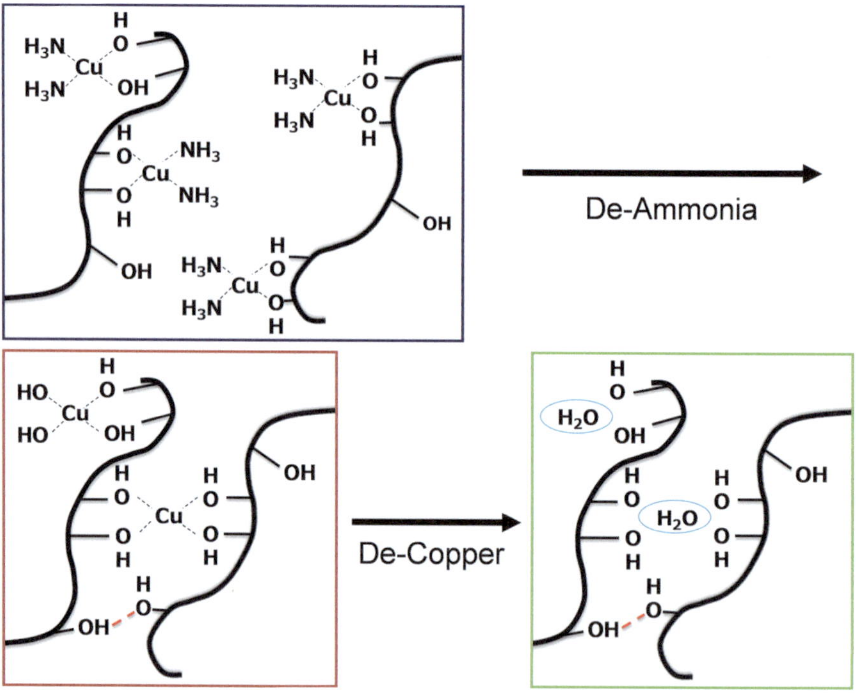

Fig. 26.3 Schematic image of coagulation and regeneration

The porous structure and relaxed amorphous region show some important characters. The cuprammonium filament has uniform and large size of pores that water, reagent, dyes, etc. can be absorbed quickly and excessively into the filament.

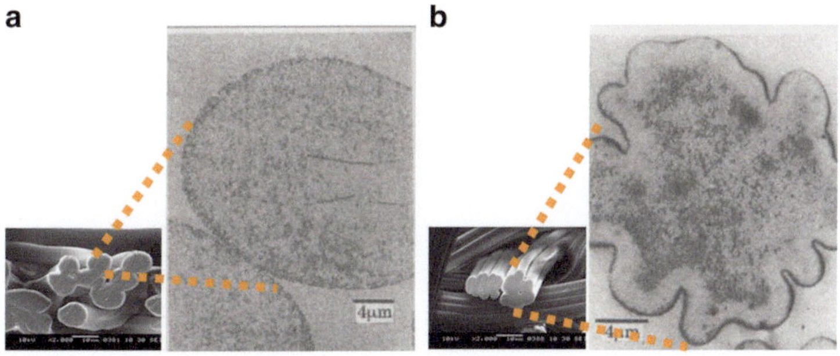

Fig. 26.4 Magnified view of regenerated cellulose's cross section. (**a**) Cupra(Bemliese®) (**b**) Viscose rayon

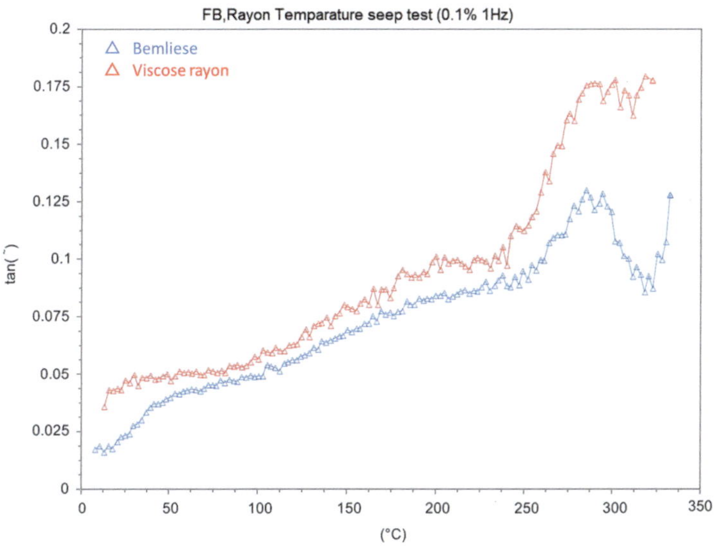

Fig. 26.5 Dynamic viscoelasticity of regenerated cellulose (2013 Asahi Kasei Analysis and Simulation center)

Regenerated cellulose is categorized into cellulose II that shows higher reactivity of hydroxyl groups than natural cellulose (cellulose I); in particular cuprammonium filament's amorphous region has less intramolecular hydrogen bond (C3–C5) and shows outstanding water absorbency and transparency (Fig. 26.8). We evaluate the consistency of intramolecular hydrogen bonds between C3 and C5 by estimating the chemical shift of C4. Cuprammonium filament C4 (Bemliese®) shows more shift upfield than that of viscose filament. It is well known that the intramolecular hydrogen bond of cellulose chain forms a seven-membered ring structure between neighboring glucose residues (Figs. 26.6, 26.7, and 26.8) [3].

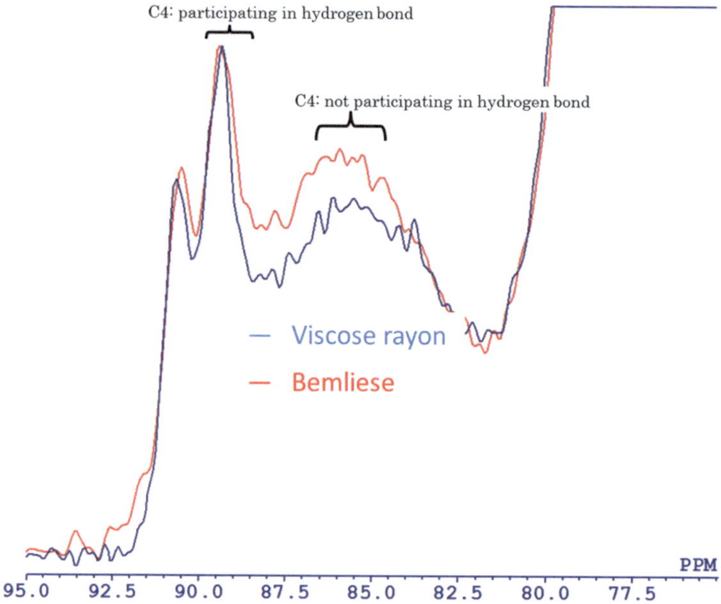

Fig. 26.6 Estimation of intramolecular hydrogen bond between C3 and C5 by ^{13}C NMR (2013 Asahi Kasei Analysis and Simulation center)

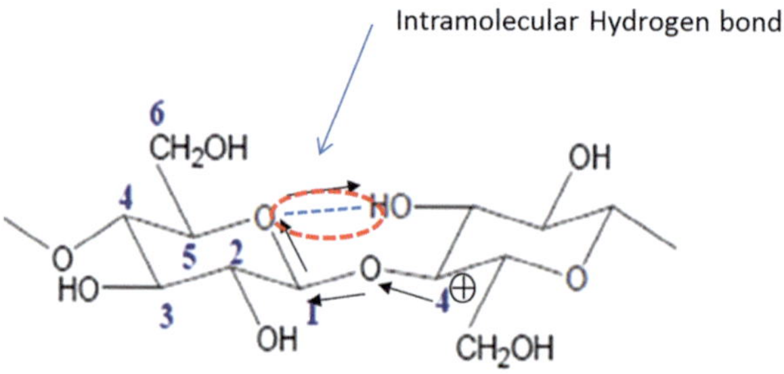

Fig. 26.7 Schematic representation of intramolecular hydrogen bond (C3–C5)

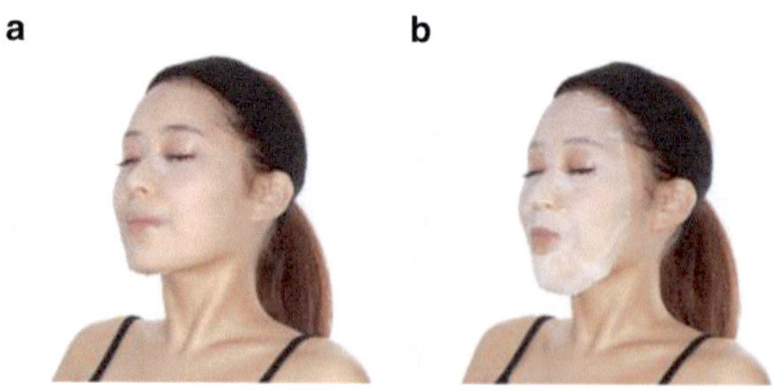

Fig. 26.8 Transparency of nonwoven (under wet condition). (**a**) Cupra(Bemliese®) (**b**) Cotton. From Bemliese®'s brochure

26.3 Wet Spunbond Process

1. **Spinning section**
   Cellulose dissolved into cuprammonium solution is discharged from the rectangular spinning apparatus that has over 100,000 holes of spinneret to hot water down onto the conveyer net. During this time, cuprammonium spinning solution is stretched about 100 times so that it can form a high orientation in the crystal region. The conveyer net is vibrated and stretches filament trace sine curves on the net so that they can laminate each other. On the conveyer net, ammonia is removed from the laminated filaments, and it forms self-bonding, thus coagulation occurs (Fig. 26.9).
2. **Refining section**
   Ingredients in the solvent have to be removed completely by diluted sulfuric acid and soft water. In this part, regeneration is completed; thus, the amorphous region has low orientation, and hydrogen bonds in the relaxed condition (on the conveyer net).
3. **Patterning section**
   The regenerated web itself has enough mechanical properties for nonwoven; in fact Bemliese® without patterning is widely applied in food/industrial applications. We have very low-pressured water needles because it is only for patterning not for entangling.

As mentioned above, cellulosic spunbond is possible only by cuprammonium solution, and its mechanical properties are settled by self-bonding at the stage of coagulation (Fig. 26.10). This process has five special features:

1. Flexibility in the design of diameter and density

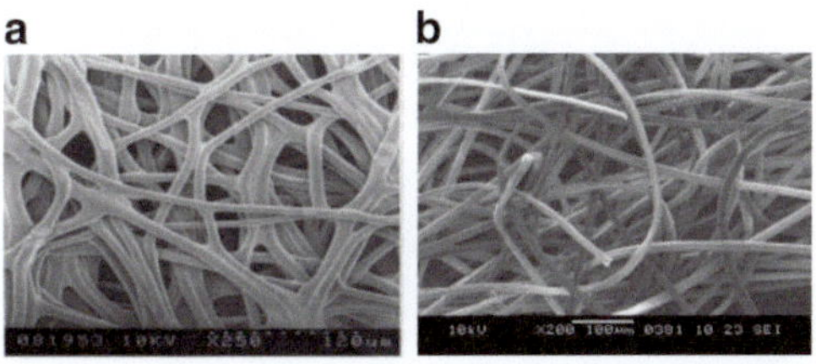

Fig. 26.9 SEM images of nonwoven. (**a**) Spunbond (Bemliese®) (**b**) Spunlace. * Without patterning

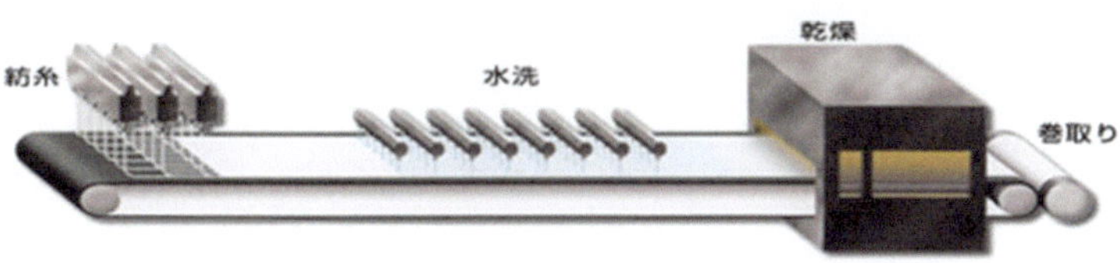

Fig. 26.10 Image of spunbond process

Due to advanced spinnability, we can choose various denier and basic weight, so that it is possible to control nonwoven's density especially in conjunction with the low-pressured water needle.

2. Flexibility in isotropy

   The filaments under condition of coagulation form self-bonding on the vibrating conveyer net, so that it is possible to control not only strength, elastic modulus, etc. but also isotropy, elongation recovery, etc. It's very advantageous especially for applications under wet condition (Cosmetics, sanitary etc.) (Fig. 26.11).

3. Variety of thickness, softness, and design

   After coagulation, the sheet is regenerated in the refining part by removal of ammonia and copper and delivered to the patterning part. Normally, the patterning part is called the "hydroentangle part" in the case of spunlaced nonwoven because their mechanical properties are obtained in this part with high pressure of water jet. In the case of spunbond "Bemliese®," high-pressure water is not necessary because its entanglement is achieved in the coagulation part so that we use very low-pressure water jet only for design and control of thickness and softness (Fig. 26.12). Of course, it is possible to produce spunbond nonwoven without water jet (no perforation).

4. Purity

   Because of Bemliese®'s continuous filament and self-bonding system and patterning under mild conditions, its fiber separation is extremely low (Fig. 26.13). Furthermore, there is no need to add surface surfactant for uniformity at sheet forming because in the spunbond process, the uniformity of sheet is obtained by uniformity of discharge from the spinneret. So Bemliese® shows very low extractives. This is the main reason that Bemliese® has been suitable for industrial wipers for a long time (Fig. 26.14).

5. Absorbency

   Thanks to the cellulose II structure, the homogeneously porous and relaxed amorphous region of Bemliese® shows enormously high water retention as much

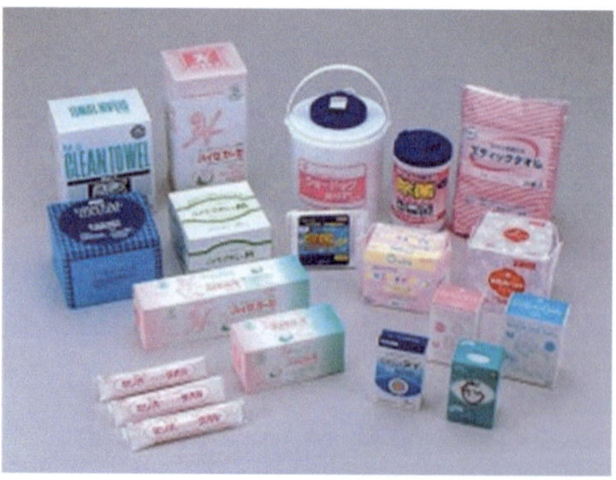

Fig. 26.11 Application for cosmetics and sanitary

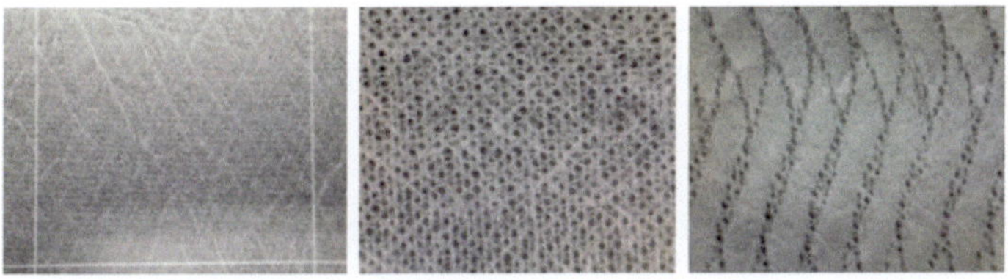

*

Fig. 26.12 Patterning variation. *Patent pending

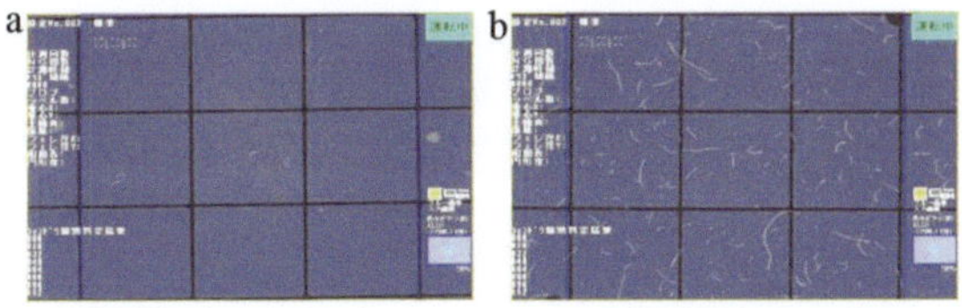

Fig. 26.13 Fiber separation under wet condition (>100 μ). (**a**) Spunbond (Bemliese®) (**b**) Spunlace

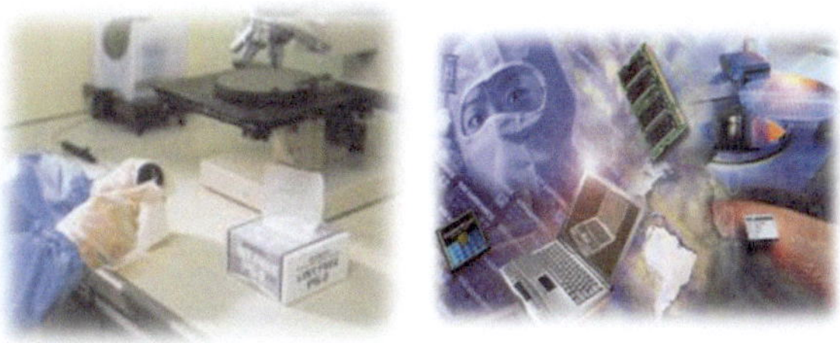

Fig. 26.14 Application for clean room

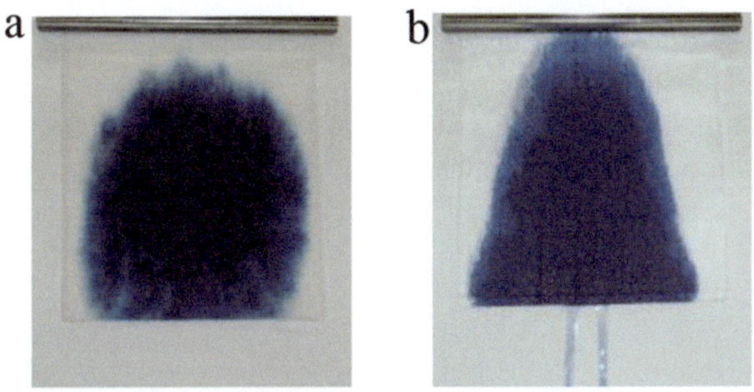

Fig. 26.15 Water retention test. (**a**) Spunbond (Bemliese®) (**b**) Spunlace. *After 15 s injection of 10 mL of blue colored water at a 45° slope (Sample weight: app.1.7 g)

as ten times its basic weight (Fig. 26.15). In addition, the cuprammonium filament's porous structure enables it to absorb water or solvent promptly and penetrate into the amorphous region of its molecular chain. So Bemliese® can be used for a long time for industrial and hospital spill control. Bemliese®'s intrinsic absorbency also shows two-dimensional adhesion on the skin so that it's especially suitable for cosmetics such as facial masks.

26.4 New Development of Cellulosic Spunbond "Bemliese®"

As mentioned, the cuprammonium solution and spunbond process have very good spinnability and controllability of filament diameter. We have been researching to enhance these features in this decade and achieved the production of a nonwoven that has fine deniered endless cellulosic filament.

Cellulosic nonwoven that has fine denier was not on the market until 2005; Bemliese® was the first product in the market. Target of fine-deniered cellulosic nonwoven was to improve tenacity, uniformity of pore size, elastic modulus (especially for cross direction), softness, speed of absorption, wipe off, whiteness, etc.

Fine-deniered Bemliese® has the finest diameter around 3–5 μm and 0.2 dt. Usually regenerated cellulosic fiber has a diameter of about 13 μm. Also regular Bemliese® has a diameter of about 10–13 μm and 2.2 dt filament (Fig. 26.16).

Now fine-deniered Bemliese® has wide spreading applications for the sanitary, industrial, and cosmetics markets [4].

26.5 Key Techniques of Microfilament

1. Optimized discharge and diameter of spinneret

   Cuprammonium solution's good spinnability enabled it to produce fine filament from large spinneret holes, so we tried finer spinneret and lower discharge of spinning solution to produce much finer filament.
2. Spinning velocity

   To obtain finer filament, it is very important to improve the flow velocity of the accompanied flow of water. Spinning apparatus was reconstructed for this much finer aim. Furthermore, not to accelerate coagulation adequate temperature of water was selected.

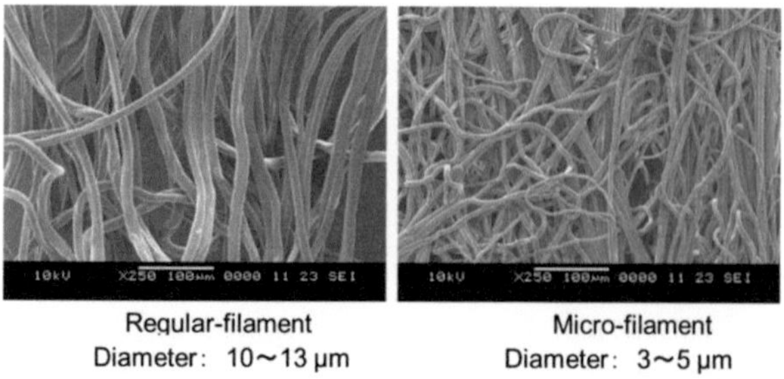

Fig. 26.16 SEM image of microfilament and regular filament

3. Improvement of line speed
   Generally speaking finer filament causes less productivity, but we tried to improve line speed without deterioration of mechanical properties and quality. There is no deterioration of productivity.

26.6 Priority of Microfilament Bemliese®

1. Fine pore size
   Fine denier comes off the improvement of the specific surface area, approximately three times. It is possible to obtain fine and uniform pored nonwoven (Fig. 26.17). Filtration and wipe off get improved, and microfilament Bemliese® is suitable for industrial and cosmetic applications.
2. Good absorbency speed
   Fine filament and fine pore size cause high-speed absorption without deterioration of retention (Fig. 26.18).
3. Isotropy
   Fine denier filaments make a lot of points of self-bonding, which make obvious improvement of mechanical properties especially at the cross direction. It makes nonwoven stable and easily handled under wet conditions, so

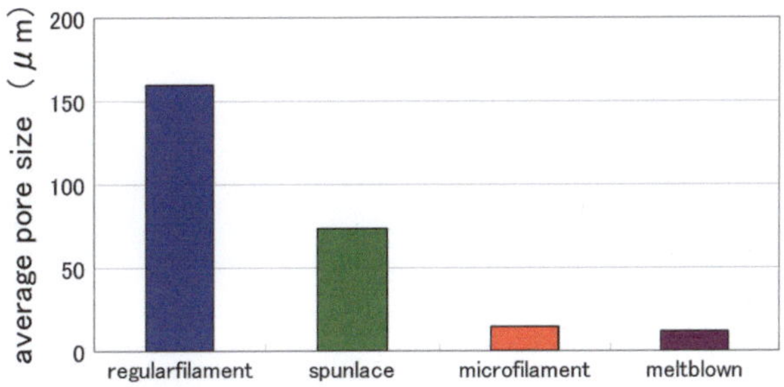

Fig. 26.17 Evaluation of pore size

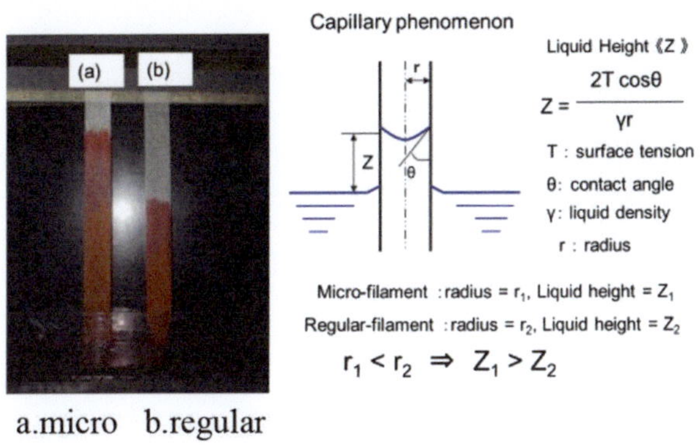

Fig. 26.18 Evaluation of absorbency. (**a**) Micro (**b**) regular

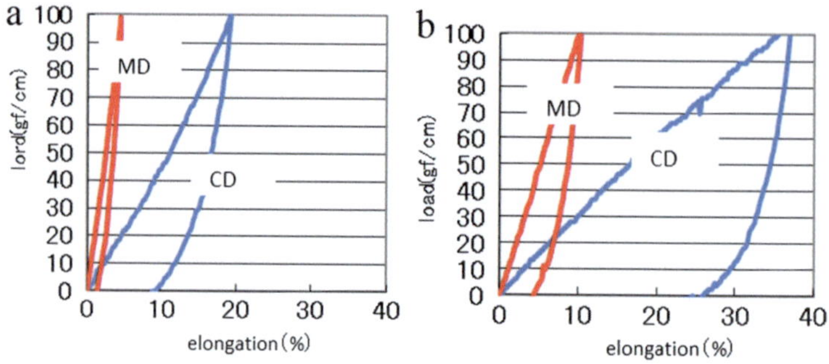

Fig. 26.19 Evaluation of elongation recovery. (**a**) Microfilament (**b**) regular filament

microfilament Bemliese® is very suitable for sanitary, industrial, and cosmetic applications (Fig. 26.19).

References

1. K. Kamide, Regenerated cellulose fiber industry. Part II Cuprammonium Rayon. pp. 63–113 (2006)
2. J. Sato, *A Study on Solid Structure of Regenerated Cellulose Fibers Related with Fibrillation* (University of Miyazaki, Miyazaki, 1999)
3. K. Kamide, Polym. J. **17**(5), 701–706 (1985)
4. JP3983088

Chapter 27
Thermoplastic Polyurethane Nonwoven Fabric "Espansione"

Hiroyasu Shimizu

Abstract "Espansione" is a melt-blown nonwoven fabric composed of thermoplastic polyurethane elastomer, which has been developed using original melt-spinning technology. "Espansione" is composed of 100 % thermoplastic polyurethane elastomer continuous fine filaments. With regard to the structure, "Espansione" has a number of physical properties such as high stretchability, elasticity, air permeability, moisture permeability, and easy handling during welding and cutting process.

A hot-melt-type adhesive polyurethane nonwoven fabric, "Espansione FF" which is composed of special thermoplastic polyurethane for hot-melt adhesive applications, has been developed by further improving the technology in the manufacture of "Espansione." No solvent is used, making it environment-friendly and human-friendly. Because "Espansione FF" uses hot-melt adhesive technology, "Espansione FF" is superior in work efficiency and offers various features such as long-lasting stretchability and air permeability after adhesion. "Espansione FF" has a wide variety of end uses.

Keywords Polyurethane • Elastomer • Nonwoven fabric • Hot-melt adhesive • Melt-blow

27.1 Introduction

Globally, the nonwoven industry is one of the grown-up fields in a textile industry. In recent years, various nonwoven fabrics manufacturing by various raw materials and manufacturing process have wide variety of end uses according to their characteristic. Technology of nonwoven fabrics has been improved significantly in nearly all available major manufacturing processes [1, 2].

"Espansione" is a melt-blown nonwoven fabric composed of thermoplastic polyurethane elastomer which KBSEIREN has developed using its original polyurethane melt-spinning technology. "Espansione" has a number of superb physical

H. Shimizu (✉)
Development Division, Hokuriku Synthetic Fiber Plant, KBSEIREN, Ltd, Osaka, Japan
e-mail: hiroyasu.shimizu@kb.seiren.com

Society of F.S. Technology, Japan (ed.), *High-Performance and Specialty Fibers*,
DOI 10.1007/978-4-431-55203-1_27

properties such as high stretchability, elasticity, air permeability, and easy handling. "Espansione" can be widely applied for many fields by utilizing its unique characteristics.

"Espansione" can be developed by itself or in combination with other materials. It is applicable to apparels, medical fabrics, sanitary products, clean engineering products, and miscellaneous applications. By further improving the proprietary technology in the manufacture of "Espansione," KBSEIREN developed a new hot-melt adhesive polyurethane nonwoven fabric, "Espansione FF."

27.2 "Espansione"

27.2.1 Manufacturing Process of "Espansione"

"Espansione" is manufactured by thermoplastic polyurethane nonwoven technology of melt-blown process. Melt-blown process is in which a thermoplastic fiber-forming substance is melted and extruded through a die having many orifices arranged in line at constant pitch. Convergent streams of high-velocity heated air (exiting from the top and bottom sides of the die nosepiece) rapidly attenuate the extruded polymer streams to form microfibers. The attenuated fibers subsequently get laid randomly onto the collecting screen made of wire, forming a nonwoven web, thus obtaining nonwoven fabric wherein the fibers collected on the collecting screen have been fused with their own heat. Additional bonding and finishing processes may be applied to nonwoven webs as necessary. It is usually used to increase the web strength and abrasion resistance. Additional physical or chemical treatment such as calendering and embossing can be applied during production.

27.2.2 Technological Features of "Espansione"

"Espansione" is composed of random webs which are made of 100 % thermoplastic polyurethane elastomer continuous fine filaments. "Espansione" offers superior air permeability and easy handling property traditionally found in melt-blown nonwoven fabrics.

"Espansione" possesses many excellent characteristics such as (1) high stretchability in all directions, (2) high elongation recovery, (3) high air permeability due to randomly webbed fine filaments with only the intersections melt bonded, (4) traps and prevents minute dust from penetrating because filaments are tightly webbed, (5) lint-free owing to continuous filaments and thermally melt bonded, (6) high coefficient of friction with high slippage resistance, and (7) welding/heat cutting easily done with a high-frequency welder or a heater. Table 27.1 shows the physical properties of "Espansione." The unit weight of fabric is from 25 to 300 g/m

Table 27.1 Physical properties of "Espansione"

Type	Weight (g/m^2)	Thickness (mm)	Breaking tenacity* (cN/25 mm)		Breaking elongation* (%)		Elongation recovery** (%)	
			Warpwise	Weftwise	Warpwise	Weftwise	Warpwise	Weftwise
UHO-50	50	0.21	1200	900	390	390	90	90
UHO-75	75	0.34	1515	1380	405	420	90	90
UHO-100	100	0.42	2070	1940	435	440	90	90
UHO-300	300	1.01	7265	6570	585	540	90	90
UYO-75	100	0.33	1700	1390	430	425	90	90
F13-75	75	0.33	1450	1410	395	405	90	90
FHF-85	85	0.36	1880	1790	405	410	91	91
F13-50	50	0.17	1830	1700	320	320	91	91

*Breaking tenacity and elongation are measured with a sample of 25 mm width
**Test method: JIS L 1096

^{2}. “Espansione” offers superior stretchability, high elongation recovery, and flexibility.

“Espansione” can be widely applied for many fields by utilizing its unique characteristics. It has been developed by applying aftertreatment to the product and using it either alone or through combinations with other materials. The main applications are medical uses (surgical dressings, elastic first-aid adhesive tapes, dental masks), clean room supplies (dust-free gloves, masks, inner caps), textiles and sports equipment (stretch padding materials, interlining, sports shoes), and industrial materials (insoles, car seat linings, synthetic leather base cloths).

Since “Espansione” is made of 100 % thermoplastic polyurethane elastomer, special attention is required as in the case of handling polyurethane spandex fiber. When used for a surface, “Espansione” may gradually yellow and weaken after long exposure to the light and NOx gases. The degree of yellowing is similar to that of polyurethane spandex fiber. “Espansione” should normally be used at temperatures below 100 °C. It may be glued to each other at 100 °C or higher. As for the chemical resistance, “Espansione” suffers the same level of deterioration as nylon in highly concentrated acid or alkali. Chlorine bleaching agent deteriorates “Espansione.” In use of organic solvent, “Espansione” is soluble in dimethylformamide, dimethyl sulfoxide, tetrahydrofuran, and methyl ethyl ketone.

The performance retention of “Espansione” when processing the nonwoven fabric in a variety of conditions is shown in Table 27.2.

Table 27.2 Performance retention of “Espansione” after treatment (light resistance, heat resistance, chemical resistance, washing resistance)

		Strength retention (%)	Elongation retention (%)
Light (Xe lamp)	20 h	85	67
	40 h	50	35
NOx gas	24 h	92	98
Chlorine	300 ppm, 30 min	82	105
Washing	10 times	100	100
	20 times	96	96
Dry heating	100 °C, 30 min	100	101
	140 °C, 30 min	97	98
Steaming	30 min	95	105

27.3 "Espansione FF"

27.3.1 Technological Features of "Espansione FF"

As the hot-melt material, those made of a resin such as ethylene-vinyl acetate (EVA), ethylene-ethyl acrylate (EEA) copolymer, polyethylene-atactic polypropylene, polyamide, polyester, polyolefin, polyurethane, or the like have hitherto been known. A hot-melt nonwoven fabric is used to laminate the nonwoven fabric with the adherend and fuse the nonwoven fabric with heating to adhere it. No solvent is used, making it environmentally friendly. Because it uses hot-melt technology, hot-melt adhesive is superior in work efficiency.

"Espansione FF" is a hot-melt-type adhesive polyurethane nonwoven fabric. For developing new technology of "Espansione FF," KBSEIREN has made further advances to its proprietary technology in the manufacture of melt-blown nonwoven fabric "Espansione." "Espansione FF" is the same structure as shown in the electron micrograph of Fig. 27.1. "Espansione FF" can be easily adhered to an iron or a heat presser at a temperature of between 100 and 200 °C, depending on the fabrics to be adhered. KBSEIREN has marketed two types of hot-melt adhesive polyurethane nonwoven fabric "Espansione FF." As shown in Fig. 27.2, "Espansione FF" was sandwiched between adherends (cotton textile fabrics or polyamide textile fabrics) followed by bonding with heat pressing. "Espansione FF" shows the good adhesive strength which was excellent like other hot-melt adhesives. Table 27.3 shows that

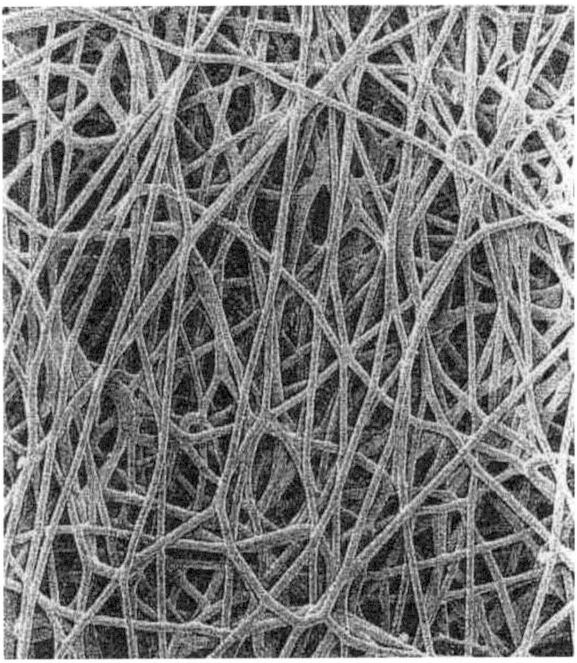

Fig. 27.1 Electron micrograph of "Espansione"

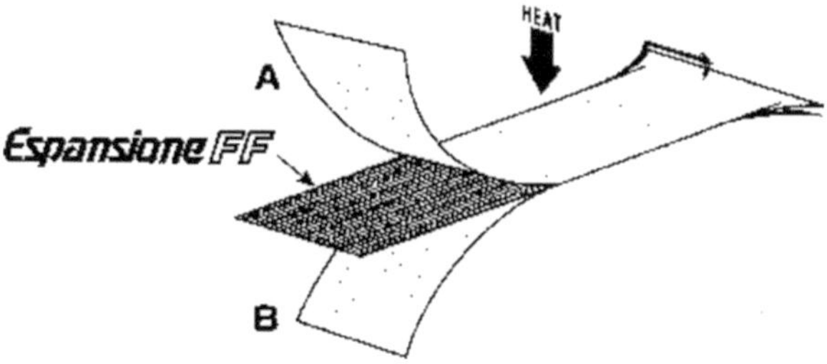

Fig. 27.2 How to use "Espansione FF"

the adhering temperature range of "Espansione FF" UEO type is from 100 to 180 °C, and for "Espansione FF" EDH type, it is from 130 to 200 °C. The nonwoven fabric width is 900–1500 mm, and it comes in weights of 25, 35, 50, 75, and 100 g/m^2. Comparison in points of stretch recovery and elastic recovery between "Espansione FF" and polyamide hot-melt nonwoven fabric shows in Table 27.4. "Espansione FF" offers various features such as long-lasting stretchability after adhesion.

27.3.2 *Adhesion Properties of "Espansione FF"*

"Espansione FF" can be used for adhesion of various materials, such as natural fibers, synthetic fibers, wood, paper, natural leather, and artificial leather. The adherends can be bonded with each other by sandwiching the nonwoven fabric and heating it at a suitable temperature. The evaluation of the adhesion temperature of various nonwoven fabrics is described in Fig. 27.3.

"Espansione FF" can be easily adhered to an iron or a heat presser at a temperature of between 100 and 180 °C, depending on the fabrics to be adhered. For this reason, an adhering temperature that does not damage the properties of the fabrics can be chosen. The adhesion temperature of between 100 and 180 °C is possible for hot-melt nonwoven fabrics in general, but it is quite difficult to choose a specific item with the suitable melting point for selected adhering temperature. Since the viscosity of polyurethane elastomer used in "Espansione FF" has a low dependence on the temperature, it is not difficult to select the adhering condition of "Espansione FF."

"Espansione FF" can be used to attach knitted fabrics which have superior flexibility such as two-way tricot fabrics and cotton circular knitted fabrics. Compared with other hot-melt adhesives, "Espansione FF" retains the suppleness after bonding. In particular, it retains stretchability even after two pieces of stretchable materials are bonded by "Espansione FF." The stress of laminated fabrics adhered by "Espansione FF" shows the low value in Table 27.5. "Espansione FF" is superior in softness and flexibility, and the adherend materials retain their characteristic after adhesion.

Table 27.3 Physical properties of "Espansione FF"

Type	Weight (g/m^2)	Thickness (mm)	Breaking tenacity* (cN/25 mm)		Breaking elongation* (%)		Elongation recovery** (%)	
			Warpwise	Weftwise	Warpwise	Weftwise	Warpwise	Weftwise
UEO-25	25	0.14	125	120	240	240	90	90
UEO-50	50	0.23	250	240	250	250	90	90
UEO-75	75	0.31	375	360	255	255	90	90
UEO-100	100	0.39	500	480	270	270	90	90

*Breaking tenacity and elongation are measured with a sample of 25 mm width
**Test method: JIS L 1096

Table 27.4 Comparison in points of stretch recovery and elastic recovery between "Espansione FF" and polyamide hot-melt nonwoven fabric

Sample	Stretch recovery ratio (%)	Elastic recovery ratio (%)
Espansione FF	85	67
Polyamide nonwoven fabric	78	11

*Adhesive material: unit weight of fabric: 50 g/m^2, adherend: cotton circular knitted fabrics

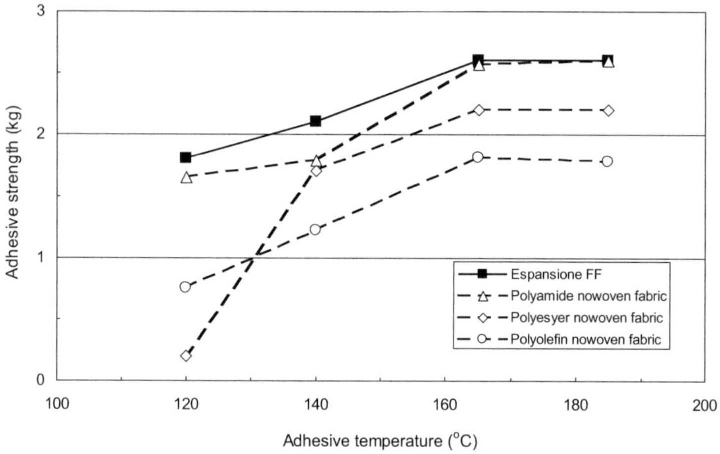

Fig. 27.3 Adhesive performance of "Espansione FF" (unit weight of fabric: 25 g/m^2)

Table 27.5 The stress at 30 % or 50 % elongation before and after adhesion

Adherend sample		Stress of two same cloths before adhesion	Stress of two same cloths after adhesion (g)			
			UEO-25	N-25	UEO-50	N-50
Cotton circular	30 % elongation	40	360	687	683	740
Knitted fabrics	50 % elongation	62	430	810	925	1845
Nylon two way	30 % elongation	268	740	769	740	3415
Tricot fabrics	50 % elongation	420	1000	1405	1740	3750

*UEO-25, UEO-50: Espansione FF (unit weight of fabric: 25, 50 g/m^2)
**N-25, N-50: adhesive polyamide nonwoven fabric (unit weight of fabric: 25, 50 g/m^2)

27.3.3 *Air Permeability of "Espansione FF"*

"Espansione FF" has air permeability in the cause of the constitution as well as other hot-melt nonwoven fabrics which are made of polyester and polyamide. On the other hand, polyurethane film has elasticity and stretchability, but air permeability is drastically lowered or completely lost. "Espansione FF" retains elasticity, stretchability, air permeability, and moisture permeability after adhesion as shown in Table 27.6.

As mentioned above, hot-melt nonwoven fabrics which consist of polyamide or polyester do not retain softness and flexibility of adherends after adhesion. "Espansione FF" excels in both suppleness and air permeability.

27.3.4 Other Properties of "Espansione FF"

"Espansione FF" has additional properties of light fastness and washing resistance. "Espansione FF" is mainly applicable to be used to adhere textile fabric; the problem of color change is undesirable for adhesive materials. Generally, polyurethane products tend to turn yellow by sunlight. "Espansione FF" has superior light fastness by modifying the polyurethane elastomer and the production process for "Espansione FF." Table 27.7 shows that little yellowing by sunlight occurs, and no deterioration due to perspiration is seen.

"Espansione FF" has good washing resistance and dry-cleaning resistance. Table 27.8 shows that the adhesion strength does not change after washing or

Table 27.6 Air permeability of hot-melt adhesive materials

Adhesive material	Air permeability ($cc/cm^2/s$)	Softness (mm)
Espansione FF	830	23
Polyamide nonwoven fabric	520	58
Polyurethane film	0	17

*The adhesive material was sandwiched and fused between polyamide two-way tricots
**Adhesive material unit weight of fabric: 25 g/m^2

Table 27.7 Change of the color b value by fade meter irradiation

Sample	Blank (0 h)	20 h	40 h
Espansione FF	0	0	0.2
General TPU	0	18.7	33.2

*Using fade meter (89 °C ± black panel temperature 3 °C)

Table 27.8 Change of ply adhesion by washing and dry-cleaning

Sample	Blank	Washing (10 times)	Dry-cleaning (once)	Dry-cleaning (10 times)
Espansione FF	9487	8820	8709	11246
Polyamide nonwoven fabric	4427	4118	4656	5251

*Adherend: cotton circular knitted fabrics
**Adhesion condition: 165 °C × 6 kg/cm^2 × 30 s

dry-cleaning. “Espansione FF” retains good properties including adhesion strength and stretchability.

“Espansione FF” requires the same instructions for use as spandex fiber does. “Espansione FF” deteriorates by acids or caustic soda in high concentration and chlorine-type bleaching agents. “Espansione FF” dissolves in organic solvent such as dimethylformamide, tetrahydrofuran, methyl ethyl ketone, etc.

Compared with solvent-type adhesive, “Espansione FF” is an environmentally friendly adhesive which uses no solvent. “Espansione FF” requires no special ventilation, making the workplace safer. And any waste and discard produced from cutting “Espansione FF” cause no environmental problems.“Espansione FF” is a promising adhesive friendly to the environment and an alternative to solvent-type adhesives.

27.3.5 Various Applications of “Espansione FF”

“Espansione FF” is widely applicable to natural fibers, synthetic fibers, wood, paper, and artificial leather. Further, adhering different materials such as waterproof moisture permeable sheets and synthetic fiber woven fabrics are possible.

KBSEIREN offers a variety of applications that exploit the pliancy and air permeability of “Espansione FF.” One example is the use of “Espansione FF” to attach knitted fabrics which have superior flexibility, such as two-way tricot fabrics, thereby creating layered fabrics which do not use the elasticity and permeability of tricot fabrics. Applications include soft girdles and shape-up wear.

Moreover, moisture-permeable films can be layered onto knitted fabrics to create fabrics that offer the advantages of both and do not lose the stretchability of knitted fabrics or the properties of moisture-permeable films.

“Espansione FF” can also be used to adhere flocked marking fabrics, thereby creating products with flexibility, which heretofore had been difficult to achieve. Using “Espansione FF” for the belt part of innerwear can completely eliminate the former twisting and slack. Previous belts were adhered by adhesives in a dot form, which made for frequent twisting and slack during use. “Espansione FF” has excellent stretchability and can be used in creating lining and interlining for apparel with superior form stability. At the same time, it can also offer maintenance of form in the manufacture of wireless brassieres and sports bras. The product can also be used as an adhesive for various emblems and marks, and even drawings photocopied onto one surface of “Espansione FF” using a common photocopier can be applied by heat adhesion to apparels such as T-shirts.

In addition, a high-temperature durable hot-melt polyurethane nonwoven fabric (EDH type) maintains the original properties of “Espansione FF.” The adhering temperature of this nonwoven fabric is 130–200 °C and will, for example, withstand the production process in the manufacture of shoes, which requires high-temperature processing in the finishing stages after adhering. Through this application, the fabric can be used for the stretching of portions requiring flexibility in the shoe material, and the material will not lose flexibility even after the adhesion process.

27.4 Conclusion

In recent years, nonwoven industry has been forecast as one of the grown-up fields in a textile industry. Various nonwoven fabric manufacturing by various raw materials has been improved significantly nearly by the development of new materials and manufacturing process technology.

"Espansione" is a melt-blown nonwoven fabric composed of thermoplastic polyurethane elastomer which has a number of unique characteristics. The high stretchability and elasticity are applied to the industry needs of providing a better fit and body. "Espansione" can be developed by itself or in combination with other materials. It is applicable to apparels, medical fabrics, sanitary products, clean engineering products, and miscellaneous applications.

Hot-melt-type adhesive polyurethane nonwoven fabric "Espansione FF" has several distinguishing characteristics derived from structure and superb physical properties. Hot-melt adhesives including "Espansione FF" have the advantage of being solvent-free, which can be said to make them suited to protecting the global environment. KBSEIREN can be expected to increase its development, production, and sales of hot-melt nonwoven fabrics, through developing new materials and through developing devices and know-how to use them.

KBSEIREN continues the development of a new product so that deployment of an elastic nonwoven fabric may be made also to the field could not be satisfied with the conventional product of a field, and it thinks that he would like to continue to try hard to develop original products and market cultivation for the purpose of getting used with a functional nonwoven fabric manufacturer.

References

1. United States Patent Publication No.6429159
2. JTN Asian Textile Business, **565**, 34 (2001)

Part VII
Fibers in Future

Chapter 28
Future Man-Made Fiber

Akira Yonenaga

Abstract The demand for synthetic fibers is increasing with the burgeoning global population and the rising level of disposable income in developing countries. Given the shortage of natural resources, concepts such as sustainability and recycling are increasingly important. The future will see the development of 'super functional' fibers: biomimicked fiber, design-driven cellulose fiber, spider silk fiber, intelligent fiber, and bio-base fibers.

Keywords Sustainability • Functional fiber • Biomimicked fiber • Cellulose fiber • Spider silk fiber • Intelligent fiber • Bio-base fiber

28.1 Introduction

In this, the final chapter of this memorial publication, we describe the textiles of the future and their production in terms of global changes in society and work life. We consider only the near future, allowing us to focus on the likely reality rather than merely speculating.

28.2 General Future Forecast

Any predictions around the future of the fiber industry must consider not only potential technologies but also the ongoing social and environmental surroundings in which it will operate [3].

The Fiber Engineering Program at Shinshu University [1] and the special issue regarding the exploration of the function and materials of bio-fibers [2] inform this chapter.

A. Yonenaga (✉)
Independent Textile Journalist, Osaka, Japan
e-mail: yonenaga@gold.ocn.ne.jp

Society of F.S. Technology, Japan (ed.), *High-Performance and Specialty Fibers*,
DOI 10.1007/978-4-431-55203-1_28

28.2.1 Social Structural Change: From Consumption to Sustainable Society

Consumption-based society is nearing its end; the way of the future involves sustainable resources and environments. This will not only be a fundamental change to our social structure but will also affect the demand for and consumption of textiles.

28.2.2 Explosive Increase in Global Population

The global population is predicted to increase from the present 6.8 billion to 9 billion in 2030. This increase will mainly come from under-developed countries, whereas developed countries will exhibit a declining population growth. This global population growth will lead to substantial increases in the demand for textiles.

28.2.3 Aging Society

Humanity as a whole is moving towards a population that comprises a growing number of elderly people, fewer children, later marriage, and international marriage. Medical and welfare services will need to adapt to meet the needs of this aging society.

28.2.4 Limited Global Capacity for Food and Natural Resources

Looking to the future of foods and natural resources, the forecast is obviously that of a scarce food supply and a worldwide increased demand for marine products. Thus, shortages in food and natural resources are likely to become a serious social problem.

28.2.5 Serious Shortage of Water Supply

Water is going to be in short supply in most under-developed countries. This will have various associated negative effects, not only on human life and industrial activities but also on the growth of plants and forests.

28.2.6 Multi Polarized Society: Economic Bloc and Resource Nationalism

Trends toward economic blocs could intensify, and the protection of natural resources may lead to pronounced resource nationalism.

28.2.7 Government Conversion: Localization and Autonomic Dispersion Style

Governments are likely to localize systems to concentrate interests and autonomic dispersion systems, which will in turn affect more complicated industrial management systems. These changes to social structures will undoubtedly affect the supply balance of textile fibers. Under these various changes, foreseeable demands of textile fibers are basically upward and will be tighten in future.

28.3 Forecast of Future Fiber Trend

One must understand textile dynamics before making conclusions about the development of textiles in the future. Assuming the long-term average annual growth rate remains unchanged, the textile market would account for 106 million tons in 2020 and nearly 139 million tons in 2030, corresponding to an increase in average per capita consumption (Fig. 28.1) from12.2 kg in 2012 to 13.9 kg in 2020 and 16.7 kg in 2030 [4]. Thus, this picture naturally portrays an increasing global demand for textile fibers.

28.3.1 Coping with Increased Demands

Our ability to increase the production of natural fibers is evidently limited, meaning the only way forward is to increase the production of synthetic fibers. However, this reality is also constrained by the availability of raw materials, particularly raw fossil material, which continues to both rely upon and disregard the environment. Any increase in fiber production will necessitate serious study both of the availability of raw material and of how best to create a sustainable society.

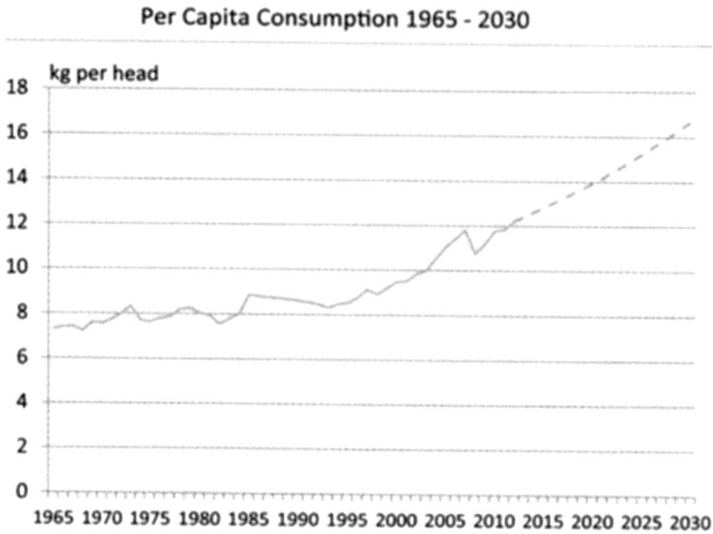

Fig. 28.1 Per capita fiber consumption trends (Source: Fiber Journal October 2013)

28.3.2 Decreasing the Costs of Fiber Production

Cost competitiveness will be an essential component of the future survival of synthetic fibers. Developing countries have the advantage of being able to provide cheaper labor and services, but the cost of raw materials applies equally to any country. Therefore, news from the USA that shale gas technology was a reality was exciting, and it will certainly assist in the cost-competitive creation of synthetic fibers.

Raw materials from shale gas, such as ethylene and ethylene glycol, will provide a cost advantage of more than 40 %, providing practical support to the survival of synthetic fibers (Fig. 28.2), via not only decreases in the costs of raw materials but also savings applied throughout the manufacturing process.

28.3.3 Development of Recycling Technology

The ability to recycle synthetic fibers is already broadly developed and routinely used; it is paving the way for the production of specific materials, such as polylactic acid (PLA) and bio-fibers. The Japanese recycling specialist, Takayasu Co., has reported that super-fine recycled polyester fibers <1.0 dtex. have been developed for automotive nonwoven fabric [6]. Diversification of recycled synthetic fibers will be a foundation material source.

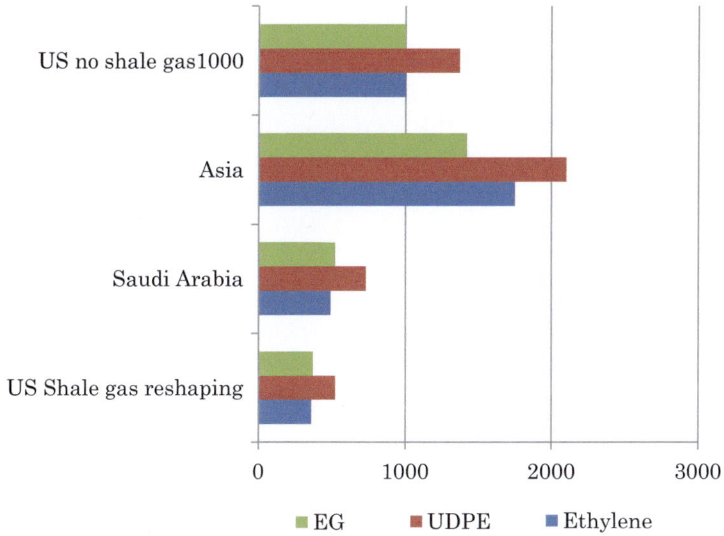

Fig. 28.2 Shale gas reshaping the US chemical industry (*horizontal bar* unit: $/ton) (Source: Management Sensor P187-8/2013) [5]

28.3.4 Expansion of Man-Made Fiber Areas

Cellulose and natural resources such as kenaf, kapok, bamboo, and banana will attract attention in terms of sustainability. However, the availability of these fibers will be limited.

28.3.5 Development of Biomass Fiber

See Sect. 28.4.5 for details on the development of biomass fiber.

28.4 Future Super-Functional Fiber

It is essential that textiles perform well. As such, the laboratory is often the starting point in the development of textiles, and this provides further potential to explore new fibers in the future. Various fields have explored high-functioning and super-performance fibers such as high-temperature resistant, flame-retardant, super-high strength, and anti-microbial fibers. However, consumers and markets expect fibers with further sophisticated functions and enhanced qualities that will survive in the market place. The super-high functional synthetic fibers of the future will provide various pictures of the potential to furnish human life.

28.4.1 Biomimicked Fiber

Synthetic fibers were originally developed to mimic natural fibers, for example, viscose rayon yarn to imitate silk yarn. Such products have since met many challenges, and further development of more sophisticated functions is expected in the future [7].

1. Pine core: Smart breathing textile
   Academic organizations in the USA and Japan have reported the development of fibers that imitate the botanical structure of pine cones, which open and close in response to moisture levels. Thus, this product can alleviate the feeling of dampness by increasing the air permeability of textiles and yarns as moisture builds up around them. Briefly, the developers report that the fabric remains dryer for longer and dries rapidly in extreme conditions [8].
2. Morpho's wings: "Morphotex" colorful textile
   The colors of the beautiful South American Morpho butterfly change depending on the angle of light from which they are viewed because of the complex scale structure on the wings. Teijin Fibers Limited, Japan, mimicked this with a heavily engineered filament fiber called Morphotex. It consists of no less than 60 separate nanolayers in a 15–17 μm structure. There will be interest in enhancing this unique structure for future development (Fig. 28.3).
3. Lotus effects: Perfect waterproof textile
   Various textile coatings have imitated the "lotus effect". The papillae on the lotus leaf ensure that only 2–3 % of its surface comes into contact with water droplets. This contact is confined to the outermost tips of papillae, meaning the adhesive force that would otherwise cause a droplet to spread is also minimal.

Fig. 28.3 Morpho's wing (Source: Future Materials 2/2013)

Fig. 28.4 Lotus effect (Source: Future Materials 2/2013

Instead, the surface tension forces of the water prevail and invariably cause the droplet to form a spherical globule, leading the water to simply roll off. Textiles with these properties are expected to be produced in the near future (Fig. 28.4).

4. Gecko's feet: adhesive cloth

   The gecko is able to stick without glue to flat surfaces by the soles of its feet, which have millions of tiny hairs with split ends. At the tip of each split is a mushroom-shaped cap less than one-thousandth of a millimeter across. These ensure the gecko's toes continuously maintain very close contact with the surface. Many have attempted to replicate this in bonding and adhesive systems, and future applications in building and in household use are expected.

5. Polar bears: thermal control

   Polar bears are superbly insulated by up to 10 cm of blubber, in addition to their hide and their fur; they overheat at temperatures above 10 °C and existing happily at −50 °C. The polar bear's fur consists of a dense layer of 'under fur' and an outer layer of guard hairs that appear white to tan but are actually transparent. The guard hair is 5–15 cm long over most of the bear's body. ITV Denkendorf, Germany, are creating a textile that imitates these properties, and expect to contribute to sophisticated piping, heat capture, and heat storage systems.

28.4.2 Design-Driven Cellulose Fiber Products

The largest natural resource for textile fibers is cellulose from forest wood, which accounts for >60 % of potential textile fiber uses. However, the tight binding of the molecule means it is currently difficult to utilize properly. Elaborate development

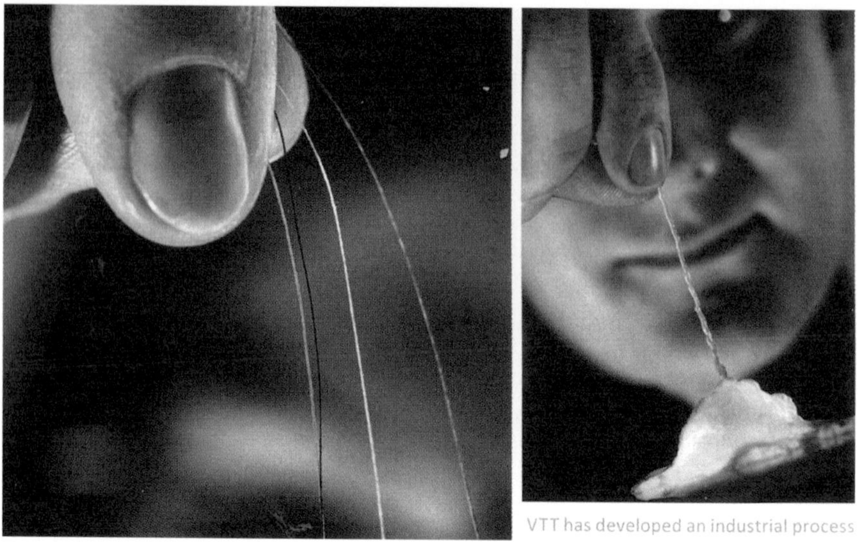

Fig. 28.5 Cellulose fiber yarn made directly from pine fibers (Source: Future Materials P286/2013)

work to use a huge reserve of buried cellulose has begun. State-of-the-art cellulose processing technologies could generate significant production value for future uses of valuable buried natural cellulose resources. A project currently being developed involves the replacement of cellulose wet spinning with extrusion technology, which will resolve the issue of environmental harm [9] (Fig. 28.5). This research project provides an opportunity to use design methods to turn raw cellulose into products and business projects.

28.4.3 Spider Silk Fiber

Artificial spider silk represents another great vision soon to become a reality. The fiber exhibits extensive physical properties that are expected to find applications in the automotive, industrial, and other fields. The latest news has come from Japan-based Spiber Inc., a venture business firm out of Keio University that has developed a mass production process for artificial spider silk. In collaboration with Spiber Inc., automotive parts manufacturer Kojima Press Industry constructed a pilot method to supply 100 kg per month of the yarn in 2013. The process uses raw material from a microbiological protein with the same components and characteristics of spider silk, but with a revised micro-molecular arrangement to enable it to be used with other materials (Fig. 28.6).

At this stage, annual production is 1.2 tons, which is expected to increase to 10 tons by 2015. The new artificial spider silk yarn has high tensile strength and

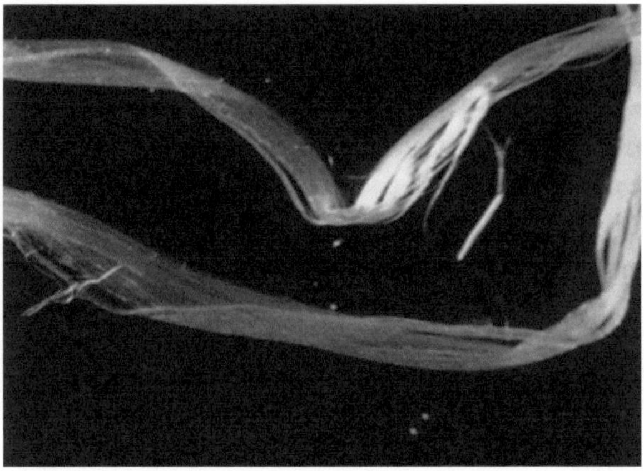

Fig. 28.6 Spiber Inc. artificial spider silk (Source: Nonwovens Report International 4/2013)

elastic properties. Spiber Inc. already owns 16 patents, and the spider silk yarn is regarded as the future of fiber. Meanwhile, recombinant spider silk has also reached the laboratory stage in the USA and Europe. The scale is still at the 50-kg stage, but excellent tensile strength could lead to military uses such as in the creation of ballistic cloth [10, 11].

28.4.4 Intelligent Fiber: Semi Conductor in Fiber with Light-emitting Diode (LED)

The intelligent textile industry is currently focusing on multi-functional textile products, the electronic functionality of which is intended for various medical, protective, military, and sports applications. Electro-conductive fiber for use in electronic textiles (so-called E-textiles) has already appeared on the market. In the near future, lighter and more effective functional semi-conductive chips will be processed directly into the heart of the fiber to be processed into electronic textiles.

Nottingham Trent University (NTU) in the UK is reportedly undertaking intensive development of textile fibers containing semi-conductor chips (Fig. 28.7). This new development will possibly lead to the creation of wearable and washable computers within garments that can easily function to retain possible communication with outside world and to control every possible textile function. The researchers at NTU hope for medical applications such as checking humidity and temperature, as well as more sophisticated chips that can conduct chemical analysis and monitor perspiration.

The researchers have highlighted this electronic integration using light-emitting diodes (LEDs) in a demonstration garment to enable people to see the technology at work (Fig. 28.8). LED textiles can also result in variations of washable and wearable computers that can monitor vital signs of wellbeing, provide intelligent textiles for military, such as invisibility cloaking capabilities, and create flexible

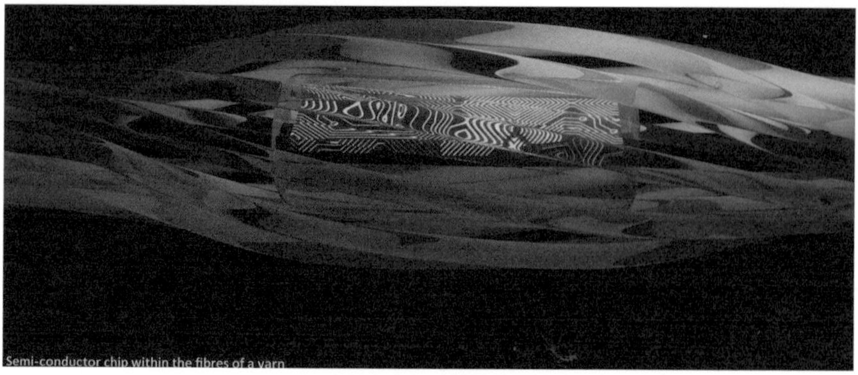

Fig. 28.7 Semi-conductor chip within the fiber of a yarn (Source: Future Materials P10, 4/2013)

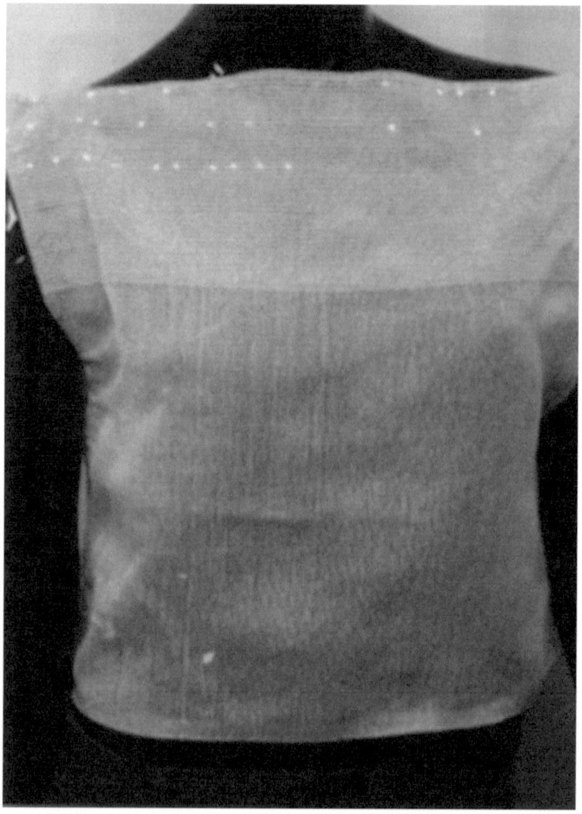

Fig. 28.8 Prototype light-emitting diode (LED) garment (Source: Future Materials P11, 4/2013)

and comfortable displays. In the future, we also expect it will be possible to introduce radio-frequency identification (RFID) chips into yarn so manufacturers can follow the garments all the way from production to retail. In addition, the technology could allow the military to monitor the atmosphere to identify harmful chemicals [12].

Various types of highly functional and multi-functional fibers, such as phase-change material (PCM), fiber robotics, and textiles with built-in bacterial control, smart color, and optical electronics, are also under development and expected to appear on the market. Super-performance fibers such as carbon fibers and aramid fibers have already been marketed substantially in limited applications on a commercial basis. These fibers will also certainly maintain a share of the future.

28.4.5 Sustainability of Future Fiber: Bio-base Fibers

Sustainability has become an increasingly important topic all the way along the textile chain. Against a background of scarce resources, understanding how to secure and handle resources and generate—besides using nature—raw materials for the production of fiber in the future is crucial. The most interesting and serious approaches focusing on sustainability—recycling, biopolymers, and sustainable natural resources—are common in the world of textiles. The resources may seem unlimited, but fossil resources are not. The concerns of both textile producers and customers have resulted in a drive for sustainable textiles. Consequently, a large number of projects intending to increase the use of biopolymers for fibers and textiles are underway [13].

Bio-base polymers are classified as either natural high-molecule polymers or synthetic polymers. The former comes primarily from polysaccharide from natural resources. Simply refining the material from its natural state is difficult because of the hard binding; however, many projects are currently developing cellulose nano fiber. The current development of new polymers from natural resources using microbes is increasing, as is the creation of synthetic biopolymers via fermentation. Representative bio-base polymers under development are shown in Table 28.1.

Poly(trimethylene terephthalate) (PTT) can be produced by standard condensation polymerization of 1,3-propanediol (PDO) and terephthalic acid. DuPont has commercialized PTT under the trade name of Sorona® with its bio-content of 35.9 %, using bio-PDO as a raw material. Although the thermal properties (the melting point and glass transition temperature) of PTT fall between PET and PBT, the molecular structure of PTT is quite unique kink (not linear) mode which provides a soft touch with comfortable stretch and recovery properties to the PTT fiber that are suitable for carpet and textile applications [14].

Recently, polylactic acid (PLA) has been highlighted as a carbon-neutral bioplastic because of its availability from agricultural renewable resources. PLA

Table 28.1 Representative bio-based polymers under development

Class	Polymer	Monomer	Synthetic method of monomer
Polyesters	PTT	1, 3-Propanediol	Bio-conversion from glucose/ glycerin
	Copolyester	Succinic acid	Fermentation
		Butanediol	Fermentation
	PLLA	L-Lactic acid	Fermentation
	sc-PLA	L- and D-Lactic acid	Fermentation
	PBS	Succinic acid	Fermentation
	PHB	Bacterial polyester	Biosynthesis by genetic engineering
Polycarbonates	PC	Isosorbide	Chemical conversion from sorbitol
Polyamides	Nylon 54	Succinic acid	Fermentation
	Nylon 4	γ-Aminobutyric acid (2-Pyrrolidone)	Decarboxylation of glutamic acid
	Nylon 610	Sebacic acid	Chemical synthesis from castor oil via ricinoleic acid
	Nylon 11	11-Aminoundecanoic acid	Chemical synthesis from castor oil
Acrylics	Butyrolactones	Tuliparin	Plant-based
		MMT	Chemical conversion from levulinic acid
Polyurethanes	PU	Plant oil-based polyols	Chemical conversion from castor oil

Source: H. Nakajima, Y. Kimura, "Bio-Based Polymers", Y. Kimura *ed.*, CMC Publishing Co., Ltd., Japan, pp. 1–23 (2013)

is also biodegradable when exposed in biologically active natural environments and is compostable under very specific conditions of high temperature (> 60 °C) and high humidity (>80 % RH), typified by the conditions for composting.

Recent advances in the fermentation of glucose obtained from sugar beet, sugar cane or corn have led to a dramatic reduction in the cost of manufacturing the lactic acid necessary to produce PLA polymers. In addition, the technology to produce PLA economically on a commercial scale has been developed worldwide by several companies. Notably, NatureWorks LLC currently operates the world's largest commercial plant for PLA (Ingeo®) with a production capacity of 140 000 tons per year at Blair, Nebraska, USA. NatureWorks has developed a low-cost continuous ring-opening polymerization (ROP) process that is the polymerization of the cyclic dimer of lactic acid, lactide, for the production of PLA.

Lactic acid is a well-known, simple α-hydroxy acid with an asymmetric carbon atom, which is synthesized by the bacterial fermentation of carbohydrates such as sugar from starch. Lactic acids are optically active compounds, including L- and D-forms. PLA polymers that are currently commercially available are ordinarily called poly(L-lactic acid), PLLA, although they are random copolymers containing large amounts of L-units and small amounts (0.5–20 %) of D-units in the polymers.

The melting point of PLA increases with decreasing content of D-isomer, ranging from about 130 to 180 °C.

PLA is a crystalline polymer of aliphatic polyester with thermoplastic processability. Therefore, it can be melt-spun into various types of fibers and nonwovens by conventional melt-spinning machines, and its yarn properties are relatively similar to those of PET. However, PLA fibers are highly functional, coupled with the intrinsic characteristics of PLA such as biodegradable/compostable, bacteriostatic, flame-retardant, weather-resistant and moisture management (fast wetting/drying performance) properties, when compared with PET fibers [15]. Furthermore, the environmental impact study by LCA (Life Cycle Assessment) revealed that the total CO_2 emission of PLA from cradle to grave (feedstock + processing + disposal) is the lowest (3650 kg/ton) among existing conventional plastics/fibers including PET (6443 kg/ton) and viscose rayon (14,680 kg/ton) [16].

Unitika Ltd.. Japan, is a leading company manufacturing the PLA fibers/nonwovens (Terramac®) in the world. Potential applications of these fibers and nonwovens include geotextiles (vertical drain sheet, sand bag, erosion protect and slope stabilization), agricultural/horticultural products (plant cover, plant pots, strings), home furnishings & clothing (towel, bedding, furniture wadding and filling, tea bag filters, casual wear), hygienic products (wipes, disposable diapers, personal care products) and industrial uses (air/liquid filters, cabin parts of vehicles) etc.

The vision of producing sustainable raw materials equipped with new functionalities is already within reach. The advent of genetic engineering means technology has reached a stage where the creation of sustainable biomaterials using genetically modified microorganisms instead of plant-based resources or oil is possible [17].

Gevo Inc. is leading the way in developing and implementing renewable technology to produce bio-based chemical products and next-generation biofuels that replace petrochemicals, with a focus on isobutanol as a platform chemical. Gevo's technology team has developed proprietary yeast to efficiently convert fermentable sugar into isobutanol through synthetic biology and metabolic engineering.

It is fascinating that we can now create not only biotechnology fiber but also the raw materials for the production of fiber.

In the post-fossil world of the future, it is likely that textiles will be made via biotechnology using bacteria or fungi. Current research projects are focused on the wet-spun biopolymer fibers of alginate and chitosan, produced with biotechnology. By varying fermentation conditions, nutrition media, and biopolymer isolation protocols, the characteristics of the raw materials can be adjusted, which in turn directly affects the properties of fibers made from those biopolymers.

In addition, modern biotechnology allows the genetic modification of microorganisms, thus influencing polymer output and characteristics from the very beginning [18] (Fig. 28.9). Thus, in the near future, biotechnology and textiles will combine in the use of microorganisms to synthesize fibers and yarns.

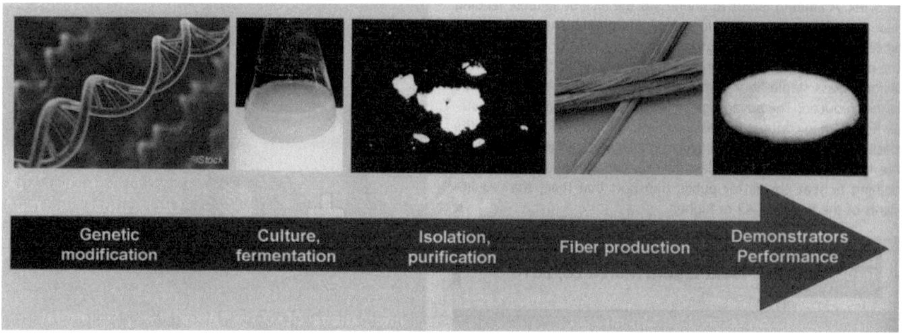

Fig. 28.9 Biopolymer from bacteria—adjustment of materials from the beginning (Source: Technical Textiles P E224 5/2013)

28.4.6 Challenges for Fiber Producers in a Sustainable Future

Japanese fiber producers are being challenged to increase their already substantial efforts in the construction of a sustainable society.

Toray Industries has set up a broad brand "Ecodear," which encompasses the company's bio-mass development projects. Furthermore, Toray has not only marketed PLA and PTT fiber but also has been jointly developing fully bio-based PET fiber with Gevo, using terephthalic acid via Gevo's renewable paraxylene from bio-based isobutanol and commercially available renewable mono ethylene glycol (MEG); that is a world first fully renewable PET fiber -i.e., all of the carbon in this PET is from renewable feedstocks. Nylon has made the transition from sebacic acid to make nylon 610, and Toray has been collaborating with the giant food manufacturer Ajinomoto Co., Ltd. to develop bio-base nylon 56. These bio-base synthetic fibers are expected to arrive on the commercial stage in the near future [19].

"Asahi Kasei Corp. works under the slogan "healthy life and comfort living under co-living with environment"." This company has created high-grade artificial leather using chemical recycling and excellent energy-saving processes. Teijin Limited has developed a complete recycling system for polyester fiber: "Eco circle" recycles the polyester chemical component to reduce consumption of chemical materials, energy, and carbon dioxide emission. Kuraray Group's "Ecotalk" brand is based on carbon dioxide reduction products and typical sustainable artificial leather, "Tirrenina," which uses no solvents and a completely closed recycling system based on the "Clarino Advanced Technology System" (CATS) [20].

In the fiber and textile business, Unitika enhances the applications of industrial fibers/nonwovens by further promoting a shift to bio-based materials. In addition to PLA fibers/nonwovens (Terramac®) as above-mentioned, bio-polyamide Nylon11 fiber (Castron®) has been developed in collaboration with Arkema. Furthermore, Unitika has developed a bio-based (bio-content: 56 %) high performance polyamide 10T (XecoT®) with high heat resistance over 300 °C and low moisture absorption, using bio-based 1,10-decanediamine from castor oil via sebacic acid, as a super engineering plastic/fiber for the next generation.

Rayon fiber is referred to generically as a regenerated cellulosic fiber commonly derived from wood pulp via the chemical manufacturing process. Rayon is the first man-made fiber, having the history of three generations of technology/product called viscose, modal or lyocell respectively.

Viscose rayon fiber, a typical sustainable fiber, previously had a negative image concerning its environmental burden and costs compared with other synthetic fibers. However, the product has made a comeback because of efforts to find environmental solutions and a renewed and positive assessment of its position in a sustainable future.

The third generation rayon fiber, Lyocell (Tencel®), manufacturing process is an extremely environmentally friendly process when compared with the first generation viscose rayon, in which the solvent is recovered and reused up to 99.8 % and the remaining emissions are broken down in biological water treatment plants although it uses a substantial amount of energy, and uses an organic solvent of petrochemical origin.

These days, viscose rayon is a leader of sustainable material for the textile industry, and its future is certain. Lenzing Corporation, Austria, is a leading rayon fiber producer with the capacity to create more than one million ton cellulose rayon fibers worldwide, and various professional rayon fiber producers all over the world will follow this trend. Japanese rayon fiber manufacturers—Daiwabo and Omikenshi—are developing a variety of modified viscose rayon fibers based on sustainable systems.

Daiwabo has already created rayon fibers with a variety of special qualities, such as heat ray shields, ultraviolet (UV) cut-off incorporated ceramics, heat retention qualities, and deodorants, under the company's eco-friendly line up. Further multi-functional fibers are expected to appear on the market in the near future.

Omikenshi Co., Ltd not only develops functional viscose rayon fibers but is also using natural cellulosic materials such as kenaf and jute—fully advanced sustainable fibers of annual plants— under the brand "Li Terra". Kenaf fiber has already appeared on the market, and other practical sustainable fibers will follow.

28.4.7 The Outlook for Textile Fibers

Synthetic fibers have progressed to the present day supported by substantial innovative technical developments. Various line-ups such as super-functional fibers, nano-fibers, and sustainable series of fibers have appeared on the market, and further advances can be expected. Manufacturers of synthetic fibers will pursue both fashionable and high-function fields to ensure the future of textile applications. The market forecasts that fiber applications will aim at industrial, medical, and living fields in the future (Fig. 28.10), and that, with diversification and high-functioning applications, more technological high-grade fiber will be developed in the near future. Practical applications for high-functioning fiber will be seen in

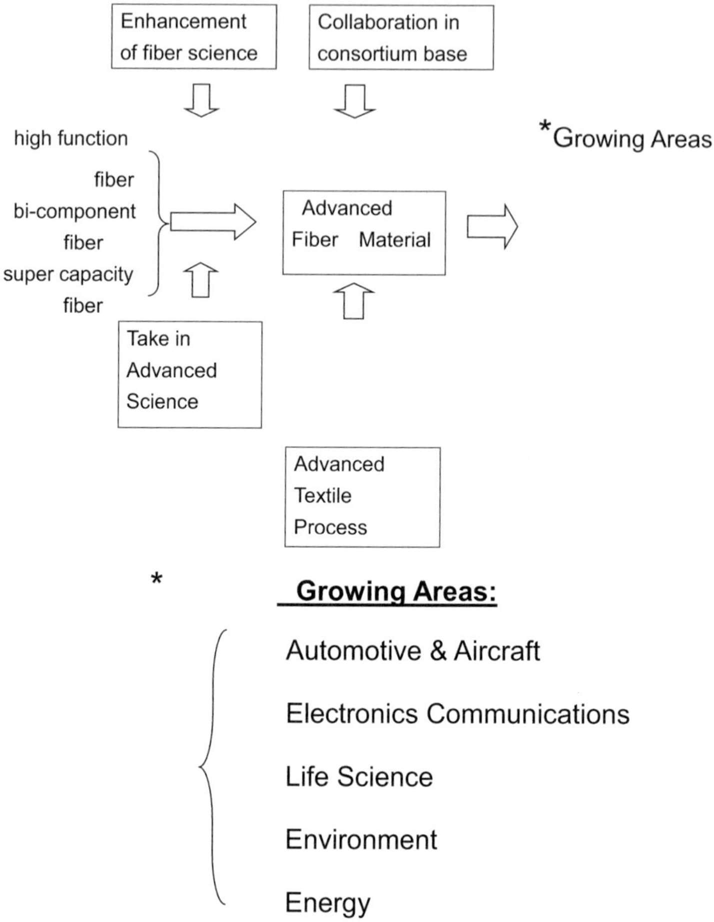

Fig. 28.10 Advanced fiber materials catch up in growing areas

lightweight transportation materials, medical materials, electric battery/condenser separators, electronic parts, E-textiles, and in the building and construction, marine, and airspace fields. Moreover, new industrial and life science applications will invite the creation of further innovative technical fibers.

In the future, technical fiber developments will extend to searches for new polymers, super high-performance fibers, comfortable fibers, and high-functioning fibers. New applications will be developed for these new materials.

Using innovative advances, such as nano-technology and bio-technology, as well as super-high-performance capacities, fiber technology is expected to have a future role in areas such as transportation, life sciences, electronics, environment, and energy fields, in particular.

In conclusion, we consider the global textile and fiber industry has the potential for a positive future, supported by growing consumer demand for fiber and increasingly swift and advanced technical developments [21].

References

1. Fiber Engineering (Shinshu University 21 Century COE Program), ed. by H. Shirai, K. Yamaur (Maruzen Co., Ltd., 2005)
2. Special issue, future exploration by bio-fibers function & materials (March 2009)
3. H. Hoshiro, Future aspect of textile fiber industry, (March, 2010)
4. A. Egelhard, Analysis of per capita fiber consumption, Fiber J. (October 2013)
5. Y. Fukuda, Management sensor P18 7-8/2013
6. http://www.takayasu-rf.co.jp
7. A. Wilson, Mirror to nature, future materials, (November, 2012)
8. Tara Hounslea, Future materials, 2/2013
9. Future materials, 6/2013
10. Nonwovens report international, 4/2013
11. Future materials, 2/2013
12. Future materials 4/2013
13. B. Defraye, Man-made fiber year book, (2013)
14. M. Kaku, Poly(trimethylene terephthalate, PTT), in *Bio-Based Polymers*, ed. by Y. Kimura (CMC Publishing Co., Ltd., Japan, 2013), pp. 86–94
15. M. Mochizuki, Poly(lactic acid) fibres, in *Handbook of Fibre Structure*, ed. by J. Hearle, S. Eichhorn, M. Jaffe, T. Kikutan (Woodhead Publishing, Cambridge, 2009), pp. 257–275
16. T.H. Erwin, K.R. Vink, K.R. Rago, D.A. Glassner, P.R. Gruber, Polymer degradation and stability **80**, 403 (2003)
17. H. Nakajima, Y. Kimura, in *Bio-Based Polymers*, ed. by Y. Kimura (CMC Publishing Co., Ltd., Japan, 2013), pp. 1–23
18. T.R. Hammer, Technical textiles P E224 5/2013
19. Toray annual report 2013
20. Seni-news, Special issue 9/12/2013
21. K. Shimomura, Outlook of textile fiber industry, (20/11/2013)

MIX
Papier aus verantwortungsvollen Quellen
Paper from responsible sources
FSC® C105338

If you have any concerns about our products,
you can contact us on
ProductSafety@springernature.com

In case Publisher is established outside the EU,
the EU authorized representative is:
Springer Nature Customer Service Center GmbH
Europaplatz 3, 69115 Heidelberg, Germany

Printed by Libri Plureos GmbH
in Hamburg, Germany